Albert Kobina Mensah
**Soil Pollution and Remediation**

# Integrated Global STEM

Edited by
Robert Krueger, Wole Soboyejo and Anita Wattson

## Volume 4

Albert Kobina Mensah

# Soil Pollution and Remediation

Risk Assessment, Phytoremediation, Revegetation

**DE GRUYTER**  The WPI Press

**Author**
Albert Kobina Mensah
Soil Research Institute
Council for Scientific and Industrial Research
Academy Post Office PMB
Kwadaso, Ghana
albertkobinamensah@gmail.com

ISBN 978-3-11-166038-7
e-ISBN (PDF) 978-3-11-166204-6
e-ISBN (EPUB) 978-3-11-166348-7

**Library of Congress Control Number: 2024950670**

**Bibliographic information published by the Deutsche Nationalbibliothek**
The Deutsche Nationalbibliothek lists this publication in the Deutsche Nationalbibliografie;
detailed bibliographic data are available on the internet at http://dnb.dnb.de.

© 2025 Walter de Gruyter GmbH, Berlin/Boston, Genthiner Straße 13, 10785 Berlin
Cover image: Karasmake/iStock/Getty Images Plus
Typesetting: Integra Software Services Pvt. Ltd.

www.degruyter.com
Questions about General Product Safety Regulation:
productsafety@degruyterbrill.com

My dearest wife, Priscilla Hattoh Mensah, and son, Aseda Kojo Mensah.

# Preface

The process of mineral extraction results in substantial damage to the topsoil, which leads to soil degradation in the form of deterioration of the soil structure, susceptibility to soil erosion, excessive leaching of nutrients, soil compaction, decrease in soil pH, accumulation of heavy metals in soil, depletion of organic matter, reduced accessibility of nutrients for plants, diminished capacity for cation exchange, decline in microbial activity, and ultimately, a consequent decline in soil fertility. Effective management of topsoil is indispensable in the execution of a reclamation strategy, as it serves to minimize nutrient depletion and ultimately expedite the process of restoring soil health and quality.

Ghana is among the top ten gold-producing countries in the world, and its actions toward achieving environmental sustainability in the mining sector must be shared with the world. There are some great success stories as well as challenges in mining sector sustainability from Ghana's case, which are left undocumented and are limited in investigations in a scientific book. Such enviable feats chalked up by some mining companies must be documented so that lessons can be borrowed for replication in restoring similar degraded mining sites elsewhere across the globe. Additionally, companies can learn from the success stories and challenges encountered in mine land reclamation and revegetation in this book.

Revegetation may present a sustainable option for the reclamation and restoration of mine soil degradation. The restoration process involves many strategies aimed at improving the quality of soil, such as augmenting the quantity of soil organic matter, enhancing nutrient availability, increasing cation exchange capacity, stimulating biological activities, and optimizing the physical qualities of the soil.

Researchers, scientists, and consultants in the field of soil pollution and remediation have conducted a great deal of study using a variety of techniques and approaches. However, fragmented reporting of techniques and results has resulted from the documentation and dissemination of success stories, challenges, and findings mostly through individual technical reports and publications in scholarly journals. This book provides an in-depth analysis of the many scientific methodologies used to identify environmental risks related to potentially toxic elements (PTEs) in mining sites and revegetation as a strategy for ameliorating contaminated and degraded mining sites. The book covers the application of these methods in identifying soil-human health risks and planning toward the reclamation of such derelict ecosystems.

The book combines reviews of relevant literature, laboratory investigations on PTEs from representative mine-contaminated soil and spoil samples, as well as appraisals of case studies on successful reclamation and revegetation of mine-degraded lands. Applications of the total element concentration method, size fractionation experiments, sequential extraction analyses, risk assessment indices, geospatial analysis, redox chemistry experiments, synchrotron radiation science, incubation experiments,

https://doi.org/10.1515/9783111662046-202

and pot experimental trials in soil remediation works were documented first hand in a single piece in this book.

The book is organized into 22 chapters, each dedicated to soil contamination caused by mining and revegetation as a sustainable solution. The initial parts of the book deal with various techniques for identifying soil-human health risks. They include topics such as the consequences of heavy metal presence and build-up, the sources from which heavy metal pollutants originate, and the possible hazards they bring to plant, human, and soil health. The second parts begin with the concept of mining sector sustainability and explore revegetation as a strategy for reclaiming and remediating mining-contaminated lands, with the objective of restoring ecosystem functionality, improving soil characteristics, and cleaning metal-contaminated soils. The book may serve as a valuable resource for individuals occupying various professional roles and engaging in academic pursuits, such as project officers operating within the environmental, safety, and health divisions of mining enterprises, consultants specializing in land reclamation, lecturers specializing in environmental and soil sciences, students, and individuals with a strong interest in environmental protection.

*Albert Kobina Mensah, Ph.D.*
*Soil Research Institute-Kwadaso, Kumasi-Ghana*
*11 November 2023*

# Aknowledgments

My deepest gratitude goes to all individuals who, in one way or another, contributed to making this book a success. I express many thanks to my professors in Germany who trained me in their laboratories in Bochum and Wuppertal. Special mentions go to Professor Bernd Marschner and Professor Joerg Rinklebe for training me in soil pollution and remediation during my Ph.D. in Germany.

https://doi.org/10.1515/9783111662046-203

# Contents

Albert Kobina Mensah

# Chapter 1
# Evolution of land reclamation practices and introducing mine land degradation and revegetation in Ghana

**Abstract:** This book explores the impact of mining activities on soil quality in Ghana, focusing on the strategies used for soil rehabilitation through revegetation. Surface mining, particularly blasting, leads to reduced organic matter content and presents challenges for vegetation establishment. The resulting soils have low nutrient concentrations and altered pH levels, and contain toxic elements that can be toxic to plant growth and human health. The Environmental Protection Agency (EPA) in Ghana requires environmental impact assessments and remediation efforts from mining corporations. Revegetation is an effective strategy that introduces organic matter into the soil to eliminate or reduce contaminants, improving soil quality, protecting public health, and returning degraded mine lands to usable states. The book provides a critical review of works, studies, projects, demonstrations, and experimental trials on revegetation.

## 1.1 Introduction

Land reclamation in mining regions has evolved significantly over the past century, driven by increasing environmental awareness and advancements in ecological science. Historically, mining operations prioritized resource extraction, leading to severe landscape degradation and ecological damage. However, pivotal moments, such as the publication of Rachel Carson's "Silent Spring" and the enactment of the Surface Mining Control and Reclamation Act, catalyzed a shift toward structured reclamation practices. Today, innovative techniques like phytoremediation and ecosystem restoration are being implemented, particularly in regions like Ghana, where mining has profoundly impacted the environment. This chapter explores the evolution and current state of land reclamation practices, emphasizing their importance for sustainable development and ecological integrity.

**Albert Kobina Mensah**, Soil Research Institute, Council for Scientific and Industrial Research, Kumasi, Ghana, e-mail: albert.mensah@rub.de, albertkobinamensah@gmail.com, ORCID: https://orcid.org/0000-0001-5952-3357

https://doi.org/10.1515/9783111662046-001

## 1.2 Historical developments in land reclamation

Land reclamation, particularly in mining regions, has a long history rooted in the need to mitigate the environmental degradation caused by extensive resource extraction [536]. The historical development of land reclamation practices has been shaped by increasing awareness of environmental impacts, advancements in ecological science, and evolving regulatory frameworks [504].

In the early stages of industrial mining, particularly during the nineteenth and early twentieth centuries, land reclamation was not a priority. Mining operations were primarily focused on resource extraction with little regard for the environmental consequences [152]. The result was widespread landscape degradation, characterized by barren lands, water pollution, and the accumulation of mine tailings – byproducts of the extraction process that often-contained hazardous materials [255]. The environmental impacts of these unregulated activities were severe, leading to the loss of arable land, destruction of habitats, and contamination of water bodies [537].

The turning point in the history of land reclamation came during the midtwentieth century, as environmental awareness began to rise globally. The publication of Rachel Carson's Silent Spring in 1962, which highlighted the detrimental effects of pesticides, sparked broader environmental movements that brought attention to various forms of ecological damage, including those caused by mining [104]. This period marked the beginning of more structured and science-based approaches to land reclamation [202].

One of the earliest significant developments in land reclamation was the introduction of the Surface Mining Control and Reclamation Act (SMCRA) in the United States in 1977. The SMCRA was one of the first laws to mandate that mining companies restore the land to its original state or to a condition that is environmentally stable and suitable for future use [390]. This legislation set a precedent for other countries, encouraging the adoption of similar regulations worldwide.

## 1.3 Evolution of reclamation techniques in response to environmental challenges

As the environmental impact of mining became more apparent, reclamation practices evolved to address specific challenges such as soil degradation, water pollution, and loss of biodiversity [90]. Early reclamation efforts primarily involved regarding the land and planting fast-growing vegetation to stabilize the soil and reduce erosion [116]. However, these practices were often superficial and did not address the underlying issues such as soil contamination and the long-term sustainability of the ecosystem [227].

The late twentieth and early twenty-first centuries saw significant advancements in reclamation techniques, driven by a deeper understanding of ecological processes and the development of more sophisticated technologies [260]. Reclamation practices began to incorporate soil amendments, such as the addition of organic matter and lime, to improve soil fertility and pH levels [453]. These amendments helped to create conditions more conducive to plant growth and supported the reestablishment of native vegetation [116].

Another milestone in the evolution of reclamation practices was the development of phytoremediation techniques. Phytoremediation involves the use of plants to remove, stabilize, or neutralize contaminants in the soil, such as heavy metals [417]. This method gained popularity in the late twentieth century as a cost-effective and environmentally friendly approach to dealing with contaminated soils in mining regions [23]. Plants like Indian mustard and poplar trees have been used to absorb heavy metals from the soil, reducing the environmental risks associated with mining activities [397].

In recent years, the concept of "ecosystem reclamation" has gained prominence. This approach goes beyond merely stabilizing the land and focuses on restoring the entire ecosystem, including the soil, water, vegetation, and wildlife [114]. Ecosystem reclamation involves the reconstruction of natural habitats and the reintroduction of native species, aiming to create a self-sustaining environment [210]. This holistic approach recognizes the interconnectedness of various ecological components and seeks to restore ecological balance [74].

## 1.4 Key milestones and their influence on current best practices

Several key milestones have shaped the evolution of land reclamation techniques and influenced current best practices. The introduction of legislation such as the SMCRA provided a legal framework that required mining companies to take responsibility for environmental restoration [390]. This led to the establishment of standards and guidelines for reclamation that have been adopted globally [538].

The development of scientific methods for soil analysis and the use of remote sensing technologies have also been significant milestones. These advancements have allowed for more precise assessment of soil conditions and more effective monitoring of reclamation progress [201]. Remote sensing, in particular, has enabled the large-scale evaluation of reclamation efforts, providing valuable data on vegetation cover, soil erosion, and water quality over time [273].

The growing emphasis on sustainability and the integration of ecological principles into reclamation practices have further refined current best practices. Today, land reclamation is not only about restoring the land's appearance but also about ensuring that the reclaimed land can support biodiversity, contribute to carbon sequestration, and provide ecosystem services such as water filtration and soil stabilization [505].

The evolution of land reclamation practices reflects a broader shift toward sustainable development and environmental stewardship. The milestones achieved in the field have laid the foundation for current practices that prioritize ecological integrity and long-term sustainability, ensuring that reclaimed lands can be safely and productively used for future generations [762].

## 1.5 Mining and land degradation in Ghana

Numerous research studies have provided evidence suggesting that the soils in Ghana are significantly affected by mining activities, particularly those arising from surface mining operations [e.g., 324]. The mining procedure employs large-scale equipment and incorporates explosive techniques for extraction purposes. The process of blasting in isolation could present detrimental effects on soil organisms that provide benefits to the soil, as well as affect the stability of soil aggregates. Over time, this can lead to a reduction in organic matter content within the soil. The soil, which is also the newly formed substrate or growth media, can pose challenges for vegetation establishment.

Numerous sites in Ghana that have been subjected to mining activities are susceptible to erosion, mostly attributable to the absence of preexisting vegetation, the prevalence of fine, scattered particles, and the presence of steep slopes. Consequently, substantial gullies and pits are formed as a result of these mining operations. The substrates, sometimes referred to as overburden or mine spoils, are a prevalent feature found in virtually all gold mining areas in Ghana. The majority of mine substrates have very low concentrations of macronutrients, particularly nitrogen (N), phosphorus (P), and potassium (K) [e.g., 332]. The issue of low pH is a significant challenge in waste materials that contain iron pyrites Mensah et al., [323]. When these materials undergo weathering, they produce sulfuric acid, resulting in low pH values if the waste lacks the ability to neutralize acidity. The presence of some metals, such as aluminum and zinc, in acidic waste can pose a substantial challenge to the growth of plants due to their toxic nature. [539] documented the pH levels in the soils of a specific mining location known as Prestea/Bogoso in the western region, revealing a significant acidity with a pH as low as 3.96. The presence of a low pH is a distinguishing feature observed in most gold mine spoils [335].

Many soils impacted by gold mining activities undergo restoration in accordance with the mandate of the Environmental Protection Agency (EPA). Conversely, many soils, particularly those arising from unlawful small-scale mining activities, are left devoid of any remediation efforts. Illustrative instances can be observed in the expansive mining zones located in the western part of Ghana, wherein numerous instances of unauthorized mining operations are prevalent. The process of mining results in the removal of vegetation and subsequently causes the depletion of certain plant nutrients from the area. In addition, the procedure involves the utilization of bulldozers and other heavy machines to remove the top layer of soil, after which the collected soil samples are sent to the laboratory for the specific objective of mineral extraction. The shallow depth of topsoil, typically around 20 cm, contains the majority of plant-accessible nutrients. Consequently, the scraping action of bulldozers during land clearing diminishes soil fertility and productivity. This process exposes the unfavorable subsurface, which is inappropriate for agricultural development [540]. Figure 1.1 shows an abandoned but a contaminated site in Ghana.

Moreover, the implementation of techniques that eliminate the vegetative material inhibits the incorporation of organic matter into the soil. According to [541], the depletion of organic matter has been identified as a contributing factor to the decline in soil fertility, degradation of soil structure, reduced water retention capacity, and diminished biological processes within the soil.

In the context of Ghana, many forms of industrial growth and expansion, including but not limited to manufacturing, construction, agriculture, and tourism, have contributed to an increased demand for the establishment of official legislative measures to support environmental impact assessments. As a result, the legislative body passed Act 490 in December 1994, which served to formally establish the EPA within the legal framework. According to [542], the Environmental Protection Council (EPC) in Ghana was granted increased authority under the legislation, leading to its transformation into the EPA. This expansion of powers aimed to facilitate the promotion of ecologically sustainable development. As stated in Section 12 (1) of the Act, the EPA has the authority to request, through written notice, that individuals responsible for any undertaking[1] that is deemed by the Agency to have or potentially have a negative impact on the environment submit an environmental impact assessment to the Agency. This assessment should include the necessary information and be submitted within the time frame specified in the notice (EPC, 1994, cited in [728]). The legislation according to the 27 principles of sustainable development (see Agenda 21 document)[2] is called the principle number 15, and it is referred to as the precautionary principle. Therefore, the legislation provides EPA with the necessary legal authority to require project proponents to carry out and submit studies on the environmental impact assessment for approval.

When the EPA determines that the actions of a particular endeavor present a significant risk to the environment or public health, the agency has the authority to issue an enforcement notice to the individuals accountable for the endeavor. This notice

---

1 In Ghana's EPA, an undertaking is any activity that is envisaged to have a potential impact on the environment and the social norms and values of the people where the project is taking place.
2 https://sustainabledevelopment.un.org/content/documents/641Synthesis_report_Web.pdf

mandates that the responsible parties undertake measures that deemed necessary by the agency to prevent or halt the said activities [728].

The EPA mandates that all mining corporations undertake remediation efforts to restore and rehabilitate lands that have been adversely affected during the extraction process, as stipulated in the relevant regulatory framework. Thus, actions must be taken by the corporation to restore the quality of the land and soil that got denatured during the operational phase of the mine's project life cycle. An integral component of the reclamation and rehabilitation process involves the implementation of revegetation strategies on the formerly mined areas.

**Fig. 1.1:** A contaminated mine-degraded site in Ghana.

## 1.6 Effects of mining activities on the physical, chemical, and biological properties of soil

The process of mining has a detrimental impact on the visual appeal of the natural surroundings, as well as on various soil elements, including soil horizons, soil structure, soil microbial populations, and nutrient cycles. These components play a vital role in maintaining a balanced and thriving ecosystem. Consequently, mining activities lead to degradation and loss of both vegetation and cause soil compaction [543]. The overburden dumps exhibit various unfavorable characteristics, including heightened bioavailability of metals, increased sand content, insufficient moisture, heightened compaction, and relatively low organic matter content. According to Ghose [179], the discharge of acidic waste may result in the release of salt or the presence of sulfidic substances, leading to the generation of acid mine drainage. The potential impacts of mine wastes are diverse, encompassing soil erosion, contamination of air and water, toxicity, geoenvironmental disasters, reduction in biodiversity, and finally, economic decline [544, 545].

The extraction activities conducted at many open-cut coal mines located in central Queensland, Australia, led to the deposition of tertiary-origin spoil materials, which consisted of waste rock, on the surface [546]. tertiary sediments that have undergone significant weathering can exhibit high levels of sodium and varying degrees of salinity. As a result, these sediments can develop surface seals that impede water infiltration and decrease the amount of water available to plants. After the process of drying, it is common for these substances to develop robust crusts that hinder the emergence of seedlings. The aforementioned materials exhibit a lack of various nutrients, most notably nitrogen and phosphorus [546].

Irrespective of the specific type of overburden employed, mine soils generally exhibit low levels of plant-accessible nitrogen (N) and phosphorus (P), potentially impeding the establishment of trees. Therefore, it is typically necessary to apply fertilizer at a certain stage of the cycle in order to achieve accelerated growth of trees that are planted on reclaimed mine sites. Mineral extraction causes significant degradation to the topsoil. The authors, Sheoran et al. [453], have observed that physical disturbance to the topsoil during activities such as stripping, stockpiling, and reinstatement can lead to significant nitrogen transformations and movements, ultimately resulting in major loss.

The presence of contaminants in soils is often associated with a decrease in organic matter content [547]. This decline can be attributed to the inhibitory effects of elevated metal concentrations, which negatively impact soil's biological processes, hamper vegetation growth, and impede the synthesis of organic matter. Dutta and Agrawal [136], posit that the diminished presence of organic carbon in mine spoil can be attributed to the disruption of ecosystem functioning, depletion of the soil's organic pool, and the loss of the litter layer during mining, which serves as a crucial storage and exchange site for nutrients.

The process of gold mining has significant repercussions on the ecosystem, leading to the degradation of soil and water quality due to the buildup of heavy metals. This, in turn, adversely affects the aquatic habitats of several fish species. The activity also has an influence on the water supply volume in mining towns, leading to a rise in the cost of water treatment. This, in turn, has a detrimental effect on the economic well-being of the people living in these areas who depend on water resources. The long-lasting social and environmental effects of mining in host communities can be traced to a number of flaws, such as inadequate community participation or involvement of project-affected individuals, difficult and time-consuming registration processes for small-scale mining businesses, inadequate mining policies, and lax enforcement of mining regulations as per Mensah et al. [324], and Mensah and Tuokuu [323].

## 1.7 Mining and soil pollution with potentially toxic elements: an introduction

The issue of soil pollution is of international significance and necessitates thorough investigation. More than 20 million hectares of land worldwide are affected by heavy metal contamination, including that of As, Cd, Cr, Hg, Pb, Co, Cu, Ni, Zn, and Se [276]. Soil contamination is typically determined through a comparative analysis of element concentrations in contaminated soils and uncontaminated soils. In essence, this indicates that the aggregate concentrations of the elements detected in the current soil surpass the predetermined background levels, geobaseline, or regulatory standards. Put simply, this phenomenon arises due to the buildup of deleterious elements in the soil beyond the designated thresholds, particularly in comparison to the natural surroundings [214, 376, 509]. Hazardous elements can manifest from various sources, including inorganic origins such as potentially toxic elements (PTEs) extracted from mining operations and radioactive residues including thorium and uranium ([879, 1044]). Agricultural practices that involve the application of pesticides and fertilizers can also contribute to the presence of hazardous elements [96]. Organic substances like petroleum can also contribute to the occurrence of hazardous elements [975].

The environmental health and human health are significantly impacted by the presence of inorganic PTEs in soil, primarily due to their intrinsic resistance to microbial decomposition [360]. Given the circumstances, the persistence of these compounds in the soil is prolonged unless proactive measures are implemented to eliminate or diminish their concentrations, as well as the corresponding hazards, via diverse remedial approaches [85, 509].

Elements with the capacity to inflict significant harm include but are not limited to Se, As, Cd, Cr, Hg, Pb, Co, Cu, Ni, and Zn. These elements also pose substantial risks to animal populations, critical environmental resources including the food chain, surface and groundwater sources, and flora and fauna. When the average concentrations of these elements in soil surpass the allowable threshold levels, these hazards ensue. It is important to highlight that a considerable quantity of locations, exceeding 2.8 million, have been recognized as potentially susceptible to soil contamination across the European Union. Furthermore, empirical data indicates that approximately 19% of agricultural soils in China contain detrimental contaminants in excess of the prescribed threshold limits.

There is a wealth of literature concerning the environmental and social consequences that are inherent in the process of gold mining in the host communities of Ghana [323, 324]. An investigation conducted by Armah et al. [43], and Mensah et al. [321], revealed that unregulated arsenic (As) discharge negatively impacts soil quality, food chain, surface and groundwater sources, and soil. In a similar vein, Addai et al. [6], documented elevated levels of heavy metals, including arsenic, lead, cadmium, and mercury, in lettuce cultivated on abandoned mine fields in the Ashanti region of Obuasi, Ghana.

The tailings dam is a topographical feature that results from surface mining techniques utilized for the extraction of minerals. The scholarly literature suggests that the yearly worldwide output of tailings could vary between 5 and 7 billion tons [140, 285]. The primary factor contributing to the environmental consequences of surface mining in the vicinity is the construction of tailing dams, as stated by Lin et al. [896] in reference to mineral mining. These dams are significant due to the substantial volumes of material that are produced. Mine tailings consist of residual fine-grained particles, which commonly span a size range of 1–600 µm, which persist subsequent to the extraction of valuable minerals from the ore in the course of mining activities. Additionally, processed water containing dissolved metals and chemicals used in the mineral extraction process is present in the tailings [140].

Tailings dams are engineered structures that are purposefully engineered and constructed as containment facilities within gold mining operations. Their primary function is to store waste materials that are produced during the ore extraction and processing stages. The dams have the potential to store a wide range of hazardous chemicals, including toxins and components, that could endanger human and animal health and pose ecological risks to the environment.

Soil contamination by PTEs can arise from anthropogenic and natural origins, contaminating agricultural crops, land, surface water, and the atmosphere. Intentional or unintentional, the ingestion of these toxic substances through contact with the soil may lead to a range of adverse health effects. A correlation has been identified, as reported by [1045], between the presence of mercury (Hg) and cadmium (Cd) and the onset of renal and pulmonary complications. Additionally, research findings indicate that renal complications with toxicological ramifications may result from exposure to these chemicals in high concentrations. These health concerns can significantly impact individuals' overall well-being, resulting in substantial repercussions for their economic and social means of subsistence.

Evaluation of potential risks to human health and safety posed by the presence of PTEs in mine tailings and spoils is a significant international concern. The book delves into the potential health ramifications associated with PTEs discovered in soil, which subsequently contribute to food chain pollution. Abandoned mine land that is utilized by farmers for food crop cultivation has the potential to introduce contaminants into the food chain.

A risk assessment was conducted in this book to investigate the potential hazards that PTEs, which are found in abandoned and active mine wastes and tailings as well as in soils adjacent to gold mining sites, may pose to human health and the environment. The research was carried out by utilizing established methodologies that have gained widespread acceptance within the scientific community. Furthermore, the study integrated globally recognized indices that evaluate soil contamination and the corresponding health hazards linked to inhalation and ingestion of contaminated soil.

## 1.8 Heavy metals or potentially toxic elements: definitions, history, and debates

Certain heavy metals, including zinc, iron, and cobalt, are considered essential nutrients due to their vital functions in the nutrition of plants, animals, and humans; thus, they are called essential heavy metals. Additional examples include indium, ruthenium, and silver, all of which are relatively innocuous. On the contrary, specific forms or quantities of certain heavy metals – such as arsenic, cadmium, mercury, and lead – can induce severe toxicity; these heavy metals are therefore classified as nonessential. Heavy metal poisoning may arise from a multitude of sources, encompassing, but not limited to, the following: industrial waste, treated timber, paints, mining tailings, and smelting.

Particularly in studies of pollution impacts, the terms "heavy metals" and "potentially toxic elements" are frequently and occasionally used interchangeably in environmental sciences. There has been considerable interest among nonscientific members of the public regarding the definition of a heavy metal and, by extension, which elements are categorized as light in scientific literature. Even within the scientific community, there is disagreement regarding the definition of heavy metals and whether or not they are all metals. There are those who argue that the phrase "heavy metals" might be misleading due to the fact that not all of the metals enclosed in brackets and labeled "heavy metals" are, in fact, metals.

Therefore, these proponents believe that the correct term to use in order to avoid confusion would be potentially toxic elements. PTEs may then be used to include all elements that are needed in trace or smaller amounts. Their occurrence in larger quantities may be harmful to the food chain, the ecological system, the surrounding environments, surface and groundwater, and so on. Thus, it might be better to refer to them as PTEs. In their larger quantities, PTEs cumulatively present detrimental health consequences to both humans and animals.

Gustin et al. [834] provided convincing arguments in their article, titled "The term 'heavy metal(s)': history, current debate, and future use." They argue that "heavy metals" are used for a collection of predominantly metallic elements that exist naturally in the environment in small amounts (commonly less than 1 mg/kg). These elements are derived from the transition metals or d-block of the periodic table of the elements. They include cadmium (Cd), copper (Cu), lead (Pb), zinc (Zn), mercury (Hg), the metalloid arsenic (As), the semimetal thallium (Tl), and the post-transition element antimony (Sb). All of these elements exist in their pure elemental form, which is characterized as "heavy."

They are usually distinguished from "light" metals like aluminum (Al) and the metallic forms of alkaline and alkaline earth metals such as calcium, magnesium, potassium, and sodium. The light elements are often categorized as main or major elements because they are required by plants, animals, and humans in larger quantities, and as such their existence in large amounts does not pose deleterious health implications. In other instances, the light metals are also called essential elements or nutrients in plant nutrition.

The phrase "heavy metal" is commonly a collective term used to describe a set of metals and metalloids that have an atomic density greater than 4 g/cm$^3$ or are at least 5 times denser than water [1047]. These elements are characteristically referred to as trace elements since they exist and are required in very slight concentrations in biological systems. Heavy metals are defined as naturally occurring metals with an atomic number larger than 20 [1046]. In scientific literature and in medical geology, heavy metals are a considerable group of contaminants and pollutants that negatively impact the environment in terms of air, water, sediments, soil, plants, food crops, and humans.

Pourret [1048] proposed the names "metal," "metalloid," or "trace metal." Pourret and Bollinger [1049] introduced the terms "potentially toxic metals/elements" or "trace metal/elements" depending on the situation. However, in 2019, Pourret and Hursthouse specifically suggested PTEs as the preferred alternative phrase. Thus, in modern times, the term "PTEs" appears to be dominating in different research works, different countries, and among professions and disciplines. These fields and works include those in engineering, science communication, policymaking, environment, ecotoxicology, medical geology, public health, mining, environmental engineering, sanitation and water sciences, and civil engineering (see Fig. 1.3).

Pourret and Hursthouse [394], in their letter, "It's time to replace the term 'heavy metals' with 'potentially toxic elements' when reporting environmental research," contends that the designation "heavy metals" lacks precision, is deceptive, and lacks significance in the context of environmental research and public health studies. Their article proposes substituting the word "potentially toxic elements" instead. They argued that the designation "heavy metals" is determined on subjective criteria, such as density or molar mass, and does not accurately represent the chemical speciation, bioavailability, or toxicity of the elements. Their letter examined the literature and historical background of the phrase "heavy metals" and its disputes and objections, while presenting illustrations and data of its application in various research fields

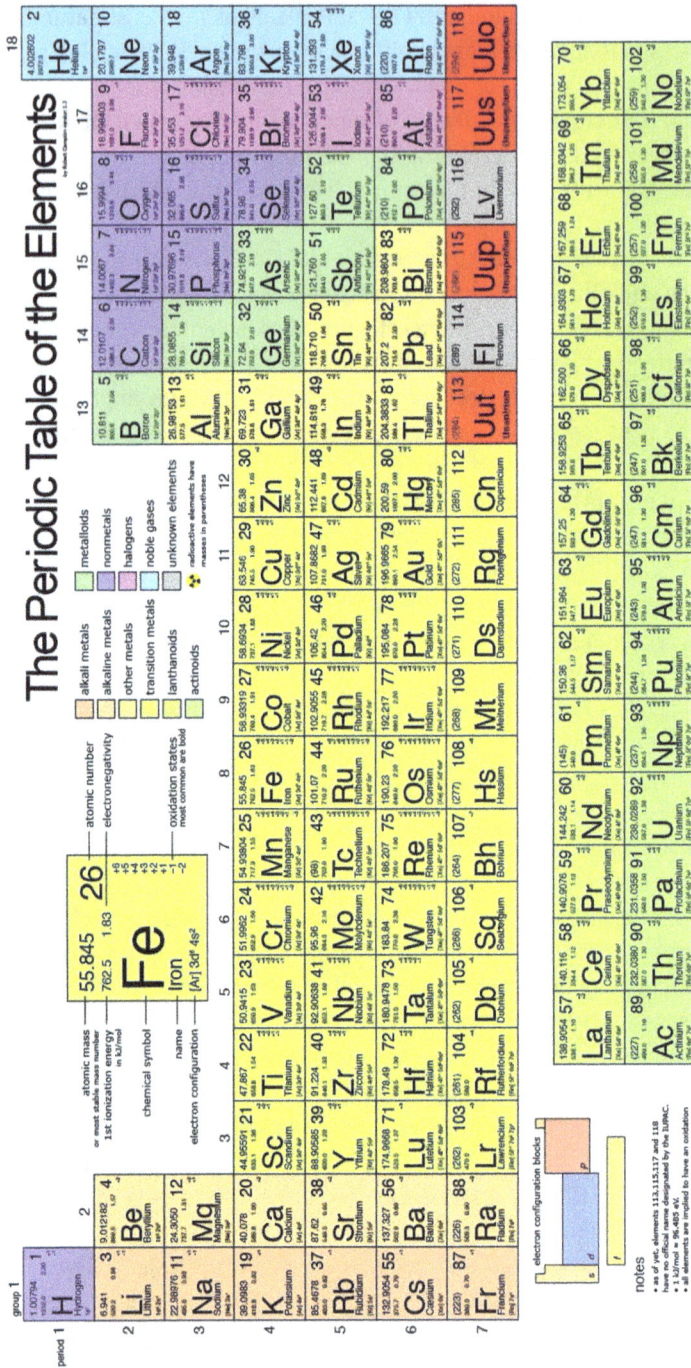

**Fig. 1.2:** The periodic table showing heavy metals and metalloids.[3]

---

3 Accessed on March 08, 2025 via: https://hi-static.z-dn.net/files/d4d/5bd4b6cea3f5ede67dcc4b7b5ab5542.jpg

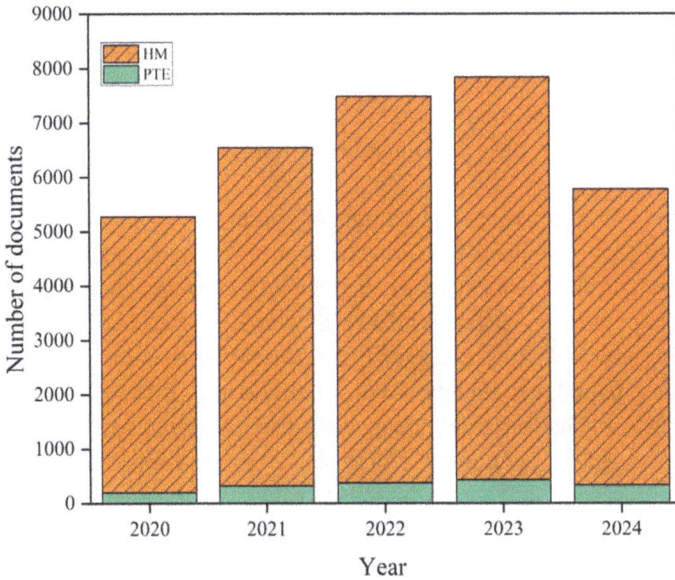

**Fig. 1.3:** Proportions of publications by research disciplines using the term "heavy metals" in their titles. Source: Author's own work.

and academic publications. They therefore suggested that researchers should employ widely accepted definitions and provide explicit nomenclature for elements, either individually or as a collective of metals or metalloids. Further, they recommended that the term "PTEs" should be used in environmental studies that focus on contamination and possible toxicity.

To put it succinctly, the matter was recently resolved by Gustin et al. [834], who argued that a universally accepted, rigid scientific definition of a heavy metal or PTE does not exist. They further stated that regardless of whether authors opt to employ the term or not, the extent of components included in the study should be explicitly stated in the study, research, and technical report at an early stage. In other words, the term in question ought to be specified or disclosed during the preliminary stages of project planning, technical report composition or drafting, or when presenting research findings in a documentary, scientific study, or documentary.

## 1.9 Introducing the concept of restoration of mine degraded and contaminated sites

Revegetation is an effective phytoremediation strategy for the elimination or reduction of toxic elements and pollutants from soil or water, and it includes the introduction of organic matter into the soil. Phytoremediation is a plant-based remediation strategy that aims at cleaning the soil from contaminants and to restore the functionality of a degraded or contaminated ecological system. Tonelli et al. [481], trace the historical development of phytoremediation and conceptualize it as an efficient, well-accepted, low-cost, eco-friendly strategy to contribute to the restoration of environments.

These efforts, hence, aim at improving and ultimately restoring the quality of the degraded lands. The land or the soil may be restored or improved by reducing or eliminating the transfer of metals and metalloids into food chain to safeguard food quality and protect public health of humans. Such remedial action by reducing ecotoxicology of the contaminated mining sites is what is termed as *remediation*. When certain aspects of the degraded land are repaired (e.g., filling gullies to control erosion), it is called land or soil *rehabilitation*. The whole concept of taking actions aimed at ultimately restoring the quality of the degraded soil or land refers to as land or soil *reclamation*. In the mining industry, the term "restoration" is used interchangeably with rehabilitation and refers to the process of returning the ecosystem to its previous state [1050]. Alternatively, restoration may involve the establishment of a new ecosystem when the changes in both living and nonliving components have been excessively severe [1051, 1052]. In essence, this refers to a procedure wherein the environmental effects of mining activities are mitigated through the restoration of a durable land surface, subsequently accompanied by the reintroduction of vegetation or the establishment of an alternative land utilization on the rehabilitated landform. The land may thus be reclaimed for various purposes such as research and education, agriculture and forestry, recreation, aquaculture, and tourism. These are referred to as *post-mining land use*. Many examples of post-mining land use are outlined in Mborah et al. [303].

The objective of using revegetation to recover the soil quality and its health is grounded in science in that there is a strong correlation between vegetation cover and the subsequent richness of topsoil. This is further demonstrated by [548] when he reported that trees have a direct impact on both soil fertility and productivity. The author posited that the utilization of trees for the restoration of degraded lands has been a longstanding practice. This is attributed to trees' ability to thrive in challenging climatic and soil conditions, as well as their potential for mitigating soil erosion. Furthermore, the author noted that the degradation of land is often accompanied by a decrease in organic matter, a significant portion of which is contributed by the foliage and roots of trees.

The presence of organic matter in soil serves the dual purpose of providing essential nutrients to plants and contributing significantly to the preservation of soil's physical, chemical, and biological characteristics [557]. According to [557], the primary source of soil nitrogen, phosphorus, and sulfur in tropical soils is organic matter. Vegetation significantly contributes to erosion management in various contexts, including gullied regions, landslides, sand dunes, building sites, road embankments, mine spoils, and pipeline corridors [341].

**Fig. 1.4:** A sign post showing an area earmarked for restoration project in the Ashanti region of Ghana.

# 1.10 Current state of land reclamation in mining areas in Ghana

## 1.10.1 Commonly used land reclamation practices in Ghana

Mining is a significant economic activity in Ghana, contributing substantially to the country's GDP and employment [44]. However, the environmental impact of mining, particularly in terms of land degradation, has been profound. The extraction of minerals such as gold, bauxite, and manganese has led to the destruction of vast tracts of land, resulting in soil erosion, loss of vegetation, and the formation of mine tailings that pose risks to both the environment and human health [208]. In response, land reclamation has become a critical component of environmental management in mining-affected areas of Ghana [254].

*Regrading and reshaping of landforms*: One of the most commonly used reclamation practices in Ghana involves the regrading and reshaping of mined landforms to restore the landscape to a more natural and stable state [48]. This process typically includes the leveling of spoil heaps and the filling of open pits to reduce the risk of erosion and to create a foundation for future land uses [83]. Regrading is often the first step in the reclamation process and is essential for creating a surface that can support subsequent restoration efforts [335].

*Vegetative cover and reforestation*: Following regrading, the establishment of vegetative cover is a critical practice in land reclamation. In Ghana, reforestation efforts often involve the planting of fast-growing tree species, such as Acacia, Teak, and Cassia, which are known for their ability to stabilize soils, enhance organic matter, and improve soil structure [335]. These species are typically chosen for their resilience to harsh environmental conditions and their ability to quickly provide a protective cover that minimizes soil erosion and promotes water infiltration [7, 369].

In addition to fast-growing species, there has been a growing emphasis on the use of native and indigenous species in reforestation efforts. This shift is driven by the recognition that native species are better suited to the local environment and can support the restoration of local biodiversity [49]. The Ghanaian government, in collaboration with mining companies and nongovernmental organizations (NGOs), has initiated several reforestation projects aimed at restoring native forest ecosystems in degraded mining areas [45].

*Soil amendments and fertilization*: To address soil degradation, particularly in areas where mining has significantly altered the soil chemistry and structure, the application of soil amendments and fertilizers is a common practice. In Ghana, amendments such as lime, compost, and organic fertilizers are used to improve soil pH, enhance nutrient availability, and increase organic matter content [17]. These amendments help create a more hospitable environment for plant growth, thereby accelerating the reclamation process [370, 479].

*Phytoremediation*: Phytoremediation, the use of plants to extract, sequester, or detoxify contaminants in the soil, is also gaining traction in Ghana. This practice is particularly relevant in areas where mining activities have resulted in the contamination of soils with PTEs such as arsenic, lead, and mercury [84]. Certain plants, such as vetiver grass and Indian mustard, have been identified for their ability to uptake heavy metals and are being used in experimental reclamation projects to reduce soil toxicity and rehabilitate contaminated sites [16, 524].

## 1.10.2 Successes and challenges of land reclamation strategies

*Successes in land reclamation*: In Ghana, several land reclamation projects have achieved notable successes, particularly in the restoration of vegetative cover and the stabilization of degraded landscapes [47]. Reforestation efforts have resulted in the reestablishment of forested areas that provide habitat for wildlife, improve local microclimates, and contribute to carbon sequestration. Additionally, the use of soil amendments has proven effective in restoring soil fertility, enabling the cultivation of crops on previously degraded lands [369].

One of the key successes of reclamation efforts in Ghana has been the rehabilitation of mined lands for agricultural use. In some regions, reclaimed lands have been converted into productive farmlands, supporting the livelihoods of local communities [335]. This success is particularly significant, given the importance of agriculture to Ghana's economy and food security [369]. The integration of agroforestry practices in reclamation projects has further enhanced the sustainability of these efforts, allowing for the simultaneous restoration of ecological functions and the generation of economic benefits [1061].

*Challenges in land reclamation*: Despite these successes, land reclamation in Ghana faces several challenges. One of the most significant challenges is the limited enforcement of environmental regulations. While laws such as the EPA Act require mining companies to rehabilitate degraded lands, enforcement is often weak, and many companies fail to comply fully with reclamation requirements [206]. This has led to inconsistencies in the quality and extent of reclamation efforts across different mining regions.

Another challenge is the persistence of PTEs in reclaimed soils. Even after the application of phytoremediation and other reclamation techniques, some PTEs remain in the soil, posing ongoing risks to environmental health and agricultural productivity [23]. The effectiveness of phytoremediation is often limited by factors such as plant selection, soil conditions, and the extent of contamination, making it difficult to achieve complete remediation [417].

Additionally, the sustainability of reclamation efforts is often threatened by socioeconomic factors. In many mining-affected communities, there is a lack of awareness and education about sustainable land use practices, leading to the degradation of reclaimed lands through activities such as illegal mining, overgrazing, and unsustainable farming practices [1053]. These activities can undo the progress made through reclamation efforts, resulting in a cycle of degradation and restoration [208].

Finally, the high cost of land reclamation remains a significant barrier. Comprehensive reclamation projects require substantial financial resources for activities such as soil amendment, reforestation, and long-term monitoring [479]. In many cases, mining companies, particularly smaller operators, may lack the funds necessary to carry out effective reclamation. This has led to a reliance on government and donor funding, which may not always be sufficient or sustainable in the long term [58].

## 1.11 Regional variations in reclamation success

The success of land reclamation strategies in Ghana varies across different geographic regions, reflecting the diversity of environmental conditions, mining practices, and socioeconomic contexts [83]. For instance, in the western region, which is heavily affected by gold mining, reclamation efforts have focused on restoring large-scale open-pit mines, with varying degrees of success [335]. In contrast, in the Ashanti region, where both large-scale and small-scale mining activities are prevalent, reclamation efforts are often challenged by the high prevalence of illegal mining, which hinders the long-term sustainability of restoration projects [58].

In some regions, the involvement of local communities in reclamation projects has been a key factor in achieving success. Community-based reclamation initiatives, supported by NGOs and government agencies, have demonstrated that local participation can enhance the effectiveness of reclamation efforts by ensuring that projects are tailored to the needs and priorities of the communities [487]. However, in other regions, the lack of community engagement has led to conflicts over land use and the neglect of reclaimed lands [208].

## 1.12 Understanding soil quality indicators in reclaimed lands

### 1.12.1 Critical soil quality indicators in land reclamation

Assessing the success of land reclamation projects, particularly in mining-affected areas, requires a comprehensive evaluation of soil quality. Soil quality indicators are vital metrics that provide insight into the physical, chemical, and biological properties of soils, reflecting their ability to support vegetation, maintain ecological functions, and sustain long-term land use [238]. The most critical soil quality indicators used in land reclamation projects include soil pH, organic matter content, soil structure, nutrient availability, and microbial activity [128, 484].

Soil pH is a crucial indicator of soil quality, as it directly affects the availability of nutrients to plants and the activity of soil organisms [91]. In reclaimed lands, soil pH often deviates from the optimal range (typically 6.0–7.5 for most crops), particularly in areas where mining activities have led to acidification or alkalinization [231]. The measurement of soil pH is typically conducted using a pH meter or colorimetric methods in soil extracts. Interpreting soil pH involves understanding its impact on nutrient solubility and potential toxicities; for instance, extremely low pH can increase the solubility of toxic metals like aluminum, while high pH can lead to nutrient deficiencies [153].

Organic matter content is a critical indicator of soil health and fertility, influencing water retention, soil structure, and nutrient cycling [262]. In reclamation projects, the restoration of organic matter is often a key goal, as it enhances soil resilience and supports the reestablishment of vegetation [459]. Organic matter is measured through methods such as loss on ignition or wet oxidation, where the amount of carbon in the soil is quantified [351]. Higher organic matter content typically correlates with improved soil fertility, better moisture retention, and increased biological activity, all of which are essential for the successful reclamation of degraded lands [185].

The soil structure refers to the arrangement of soil particles into aggregates, which influences aeration, water infiltration, and root penetration [94]. A good soil structure is indicative of a healthy soil environment that can support robust plant growth [480]. In reclaimed lands, soil structure is often compromised due to compaction and the disruption caused by mining activities. The soil structure is assessed through visual examination and physical tests such as aggregate stability tests, which measure the resistance of soil aggregates to disintegration when exposed to water [240]. Well-aggregated soils with stable structure are more resistant to erosion and are better suited to sustaining long-term vegetation.

The availability of essential nutrients such as nitrogen (N), phosphorus (P), and potassium (K) is a critical determinant of soil quality in reclaimed lands [91]. Nutrient availability is often reduced in disturbed soils due to the loss of topsoil and organic matter [422]. In reclamation projects, soil tests are conducted to measure the concentrations of available nutrients, typically through chemical extraction methods [212]. The interpretation of these tests involves comparing the nutrient levels to established benchmarks for optimal plant growth. Ensuring adequate nutrient availability is crucial for supporting the establishment and growth of vegetation in reclaimed lands [401].

Soil microbial activity is a key indicator of soil biological health, reflecting the presence and activity of beneficial microorganisms that drive nutrient cycling, organic matter decomposition, and soil structure formation [383]. In reclaimed lands, the microbial activity can be suppressed due to soil disturbance and contamination [54]. Measuring the microbial activity often involves assessing the soil respiration rate, enzyme activities, or microbial biomass carbon [463]. High microbial activity is associated with healthy, functioning soils that are capable of sustaining plant growth and ecosystem recovery [219].

## 1.13 Correlation between soil quality indicators and long-term sustainability

The correlation between soil quality indicators and the long-term sustainability of reclaimed lands is a critical area of focus in land reclamation research. Sustainable reclaimed lands are those that can maintain their productivity, ecological function, and resilience over time, without requiring continuous intensive management [260]. Understanding these correlations is essential for developing reclamation strategies that ensure long-term soil health and ecosystem stability [1054].

Maintaining a balanced soil pH is crucial for the long-term sustainability of reclaimed lands. Soils with a stable pH within the optimal range support better nutrient availability and reduced metal toxicity, which are essential for the sustained growth of vegetation [299]. Long-term studies have shown that soils with well-managed pH levels tend to have higher resilience to environmental stresses, such as drought and extreme weather events, making them more sustainable over time [1055, 1056].

Organic matter plays a pivotal role in building soil resilience. Soils with higher organic matter content are better equipped to retain moisture, support microbial communities, and maintain nutrient cycling, all of which contribute to long-term sustainability [887]. In reclaimed lands, the continuous addition of organic amendments, such as compost or manure, can help to build and maintain organic matter levels, leading to more sustainable ecosystems [185]. The positive feedback loop between organic matter and soil structure also enhances the ability of soils to resist erosion and degradation [262].

The development of good soil structure is essential for the long-term stability and sustainability of reclaimed lands. Well-structured soils promote efficient water infiltration and root penetration, reducing the risk of erosion and improving plant establishment [94]. Over time, soils with stable aggregates are less likely to suffer from compaction or surface crusting, which can hinder plant growth and lead to land degradation [480]. Thus, the soil structure is closely linked to the sustainability of reclaimed ecosystems.

The availability of nutrients is directly linked to the productivity of reclaimed lands. Sustained nutrient availability ensures that vegetation can thrive without the need for excessive fertilization, which can be both economically and environmentally costly. Over the long term, soils that can maintain adequate nutrient levels through natural processes, such as nitrogen fixation or organic matter decomposition, are more likely to support sustainable ecosystems. Conversely, nutrient-deficient soils may require ongoing intervention to prevent land degradation and loss of productivity.

Microbial activity is a strong indicator of the biological health and sustainability of reclaimed soils. Active microbial communities are essential for nutrient cycling, organic matter decomposition, and the formation of soil structure [383]. In the long term, soils with high microbial activity are more resilient to disturbances and are better able to recover from environmental stressors [219]. The presence of a diverse and active microbial community is therefore a key factor in ensuring the sustainability of reclaimed lands.

In summary, soil quality indicators such as pH, organic matter content, soil structure, nutrient availability, and microbial activity are critical in assessing the success of land reclamation projects. These indicators not only reflect the current health of the soil but also provide insight into its potential to sustain vegetation and ecological functions over the long term. A balanced and holistic approach to managing these indicators is essential for ensuring the long-term sustainability of reclaimed lands, particularly in mining-affected areas, where soil degradation and contamination pose significant challenges.

# 1.14 Comparing reclaimed and undisturbed soils

## 1.14.1 Differences in pedological characteristics

Reclaimed soils, particularly those in post-mining landscapes, exhibit significant differences in pedological characteristics compared to undisturbed soils. These differences arise from the extensive disturbance caused by mining activities, which often involve the removal of topsoil, disruption of soil structure, and exposure of subsoil or parent material [762]. The process of reclamation, while aimed at restoring soil functionality, often results in a soil profile that is markedly different from that of undisturbed lands [934].

One of the most pronounced differences between reclaimed and undisturbed soils is soil texture and structure. In undisturbed soils, natural processes such as weathering, organic matter accumulation, and bioturbation by soil organisms lead to the development of well-defined soil horizons with a stable structure [971]. In contrast, reclaimed soils often lack these distinct horizons due to the mixing of soil layers during mining and reclamation activities (Shukla et al., 2003). Reclaimed soils may exhibit a more uniform texture, often dominated by coarse fragments if the reclamation process involved the use of overburden or spoil material [453].

The structure of reclaimed soils is also typically less stable compared to undisturbed soils. The compaction that occurs during mining and reclamation can lead to poor aggregation and reduced porosity, limiting water infiltration and root penetration [177]. Over time, this can result in increased surface runoff and erosion, further degrading the soil [762].

Soil organic matter (SOM) is a critical component of soil health, influencing nutrient availability, water retention, and soil structure. In undisturbed soils, the accumulation of organic matter through plant litter and root decay supports a rich microbial community and robust nutrient cycling [383]. However, reclaimed soils often have significantly lower organic matter content due to the loss of topsoil and the disruption of natural organic inputs [454].

The nutrient content of reclaimed soils is also typically lower than that of undisturbed soils. Essential nutrients such as nitrogen, phosphorus, and potassium are often depleted during the mining process and may not be fully restored through reclamation efforts [453]. Even when fertilizers or soil amendments are applied, reclaimed soils may struggle to retain these nutrients due to poor structure and low organic matter levels [1057].

The pH of reclaimed soils can vary widely depending on the type of mining activity and the materials used in reclamation. In some cases, reclaimed soils may be more acidic or alkaline than undisturbed soils due to the exposure of sulfur-rich minerals (leading to acid mine drainage) or the application of alkaline materials during reclamation [226]. These shifts in pH can alter the availability of nutrients and the mobility of PTEs, making the chemical environment of reclaimed soils quite different from that of undisturbed soils ([762, 1057]).

Undisturbed soils typically support a diverse and active biological community, including bacteria, fungi, earthworms, and other soil fauna. These organisms play a crucial role in nutrient cycling, organic matter decomposition, and soil structure formation [1058]. In contrast, reclaimed soils often exhibit reduced biological activity and biodiversity due to the loss of organic matter, changes in soil structure, and the presence of contaminants [197]. The slow recovery of soil biota in reclaimed soils can limit the restoration of ecological functions and delay the establishment of a stable, self-sustaining ecosystem [344].

## 1.15 Evolution of soil properties over time

The physical, chemical, and biological properties of reclaimed soils evolve over time as natural processes such as weathering, organic matter accumulation, and microbial colonization gradually improve soil conditions [454]. However, the rate and extent of this evolution can vary widely depending on the initial conditions of the reclaimed soil, the reclamation practices employed, and the environmental context ([483, 762]).

Over time, the physical properties of reclaimed soils can improve as organic matter is added, either through natural litterfall or through reclamation practices such as mulching and composting [887]. This organic matter helps stabilize soil aggregates, increasing porosity and reducing compaction [94]. The development of a more stable soil structure enhances water infiltration and root penetration, which are critical for the establishment of vegetation and the prevention of erosion [762].

However, the physical recovery of reclaimed soils can be slow, particularly if the soil has been heavily compacted or if the reclamation process involved the use of suboptimal materials [177]. In some cases, physical amendments such as deep ripping or the addition of soil conditioners may be necessary to accelerate this recovery [929].

The chemical properties of reclaimed soils, including pH, nutrient content, and the presence of contaminants, also evolve over time. The addition of organic matter and the growth of vegetation can help buffer soil pH, bringing it closer to the optimal range for nutrient availability [299]. Nutrient levels may gradually increase as organic matter decomposes and releases nutrients, although this process can take years or even decades to fully restore soil fertility [762, 1057].

The fate of PTEs in reclaimed soils is a critical concern. Over time, some PTEs may be immobilized through processes such as adsorption to clay particles or incorporation into stable organic matter, reducing their bioavailability [304]. However, the persistence of PTEs in the soil can pose long-term risks to plant growth and environmental health, and their management remains a significant challenge in land reclamation [9, 230].

The biological recovery of reclaimed soils is perhaps the most complex and variable aspect of soil evolution. Microbial communities are often the first to colonize reclaimed soils, and their activity is essential for nutrient cycling and organic matter decomposition [219]. Over time, as organic matter accumulates and soil structure improves, these microbial communities can become more diverse and stable [344]. The gradual increase in microbial diversity and stability is a key indicator of successful biological evolution in reclaimed soils [1065].

The establishment of higher trophic levels, such as soil fauna and plant roots, depends on the recovery of microbial communities and the overall improvement of soil conditions [1021]. The return of soil biodiversity is a key indicator of the success of reclamation efforts, as it signifies the restoration of ecological functions and the development of a self-sustaining ecosystem [1058, 1065].

In summary, reclaimed soils differ significantly from undisturbed soils in terms of their physical, chemical, and biological properties. These differences are primarily the result of the disturbances caused by mining activities and the challenges associated with reclamation [762]. Over time, reclaimed soils can recover many of these properties, but the rate and extent of recovery depend on a variety of factors, including the initial soil conditions, the reclamation techniques used, and the environmental context [1057]. Understanding these differences and the processes of soil evolution is critical for improving reclamation practices and ensuring the long-term sustainability of reclaimed lands [453].

# 1.16 Long-term outcomes of land reclamation efforts

## 1.16.1 Sustainability of reclaimed lands: insights from long-term studies

Long-term studies of land reclamation efforts are crucial for understanding the sustainability of these projects, particularly in terms of soil quality and the persistence of PTEs. These studies provide valuable insights into how reclaimed lands evolve over time, highlighting both the successes and challenges of restoring ecosystems affected by mining and other industrial activities [762].

## 1.16.2 Changes in soil quality over time

Soil quality is a key indicator of the success and sustainability of land reclamation. Long-term studies have shown that soil quality in reclaimed lands can improve significantly over time, particularly with the continuous addition of organic matter, the establishment of vegetation, and the implementation of proper soil management practices. However, the rate and extent of improvement can vary depending on the initial condition of the soil, the reclamation techniques used, and the environmental context [197].

For instance, studies conducted in reclaimed mining areas in Canada and Australia have shown that SOM content, nutrient availability, and soil structure can gradually recover to levels comparable to those of undisturbed soils over a period of several decades. These improvements are often attributed to the successful establishment of plant cover, which contributes to organic matter accumulation, root development, and the stabilization of soil structure [184].

However, not all reclaimed lands exhibit such positive outcomes. In some cases, soil quality remains compromised, particularly in areas where the original soil profile was severely disrupted or where contamination from PTEs persists. For example, research in former coal mining regions of the United States has found that, even after decades of reclamation, some soils still exhibit low fertility, poor structure, and limited biological activity. These issues are often linked to incomplete restoration of the soil profile and the challenges of managing residual contamination [454].

## 1.16.3 Persistence and changes in PTE concentrations

The behavior of PTEs in reclaimed soils is a critical factor influencing the long-term sustainability of these lands. Long-term studies have revealed that while some PTEs may become immobilized or less bioavailable over time, others may persist in the soil, posing ongoing risks to the environment and human health [9].

For example, research in reclaimed lands affected by lead and zinc mining in Europe has shown that, although surface PTE concentrations may decrease due to leaching, deep soil layers can remain heavily contaminated. This vertical migration of PTEs can pose risks to groundwater quality and may lead to recontamination of surface soils under certain conditions, such as erosion or land use changes [385]. Additionally, the effectiveness of natural attenuation processes, such as the binding of PTEs to organic matter or clay particles, can vary. In some long-term reclamation projects, PTEs have been found to remain bioavailable, particularly in acidic soils where metal solubility is high. This ongoing bioavailability poses a risk to plants and soil organisms, potentially undermining the sustainability of the reclaimed ecosystem [25].

## 1.16.4 Long-term vegetation dynamics

The establishment and succession of vegetation on reclaimed lands play a critical role in the long-term stabilization and sustainability of these ecosystems. Long-term studies have shown that successful vegetation cover can lead to improvements in soil quality, enhanced biodiversity, and the creation of self-sustaining ecosystems. However, achieving such outcomes often requires careful selection of plant species, ongoing management, and, in some cases, periodic intervention [125].

In some reclaimed mining areas, initial plantings of fast-growing, hardy species have been followed by the gradual colonization of native species, leading to more diverse and resilient plant communities. This natural succession process is essential for restoring ecological functions and supporting wildlife. However, in other cases, reclaimed lands have struggled to maintain vegetation cover, particularly in regions with harsh climates or where soil conditions remain poor. The failure of vegetation to establish or persist can lead to soil erosion, loss of organic matter, and a decline in overall ecosystem health [210].

# 1.17 Justification and significance of the book

To ensure the long-term sustainability of the mining industry, it is crucial to gradually restore and rehabilitate polluted and damaged mining sites throughout the active phase and at the end of the mine's project life cycle. Furthermore, it is imperative to implement sustainable methods to rectify the harm inflicted on the land during mineral extraction, restore the mine sites, and guarantee the secure reutilization of the land. Inevitably, implementing such measures is necessary to ensure the preservation of the ecosystem and the well-being of both humans and animals in mining regions. There is an urgent need for gentle remediation studies to address the possible harm posed by contaminated abandoned gold mining spoils [332]. One of the methods used is "green remediation," which has been recognized as a mild in situ remediation technique [1059].

Thus far, a multitude of physical and chemical approaches have been utilized or suggested to restore or recover the integrity of polluted and deteriorated mine sites. The physical approach mainly entails the filling of excavated holes, reshaping the terrain, and later removing the contaminated soil from the surface and replacing it with clean uncontaminated topsoil. In addition, the physical approach employs containment techniques such as constructing barriers, caps, and liners to prevent the migration and transmission of hazardous compounds into the adjacent ecosystem and mining communities [515]. The technique has also incorporated chemical treatments with oxides (e.g., iron oxides) to stabilize contaminants, subsequently reduce their solubility, and thereby reduce their environmental and human health effects [251, 376, 509].

The application of physical and chemical stabilization techniques in the restoration of mine-affected lands is a subject of ongoing debate in the scientific literature. These procedures have the potential to either hinder or improve the movement of PTEs in contaminated sites ([376, 1060]). As an illustration, Mensah et al. [332], observed that the utilization of compost and manure on an abandoned mine site contaminated with As in Ghana resulted in a 332% and 315% increase in the bioavailable

As content, respectively. According to a study by [1061], a field with arsenic contamination (As) showed an increase in soil organic carbon content after the application of cow dung manure. This, in turn, helped to reduce the availability of As in a paddy field located in Pakistan. Alternatively, alterations in soil pH and redox conditions might lead to the dissolution of organic carbon and metal oxides utilized in chemical treatments. They might then release these substances when PTEs stick to or are absorbed by their surfaces, which could pollute groundwater and contaminate the food chain [150, 268, 549]. In addition, Yang et al. [520], observed that applying biochar to a polluted area during floods abetted the solubility of arsenic (As), leading to increased mobilization and release of the contaminant.

Therefore, the successful implementation of these physicochemical methods to remediate vast areas of severely polluted land may require significant financial investment and specialized knowledge in diligent monitoring, ongoing quantification, and evaluation. At the same time, the feasibility of using these methods to restore large portions of polluted and deteriorated mining sites may not meet the criteria for socio-environmental and economic sustainability for the mining corporation, local residents, and surrounding ecosystems.

Phytoremediation through revegetation, which uses plants to restore polluted and damaged mine sites, could be a low-cost way for the mining industry to achieve long-term sustainability for the host mining community, its environment, and its economy. For example, the strategy could be used to address heavy metal contamination issues in gold mine soils ([360, 964]). Therefore, by reducing food chain contamination, limiting their mobility, reducing metal transfer into surface and groundwater, and guaranteeing a metal-contaminated-free environment, mine land revegetation could lessen the negative effects of PTEs on human health. Furthermore, revegetation may aid with carbon sequestration, regulate wind and water erosion, guarantee water quality, lessen runoff from contaminated mining waste, preserve and improve soil fertility, and conserve soil moisture.

In Ghana, revegetation is still a developing environmental engineering science that requires more in-depth investigation and study to expand its applicability. In addition, the pros and cons of various case studies, experiments done in the field and in a greenhouse, and applications in restoring mine-damaged sites from different mining companies in Ghana need to be documented and carefully evaluated. By doing so, the success stories and positive lessons can be upscaled and the challenges rectified to clean and reclaim the country's many polluted and degraded mine sites. To this end, I:

i.   provide the most recent review on the current status and progress of revegetation by mining companies and research institutions; and
ii.  critically appraise the science of revegetation for the restoration of mine-degraded sites in Ghana to ultimately achieve sustainability in the mining sector.

This book provides critical literature reviews regarding the reestablishment of vegetation in areas affected by gold mining waste materials, such as mine spoils and tailings.

A comprehensive analysis was conducted on the rehabilitation of deteriorated and polluted mining areas in Ghana. These literary works were published as articles in the academic databases "Web of Science," "Science Direct," "Springer," "Taylor and Francis," and "Frontiers." Studies and articles explored various topics related to mining, including the effects of mining on soil quality, the process of restoring mining sites, the regulatory frameworks governing the mining sector in Ghana, the sustainability of the mining sector, phytoremediation (using plants to clean up contaminated soil), land re-habilitation, phytoavailability (the availability of nutrients to plants), and plant uptake (the absorption of nutrients by plants).

It also presents a literature analysis on the regulatory agencies that are responsible for monitoring, supervising, and enforcing environmental restoration and sustainability efforts carried out by different mining enterprises. This provides a thorough evaluation of both scientific and gray literature on the use of plants to restore and clean up deteriorated and contaminated mine sites.

## 1.18 Conclusions

This chapter introduces the topic of mine land degradation and revegetation in Ghana, and provides the background and objectives of the book. The chapter discusses the impact of mining activities on soil quality, the strategies employed for soil rehabilitation through revegetation, and the regulatory framework governing the mining sector's sustainability in Ghana. The chapter highlights the importance of restoring degraded mine lands to ensure productive use, environmental protection, and social justice. It also sets the stage for the subsequent chapters, which cover various methods, techniques, and case studies related to mine land reclamation and revegetation.

Albert Kobina Mensah

# Chapter 2
# Methods used in soil and human health risk assessment

**Abstract:** The chapter discusses the determination of total element contents in soil and environmental media as a crucial step in identifying potential risks associated with toxicity. Various laboratory methods and protocols are used to determine the total element or contaminant contents, such as the nitric acid digestion method or aqua regia. Soil characterization is also conducted to assess physical, chemical, and biological parameters. The chapter further highlights the environmental and human health risks associated with gold mining, including water pollution, cyanide spillage, soil pollution, and various diseases. A case study on arsenic contamination in a gold mine tailing is presented, showing the distribution of total arsenic content in different samples. The chapter concludes by discussing the use of risk assessment indices to evaluate soil contamination and human health risks, such as contamination factor, enrichment factor, pollution load index, geo-accumulation index, hazard quotient, and hazard index. These indices are important for assessing pollution, planning remediation, and ensuring compliance with regulations.

## 2.1 Determination of the total element contents

The total elemental content (TEC) technique serves as a valuable method for evaluating the distribution of contamination and the dispersion of pollutants originating from contaminated sites. It serves as a valuable tool for evaluating environmental and human health issues. Furthermore, TEC provides relevant data concerning the nutrient composition found in geophagic substances. Furthermore, the TEC is relevant to theories linking geophagy to potential contamination and its implications for human health concerning PTEs. Thus, assessing the total pseudo-trace element content is a vital initial step and fundamental for identifying potential toxicity risks associated with an element. Before any full-scale remediation strategy could be planned for any soil or environmental media, initial characterization of the media under investigation is a prerequisite requirement. Thus, determination of total pseudo-trace element contents is a very important initial step and key to identifying a potential risk associated with the toxicity of an element. Soil characterization, determination of

**Albert Kobina Mensah,** Soil Research Institute, Council for Scientific and Industrial Research, Kumasi, Ghana, e-mail: albert.mensah@rub.de, albertkobinamensah@gmail.com, ORCID: https://orcid.org/0000-0001-5952-3357

https://doi.org/10.1515/9783111662046-002

soil's physicochemical properties, as well as determination of the total element contents in the contaminated soils are carried out in experiments using various laboratory methods and protocols. For example, the contents of total elements or contaminants are determined using the nitric acid digestion method or aqua regia. In the nitric acid digestion method, the digestion process is carried out at 120 °C for 15 min. This approach utilizes 0.25 g of soil plus a mixture of 10 mL concentrated nitric acid and 10 mL deionized water, as specified by the [574]. After extraction, the total elements in the samples' filtrates are determined either with the atomic absorption spectrophotometer or with the inductively coupled plasma optical emission spectrophotometer (ICP-OES).

Soil characterization is done for physical, chemical, and biological parameters such as soil pH, electrical conductivity, total carbon and nitrogen, texture, moisture content, exchangeable cations, cation exchange capacity, available P, and total element contents. For the soil pH, it is carried out with a pH meter in conjunction with a 1:5 ratio of soil to 0.01 M $CaCl_2$ solution. Procedures according to Wright and Bailey (2001) can be used to calculate the total carbon, nitrogen, as well as sulfur. This is done after going through the extraction procedure for C and N. C, N, and S are determined using the elemental analyzer.

In other jurisdictions, C is determined via the Walkley Black method, but all these depend on the protocol at the accredited laboratory or what the research team uses or prefers in their labs. The laser scattering approach developed by [550] can be employed for determining the texture. This method gives different peaks representing the specific contents of the clay, sand, silt, and others. Thereafter, the texture of the sample is predicted using the textural triangle.

For the characterization of soil amendments and their properties (with the exception of the texture), the same procedures used for the soil are employed but different weights may be required depending on the protocol used in the laboratory. The results are displayed either in the form of tables or in graphs. Tables 2.1 and 2.2 give examples of the characterization or the physicochemical properties of the soil and amendments used in one phytoremediation project by Mensah et al. [332].

## 2.1.1 Application of the total element contents method in identifying environmental risks

Gold mining presents varied environmental and human health threats in the areas where the mining takes place. Water pollution and increased cost of water treatment, cyanide spillage from mine tailings, collapse of tailings, pollution of aquatic biodiversity, mercury pollution, depletion of fertile lands, food crop contamination, soil pollution, and many neurological, genetic, and skin diseases are common in mining areas. These emanate from unsustainable mining practices such as poor management of mine wastes and abandonment of mine lands at the closure phase of the mineral extraction project's life cycle. Consequently, these affect the health of residents in mining communities and deprive them of their livelihoods.

To investigate this hypothesis in detail, I first collected samples from an abandoned mine tailing (10 individual sampling points and bulked together into one) that measures 126,000 m$^2$. Initial analyses of the samples indicated total As content at 1,809 mg/kg, while other heavy metals in the mining tailing were either below or equal to the recommended threshold. Thus, As pollution is a major problem in the mining region, and therefore studying it in detail and devising a sustainable remediation or mitigation strategy will be worth pursuing. This led to the study of arsenic (As) in detail. The initial elemental concentrations are shown in Fig. 2.1. The detailed follow-up of total As contents in the active, abandoned top soil, and abandoned profile mine spoils are shown in Fig. 2.2.

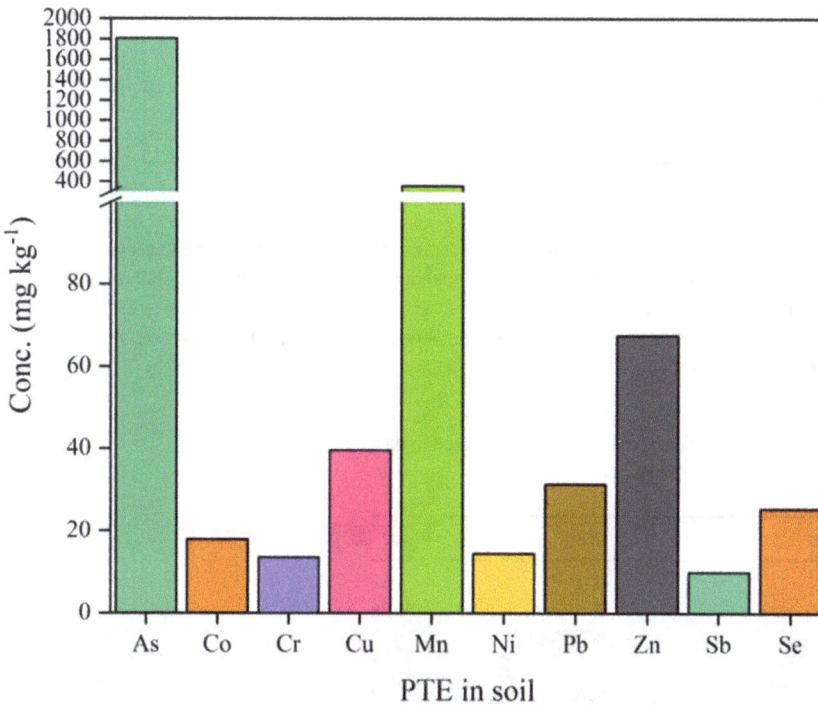

**Fig. 2.1:** Initial total element concentrations in a gold mine tailing in southwestern Ghana.

### 2.1.1.1 Conclusion and implications of graph in Fig. 2.2

The risk of pollution was in the following order: active mine spoil < profile < abandoned mine spoil. Thus, even though both mine spoils contained As contents in many folds above the recommended thresholds, the extent of contamination and risks were more pronounced in surface soils and profiles. The higher As content in the surface soils

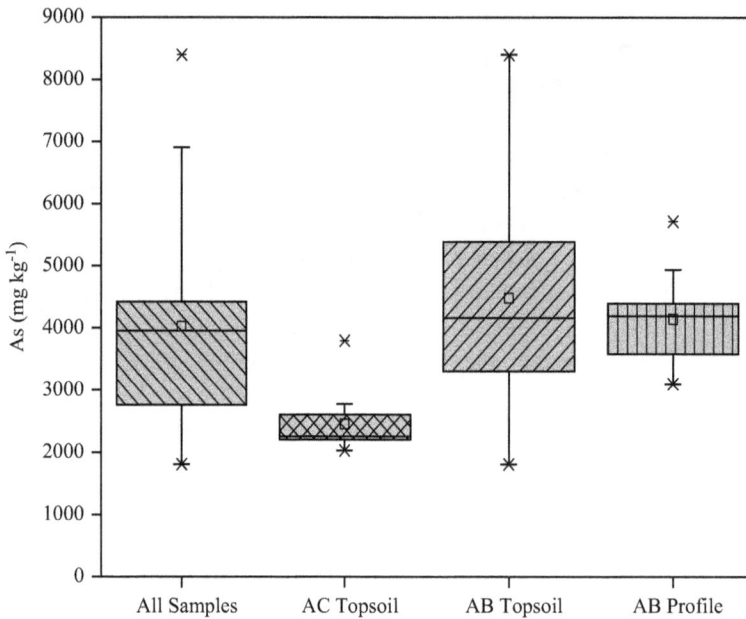

**Fig. 2.2:** A box plot showing the distribution of total content of As in all samples ($n$ = 51): active (AC topsoil; $n$ = 10), abandoned topspoil (AB topsoil; $n$ = 31), and abandoned profile mine spoils (20–100 cm depth, AB profile $n$ = 10). Asterisk marks (lower and upper) in the range represent 1% and 99% percentiles, respectively. Asterisks outside the range represent outliers. Small rectangular boxes represent the mean. Source: Mensah et al. [321].

**Tab. 2.1:** Physicochemical properties of a contaminated mine spoil soil in Mensah et al. [331].

| Parameter | Unit | Value |
|---|---|---|
| Sand | % | 62 |
| Silt | % | 36 |
| Clay | % | 3 |
| pH (CaCl$_2$) | | 6.4 |
| EC | dS/cm | 1.9 |
| Total C | % | 1.2 |
| Total P | mg/kg | 13.4 |
| *Total element content* | *mg/kg* | |
| Fe | | 23,200 |
| Ca | | 8,994 |
| Mg | | 5,502 |
| As | | 4,282.6 |

**Tab. 2.2:** Physicochemical characterization of compost and manure used in Mensah et al. [322].

| Parameter | Unit | Compost | Manure |
|---|---|---|---|
| pH | | 6 | 7 |
| EC | dS/cm | 0.009 | 0.023 |
| Al | | 5,137 | 2,257 |
| Ca | | 15,880 | 88,646 |
| Fe | | 8,734 | 1449 |
| K | mg/kg | 3,161 | 17,859 |
| Mg | | 4,664 | 5,432 |
| Na | | 1,204 | 2,847 |
| P | | 1,126 | 9,521 |
| Mn | | 471 | 392 |
| Zn | | 166 | 234 |

might be due to many years of accumulation and stockpiling of the mining waste materials. Moreover, the extent of pollution was higher in the abandoned mining spoil than in the active mine spoil. It was assumed that these differences might have arisen from difference in either ore characteristics, historical extraction procedures [321], age of the mine spoils [37], or different tailing depositional times as reported in Edraki et al. [140].

## 2.2 Identifying soil and human health risks using risk assessment indices

The results generated from the total potentially toxic element (PTE) contents determined via the nitric acid digestion or aqua regia will enable one to do a calculation and carry out a risk assessment in terms of soil contamination and human health impacts of the mining spoils on the people living in the mining areas. The calculations can be extended to human health impacts and risk assessment using average daily dose (ADD), hazard quotient (HQ), and hazard index (HI). The assessment can further be carried out for carcinogenic and noncarcinogenic risks. Various equations used in the estimation of risks are mentioned below.

### 2.2.1 Soil contamination assessment indices

The degree of soil contamination associated with PTEs of a contaminated field or site is determined or quantified using four indices: contamination factor (CF), enrichment factor (EF), pollution load index (PLI), and geo-accumulation index ($I_{geo}$). Additionally, noncarcinogenic human health risk associated with possible soil ingestion of PTEs is calculated using HQ and HI. These indices have successfully been employed in past studies for monitoring the extent of PTE contamination and assessing human health risk of soils along rivers in Germany [405], in garden vegetable soils [37], in children's playgrounds in Beijing [225], in temperate and arid regions in Germany and Egypt [440], around a lead/zinc smelter in southwestern China [271], around an industrial area in Greece [35, 37], around abandoned iron ore mines in north central Nigeria [551], and in mine tailings in Ghana [321].

Soil contamination assessment indices are important for the following reasons:
- They help assess pollution and locate polluted areas.
- These indices help analyze soil and sediment pollution and establish effective remediation techniques to preserve human health and the environment.
- Regular monitoring and assessment of soil CFs are essential for maintaining soil health, protecting the environment, and ensuring the long-term sustainability of land use practices.
- These indices aid in assessing ecological effects. Here, the contaminated soil can disrupt natural ecosystems by affecting the health of soil-dwelling organisms, which are the foundation of terrestrial food webs. This can have cascading effects on higher trophic levels and ecosystem functioning.
- These indices aid in assessing persistence and planning remediation. Different contaminants have varying degrees of persistence in the soil, with some remaining for years or even decades. Therefore, doing a soil contamination assessment will assist to plan specific mitigation and remedial measures for controlling continuous pollution associated with PTEs.
- Monitoring helps identify the extent of contamination and informs decisions about remediation efforts to restore soil health.
- Abet in legal and regulatory compliance. Many regions have regulations in place to limit soil contamination, especially in industrial and agricultural areas. Monitoring soil contamination is necessary to ensure compliance with these regulations and avoid legal issues.

### 2.2.1.1 Soil contamination factor (CF)

To monitor soil health in the context of contamination, various techniques can be used, including soil testing, chemical analysis, and biological assessments (Fig. 2.3). It is defined as the ratio of the concentration of the toxic element (TE) in an environmental medium, for example, soil or sediment to the background concentration of the same element in the soil or in the sediment. In brief, the CF is the sample pollutant concentration divided by its background concentration in uncontaminated soils. Higher CF values indicate a higher extent of pollution. The soil CF is calculated using the following equation:

$$CF = C_s / C_{ref} \tag{2.1}$$

where CF represents the contamination factor of soil, $C_s$ is the total element content in soil (mg/kg), $C_{ref}$ is the reference/background value of the element in uncontaminated soil (average values of world's soil were obtained from Kabata-Pendias [230]).

The reference background values for various PTEs in mg/kg, also reported in Fig. 2.1, can be used as follows: As = 6.8; Al = 20,000; Cd = 0.2; Cr = 59.9; Cu = 38.5; Fe = 20,000; Ni = 29, Pb = 27; Ti = 0.57; V = 90; and Zn = 70. The reference value for Ti is obtained from Saha et al. [414], that of V from Li et al. [271], and those of Al and Fe are obtained from Shaheen et al. [440].

This index further classifies the degree of soil contamination using the following categories: low contamination (when CF < 1), moderate contamination (CF = 1–3), considerable contamination (CF = 3–6), and very high contamination (CF ≥ 6) [271, 321, 405, 440].

## 2.2.1.2 Pollution load index (PLI)

The pollution load index serves as a criterion to evaluate instances of multi-element contamination at a place. There are instances where the scope of the site cannot be ascribed to a single element. Consequently, the pollution may originate from many instances of elemental contamination.

The soil contamination index is utilised for human risk assessment in such instances. This occurs by aggregating all the averages of the contamination variables and elevating the final output to the exponent of one divided by the number of elements deemed relevant to the pollution at the affected areas.

In mining land reclamation, there are instances where the degree of pollution may be ascribed to a singular source. Particularly in mine tailings, one can infer numerous instances of elemental pollution at the site, which may encompass contamination from various sources, including arsenic as the predominant element, along with silver, mercury, cyanide, antimony, selenium, titanium, and others.

PLI is the product of individual CF values of all studied PTEs, as per Rinklebe et al. [405], and [552]:

$$PLI = (CFs_1 \times CFs_2 \times \cdots \times CFs_n)^{1/n} \tag{2.2}$$

where $CFs_1$, $CFs_2$, $CFs_n$ are the contamination factors of elements 1, 2, . . ., $n$ under consideration. PLI > 1 indicates substantial PTE soil contamination of the area or site under study or consideration.

## 2.2.1.3 Enrichment factor (EF)

Enrichment factor (Fig. 2.4) helps to ascertain whether PTE pollution in the soil is geogenic or anthropogenic in origin. It assumes that the contents of elements such as Al and Fe occur in a natural medium and are mostly geogenic; hence, they are used as normalizer elements in calculating the enrichment [271, 440]. Therefore, enrichment of the soil with higher contents of PTEs (higher EF values) is an indication of pollution from anthropogenic origin. It is usually argued in other jurisdictions that EF values greater than 1.5 may give a clue that the source of pollution from the contaminated site is more anthropogenic than geogenic. The EF values categorize contamination as minor (if EF = 1.5–3), moderate (EF = 3–5), severe (EF = 5–10), and very severe (EF > 10) as per Antoniadis et al. [36, 37]:

$$EF = (C_s/Al_s)/(C_{ref}/Al_{ref}) \tag{2.3}$$

where Als is the total content of Al in the contaminated soil, and $Al_{ref}$ is the total content of Al in the uncontaminated/background reference soil [e.g., 271, 414, 440].

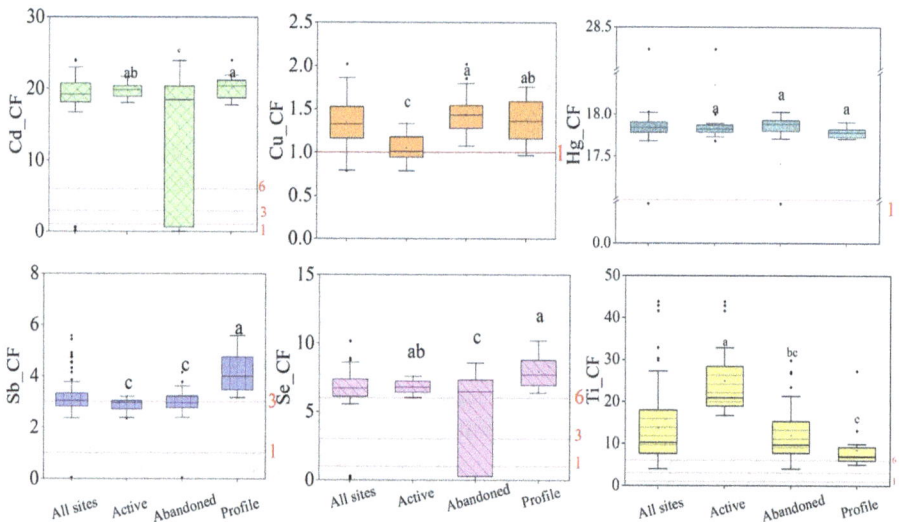

**Fig. 2.3:** Soil contamination factor (CF) of PTEs in the active, abandoned, and profile pits in As-contaminated gold mine tailings. Asterisk marks (lower and upper) in the range represent the 1% and 99% percentiles, respectively. Asterisks outside the range represent outliers. Small rectangular boxes represent mean values. The added horizontal red lines indicate the threshold above which there is contamination.

### 2.2.1.4 Geo-accumulation index ($I_{geo}$)

Geo-accumulation index as per Klubi et al. [245], is given as follows:

$$I_{geo} = \log 2\left(\frac{C_S}{1.5C_{refs}}\right) \tag{2.4}$$

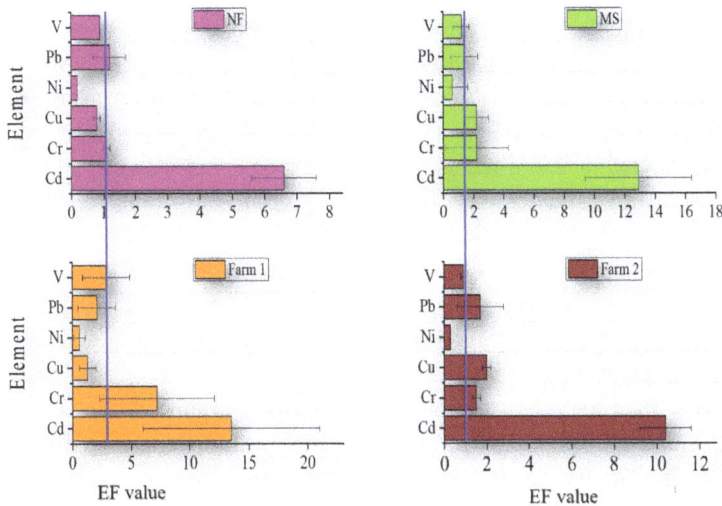

Fig. 2.4: Soil enrichment factor (EF) of selected PTEs in the soils and farms near the mine spoil. NF = natural forest; MS = mine surrounding. Farm 1 = Peasant farm under cassava cultivation; Farm 2 = Subsistence farm consisting of cocoa, plantain and cocoyam. The added horizontal red lines indicate the maximum allowable EF limit.

## 2.2.2 Human health risk assessment indices

### 2.2.2.1 Hazard quotient (HQ) and hazard index (HI)

The HQ and HI are two measures used to evaluate the potential health risks posed by exposure to environmental contaminants. They are widely used in environmental health risk assessment and toxicology to evaluate the potential health effects of exposure to pollutants in the environment. Both the HQ and HI are important tools in environmental health risk assessment and toxicology, and they are used in conjunction with other measures to evaluate the potential health risks posed by exposure to environmental contaminants. However, it is important to note that these measures are based on a limited amount of information and should be used with caution in making decisions about human health risks. Further investigation and data collection are often necessary to fully understand the potential health effects of exposure to environmental contaminants.

The soil-to-human health risk assessment (SHRA) methodology, according to [553], can be applied to conduct a detailed health risk assessment of any residential or commercial/industrial scenario and is recommended as a viable initial tool in planning remediation of contaminated sites. The reported limitations (e.g., [453, 554]) of SHRA are that it is based on rigid assumptions that are not unique for every exposed human, does not consider preexisting health problems, and lacks certainty.

Recognizing these limitations, SHRA is the single most important approach to identify possible health risks deriving from PTE exposure and it is the most adopted method in the literature in assessing the human exposure to soil PTEs ([453, 1012]). Eventually, SHRA indicates the possibility of health risk associated with exposure to PTE. Thus, such studies provide a necessary initial step of assessing health risk, aid in addressing problems concerning the risk of metal pollution and human health outcomes associated with mining, and abet in proposing more sustainable approaches toward reclamation of multielement-contaminated mining spoils in gold mining-dominated regions.

1.  HQ: This is a measure of the potential health risk posed by exposure to a single contaminant. It is calculated by dividing the estimated exposure dose of a contaminant by its reference dose, which is the level of exposure considered safe for human health. The HQ is a unit less value, and values greater than 1 indicate that the exposure dose is higher than the reference dose and may pose a health risk. The HQ is useful in determining the potential health risks associated with exposure to specific contaminants and in identifying contaminants that may require further investigation.

2.  HI: This is a measure of the potential health risk posed by exposure to multiple contaminants. It is calculated by summing the HQs for all contaminants of concern in a particular exposure scenario. The HI is also a unit less value, and values greater than 1 indicate that the combined exposure to multiple contaminants may pose a health risk. The HI is useful in evaluating the overall health risk posed by exposure to multiple contaminants and in determining the need for further investigation and risk management.

## 2.2.3 Carcinogenic and noncarcinogenic human health risk assessments

### 2.2.3.1 Noncarcinogenic health risk assessment

Noncarcinogenic soil risk assessments are conducted to evaluate the potential health risks associated with exposure to contaminants in soil that do not primarily cause cancer but may still pose health concerns. These assessments focus on non-cancer health effects, such as acute and chronic toxicity, developmental effects, reproductive effects, and other adverse health outcomes. Noncarcinogenic soil risk assessments are essential for protecting human health and the environment from the adverse effects of exposure to hazardous substances in soil. These assessments help guide regulatory decisions and remediation efforts to mitigate risks effectively.

### 2.2.3.2 Importance of soil-to-human health risk assessment: applications and significance of past findings

Noncarcinogenic health risk assessments are usually calculated for three groups of people (children, adult males, and adult females) exposed via three possible routes: *ingestion* of the contaminated soil, *inhalation* of the contaminated soil, and *dermal contact* with the contaminated soil. Children are mostly vulnerable to these risks of contamination via all three possible routes. For instance, contaminated abandoned mining sites (i.e., spoils and tailings dam) are often used by children as playgrounds, and in turn they eat the contaminated samples either by coming into contact with their skin, inhaling the contaminated substances, or accidentally ingesting the contaminated samples orally. Additionally, children are most often vulnerable because they have the tendency to eat and come into contact with the contaminated soil when they are crawling. Various studies have reported higher HQ and HI values for children, indicating that they are mostly at greater health risks [e.g., 319, 321, 340, 405, 447].

Women are also often at risk because they use abandoned minefields as pathways to farms, which are mostly contaminated with arsenic, cadmium, lead, mercury, and titanium [e.g., 316, 319, 321]. Additionally, women intentionally or unintentionally consume contaminated soil. The practice of intentionally eating soil is called geophagy. In this regard, [555] found that many women in Tanzania were at higher risk of contracting cancer from eating contaminated soil. Such practices are common among pregnant women, and it is a common cultural and traditional practice in some communities.

Geophagy may be driven by nutritional deficiencies, especially in regions where access to essential minerals or nutrients is limited. Soil or clay may contain minerals like iron, calcium, or magnesium that are lacking in the diet. These practices make women vulnerable to the health impacts of PTEs. Such practices are common in Latin America and Eastern Africa [96, 242]. Geophagy is not limited to humans. Various animal species, including birds, mammals, and insects, engage in geophagy. It is believed that animals may consume soil or clay to aid in digestion, detoxify harmful substances, or obtain essential minerals.

Hou et al. [215], conducted a thorough examination of research pertaining to the presence of PTEs in soil samples in close proximity to gold mining sites spanning from 2001 to 2022. The study revealed that lead (Pb) and mercury (Hg) posed a noncarcinogenic hazard, specifically to children residing near the mine, but not to adults. Conversely, arsenic (As) posed a noncarcinogenic risk to both children and adults living in close proximity to the mine. The carcinogenic hazards associated with the elements As, Cd, and Cu were found to exceed the tolerable level. Researchers determined that gold mining activities have resulted in significant soil pollution in the surrounding areas, hence posing potential ecological hazards. Moreover, this contamination posed both noncarcinogenic and carcinogenic risks to individuals of all age groups residing in close proximity to the mine site.

Children often have a higher soil and human health HI compared to adults for several reasons (some of these reasons are outlined in Hou et al. [215]):

- Behavioral factors: Children are more likely to engage in behaviors that increase their exposure to soil contaminants. They often play on the ground, put objects and hands in their mouths, and have a higher likelihood of ingesting soil accidentally. These behaviors can lead to greater exposure to soil contaminants.
- Body size: Children have smaller body sizes and lower bodyweights compared to adults. As a result, the same level of exposure to contaminants in soil can have a more significant impact on a child's body on a per-unit-weight basis. Their developing organs and systems may also be more vulnerable to the toxic effects of certain substances.
- Developmental factors: Children's bodies and organ systems are still developing, which can make them more susceptible to the adverse effects of soil contaminants. Their developing nervous, immune, and endocrine systems may be particularly sensitive to environmental toxins.
- Higher breathing rates: Children generally have higher breathing rates per unit of bodyweight than adults. This means that if there are airborne contaminants associated with the soil, such as dust particles or volatile chemicals, children may inhale more of them relative to their bodyweight, increasing their exposure.
- Hand-to-mouth activity: Children commonly engage in hand-to-mouth activities, including touching contaminated surfaces or soil and then putting their hands or objects in their mouths. This behavior can lead to the ingestion of soil contaminants that may not affect adults in the same way.
- Duration of exposure: Children may spend more time outdoors and have longer exposure durations to soil contaminants, especially if they live near contaminated sites or in areas where soil contamination is prevalent.
- Immature detoxification processes: Children may have less developed detoxification mechanisms in their bodies compared to adults. This can result in slower clearance of certain contaminants, allowing them to accumulate to higher levels in the body.
- Lack of awareness: Children may not be aware of the risks associated with soil contamination, and they may not take precautions to minimize their exposure.
- Dietary habits: Children's dietary habits, including consumption of fruits and vegetables grown in contaminated soil, can contribute to their exposure to soil contaminants.

## 2.2.4 Calculating average daily dosages, hazard quotients, and hazard index for noncarcinogenic risks

Three routes of PTE exposure to the human body are usually explored: (1) ingestion, (2) dermal absorption, and (3) inhalation of air-borne particles.

**Example 1 (exposure route: ingestion):** We aimed to quantify the soil-to-human health impacts of PTEs associated with soil ingestion of the contaminated soils in an abandoned gold mining spoil.

**Methods:** This involved three main steps: first, we calculated the ADD of possible soil ingestion of PTEs; this was followed by the calculation of HQ; and then we calculated HI. HI is the summation of HQs of all the studied PTEs. It is assumed that an adverse health impact exists when HQ and HI are above 1, while adverse health effects are unlikely at HQ and HI of less than, or equal to, 1 [e.g., 271, 333, 405]. The calculations are as follows:
ADD, calculated in mg/kg bodyweight per day:

$$ADD = C_s \frac{(IR \times EF \times ED \times 10^{-6})}{(BW \times AT)} \tag{2.5}$$

where IR is soil ingestion rate (children = 200; adults = 100 mg dust/day); EF is exposure frequency (children = 350; adults = 250 days/year); ED is exposure duration (children = 6 years; adults = 25 years); BW is bodyweight (children = 15 years; adult males = 68 kg; adult females = 58 kg); AT is average time (which is 6 × 365 days = 2,190 days for children, and 25 × 365 days = 9,125 days for adults); and $10^{-6}$ is for unit conversion. Values are used and reported in previous studies [e.g., 41, 86, 321, 405, 440].
Calculation of HQ was then done:

$$HQing = \frac{ADD}{RfD} \tag{2.6}$$

where RfD is the PTE oral reference dose as follows: As = 0.0003; Al = 1; Cd = 0.001; Cr = 1.5; Cu = 0.04; Fe = 0.7; Ni = 0.02, Pb = 0.0035; Ti = 0.0003; V = 0.28; and Zn = 0.3 (values in mg PTE per kg bodyweight per day, as obtained from [271, 271, 405, 440, 447]). RfD values for Ti were obtained from Saha et al. [414], and that of V from Li et al. [271].

**Findings:** The resultant HI values for all sites were 0.7–134.56 for children, 0.05–10.6 for adult males, and 0.13–12.77 for adult females (Fig. 2.2). In all fields, HI was in the following decreasing order: children > adult females > adult males. The high HI values, especially for children in the mine surrounding, the *Pueraria* sites, and farms 1 and 2, surpassed by far the critical threshold of 1, indicating very high health risk implications associated with PTEs for children in the mining communities. Similar results have also been found in other similar works [e.g., 35, 340].

**Conclusions and implications:** These high HI values were largely contributed by As (range: 63.2–88.1%), displayed in fig. 2.5, indicating very high health risk implications associated mainly with As for women and children in the mining areas. In conclusion, I speculate that women and children in the mining areas may be susceptible to health-related problems. Women and children may thus be exposed to As-related health threats such as that of Buruli ulcer, other skin diseases, genetic disorders, neurological problems, birth deformities in newborn babies, and cognitive dysfunctions among children. In this regard, [556] found that cases of Buruli ulcer were greater in As-enriched drainages and farmlands in mining districts in Ghana. Additionally, cases of bladder and lung cancers, reproductive outcomes, and declined cognitive function are consistently reported among As-exposed populations in Latin America [e.g., 96, 242].

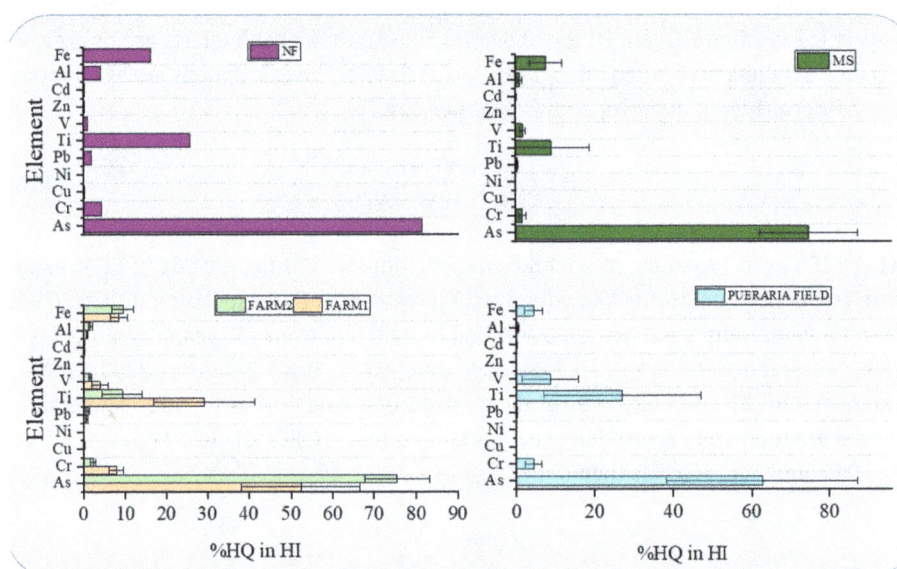

**Fig. 2.5:** Mean percentage contributions of hazard quotients of the studied PTEs to the hazard index of the various mining environments. NF, natural forest; MS, mine surrounding; PF, *Pueraria* field.

**Example 2:** We collected 20 samples from an active mine spoil, 62 samples from abandoned mine spoil, and 20 samples from 1 m depth soil profile dug at the abandoned mines [see 316]. We air-dried the samples for 48 h and later milled, homogenized, and sieved via 2 mm sieve. We characterized them for their basic physicochemical properties using standard methods as per [557].

We extracted the total content of Al, Cd, Cu, Hg, Ti, Fe, Mn, Se, and Sb using the microwave digestion method as described by the [558]. Contents of PTEs in the soil were digested and measured with ICP-OES (Spectroblue, Ametek Materials Analysis Division).

For the details of the procedure, quality control measure, quality assurance, standard solutions, standard reference materials, blanks, as well as replicate measurements were employed. We then evaluated the topsoil (0–20 cm) contamination as well as the profile (100 cm) with PTEs using the following indices as per previous studies [e.g., 35, 319, 321].

The HI of the elements for children were as follows: active = 1,681.81; abandoned = 1,443.21; and profile = 1,921.64. All adult female HI were: active = 155.3; abandoned = 133.3; and profile = 177.5. In addition, HI values for adult males were: active = 132.5; abandoned = 113.7; and profile = 151.4. High HI values in the mine spoil exceeded the far-critical threshold of 1, indicating very serious health risks associated with PTEs for children and adults in the communities (Fig. 2.4). Children > adult females > adult males were the HI calculated in decreasing order; similar results have also been found in prior investigations (e.g., Antoniadis et al., [35]).

**Example 3 (exposure route: dermal contact):** In order to determine the probability of dermal contact, we first multiplied the PTE concentration in the mine spoil by the exposed skin area, the adherence factor, the dermal contact factor, the frequency of exposure, and the duration of exposure. Subsequently, the result was divided by the body-weight and the mean time. The following equation gives a summary of the calculations:

$$\text{ADDderm} = \frac{C_s \times \text{SA} \times \text{AF} \times \text{ABS} \times \text{EF} \times \text{ED}}{\text{BW} \times \text{AT}} \tag{2.7}$$

where ADDderm is the average daily dose of dermal contact (mg/kg per day), $C_s$ is the PTE concentration in the mine spoil (mg/kg), SA is the exposed skin surface area (cm²), AF is the adherence factor (mg/cm² per day), and ABS is the dermal contact factor (no unit).

Exposure via inhalation of contaminated particles in soil and dust is calculated using the following formula:

$$\text{ADDih} = \frac{C_s \times \text{ihR} \times \text{EF} \times \text{ED}}{\text{PEF} \times \text{BW} \times \text{AT}} \tag{2.8}$$

where IhR is the inhalation rate (m³/day), EF is the exposure frequency (day/year), and PEF is the particle emission factor (m³/kg).

**NB:** Subsequently, HQs and HI via each exposure pathway are calculated using the above formula as used in the soil ingestion equation, but here the ADD is replaced with the value found for the exposure pathway.

**Example 4 Introduction and study objectives (Fig. 2.7):** Soils may be ingested by humans either deliberately or inadvertently. This activity is referred to as geophagy. It entails the intentional consumption of terrestrial substances, such as clay and various soil kinds. Clayey soils are typically favoured due to their pliability and the presence of minerals and elemental aggregates, such as iron. The practice is predominant among pregnant women in Ghana. The study examined the health effects of geophagic clays in eight marketplaces, focusing on nine PTEs (As, Cd, Cr, Cu, Fe, Mn, Ni, Pb and Zn).

**Methods:** We collected 40 samples of geophagic white clay from eight markets in Kumasi, Ashanti region of Ghana, for analysis. Samples were dried, crushed, and sieved to prevent contamination. pH and electrical conductivity were measured, and potentially toxic elements (PTEs) were extracted and analyzed using Atomic Absorption Spectrophotometry. Soil contamination was assessed using contamination factors, Pollution Load Index, and Geo-accumulation index. Human health risks from PTEs were evaluated through average daily dosage calculations for ingestion and dermal contact. Quality control involved standard reference materials and statistical analyses, including ANOVA and Pearson correlation, to assess relationships between PTEs and soil factors.

**Results and discussion:** The study highlighted the impact of soil carbonate content on soil properties, pH, structure, and nutrient availability. It revealed varying concentrations of potentially toxic elements in geophagic clays across markets, with some markets showing higher levels linked to gastrointestinal and neurological disorders. Arsenic and manganese were detected in several markets, while cadmium exposure was notably higher in both children and women consumers. Lead levels were significantly elevated in some of the markets, causing severe health issues, and zinc was predominantly high and lead to deficiencies and related health problems.

**Conclusions:** The distribution of these PTEs indicates potential environmental and public health risks. Areas with high levels of Cu, Pb, and Cd are particularly concerning due to their severe health impacts. Regular monitoring and mitigation strategies are essential to reduce exposure and protect public health. The study provides a comprehensive assessment of the health risks associated with the consumption of geophagic clays in Ghana, particularly among vulnerable populations such as pregnant women and children.

## 2.2.5 Carcinogenic health risk assessment

Assessing the risk of carcinogenic soil-to-human health is a critical component of environmental and public health management. Carcinogenic soil refers to soil contaminated with substances that have the potential to cause cancer in humans. Conducting a risk assessment involves evaluating the exposure to these carcinogens and determining the likelihood of adverse health effects. Carcinogenic soil risk assessments are complex and require expertise in toxicology, environmental science, and risk analysis. They are crucial for protecting human health and guiding appropriate actions to manage and remediate contaminated soil sites.

Cancer risk is based on the human intake values, DIng, DDerm, and DInh, calculated for the soil-to-human pathways as shown earlier.

Soil-related CR for each TE separately, CRSi (unit less), is calculated as follows:

$$CRSi = CRIng + CRDerm + CRInh \qquad (2.9)$$

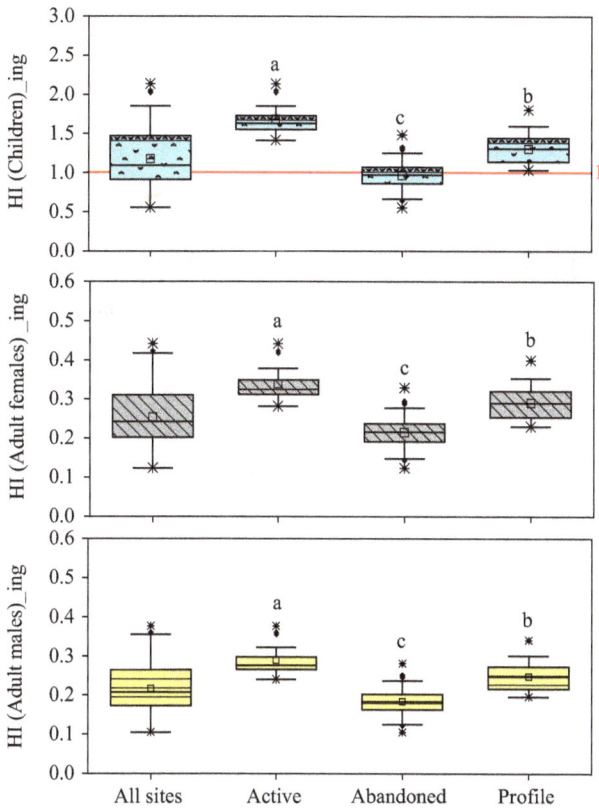

**Fig. 2.6:** Noncarcinogenic soil ingestion hazard index (HI) of PTEs from the active, abandoned, and profile. Asterisk marks (lower and upper) in the range represent 1% and 99% percentiles, respectively. Asterisks outside the range represent outliers. Small rectangular boxes represent mean values. The added horizontal red lines indicate the critical HI point.

where CRIng, CRDerm, and CRInh (all unit less) are cancer risks related to soil ingestion, soil dermal absorption, and inhalation of air particles, respectively. These values are calculated as follows:

$$CRIng = DIng \times OSF \qquad (2.10)$$

where OSF is the oral (e.g., ingestion) slope factor (mg TE/kg BW per day)$^{-1}$. The values of OSF for various TEs as used in this study are reported in Antoniadis et al. [35]:

$$CRDerm = DDerm \times DSF \qquad (2.11)$$

where DSF is the dermal slope factor (mg TE/kg BW per day)$^{-1}$, calculated as follows:

$$DSF = OSF/ABSGI \qquad (2.12)$$

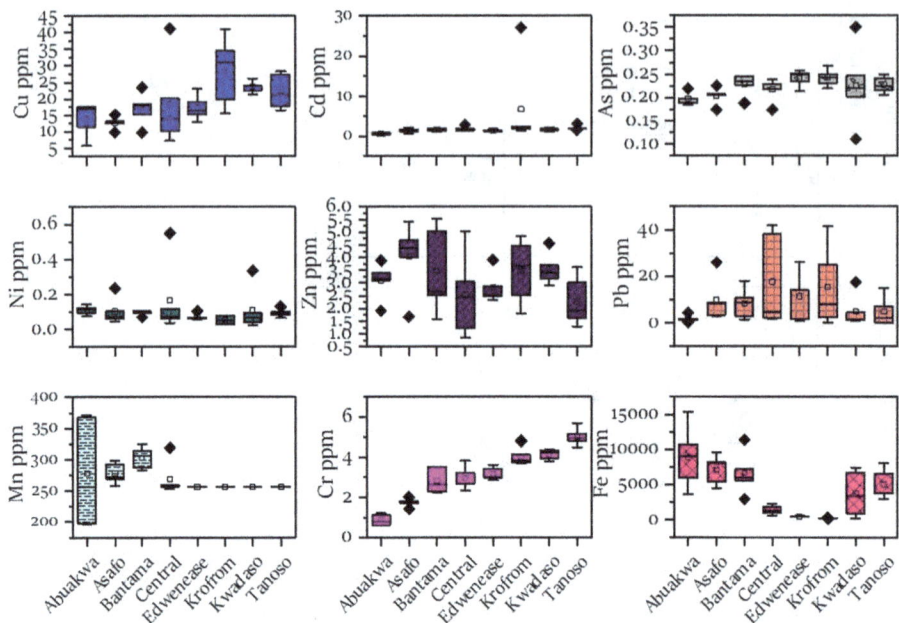

**Fig. 2.7:** Levels of PTEs in geophysical clays collected from 8 markets in Ghana consumed by mostly women. This practice is called geophagy.

$$CRInh = DInh \times ISF \qquad (2.13)$$

where ISF (mg TE/kg BW per day)$^{-1}$ is the inhalation slope factor, calculated as follows:

$$ISF = 70 \, kg \, BW / \left( UR \left[ \mu g \, m^3 \right] \times 20 \left[ m^3 d^1 \right] \times 10^3 [mg^1] \right)$$

where UR is the unit risk ($\mu g$ TE/m$^3$ air), and is used in various other similar works as shown in Antoniadis et al. [35].

The combined cancer risk of all carcinogenic TEs, CRS (unit less), was then calculated as follows:

$$CRS = (CRSi)$$

## 2.2.6 Key steps in assessing the risk of carcinogenic and noncarcinogenic soil-to-human health risks

1. Identification of carcinogenic and non-carcinogenic contaminants:
   - Identify and characterize the specific carcinogenic substances present in the soil. Common carcinogenic contaminants in soil include heavy metals (e.g., arsenic, cadmium, and lead), organic chemicals (e.g., polycyclic aromatic hydrocarbons or PAHs), pesticides, and asbestos.
2. Exposure assessment:
   - Assess how people can come into contact with the contaminated soil. This includes evaluating potential exposure pathways, such as inhalation of airborne contaminants, dermal contact, and ingestion of soil.
   - Estimate the extent and frequency of exposure for different population groups, including residents, workers, and recreational users of the area.
3. Toxicity assessment:
   - Determine the toxicity and carcinogenicity of the identified contaminants. This involves reviewing available scientific literature and toxicological studies to understand the health risks associated with exposure to these substances.
   - Establish reference doses or cancer slope factors, which are used to quantify the risk of cancer associated with exposure to a particular carcinogen.
4. Risk characterization:
   - Combine the exposure assessment data with the toxicity assessment data to quantify the risk of cancer for exposed individuals. This is typically expressed as a probability or an HQ.
   - Consider various factors, such as the duration and intensity of exposure, as well as individual susceptibility, when assessing the risk.
   - Evaluate the cumulative risk from multiple contaminants if more than one carcinogen is present in the soil.
5. Risk management and mitigation:
   - Based on the risk assessment, develop risk management strategies to reduce or eliminate exposure to carcinogenic soil. These strategies may include remediation efforts to clean up the contaminated soil, restricting access to the contaminated area, or implementing protective measures for workers.
   - Establish clean-up goals or remediation standards that aim to reduce contaminant levels in the soil to acceptable levels.
6. Communication and public awareness:
   - Communicate the results of the risk assessment to affected communities, stakeholders, and regulatory agencies.
   - Engage with the public to ensure that they are aware of potential risks and understand any precautionary measures they should take.
7. Regulatory compliance:

   – Ensure compliance with local, state, and federal regulations related to soil
     contamination and carcinogenic substances.
   – Adhere to guidelines and standards set by environmental agencies when con-
     ducting risk assessments and implementing remediation measures.
8. Monitoring and follow-up:
   – Regularly monitor the site to assess the effectiveness of remediation efforts
     and track changes in soil contamination levels.
   – Update risk assessments as new data become available or as site conditions
     change.

## 2.3 Conclusions

In conclusion, the use of indices is crucial for identifying polluted mining sites, developing effective remediation strategies, and ensuring compliance with regulations.
These indices serve as valuable tools in environmental management, allowing for targeted actions to be taken in areas where soil contamination and pollution is prevalent. By utilizing these indices, decision-makers can prioritize resources and efforts
toward areas that require immediate attention. Furthermore, these indices aid in the
planning and implementation of remediation measures, enabling the restoration of
polluted locations and the protection of human health and the environment. Ultimately, the use of indices plays a vital role in promoting sustainable practices and
safeguarding our ecosystems.

Albert Kobina Mensah

# Chapter 3
# Identifying risks using sequential extraction analyses, size fractionation, and acid neutralization capacity experiments

**Abstract:** The chapter focuses on the application of sequential extraction methods to assess the mobilization and remediation of arsenic (As) in heavily gold-mine-contaminated sites. The experiment tested various soil amendments, including biochar, compost, iron oxide, manure, and NPK fertilizer, to reduce the bioavailability of As and improve soil quality. The results showed that iron oxides were the most effective amendment in reducing the bioavailability of As, attributed to their sorption effect and positive charges. The chapter also found that soils with higher clay content had higher proportions of As, and the potential mobility of As was higher in the fine fraction compared to the coarse fraction. The findings have implications for the remediation of As in mining spoils and highlight the importance of pH and environmental conditions in controlling As mobility.

## 3.1 Introduction

The determination of the total element content of potentially toxic element (PTE) has been plagued by a number of drawbacks. In spite of the fact that it is a first point of call that characterizes a study site, it does not represent the actual toxicity, availability, or mobilization of the PTE being discussed. Depending excessively on such an approach may result in either an overestimation or an underestimation of the level of toxicity associated with the PTE. Using pseudo-total PTE to evaluate the level of contamination and, consequently, the pollution of an environmental medium is another one of the many drawbacks of this environmental risk assessment method.

Moreover, total PTE concentrations are the most significant contaminant in soils, but available concentrations should receive more attention. Most PTEs exist as available fractions that humans can easily ingest and bio-accumulate in diverse tissues and organs. This raises the risk to human health. More importantly, it is interesting to know how much of these TEC will become accessible and available, as this will have an impact on human health if contaminated soils, for example, are consumed. To give you an example, let us imagine that a mining site or an arable field has a concentration of 90 mg/

**Albert Kobina Mensah,** Soil Research Institute, Council for Scientific and Industrial Research, Kumasi, Ghana, e-mail: albert.mensah@rub.de, albertkobinamensah@gmail.com, ORCID: https://orcid.org/0000-0001-5952-3357

https://doi.org/10.1515/9783111662046-003

kg of arsenic or 25 mg/kg of cadmium. Would we consider this facility to be contaminated? Does it inevitably suggest that there are concerns regarding the health of the soil or the ecosystem surrounding the field? On the surface, looking at the data, it is evident to assume and point out that the sites are contaminated or polluted. The average amount of arsenic in the world's soil is 6.83 mg/kg, and that of cadmium is 0.02 mg/kg.

In this circumstance, producing a correct answer may require first separating the elements into their respective binding forms or geochemical fractions. To achieve precise results, it is essential to understand the percentages of elements that are mobile, potentially mobile, and the proportion associated with sulfides and silicates known as the residual fraction of the elements. It is possible that larger quantities of As or Cd are bonded to the residual fractions, indicating that these elements are retained in forms that do not pose environmental hazards. Typically, components associated with the residual fractions are not accessible for contamination.

In other cases, the biogeochemical conditions and characteristics that are present at the location will determine the percentage of the element that actually becomes a cause for concern regarding environmental risks or contamination of the food chain. The actual toxicity of the elements at the site will depend on the biogeochemical variables present in that environment. The variables under consideration encompass fluctuations in redox chemistry, the dominant pH levels present at the location, and additional factors including organic carbon content, site-specific water regimes, the existence and type of anions (e.g., chlorides, phosphates, sulfates, nitrates, and carbonates), as well as the concentrations and variations of sesquioxides like aluminum, iron, and manganese.

Thus, the total contents are not good indicators of bioavailability and do not reflect the true toxic effects of PTEs [513]. It is for this reason that the sequential extraction procedures (SEPs) are employed. This procedure separates PTE into different binding forms or geochemical fractions. Wenzel [513] developed a five-step methodology for extracting the geochemical fractions of As. These fractions are defined as fraction I (nonspecific sorbed As/water soluble), fraction II (exchangeable and surface/specifically adsorbed), fraction III (crystalline iron oxide fraction), fraction IV (amorphous iron oxide fraction), and fraction V (residual fraction). Fractions I and II are classified as the soluble/mobile fractions and they are most toxic in terms of ecosystem and environmental pollution and toxicity. The summation of fractions 1 to 4 gives the potential fraction. Greater percentages indicate greater environmental risks to water, groundwater, and translocation into the food chain. Figure 3.1 shows a summary of SEP in a flowchart diagram and Fig. 3.2 shows the sequentially extracted As samples ready for analysis.

Here, you calculate the percentage recovery after the extraction of the various fractions. The calculation is as follows:

$$\text{Recovery} = \frac{(\sum \text{FI} - \text{FV})}{\text{Total As}} \times 100 \qquad (3.13)$$

where ($\sum$ FI – FV) is the sum of the element extracted in the five-step sequential extraction and "Total As" is the result obtained from the nitric acid microwave digestion.

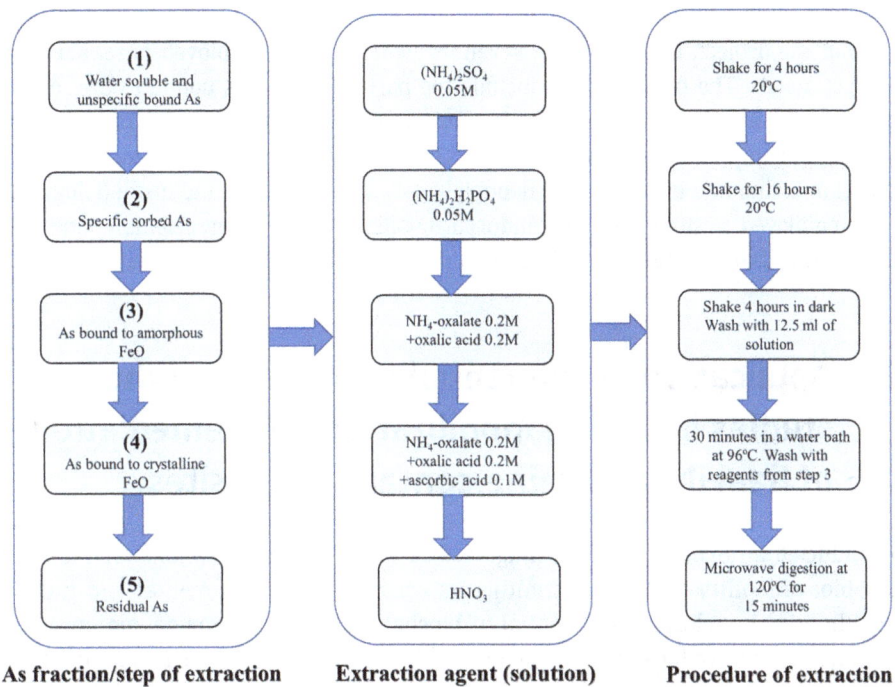

| **(1)**<br>Water soluble and<br>unspecific bound As | | $(NH_4)_2SO_4$<br>0.05M | | Shake for 4 hours<br>20°C |
| --- | --- | --- | --- | --- |
| ↓ | | ↓ | | ↓ |
| **(2)**<br>Specific sorbed As | | $(NH_4)_2H_2PO_4$<br>0.05M | | Shake for 16 hours<br>20°C |
| ↓ | | ↓ | | ↓ |
| **(3)**<br>As bound to amorphous<br>FeO | → | $NH_4$-oxalate 0.2M<br>+oxalic acid 0.2M | → | Shake 4 hours in dark<br>Wash with 12.5 ml of<br>solution |
| ↓ | | ↓ | | ↓ |
| **(4)**<br>As bound to crystalline<br>FeO | | $NH_4$-oxalate 0.2M<br>+oxalic acid 0.2M<br>+ascorbic acid 0.1M | | 30 minutes in a water bath<br>at 96°C. Wash with<br>reagents from step 3 |
| ↓ | | ↓ | | ↓ |
| **(5)**<br>Residual As | | $HNO_3$ | | Microwave digestion at<br>120°C for<br>15 minutes |
| **As fraction/step of extraction** | | **Extraction agent (solution)** | | **Procedure of extraction** |

**Fig. 3.1:** Summary of the As-sequential extraction procedure in a flowchart diagram.
Source: Author's own work

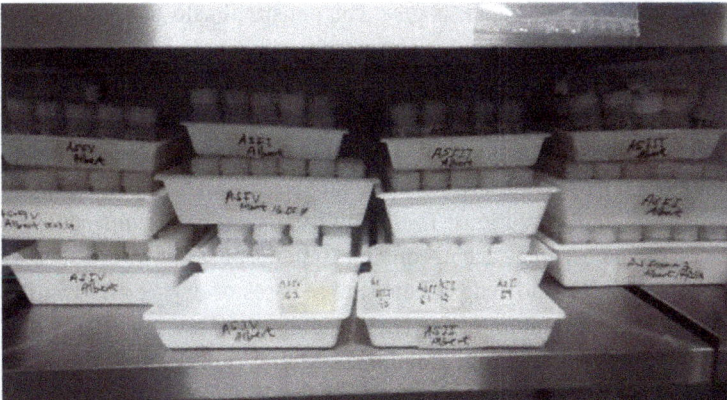

**Fig. 3.2:** Soil filtrates ready for analysis and determination of As geochemical fractions with the inductive couple plasma-optical emission spectrometry (ICP-OES). Picture taken at the Ruhr University Bochum Geography Institute Laboratory on the 14th June 2019.
Source: Author's own work

**NB**: Sometimes, one may choose to use two, three, or four steps depending on the objectives of the project. In other cases, seven or eight steps are employed to sequentially extract metals. The fractions may include the part of the metals bound to the soil organic matter and the carbonate. For instance, the seven-step approach had been used to study the impact of various amendments on immobilization and phytoavailability of nickel and zinc in a contaminated floodplain soil [436]. Similarly, the methodology has been employed to study the redox-induced mobilization of copper, selenium, and zinc in deltaic soils originating from Mississippi (USA) and Nile (Egypt) River Deltas [437].

## 3.2 Application of sequential extraction method for studies involving mobilization and remediation of heavily gold-mine-contaminated sites

I conducted an experiment by testing various soil remediation options for reducing the bioavailability of As, thus limiting its consequent environmental and human health risks. I explored the potential of biochar, compost, iron oxide, manure, and NPK fertilizer to reduce the bioavailability of As and improve the soil quality of the degraded mining site. These amendments are readily available and possess properties that make them possible to be used for soil remediation of contaminated mining sites [see, e.g., 211, 376, 509].

The amendments were applied at different rates: lowest (0.5%), moderate (2%), and highest (5%); and NPK at 0.1, 0.2, and 5.0 g/kg. The nonspecifically sorbed (readily bioavailable) and specifically sorbed As were extracted in sequence. Additionally, soil available P, total C and N, dissolved organic carbon, soil-soluble anions, and exchangeable cations were extracted after 1-day and 28-day incubation periods (Figs. 3.3–3.5). As discussed earlier, the nonspecifically sorbed As and specifically sorbed As represent the most mobile fractions and they are easily available for causing pollution due to the changing environmental conditions. These environmental conditions could be changes in the pH, dissolved organic carbon content, contents of sesquioxides such as Al/Fe/Mn, phosphorus contents, and redox chemistry in cases of drying and wetting conditions of the field. The summation of the two fractions gives the mobile fraction (represented in Fig. 3.6). The individual fractions 1 and 2, represented as water-soluble and exchangeable fractions, respectively, are shown in Figs. 3.7 and 3.8.

I observed that the most promising soil amendment was iron oxides, in which the highest dose application drastically reduced the readily bioavailable As by 93%. The reduction in As bioavailability by iron oxides was attributed to the sorption effect. In this regard, the presence of positive charges on the surfaces of iron oxides increases their binding and sorption strengths, and thus reduces the availability of As. The iron oxides used in the experiments contained greater percentages of Fe, Si, Ca, Al, Mg,

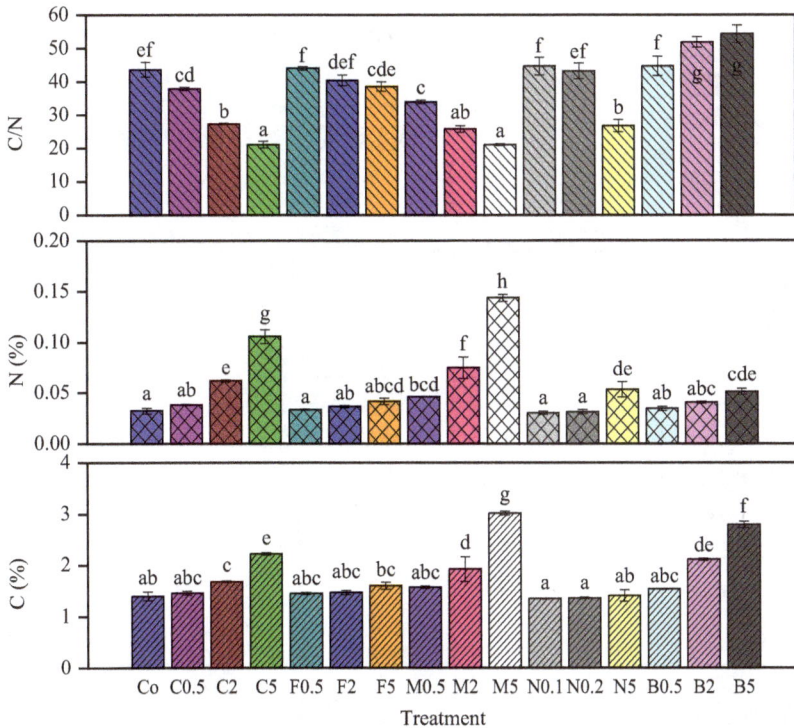

**Fig. 3.3:** Changes in the mine spoil soil total carbon, total nitrogen and CN ratio at the end of the 28-day incubation period following treatments with biochar, compost, iron oxide, manure and inorganic fertiliser. Co = control; C0.5, C2, C5 = soil treatment with compost at 0.5%, 2% and 5%, respectively; F0.5, F2, F5 = iron oxide at 0.5%, 2% and 5%, respectively; M0.5, M2, M5 = manure at 0.5%, 2% and 5%, respectively; N0.1, N0.2, N5 = NPK fertiliser at 0.1g/kg, 0.2g/kg and 5 g/kg, respectively; and B0.5, B2, B5 = rice husk biochar at 0.5%, 2% and 5%, respectively. Means with different letters differ significantly among treatments at $P < 0.05$ and same letters indicate no significant differences among means according to the Tukey's Honestly Significant Different test.
Source: Author's own work

and K, and these positive charges increase the sorption capacity, become carriers of As, and consequently reduce their mobility.

Iron oxides may also reduce the availability of As through the process of coprecipitation [520]. However, the stability of FeO highly depends on the pH of the medium. For example, under high pH media, there is a reduction in positive charges and an increase in negative charges due to the production of $OH^-$ and $CO_3^{2-}$, and there is deprotonation of Fe due to the loss of protons. In this regard, Lindsay et al. [275] reported that under moderately acidic to circumneutral pH conditions, the mobility of dissolved oxyanions such as $H_2AsO_4^-$ is restricted due to sorption onto positively charged mineral surfaces.

These findings have implications and applications in the remediation of As in Fe-As bearing dominated mining tailings and spoils. Acid mine drainage (AMD), for in-

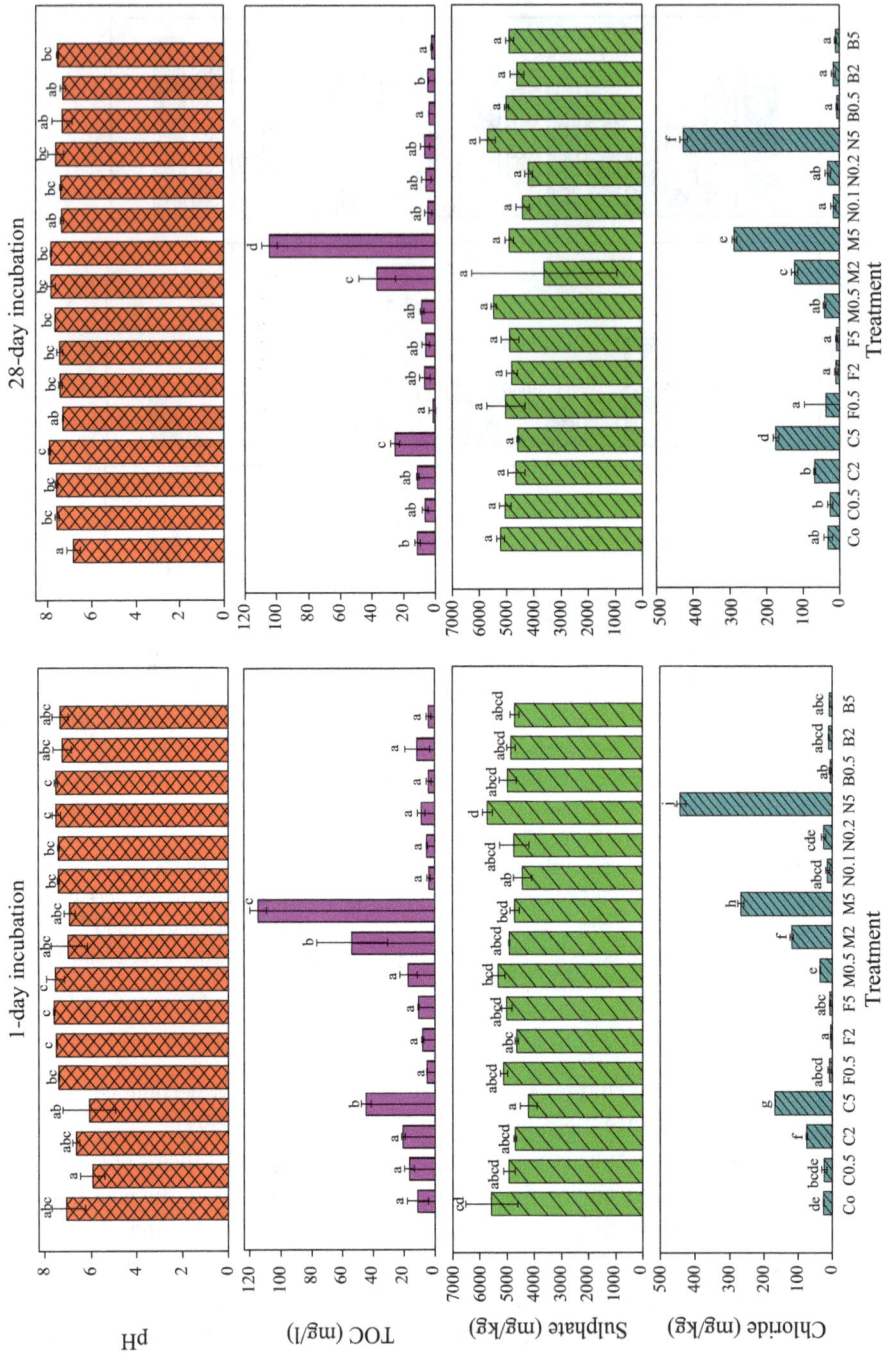

**Fig. 3.4:** Changes in the mine spoil soil pH, TOC, sulphate and chloride contents during the 1-day and 28-day incubation period following treatments with biochar, compost, iron oxide, manure and inorganic

stance, is a common environmental hazard that occurs from gold mine fields. This process creates oxidative dissolution of FeAsS and releases As into the environment, causes soil acidity, and increases the concentration of Fe. The decrease in soil pH during this process causes negative ions such as carbonates to be depleted, thus reducing the acid neutralization capacity (ANC) [275].

Lindsay et al. [275] reported that AMD develops in mine deposits where the neutralization capacity of carbonate minerals is depleted. In contrast, neutral mine drainage occurs in mine deposits where carbonate dissolution efficiently counterbalances the acid generated by sulfide oxidation. Such spoils are concomitantly characterized by high concentrations of weakly hydrolyzing substances, including Fe(II) and As [see. 275].

Thus, the application of FeO is affected by pH and changes may consequently reduce or increase the mobility of As. Mitigating the bioavailability of As in the mine sites with Fe and reducing ecological pollution may be achieved in the acidic range. Neutral to alkaline conditions may be encouraged to enhance the bioavailability for phytoextraction. This observation was confirmed by Mamindy-Pajany et al. [292], when they reported that As adsorption by FeO is effective at lower pH. We conclude that iron-rich materials can be used to reduce the bioavailability of As and mitigate the associated environmental and human health risks in the mining spoils. However, biochar, compost, manure, and inorganic fertilizer increased the bioavailability of As, which indicates that these amendments may increase the risk of As but can be used to enhance the phytoextraction efficiency of As in the gold mining spoil. Details of these experiments and findings are published by Mensah et al. [332].

## 3.3 Size fractionation experiments

Size fractionation is fundamentally performed to identify the soil separates where toxic elements may be concealed. Is it feasible that the potentially toxic element is retained within the clay, silt, or sand fractions? Consequently, size fractionation and identifying the locations of potentially toxic elements within soil separates are essential for effectively targeting the elements of interest when devising a remediation strategy.

---

**Fig 3.4:** (continued)
fertiliser. Co = control; C0.5, C2, C5 = soil treatment with compost at 0.5%, 2% and 5%, respectively; F0.5, F2, F5 = iron oxide at 0.5%, 2% and 5%, respectively; M0.5, M2, M5 = manure at 0.5%, 2% and 5%, respectively; N0.1, N0.2, N5 = NPK fertiliser at 0.1g/kg, 0.2g/kg and 5 g/kg, respectively; and B0.5, B2, B5 = rice husk biochar at 0.5%, 2% and 5%, respectively. Means with different letters differ significantly among treatments at $P < 0.05$ and same letters indicate no significant differences among means according to the Tukey's Honestly Significant Different test.
Source: Author's own work

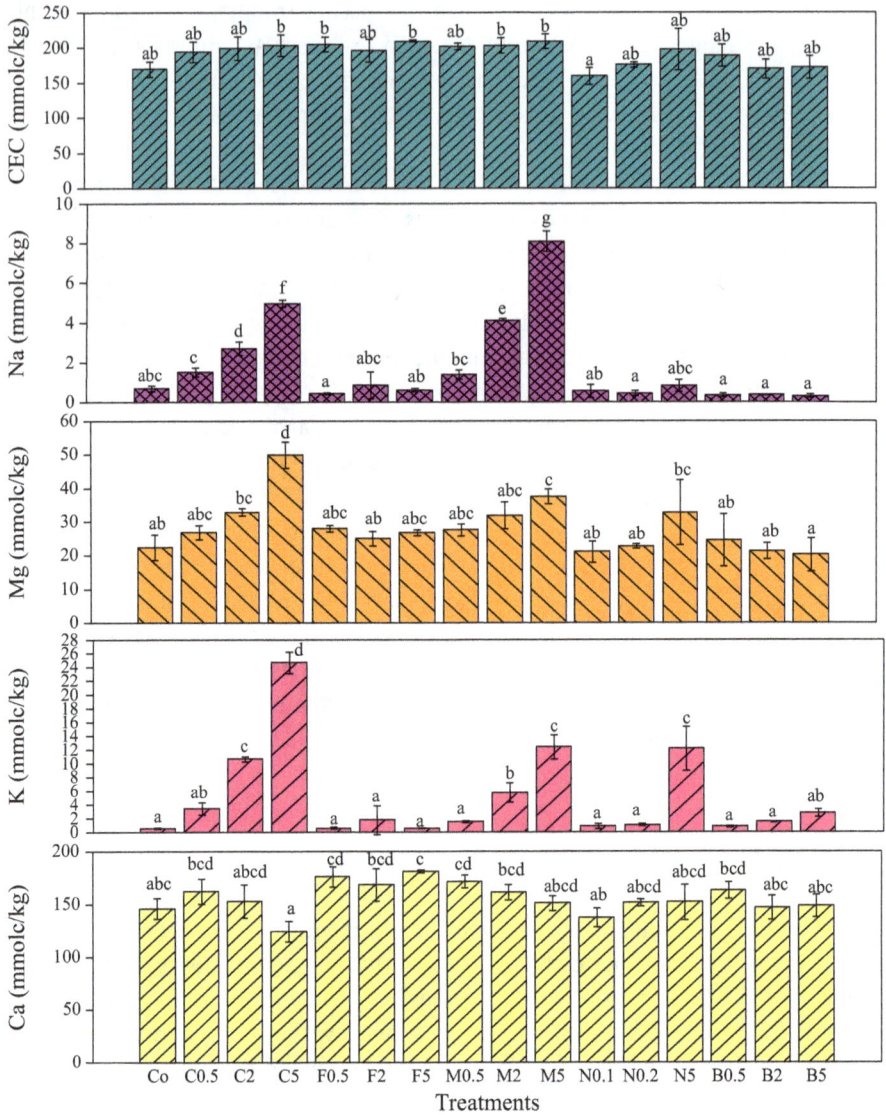

**Fig. 3.5:** Changes in the mine spoil soil exchangeable cations and cations exchange capacity at the end of the incubation period following treatments with biochar, compost, iron oxide, manure and inorganic fertiliser. Co = control; C0.5, C2, C5 = soil treatment with compost at 0.5%, 2% and 5%, respectively; F0.5, F2, F5 = iron oxide at 0.5%, 2% and 5%, respectively; M0.5, M2, M5 = manure at 0.5%, 2% and 5%, respectively; N0.1, N0.2, N5 = NPK fertiliser at 0.1g/kg, 0.2g/kg and 5 g/kg, respectively; and B0.5, B2, B5 = rice husk biochar at 0.5%, 2% and 5%, respectively. Means with different letters differ significantly among treatments at $P < 0.05$ and same letters indicate no significant differences among means according to the Tukey's Honestly Significant Different test.

Source: Author's own work

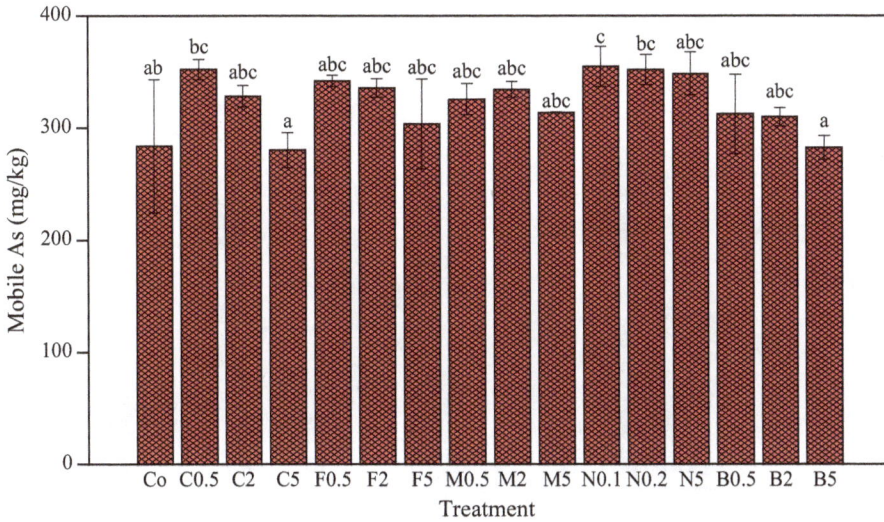

**Fig. 3.6:** Changes in the mine spoil mobile As contents during the 1-day and 28-day incubation period following treatments with biochar, compost, iron oxide, manure and inorganic fertiliser. Co = control; C0.5, C2, C5 = soil treatment with compost at 0.5%, 2% and 5%, respectively; F0.5, F2, F5 = iron oxide at 0.5%, 2% and 5%, respectively; M0.5, M2, M5 = manure at 0.5%, 2% and 5%, respectively; N0.1, N0.2, N5 = NPK fertiliser at 0.1g/kg, 0.2g/kg and 5 g/kg, respectively; and B0.5, B2, B5 = rice husk biochar at 0.5%, 2% and 5%, respectively. Means with different letters differ significantly among treatments at $P < 0.05$ and same letters indicate no significant differences among means according to the Tukey's Honestly Significant Different test.
Source: Author's own work

For instance, upon identifying that the PTE is significantly concentrated within the clay fraction, remediation strategies are formulated to specifically address the clay-sized particles. In this context, efficiency is achieved, thereby minimising the wastage of remediation efforts.

The experiments involving soil-sized fractionation contribute significantly to the precise estimation of the bioavailability of the potentially toxic elements. For instance, clays or fine fractions exhibit a larger surface area, which allows them to contain and sequester greater amounts of potentially toxic elements. In this scenario, dusts are characterised by fine particles, which consequently leads to a higher retention of particulate matter when inhaled, ingested, or when they come into contact with human skin.

It is anticipated that the potential mobile fractions are predominantly retained in the fine fractions rather than in the coarse fractions. The constituents retained within the clay fraction may subsequently be mobilised, contingent upon the prevailing environmental conditions. These actions could lead to secondary pollution effects, as the sorbed fraction within the fine particles may be released during flooding events, for example.

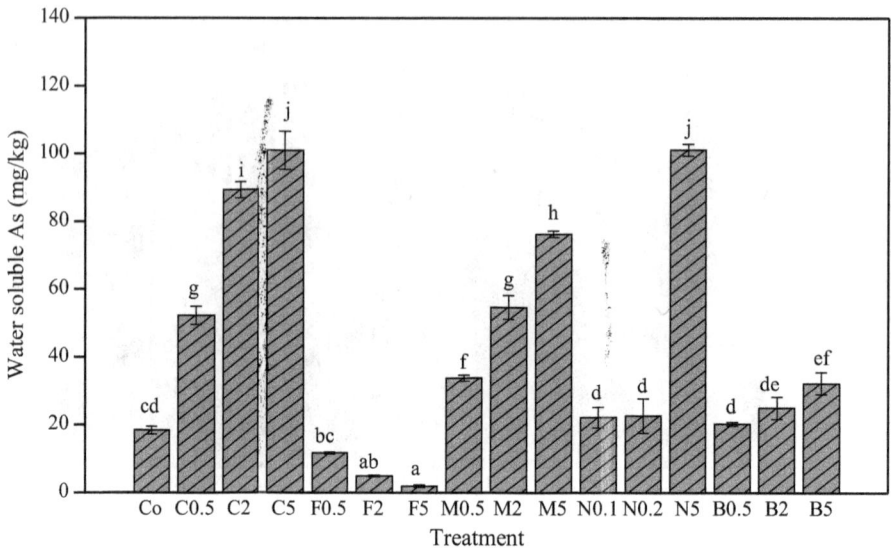

**Fig. 3.7:** Changes in the mine spoil water-soluble As contents during the 1-day and 28-day incubation period following treatments with biochar, compost, iron oxide, manure and inorganic fertiliser. Co = control; C0.5, C2, C5 = soil treatment with compost at 0.5%, 2% and 5%, respectively; F0.5, F2, F5 = iron oxide at 0.5%, 2% and 5%, respectively; M0.5, M2, M5 = manure at 0.5%, 2% and 5%, respectively; N0.1, N0.2, N5 = NPK fertiliser at 0.1g/kg, 0.2g/kg and 5 g/kg, respectively; and B0.5, B2, B5 = rice husk biochar at 0.5%, 2% and 5%, respectively. Means with different letters differ significantly among treatments at $P <$ 0.05 and same letters indicate no significant differences among means according to the Tukey's Honestly Significant Different test.
Source: Author's own work

Research indicates that soils with higher clay content contain greater proportions of As than sandy soils [37, 321]. For instance, for a more detailed analysis of As geochemical fractions, a composite sample from the abandoned, active, and profile mining spoils collected in Ghana was further separated into a coarse and a fine fraction. The coarse fraction consisted of sand, while the fine fraction consisted of silt and clay. Details of this experiment are described by Mensah et al. [321].

In Mensah et al. [321], we reported findings from an investigation on As contamination in abandoned and active gold mine spoils in Ghana: geochemical fractionation, speciation, and assessment of the potential human health risk. Here, we collected additional samples from both the abandoned mining spoil/tailing and active mining spoil/tailing. The samples were prepared, further separated into bulk, fine (silt and clay), and coarse (sand) fractions, and analyzed for total As contents and five geochemical fractions.

We observed that the total contents and potential mobility of As were higher in the fine fraction than in the coarse fraction (Fig. 3.9). This was attributed to greater proportions of $FeAsSO_4$ and $FeAsS$, and extracted As in the amorphous and residual fractions in the fine fraction relative to the contents in the coarse fraction. There was

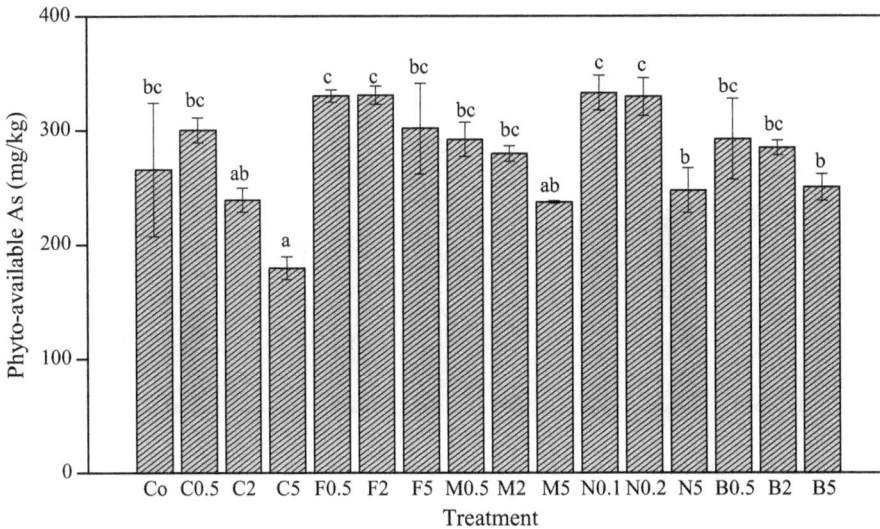

**Fig. 3.8:** Changes in the mine spoil phyto-available As contents during the 1-day and 28-day incubation period following treatments with biochar, compost, iron oxide, manure and inorganic fertiliser. Co = control; C0.5, C2, C5 = soil treatment with compost at 0.5%, 2% and 5%, respectively; F0.5, F2, F5 = iron oxide at 0.5%, 2% and 5%, respectively; M0.5, M2, M5 = manure at 0.5%, 2% and 5%, respectively; N0.1, N0.2, N5 = NPK fertiliser at 0.1g/kg, 0.2g/kg and 5 g/kg, respectively; and B0.5, B2, B5 = rice husk biochar at 0.5%, 2% and 5%, respectively. Means with different letters differ significantly among treatments at $P < 0.05$ and same letters indicate no significant differences among means according to the Tukey's Honestly Significant Different test.
Source: Author's own work

also a highly significant positive relationship between As and Fe ($r = 0.98$; $P < 0.01$), which implies that the total contents and potential mobility of As were largely a function of the content of Fe. The dominance of Fe in the fine fraction may also be due to the higher content of clay, which may increase positive charges. The presence of surface positive charges then increases the sorption capacity for As and increases the As content in the fine fraction. It is thus explained and elaborated earlier that soils with high clay content retain higher contents of As.

Consequently, the high potential mobility in the fine fraction indicates potential mobilization and release of the sorbed As under changing environmental conditions such as pH, redox potential, and changes in contents of P. For example, during flooding and rainfall, the high total As, As contained in the species (e.g., $FeAsSO_4$ and $FeAsS$), and the extracted As in the amorphous and residual fractions in the fine fraction may be liberated and become available for polluting the surrounding soils and surface and groundwater as also explained in other studies [e.g., 398].

The mobile and potential mobile fractions of As in the coarse particles of the profiles increased with depth while they decreased in the fine particles (Fig. 3.10). Here, I

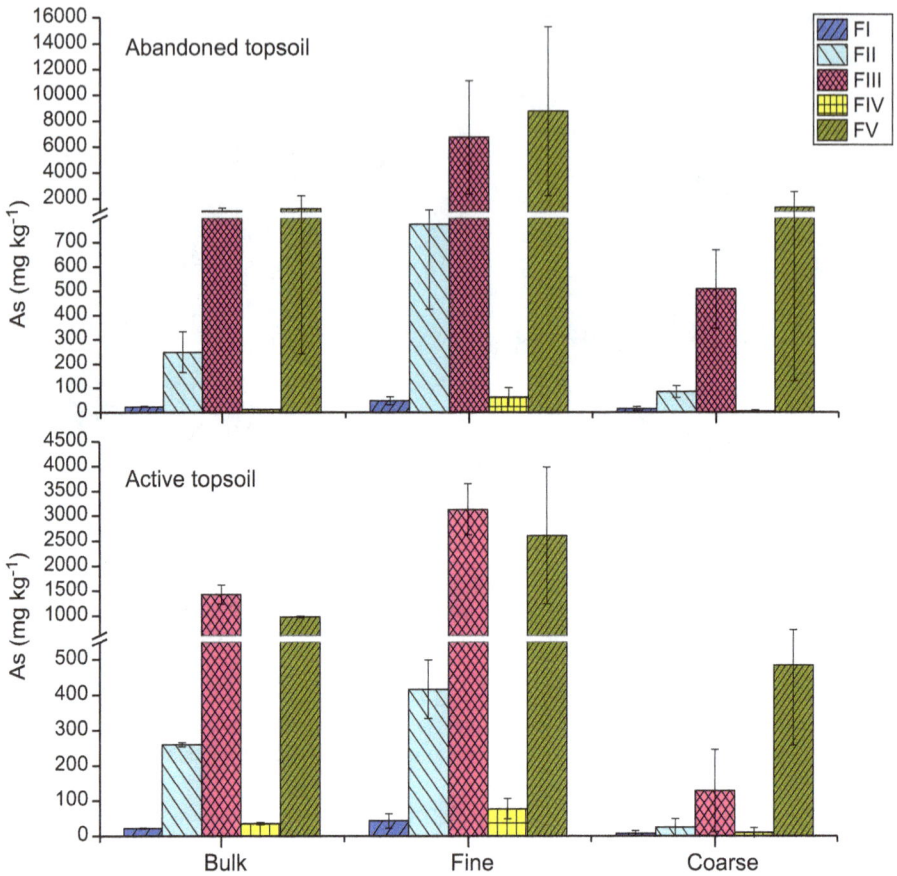

**Fig. 3.9:** Arsenic fractions in the bulk, fine and coarse fractions of abandoned and active mine spoil topsoil. Coarse fraction consisted of sand and fine fraction consisted of silt and clay.

attributed this to the effect of weathering and reduction and oxidation processes occurring in the soil's coarse particles, which are more intense on the surface than in the subsoils. The amorphous iron oxide contents, for example, may be subjected to different oxidation and reduction processes, which may lead to the release and leaching of loosely bound As from the surface soils. The released As may be lost from the soil surface or leach downwards to the soil's sublayers.

These reasons accounted for decreased As mobile and potential mobile fractions down the profiles in the fine fractions, and increased mobility in the coarse fractions down the soil profiles. In conclusion, the mining spoils pose threats to the livelihood of the residents in the mining area and remediation action should be explored to protect the environment, restore livelihoods, and safeguard the health of animals and residents.

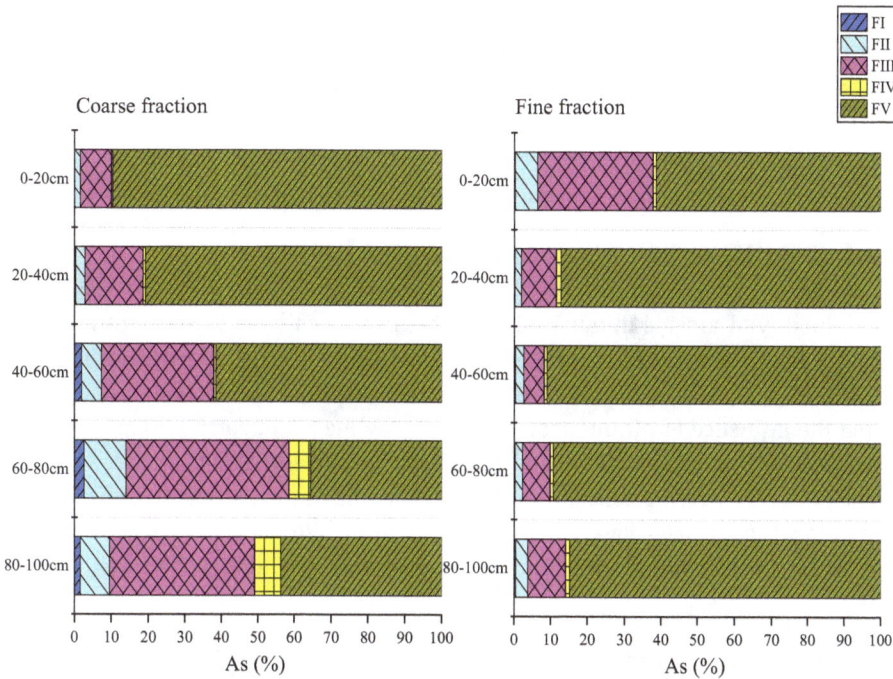

**Fig. 3.10:** Arsenic geochemical fractions in the profile of coarse and fine fractions of an abandoned mine spoils in southwestern Ghana. Details are published in Mensah et al. (2020).

## 3.4 Determination of risks using acid neutralization capacity method

This section establishes and demonstrates examples to show the solubility of PTE in relation to variations in the soil pH. Typically, elevating pH and diminishing acidity facilitate the mobilization of anionic trace elements, whereas lowering pH and elevating acidity result in decreased mobilization and, consequently, the availability of cationic trace metals. The biogeochemical effects of pH on regulating mobility and, consequently, the availability of PTEs are elaborated upon in detail in Chapter 8.

**Example 1:** The pH of the As-contaminated soil was adjusted using the ANC method [222]. ANC describes the ability of the soil to buffer acid intake within a short period. In this method, the amount of acid or base necessary to adjust the soil to a certain pH is specified. The effect of pH on As solubility was determined in a series of batch extraction tests (1:5 m/v), in which 25 g of air-dried soil was mixed with various volume doses of 1 M HCl or 0.1 M KOH. These were diluted in appropriate amounts of distilled water to obtain a total solution volume of 50 mL, which contained additions of 5 mL 0.1 M KCl as the background electrolyte. The volumes of acid added were in the range of 0.5–10 ml, and the base in the range of 0.5–5 mL, which correspond to 0–10 mmol $H^+$ or $OH^-$/25 kg of soil. The samples were shaken overnight for 22 h; thereafter, the suspensions were centrifuged and filtered. The concentrations of As, as well as pH values, in the extracts (filtrate) were determined. All the procedures were carried out in duplicate.

### 3.4.1 Results and discussion of findings

The solubility of arsenic in our study may be influenced by the concentrations and biogeochemical alterations in the mine spoil's pH, chloride ions, electrical conductivity, and exchangeable potassium and sodium ions. Consequently, we further illustrated the influence of pH on arsenic solubility by the acid neutralization capability test (Fig. 3.11). The extraction tests demonstrated that arsenic solubility rises with elevated pH levels. Arsenic concentrations were minimal at pH levels of 5.47 and 5.87, measuring 0.00 and 0.04 mg/kg, respectively. In the alkaline range, As grew constantly by a factor exceeding 10, from 1.25 mg/kg at pH 7.02 to 14.08 mg/kg at pH 10.03.

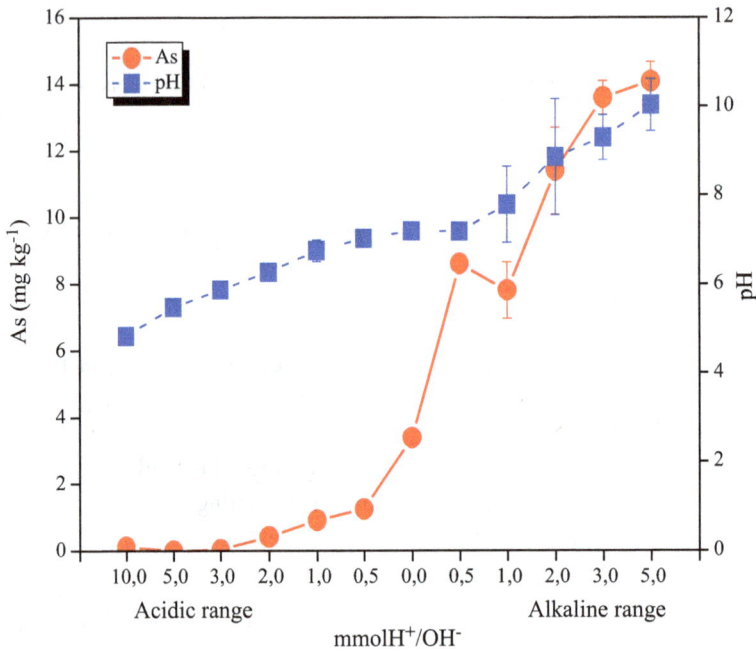

**Fig. 3.11:** Dynamics of As solubility due to pH changes (impacts of pH changes on As solubility). Source: Author's own work.

This suggests the elevated possible environmental dangers of arsenic under alkaline conditions, further corroborated by a substantial positive connection between pH and soluble arsenic ($r = 0.92$, $P < 0.01$, $n = 12$; data not shown). We suggest three primary hypotheses to elucidate As release and mobilization at elevated pH: deprotonation of Al/Fe/Mn oxides, solubilization of Fe/Mn oxides, and the formation of negative charges on soil colloids.

Arsenic compounds are known to adsorb onto the oxyhydroxides of aluminum, iron, and manganese, as well as humic substances and clay minerals [230]. At elevated pH levels, the sorption capacity of iron oxides for arsenic diminishes due to the deprotonation of iron [230]. Tack [4] indicated that since sorption onto soil colloids diminishes with increasing pH, carbonates get a larger negative charge and hydroxyl ions are produced at elevated pH levels. This occurrence can render exchangeable As accessible [336].

Gersztyn et al. [175], observed that significant quantities of arsenic were released at a pH of approximately 9. Additionally Karczewska et al. [237], discovered that As-contaminated mine tailings treated with $Ca(OH)_2$ sewage sludge exhibited a notable enhancement in As solubility within the mine tailings. Gersztyn et al. [175], indicated that soil liming with $Ca(OH)_2$, primarily a source of calcium, did not facilitate As solubilization, as calcium humates have limited solubility in alkaline environments.

**Example 2:** Evaluation of pH and mobility of heavy metal(loids) using ANC (Fig. 3.12)

**Introduction and Objectives:** Environmental concerns have emerged due to the anticipated increase in PTEs found in e-waste, including As, Pb, Cd, Co, Ni, Ti, and others, which pose significant ecotoxicological risks to various forms of life on Earth. Soil pH is a critical factor in regulating the mobility of potentially toxic elements (PTEs). This study aims to evaluate the impact of soil pH on the mobilisation of potentially toxic elements from soil contaminated by electronic waste. Samples were obtained from Agbobloshie (05°32′42.0″N 00°13′30″W), the largest electronic waste disposal site in Ghana's capital.

**Materials and methods:** Soil samples were obtained from the largest electronic waste dump site in Ghana, situated at coordinates 5° 33′ 2.95″N 0° 13′ 31.41″W. Twenty-five samples were collected from a depth of 0-20 cm, randomly distributed across the site, and combined into a bulk sample. Following a 72-hour air-drying period, the samples were ground using a mortar and pestle, homogenised, and subsequently sieved through a 2 mm sieve. A 15g sample was subsequently submitted for preliminary pseudo-total element and physicochemical properties analyses.

The pH of contaminated E-waste soil was adjusted using the acid neutralisation capacity (ANC) method as described by James and Riha (1986) and applied by Mensah et al. (2022). A series of batch extraction tests (1:5 m/v) were conducted by combining 25 g of e-waste soil with varying quantities of 1 M HCl or 0.1 M KOH. Fifty millilitres of distilled water and five millilitres of 0.1 M $CaCl_2$ were added to dilute the solutions and provide a background electrolyte. The volumes of acid added varied from 0.5 to

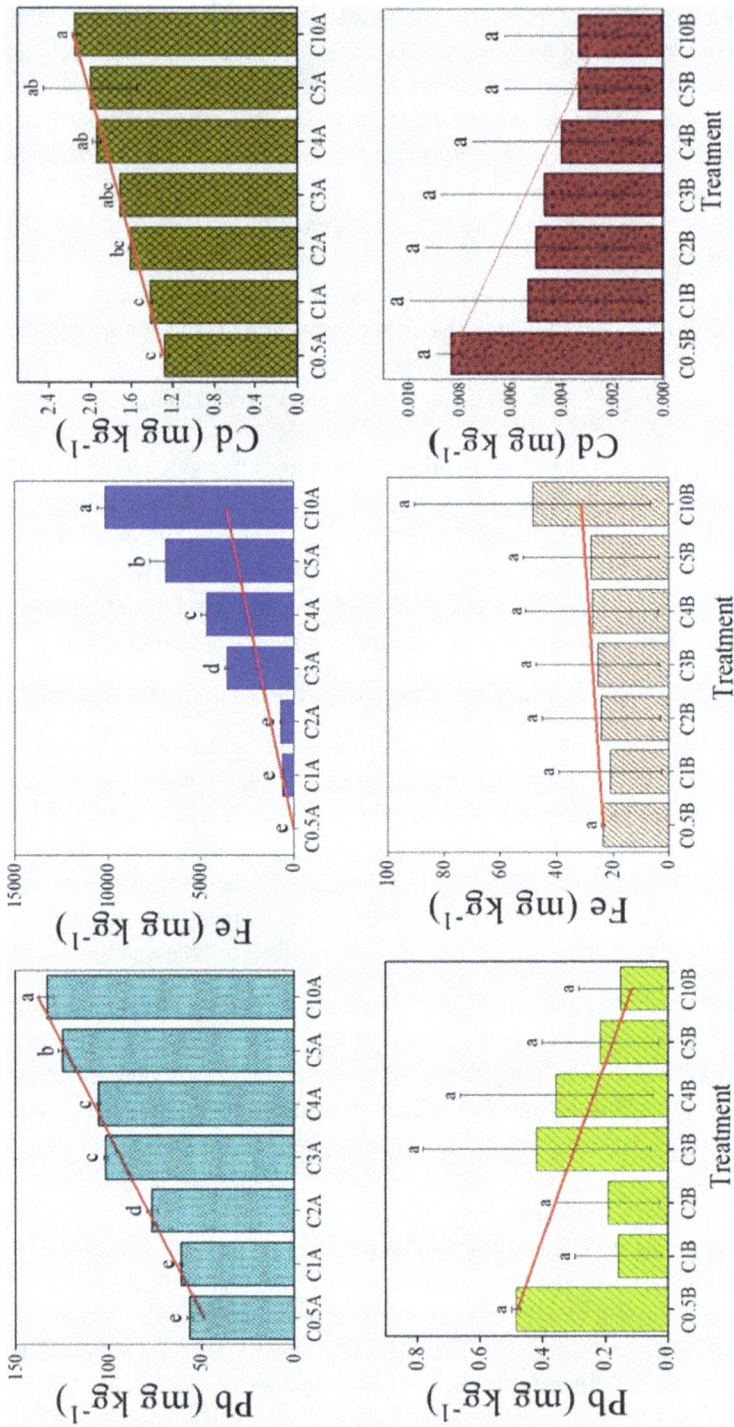

**Fig. 3.12:** Influence of PTEs on the availability of PTEs in e-wastes collected samples in Ghana. Source: Author's original work.

10 ml, whereas the base varied from 0.5 to 5 ml, corresponding to 0 to 10 mmol H+ or OH– per 25 kg of soil. The samples underwent shaking for 22 hours overnight, followed by centrifugation and filtration of the suspensions. The concentrations of potentially toxic elements (PTEs) and pH values in the filtrate were evaluated. All procedures were performed in triplicate. The ANC method assessed the following parameters: mobile PTEs, pH, EC, Mn, Al, Fe, and P as governing variables.

**Results and discussion:** All metals exhibited variations in mobility corresponding to changes in the pH levels of the e-waste soils. The trend indicated that metals exhibited increased mobility in acidic media and decreased mobility in alkaline conditions. The findings can be attributed to three primary hypotheses.

i. **Deprotonation of Al/Fe/Mn oxides:** happens at elevated pH levels: Under acidic or low pH conditions, sorption parameters such as Al, Fe, and Mn gain protons. Conversely, these elements lose their sorption capacities due to hydrolysis, which diminishes their acidic properties in alkaline media. In such scenarios, cationic PTEs become more available in acidic media, while they are more sorbed at higher pH levels.

ii. **Solubilisation of Fe/Mn oxides:** Sesquioxides lose their carrying properties as they become soluble in acidic media, and the reverse occurs in neutral or alkaline conditions. Consequently, any cationic PTEs that are precipitated, complexed, or coprecipitated at the reduced pH also become released during the process.

iii. **Development of negative charges on soil colloids:** Negative charges, including chlorides, nitrates, sulphates, phosphates, and hydroxyl ions, predominate in the soil reaction environment under alkaline conditions. At elevated pH levels, anionic potentially toxic elements exhibit increased solubility and mobility. Cationic ptes become immobile, thereby restricting their mobility in relation to pollution.

**Conclusion:** These findings illustrate the capacity of liming to correct soil acidity, as liming supplies basic cations during the process and consequently aids in buffering soil pH at higher ranges. Anionic PTEs are mobilised, while cationic PTEs become immobile under higher pH conditions.

## 3.5 Conclusions

The SEP is a useful method for studying the bioavailability and toxicity of PTEs in contaminated sites. This procedure separates PTEs into different binding forms or geochemical fractions, allowing for a more accurate assessment of their environmental risks. In the case of arsenic (As), the soluble/mobile fractions (fractions I and II) are the most toxic and pose the greatest risk to water, groundwater, and the food chain. The use of iron oxides as a soil amendment was found to be effective in reducing the bioavailability of As by sorption and coprecipitation processes. However, other amendments such as biochar, compost, manure, and inorganic fertilizer increased the bioavailability of As, indicating a potential risk but also the potential for enhanced phytoextraction of As. The size fractionation experiments showed that the fine fraction of soils, particularly those with high clay content, contained higher proportions of As and had higher potential mobility compared to the coarse fraction. This highlights the importance of considering the particle size distribution in assessing the mobility and potential risks of As in contaminated sites. Lastly, the Acid Neutralization Capacity method aided to perform an experiment and analyzed the effects of the soil pH as a key parameter that regulates the toxicity of PTEs in contaminated sites.

Albert Kobina Mensah

# Chapter 4
# Risk identification using remediation incubation experiments, redox microcosm, geospatial analyses, and synchrotron radiation science

**Abstract:** Incubation experiments are conducted in soil contamination research to understand the behavior of contaminants in soil under controlled conditions. These experiments help determine how contaminants move through soil and interact with various factors. The duration of incubation experiments varies, and researchers monitor the contaminant levels and soil properties over time. In a specific study, different soil amendments were tested for their potential to reduce arsenic (As) contamination in mining spoils. Iron oxides were found to be the most promising amendment, significantly reducing the bioavailability of As. The findings have implications for the remediation of contaminated mining soils. Redox experiments using automated biogeochemical microcosms were also conducted to study the mobilization of As under different redox conditions. These experiments aid in risk assessment and accurately designing remediation techniques for As-contaminated fields.

## 4.1 Introduction

The detailed study of the mobilization and speciation of potentially toxic elements (PTEs) in abandoned gold mine tailings and spoils in Ghana, especially during periods of significant flooding, and intense rainfall regimes has been lacking. Prior studies have primarily concentrated on identifying risks associated with surface soils as well as those found in soil profiles [321]. The existing scientific literature lacks studies on the mobilization and speciation of PTEs under different reducing and oxidizing circumstances in tropical regions such as Ghana.

Such studies will contextualize, highlight and pinpoint the significant issues arising from the investigation of ecotoxicological risks associated with potentially harmful elements in contaminated and polluted sites.

Environmental research in Ghana has suffered from an absence of advanced and high-throughput methodologies. This challenges the assessment of pollution issues re-

**Albert Kobina Mensah**, Soil Research Institute, Council for Scientific and Industrial Research, Kumasi, Ghana, e-mail: albert.mensah@rub.de, albertkobinamensah@gmail.com, ORCID: https://orcid.org/0000-0001-5952-3357

https://doi.org/10.1515/9783111662046-004

lated to potentially toxic elements (PTEs) in soil, water, food, and human samples such as blood, urine, and hair.

Overly simplistic methods that concentrate solely on the total elemental contents of potentially toxic elements (PTEs) may yield erroneous conclusions, leading to misleading and inaccurate recommendations.

Ghana plays a crucial role in the extractive industries, being recognised as one of the leading gold-producing nations globally. The environmental impacts of mineral mining are considerable. The problems associated with gold mining arise mainly from weak institutions, the increase in illegal gold mining activities, and, crucially, a deficiency in political will to enforce appropriate environmental standards and regulatory frameworks.

This chapter presents advanced strategies for accurately identifying soil-to-human health risks associated with potentially toxic elements (PTEs). Advanced methods extend beyond the simple collection of soil samples from potentially contaminated areas, the assessment of total elemental content, and the subsequent conclusion of contamination at these sites.

These approaches involve a thorough examination of the problem through incubation experiments, redox chemistry experiments on collected samples, and geospatial analyses using GIS to effectively understand the distribution of PTEs across a large area. An example involves employing geospatial analyses to delineate contamination and pollution by potentially toxic elements (PTEs) in mine tailings across an area of approximately 10,000 square meters.

This chapter presents an analysis of soil samples from Ghana's gold mine tailings, introducing synchrotron radiation as one of the high thorough-put that can be considered in appraising toxicity of PTEs in contaminated sites.

This chapter presents arguments for employing synchrotron radiation science as a sophisticated technique for the precise identification of soil-to-human health risks linked to potentially toxic elements (PTEs). These include studies that involve assessing ecotoxicological risks and environmental safety regarding toxic elements in mine-contaminated sites.

Incubation experiments are controlled investigations conducted in a laboratory or field setting to evaluate how environmental conditions impact the behavior of pollutants in soil. These experiments imitate real-field scenarios of measures used to mitigate soil pollution or fertility issues in laboratory settings. Here, various interventions are examined in laboratory incubation experiments to see which ones yield the most favorable outcomes. During incubation research, several parameters and factors such as temperature, moisture content, pH, and oxygen concentration are carefully regulated, altered, and modified. The trial is thereafter permitted or examined during a specified duration and time frame. The duration of these assessments might range from a few weeks to several years. They entail the continuous monitoring and periodic measurement of soil pollutant hazards resulting from soil remediation interventions or treatments.

Synchrotron radiation science offers sophisticated analytical instruments that improve our capacity to investigate, assess, and deal with environmental contamination. These methods support attempts to save the environment, regulate pollutants, and assess risk more successfully. In order to identify the predominant minerals in the samples, synchrotron radiation research has been used in soil contamination studies.

This chapter aims to provide discussions on how to critically appraise risk identification by employing sophisticated techniques such as redox microcosm (MC) experiments, geospatial analyses, soil incubation experimental trials, and synchrotron radiation science. The chapter demonstrates examples using data from Ghana.

## 4.2 Identifying risks using incubation experiments

Incubation experiments in soil contamination research involve controlled laboratory or field studies designed to assess the effects of various environmental factors on the fate and behavior of contaminants in soil over time in a controlled setting. These experiments are crucial for understanding how contaminants interact with soil components, microbial activity, and other factors that influence their persistence and potential for harm. Incubation studies and experiments are carried out in the laboratory to simulate the conditions in the field settings. The experimental designs are usually set to completely randomized designs because the environmental conditions are usually controlled to become homogeneous. In field settings, environmental conditions that influence outcomes are usually heterogeneous. As such, in field settings, the experimental designs are most often than not randomized complete block designs.

Incubation experiments can help scientists determine how contaminants move through soil, whether they leach into groundwater, evaporate into the atmosphere, or remain bound to soil particles. For example, a study might investigate the migration of heavy metals in soil and their potential to contaminate nearby water sources. Researchers can manipulate environmental conditions in incubation experiments to simulate real-field scenarios. Temperature, moisture levels, pH, and oxygen content are examples of parameters that can be controlled to assess their impact on contaminant behavior. These experiments can help predict how contaminants might behave under various environmental conditions.

The duration of incubation experiments can vary widely, depending on the research objectives. Some experiments may last for weeks, while others span several months or even years. Researchers monitor contaminant levels, soil properties, and microbial activity throughout the incubation period to gather data on changes over time.

For instance, using a 28-day laboratory incubation study, I tested eight different soil amendments for remediation of the As-contaminated mining spoils and tailings. These amendments included composts, rice husk biochar, corn cob biochar, NPK fertilizer, manure, sewage sludge, iron oxide, and red mud. These amendments were applied at different rates – 0.5%, 2%, and 5% (w/w) – to 300 g of the contaminated soil in an incubation jar. The NPK fertilizer was also applied at rates of 5, 0.2, and 0.1 g/kg. My aim was to explore the potential of biochar, compost, iron oxide, manure, and NPK fertilizer to reduce As bioavailability and improve the soil quality of the degraded mining site. These amendments are readily available and possess properties that make them possible to be used for soil remediation of contaminated mining sites [see, e.g., 214, 376, 509]. The amendments were each applied at different rates: lowest (0.5%), moderate (2%), and highest (5%), and NPK at 0.1, 0.2, and 5.0 g/kg. Figure 4.1 shows soil samples mixed with different rates of soil amendments and ready for incubation. Details of the experiments are published by Mensah et al. [332].

**Fig. 4.1:** Arsenic-contaminated mine spoil soil mixed with different rates of soil amendments, ready for incubation. Picture taken at the Ruhr University Bochum-Germany Geography Institute Laboratory on August 20, 2019.

**Methods:** The nonspecifically sorbed (readily bioavailable) and specifically sorbed As were extracted in sequence. Additionally, soil available P, total C and N, dissolved organic carbon, soil-soluble anions, and exchangeable cations were extracted after 1-day and 28-day incubation periods. The nonspecifically sorbed As and specifically sorbed As represent the most mobile fractions, and are easily available for causing pollution due to the changing environmental conditions.

**Results and discussion**: The most promising soil amendment was iron oxides, in which the highest dose application drastically reduced the readily bioavailable As by 93% (see Fig. 4.2). The reduction in the bioavailability of As by iron oxides was attributed to the sorption effect. In this regard, the presence of positive charges on the surfaces of iron oxides increases its binding and sorption strengths, thus reducing the availability of As. The iron oxides used in the experiments contained greater percentages of Fe, Si, Ca, Al, Mg, and K; these positive charges increase sorption capacity, become carriers of As and consequently reduce its mobility. Iron oxides may also reduce As availability through the process of coprecipitation [520]. However, the stability of FeO highly depends on the pH of the medium. For example, under high pH media, there is a reduction in positive charges and an increase in negative charges due to the production of $OH^-$ and $CO_3^{2-}$, and there is deprotonation of Fe due to the loss of protons. In this regard, Lindsay et al. [275], reported that under moderately acidic to circumneutral pH conditions, the mobility of dissolved oxyanions such as $H_2AsO_4^-$ is restricted due to sorption onto positively charged mineral surfaces.

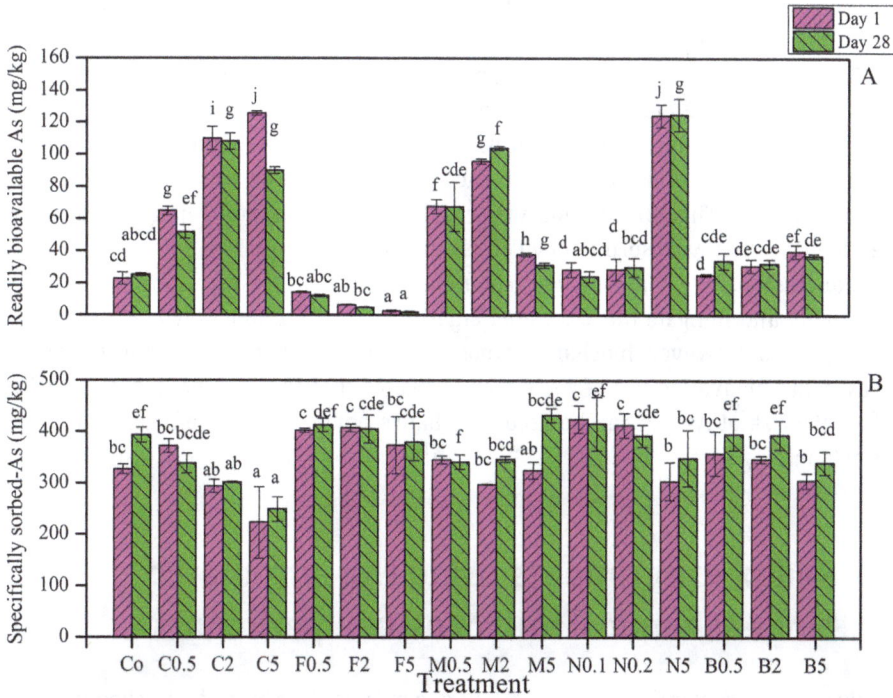

Fig. 4.2: Changes in the mine spoil nonspecifically sorbed (readily bioavailable) and specifically sorbed As contents during the 1-day and 28-day incubation periods following treatments with biochar, compost, iron oxide, manure, and inorganic fertilizer. Co = control; C0.5, C2, C5 = soil treatment with compost at 0.5%, 2%, and 5%, respectively; F0.5, F2, F5 = iron oxide at 0.5%, 2%, and 5%, respectively; M0.5, M2, M5 = manure at 0.5%, 2%, and 5%, respectively; N0.1, N0.2, N5 = NPK fertilizer at 0.1 g/ kg, 0.2 g/kg, and 5 g/kg, respectively; and B0.5, B2, B5 = rice husk biochar at 0.5%, 2%, and 5%, respectively. Charts represent the means of three replicates and error bars represent their standard deviations. Means with different letters differ significantly among treatments at $P < 0.05$, and same letters indicate no significant differences among means according to the Tukey's honestly significant difference test. Details of the results are published by Mensah et al. [332].

### 4.2.1 Implications of findings in remediation of contaminated mining soils

These findings have implications in the remediation of As in Fe-As bearing dominated mining tailings and spoils. Acid mine drainage (AMD), for instance, is a common environmental hazard that occurs in gold mine fields. This process creates oxidative dissolution of FeAsS and releases As into the environment, causing soil acidity and increasing the concentration of Fe. The decrease in soil pH during this process causes negative ions such as carbonates to be depleted, thus reducing the acid neutralization capacity [275]. Lindsay et al. [275], reported that AMD develops in mine deposits where the neutralization capacity of carbonate minerals is depleted. In contrast, neutral mine drainage occurs in mine deposits, where carbonate dissolution efficiently counterbalances acid generated by sulfide oxidation. Such spoils are concomitantly characterized by high concentration of weakly hydrolyzing substances, including Fe(II) and As [see 275].

Thus, the application of FeO is affected by pH and changes may consequently reduce or increase the mobility of As. Mitigating the bioavailability of As in the mine sites with Fe and reducing ecological pollution may be achieved in the acidic range. Neutral to alkaline conditions may be encouraged to enhance the bioavailability for phytoextraction. This observation was confirmed by Mamindy-Pajany [292], when they reported that As adsorption by FeO is effective at lower pH.

**Conclusions**: We conclude that iron-rich materials can be used to reduce As bioavailability and mitigate the associated environmental and human health risk in the mining spoils. However, biochar, compost, manure, and the inorganic fertilizer increased the bioavailability of As, which indicates that these amendments may increase the risk of As, but can be used to enhance phytoextraction efficiency of As in the gold mining spoil.

## 4.3 Redox experiments using the automated biogeochemical microcosm experiments

Mobilization of contaminants through soil depends on the physical and chemical features of both the contaminants and the site. Large areas of abandoned mining soils can be flooded periodically which alters the redox chemistry of the site, changes the species, and consequently influences the release of the toxic element. For instance, As (IV) changes into As(III) during anaerobic changes of the site, a phenomenon which increases mobility and leads to groundwater pollution [20, 577]. Additionally, reducing conditions may cause a reductive dissolution of Fe/Al/Mn oxyhydroxides and any As sorbed to these oxyhydroxides can become mobilized (Mensah, 2022). Mercury occurs on the Earth's surface in chemical forms including elemental (Hg(0)), divalent inorganic (Hg(II)) and organometallic (e.g., methylmercury) [361]. Methylmercury is the toxic species, occurs in environments and it is biomagnified in the trophic chain ([575, 361]). Gold mining using Hg-amalgamation for gold recovery has been regarded as one of the largest contributors to Hg pollution, contaminating the atmosphere, rivers, and impacting negative on the people's health [361]. Flooded and waterlogged mining fields exhibit low oxygen conditions, and such can serve as fertile grounds for Hg methylation (MeHg) sites and become hotspots for increasing the release of MeHg into adjacent streams and rivers [761].

As indicated earlier, solubility and mobilization of As under flooding or waterlogged conditions in a watershed is controlled by the soil redox chemistry [562; 563]. These flooding conditions and redox potential concomitantly influence the mobilization of As into surface water and groundwater, and affect the uptake and translocation by plants [709; 564]. Studying the mobilization of As due to varied redox settings can aid in risk assessment and in designing appropriate remediation techniques for As-contaminated fields.

Experiments on redox effects on mobilization of As were carried out using the automated biogeochemical MC under simulated laboratory conditions to mimic conditions on the field. Briefly, in this method, the redox potential levels of the soil suspension are reduced and oxidized by alternately flushing the MCs with $N_2$ and synthetic air/$O_2$. The values of redox potentials ($E_H$), pH, as well as temperature were recorded into a data logger connected via a computer to the system. Preset $E_H$ windows were set and automatically maintained with the flushing of $N_2$ (to lower $E_H$) and synthetic air/$O_2$ (to raise $E_H$). $E_H$ and pH of samples taken 6 h before the predefined $E_H$ windows were recorded. Figure 4.3 shows a typical soil MC in operation. Samples (herein re-

**Fig. 4.3:** A typical soil microcosm in operation. Picture taken during experiments at the laboratory of soil and groundwater management, Bergische University in Wuppertal, September 2018.

ferred to as suspension or soil slurry) were collected at regular intervals using syringes (Fig. 4.4). The whole filtration process took place in a glove box at an $O_2$ concentration of 0% (Fig. 4.5). Figure 4.6 shows prepared samples ready for analysis. Details of this experiment are described by Mensah et al. [332].

**Fig. 4.4:** Samples (soil slurry/suspensions) collected in syringes for centrifugation and filtration. Source: Experiments at the Laboratory of Soil and Groundwater Management, Bergische University in Wuppertal, September 2018.

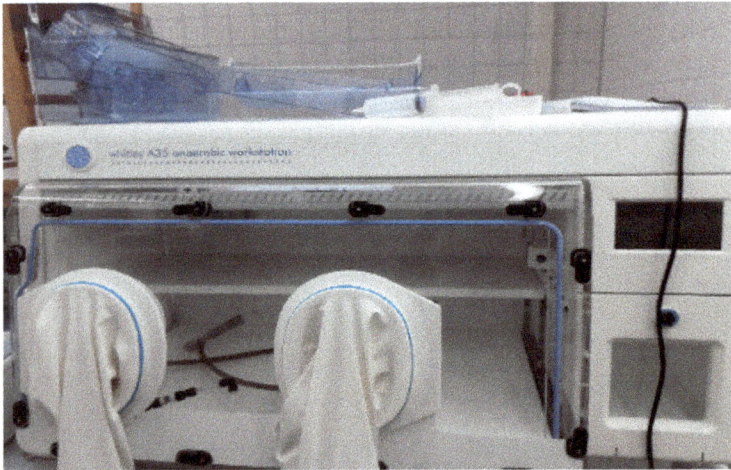

**Fig. 4.5:** Glove box for sample filtration and subsampling under anaerobic conditions. Source: Laboratory experiments in Bergische University in Wuppertal, September 2018.

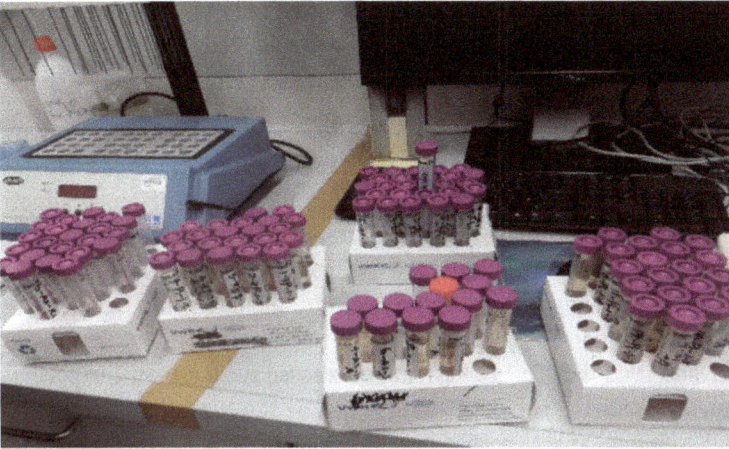

**Fig. 4.6:** Samples prepared for analysis. Source: Laboratory experiments in Bergische University in Wuppertal, September 2018.

## 4.3.1 Application of redox microcosm experiments in studies involving arsenic mobilization into groundwater and in planning remediation strategies

**Objectives and methodology**: In Mensah et al. [332], we investigated the mobilization and speciation of As in an abandoned As-contaminated gold mine spoil under reducing and oxidizing conditions. We assumed that humid tropical climates are characterized by high rainfall regimes. These events cause flooding of many urban areas, including farmlands and mining fields. Flooding and high rainfall regimes create various reducing and oxidizing conditions at mine-contaminated sites and may thus induce mobilization and release of toxic elements, including As-laden sediments and slurry into the environment. Consequently, such redox processes exacerbate pollution of nearby water courses and available groundwater, with subsequent threats to human health and livelihoods.

In summary, we explored the biogeochemistry and dynamic behavior of As in a contaminated mine spoil when it rains, or when the site is flooded or drained. This study was conducted using the novel automated biogeochemical MC experimental setup. This setup has been employed successfully in the past to study biogeochemistry and mobilization of other PTEs into groundwater [e.g., 530, 531]. The system enables the researcher to investigate in a controlled environment various biogeochemical processes that may influence mobilization and control the fate of transport of As during flooding. This equipment allows the simulation of anoxic/oxic conditions by adjusting the $E_H$ automatically with nitrogen to lower $E_H$ or synthetic air/oxygen to increase $E_H$. Contents of dissolved As, Fe, Mn, Al, and S; SUVA, DOC, and DIC; and $Cl^-$, $SO_4^{2-}$, as well as $Fe^{2+}$

were measured in the soil solution. Additionally, the selected soil samples from the lowest, middle, and highest redox potentials were selected for As K-edge X-ray absorption near-edge structure (XANES) analysis at NSRRC in Taiwan Light Photo Source.

**Findings, conclusions, and implications**: It was observed that As mobilization increased under reducing conditions while it decreased under oxidizing conditions. This mobilization was mainly regulated by the chemistry of Fe, which explained almost 100% of the data variation according to a multiple stepwise regression analysis. This was further confirmed by XANES results, where the majority of As was bound to Fe-bearing minerals such as arsenopyrite, scorodite, goethite, and ferrihydrite under both anaerobic and aerobic conditions.

It was further observed that $Fe^{2+}$ consistently increased during reducing periods and decreased during oxidizing conditions (Fig. 4.7). The implication of this is that during reductive dissolution, any As sorbed or coprecipitated with Fe mineral will be loosened and thus becomes mobilized for polluting groundwater. This hypothesis was further demonstrated by the Fe/As ratio, where lower values were observed under reducing conditions and higher values were found under oxidizing conditions. Lower values of Fe/As during reducing periods thus imply possible migration of As into watercourses during flooding commonly created by high rainfall events in tropical humid climates.

When the contaminated mining fields are drained of water and/or the floods, Fe/As values increase, and As consequently becomes precipitated/coprecipitated again due to the reintroduction of oxygen that oxidizes Fe. In conclusion, remediation measures to reduce As mobilization and associated ecological impacts should aim at limiting situations that create anaerobic conditions in the mining fields. For instance, vegetation cover or revegetation of the mining sites may improve the organic matter of the field, improve the soil structure, conserve soil moisture, reduce surface runoff, and encourage soil microbial activities, thus creating oxidizing conditions and consequently may limit As mobilization and curtail its associated health risks.

## 4.4 Spatial distribution of elements using geographical information system

Geographical information system (GIS) software provides powerful visualization tools to create maps, charts, and graphs that effectively communicate the spatial distribution of elements. You can use various symbols and color-coding techniques to highlight patterns and trends. The spatial distribution of elements using GIS is a valuable tool in various fields such as environmental science, urban planning, and natural resource management.

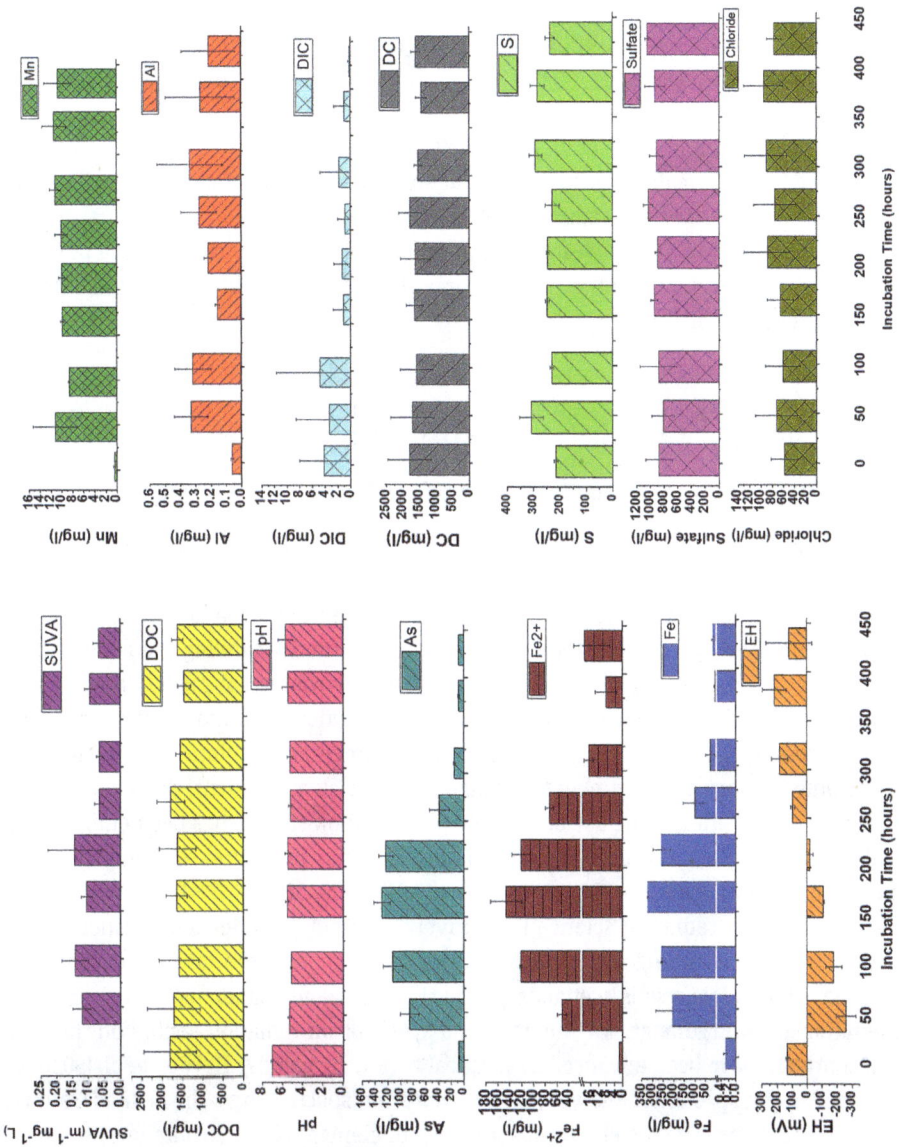

**Fig. 4.7:** Impact of different reducing and oxidizing conditions on the release dynamics of arsenic (As) as affected by redox-induced (EH) changes on pH, Fe, Fe$^{2+}$, Mn, Al, S, Cl$^-$, DOC, and SUVA in the contaminated mine spoil soil. Columns represent means, and whiskers represent the standard deviation of three replicates. Values of Fe, Fe$^{2+}$, Mn, Al, S, SO$_4^{2-}$, Cl$^-$, and DOC are in mg/L; EH in mV; and SUVA in m$^{-1}$ mg$^{-1}$ L. Source: Mensah et al. [332].

GIS allow you to analyze and visualize the distribution of elements such as land use, population, vegetation, pollution levels, and much more across geographic areas. These assist researchers and remediation projects to effectively plan and focus strategies on areas where the concentrations of metals are mostly high.

For instance, in Mensah et al. [333], we studied soil pollution with PTEs, and their associated human health risk in farms and soils near an abandoned gold mine spoil in Ghana. We collected soil samples within a depth of 0–20 cm, and individual coordinates were recorded. Details of the sample collection from individual sample points at various sites were plotted on a map to show the distribution of PTEs at the sites assessed or evaluated.

## 4.5 Element speciation experiments using synchrotron radiation science (e.g., K-edge X-ray absorption near-edge structure (XANES))

Synchrotron radiation science plays a significant role in environmental pollution research by providing powerful tools for analyzing and understanding the composition, distribution, and behavior of pollutants in various environmental media. In summary, synchrotron radiation science provides advanced analytical tools that enhance our ability to investigate, characterize, and address environmental pollution. These techniques contribute to more effective pollution management, risk assessment, and environmental conservation efforts. The use of synchrotron radiation science has been employed in soil pollution studies to ascertain the dominant minerals present in the samples. These studies have mainly been to those working on As [35, 320, 321, 332, 534].

Synchrotron radiation science is relatively new but provides a sophisticated approach to understanding the biogeochemical behavior and the fate of contaminant transport in floodplain soils, wetlands, paddy rice fields, and abandoned, active, and profile mine lands. Arguments for the need of a synchrotron radiation facility on the African continent have been advanced in a critical review by von der Heyden et al. [502].

In element speciation analysis, X-ray absorption spectroscopy, for instance, allows scientists to determine the chemical speciation of elements in environmental samples. This helps distinguish between different chemical forms of a pollutant, which is essential for understanding its mobility, toxicity, and bioavailability. In this regard, [565] reported that the bioavailable portion of As is also dependent to an extent on the speciation of the contaminant.

## 4.5.1 Applications of synchrotron radiation science in elemental speciation and remediation studies

Synchrotron radiation science can be applied to study the chemistry of soils and sediments, including the binding of pollutants to mineral surfaces. This knowledge is crucial for assessing the long-term fate and mobility of contaminants in the environment.

Synchrotron-based research helps elucidate the biogeochemical cycling of elements in ecosystems. This is essential for assessing the impacts of pollution on ecosystem health and understanding natural processes that affect pollutant distribution. Here, we conducted an experiment to quantify the impact of different redox conditions on speciation of As in an As-contaminated gold mine spoil in Ghana [332]. The selected soil samples from the lowest (−220 mV), middle (0 mV), and highest (+220 mV) redox potentials were selected for As K-edge XANES analysis. The As XANES spectra were collected at beamline TLS-07A at the National Synchrotron Radiation Research Center (NSRRC), Taiwan (https://www.nsrrc.org.tw/english/index.aspx).

Arsenic K-edge XANES spectroscopy analysis showed that sorbed As(V)-goethite, sorbed As(III)-ferrihydrite, scorodite, and arsenopyrite were the predominant As species in the mine spoil (Tab. 4.1). As(V) dominated under anaerobic conditions, and As(III) predominated under aerobic conditions, which may be attributed to either the inability of arsenate bacteria to reduce As or incomplete reduction. Consequently, the oxidized portion of As persisted during reducing conditions.

**Tab. 4.1:** Linear combination fitting results for As K-edge XANES spectra of soil samples.

| Samples | A3 (%) | A4 (%) | A5 (%) |
|---|---|---|---|
| Sorbed As(V) | 34(3) | 24(3) | 13(3) |
| Sorbed As(III) | 25(2) | 29(2) | 41(2) |
| Scorodite | 18(2) | 17(3) | 21(3) |
| Arsenopyrite | 23(2) | 30(3) | 25(3) |
| R factor | 0.011837 | 0.014286 | 0.014234 |

A3 = −220 mV Eh, A4 = 0 mV Eh, and A5 = +220 mV Eh. The numbers in parentheses indicate uncertainty, given by the Athena software. The data show the proportion of the reference spectra, which resulted in the best fit to the sample data. The normalized sum of the squared residuals of the fit is: $R\text{-factor} = \sum(\text{data-fit})^2/\sum \text{data}^2$.

The R-factor gives a clue about the strength of the fitting of the curve. The lesser the R-factor, the better the fitting, and thus the accuracy of your results. Thus, understanding the chemical and physical properties of pollutants at the molecular level can inform the development of effective remediation strategies. Researchers can use synchrotron techniques to study the effectiveness of remediation techniques, such as phytoremediation or soil amendments. For instance, in Mensah et al. [321], we employed As K-edge XANES analysis to examine the speciation of the predominant fraction of arsenic in soil samples, with the objective of assessing arsenic contamination in an abandoned and active gold mine spoils in Ghana. Understanding the mobility of arsenic (As) in the environment is crucial for comprehending its associated health hazards.

The findings of the study demonstrated the existence of two minerals containing arsenic: scorodite ($FeAsSO_4$) as the dominant and arsenopyrite (FeAsS) as a minor. These two minerals were detected and were thus the dominant species of As in the highly contaminated soil samples. This implies that the mobility and subsequent health risks associated with arsenic (As) in the region are linked to the behavior of secondary As minerals, rather than arsenopyrite, which is typically found in gold mines. In this context, it is imperative that the design of any remediation strategy targets scorodite as opposed to arsenopyrite. Details of the study findings are given in Tab. 4.2.

**Tab. 4.2:** Linear combination fitting results for As K-edge XANES spectra of soil samples.

| Sample | Scorodite | FeAsS | R-factor |
|---|---|---|---|
| AB1 | 72% ± 1% | 28% ± 1% | 0.009 |
| AB2 | 65% ± 0.8% | 35% ± 1.3% | 0.007 |
| AC | 76% ± 1% | 24% ± 1% | 0.021 |

AB1 and AB2 are selected samples from the abandoned mine spoils; AC is the selected sample from the bulk active mine spoils. Source: Mensah et al. [321]. The data show the proportion of the reference spectra, which resulted in the best fit to the sample data. The normalized sum of the squared residuals of the fit is: $R\text{-factor} = \sum(\text{data-fit})^2/\sum\text{data}^2$.

Table 4.2 indicates the percentage of each mineralogical composition of As in the mine spoil. The scorodite and arsenopyrite accounted for 65–72% and 28–35%, respectively, of total As in the selected abandoned samples, and accounted for 76% and 24%, respectively, in the active gold mine spoils. The appearance of As primary mineral arsenopyrite could be sourced from the studied gold mine, where arsenopyrite is associated with sulfide gold minerals and coexists in approximately equal proportions (50:50) with pyrite [566; 567].

## 4.6 Conclusions

The utilization of geospatial analysis, redox automated biogeochemical MC experiments, laboratory incubation trials, and synchrotron radiation science methods represents more advanced and sophisticated methodologies for thoroughly assessing and evaluating the risks of soil contamination associated with PTEs in mine-contaminated fields in Ghana. These methodologies facilitate the accurate characterization of soil-to-human health concerns prior to selecting a soil remediation approach. The aforementioned methodologies are frequently deficient among scientists in Ghana, specializing in soil sciences, environmental sciences, and public health. The examples presented and illustrated in the chapter reinforce the view that synchrotron radiation science and redox automated biogeochemical MC setup have the potential to revolutionize the process of planning and implementing an effective approach to soil remediation and revegetation of mine-contaminated and degraded sites.

Albert Kobina Mensah

# Chapter 5
# Impacts of mining on soil quality

**Abstract:** Mining activities have significant impacts on soil properties, including pH, fertility, physical composition, and moisture content. The pH of mine soil can be affected by the weathering and oxidation of rocks, resulting in fluctuations and potential acidity. Low pH levels can increase the mobility and availability of certain trace elements, while high pH levels can reduce their retention. Soil fertility is often compromised in mine spoils, with deficiencies in macronutrients such as nitrogen, phosphorus, and potassium. Organic matter plays a crucial role in providing nutrients, and the presence of metallic micronutrients can impede plant growth. The physical properties of mine soils, such as rock content and texture, can affect water and nutrient retention capacity. Soil aggregation and moisture content are also influenced by mining activities.

## 5.1 Impacts of mining on soil chemical properties

### 5.1.1 Effects on soil pH

Soil pH is a major biogeochemical factor that affects sorption and desorption, oxidation state, solubility, mobility, and thus the toxicity of soil As species [36, 81, 500]. In general, for cationic trace elements, lower pH results in higher mobility and availability, while higher pH results in increased mobility and availability for anionic species such as those of As [36].

For anionic species, the explanation is that high pH reduces the electronegativity of silicate secondary compounds and hence facilitates their availability. Moreover, at high pH, the positive charge on the surfaces of oxides diminishes, their activity and effectiveness for sorption decrease, and hence As retention reduces [568].

The measurement of soil pH serves as an assessment of the level of active soil acidity, making it a widely employed indicator in evaluating the quality of mine soil. The pH of a certain mine soil has the potential to undergo fast fluctuations due to the process of rock fragment weathering and oxidation. Pyritic minerals, characterized by the chemical formula $FeS_2$, undergo oxidation to form sulfuric acid, resulting in a significant decrease in pH. Conversely, minerals and rocks containing carbonate com-

**Albert Kobina Mensah,** Soil Research Institute, Council for Scientific and Industrial Research, Kumasi, Ghana, e-mail: albert.mensah@rub.de, albertkobinamensah@gmail.com, ORCID: https://orcid.org/0000-0001-5952-3357

https://doi.org/10.1515/9783111662046-005

pounds (Ca/MgCO₃) have a tendency to raise the pH as they undergo weathering and dissolution. Mine soils that have not undergone weathering or oxidation, and have a substantial concentration of pyritic sulfur (pyritic-S) exceeding the amount of neutralizing agents (carbonates), may experience a quick decrease in pH to a range of 2.2–3.5 upon contact with water and oxygen. When the pH of the soil decreases to a level below 5.5, the growth of legumes and forage is hindered due to the presence of toxic metals such as Al or Fe/Mn, the fixation of P, and a decrease in the number of nitrogen-fixing bacteria.

As a result, this expansion impedes the growth of plant roots and various other metabolic processes. Vegetation exhibits maximum growth potential in soils that possess a pH level within the neutral range. The optimal pH range for mine soil in order to support forages and other agronomic or horticultural purposes is said to be between 6.0 and 7.5, as documented by [569] and [570]. According to [571], the pH levels in a mining dump site located in the North Karanpura area of Ranchi in Jharkhand, India, ranged from 4.9 to 5.3. This variation in pH values suggests that the dumps exhibit an acidic nature. The acidic character observed in the area can be attributed to the geological composition of the rock there. [572] earlier reported that the bioavailability of hazardous metals such as nickel, lead, and cadmium rises when the pH level is below 5, in addition to iron.

## 5.1.2 Effects on the soil fertility

The three primary macronutrients that play a crucial role in plant growth, specifically nitrogen, phosphorus, and potassium, are commonly seen to be lacking in overburden dumps and mine spoils [573]. For instance, high contents of Al, Fe, and Mn induced by low pH values in mine spoils may bind P to the soil exchange sites and consequently reduce its availability [332]. The formation and maintenance of any plant community in newly formed mine soils, as well as many older ones, necessitate substantial amounts of fertilizer elements or modifying the soil via addition of soil amendments (e.g., manure, compost, and biochar).

Some earlier studies [e.g., 318, 326] have reported that organic matter serves as the primary provider of nutrients, including available P and K, in soils that have not been fertilized. The level of organic carbon in an overburden was found to be 0.35–0.85%. In this regard, Maiti and Ghose [902] reported that there exists a positive correlation between organic carbon and the availability of N and K, while a negative correlation is shown with Fe, Mn, Cu, and Zn.

Iron, Mn, Cu, and Zn are among the crucial metallic micronutrients that play a vital role in supporting plant growth [515]. The presence of these micronutrients in the soil can be attributed to the process of mineral weathering. The solubility of these metals is higher in acidic solutions, resulting in the formation of increased concentrations that can potentially impede plant growth [e.g., 574; 575].

# 5.2 Impacts of mining on soil physical properties

## 5.2.1 Rock content

The predominant water and nutrient retention capacity in mine soils is attributed to soil particles with a diameter smaller than 2 mm. Particles with a size above 2 mm are commonly denoted as "coarse fragments." Soils characterized by a high proportion of coarse fragments exhibit bigger pores that are unable to retain an adequate amount of plant-accessible water to support robust growth due to the potential for leaching. The composition of coarse fragments in mine spoils typically exhibits variations ranging from less than 30% to greater than 70%, which can be attributed to disparities in rock hardness, blasting methods, and spoil management practices [453].

The particle size distribution of mine soils is determined by the characteristics of their parent rocks or spoils. Over time, the rock content present on the surface of a reclaimed bench or out slopes will undergo a decrease as a result of the process of weathering, which causes the fragmentation of rock particles into smaller soil-sized particles. The presence of top soil components in rock content is generally reduced compared to spoils, resulting in improved water retention properties [577; 578]. According to [579], soil characterized by a stone content above 50% should be classified as having low quality. According to [580], it has been documented that the stone concentration of coal mine overburden dumps can reach levels as high as 80–85%. Research findings of Maiti and Ghose [902] demonstrated that the stone content in overburden dumps ranged from 35% to 65%, with a mean value of 55%.

## 5.2.2 Soil texture

The texture of soil is determined by the proportions of sand particles ranging from 2.0 to 0.05 mm, silt particles ranging from 0.05 to 0.002 mm, and clay particles smaller than 0.002 mm [453]. Soils having a sandy texture have reduced water and nutrient retention capabilities compared to soils with finer textures, such as loams and silts. Silts are characterized by their fine texture, which can lead to the formation of surface crusts. Additionally, they frequently contain elevated levels of soluble salts and exhibit poor "tilth" or consistency. The particle size distribution of soils exhibiting loamy textures is typically considered optimal. Silt loam textures are frequently observed in areas where spoils are predominantly composed of siltstones [179]. Ghose [179], in his study, found that mined soil had a maximum sand concentration of 66% and a clay percentage of only 8.6%. [581] and [582] have previously documented that the Singrauli Coalfield in India exhibits a maximum sand content of 80% and a minimum clay content of 11%. Similarly, Mensah et al. [321], in our earlier study on As contamination in active and abandoned mine spoils observed the clay content at 3%, while the spoil was predominantly sandy at a content of almost 70%. A soil with a low clay content is associated with a reduced cation exchange capability. According to Kolay [249], a higher cation exchange capacity enables the soil to accommodate larger quantities of nutrients in a single application, while also retaining adequate moisture for plant growth. In contrast, soils with lower cation exchange capacities are more prone to nutrient leaching, as applied nutrients are more easily washed away.

## 5.2.3 Soil aggregation

The process of soil aggregation plays a significant role in regulating soil hydrology, influencing soil diffusion, and determining the extent of nutrient accessibility within the soil [583]. Additionally, it has the potential to mitigate erosion and serves as a mechanism for the long-term stabilization and sequestration of organic carbon [584]. The process of aggregate structure degradation occurs as layers of soil are progressively excavated and relocated to designated areas within the site during the operation phase of the mine's project life cycle. This action results in soil compaction with consequent reduction in water holding capacity and a decrease in soil aeration. The primary determinant of macroporosity in soil, which influences drainage rate and aeration, is macroaggregate stability. Microaggregate stability exhibits greater resilience compared to macroaggregate stability due to the presence of organic matter that binds soil particles together in pores that are too small for microorganisms to inhabit.

## 5.2.4 Soil moisture content

The moisture content within a dump is a variable factor that is subject to fluctuations. This variability is impacted by several factors, including the time at which the sample is taken, the height of the dump, the presence of stones, the quantity of organic carbon, and the texture and thickness of the litter layers present on the surface of the dump [585]. According to [586], the moisture content in overburden dumps during the high summer period (May–June) was documented to reach as low as 2–3%. Maiti and Ghose [902] also add that the average field moisture content of the total mining wastes located in the North Karanpura area of Ranchi in Jharkhand, India, was reported to be 5%. Mensah et al. [332], found that the gravimetric moisture content of an abandoned mine spoil was lower than that in the manure-amended spoil in southwestern Ghana.

## 5.2.5 Soil bulk density

The bulk density of fertile natural soils typically falls within the range of 1.1–1.5 g/cm$^3$. The rooting depth in mine soils is constrained by the high bulk density. According to a study conducted by Maiti and Ghose [902], it was observed that the bulk density in overburden dumps, namely those that were 7 years old, reached a maximum value of 1.91 mg/m$^3$. [587] observed that the bulk density of soil beneath a grass sward in the UK can reach up to 1.8 mg/m$^3$. The growth of plants is directly hindered by soil compaction, as the majority of species face difficulties in successfully extending their roots through densely packed soils that have been mined [453].

## 5.2.6 Rooting depth

Mine soils that have a high bulk density (values > 1.7 g/cm$^3$) exert limited effective rooting depth (less than 2 feet). Such soils have shallow intact bedrock, and the presence of large boulders in these soils is unable to retain an adequate amount of plant-available water. As a result, these conditions are insufficient to support thriving plant communities during extended periods of drought. The consequences of such mine soils become dire, especially in areas where there is a scarcity of quality water for irrigating and maintaining planted vegetation on the degraded mine spoils. In order to provide sufficient water retention for plants during extended periods of drought, it is necessary to have a layer of loose, noncompacted soil material in which the roots of trees and shrubs used for revegetation can thrive and penetrate for available nutrients and water.

During periods of wet weather conditions, compacted zones have the potential to elevate water tables, leading to saturation and the development of anaerobic conditions within the rooting zone. In this respect, Sheoran et al. [38], reported that the frequent movement of wheeled mining machinery, such as loaders and haulers, and, to a lesser extent, bulldozers, results in the formation of densely packed areas within mining dumps. The repercussions are that the mine spoils become more compact, leading to higher bulk densities and reduction in root depth penetration. Consequently, such spoils may favor shallow cover crops such as those of grasses used in restoration projects but may impede penetration for accessible water by trees with taproot systems. In this regard, other researchers have argued that species used for mine land reclamation should have greater root length, greater agility in root growth, and superior ability for water and nutrient uptake [e.g., 332, 453].

## 5.2.7 Slope, topography, and stability

Mine soils that have been rehabilitated and possess slopes over 15% are typically deemed inappropriate for intensive land uses, such as the cultivation of vegetables or crops. However, these soils may still be acceptable for activities such as grazing livestock and reestablishing forested areas. Seasonal moisture concerns are frequently observed on broad, flat benches and fills that possess slopes with gradients below 2%.

The average slope of reclaimed modern mines tends to be significantly steeper compared to the older benches. However, the newer landforms exhibit a greater degree of smoothness and uniformity in their ultimate grade. Bench regions situated directly above undisturbed bedrock in previously mined terrains typically exhibit a rather solid nature, albeit with a potential susceptibility to slumping, particularly in close proximity to the outside periphery of the slope.

According to Sheoran et al. [453], the presence of tension cracks that are approximately parallel to the out slope suggests that the area is characterized by instability and is prone to slumping. The rise in bulk density can be attributed to a decrease in soil stability as the matrix becomes more susceptible to slaking, dispersion caused by water, and the stresses exerted by wheels, hooves, and rainfall [588]. Consequently, this phenomenon results in reduction in the level of aeration and the rate at which water infiltrates the soil, ultimately creating anaerobic environments.

## 5.2.8 Mine spoil/soil color

Mining operations involve the extraction of surface earth, which is subsequently deposited onto adjacent, unmined areas, resulting in the creation of interconnected chains of external dumps commonly referred to as mine spoil or wasteland. The cultivation of mine waste necessitates the implementation of stringent environmental conditions that are conducive for the growth of both plant and microorganism cultures. The disruption of biological functionality, coupled with the disturbance of the nutrient cycle, results in the impairment of soil system functionality.

The primary cause of this phenomenon can be attributed to the presence of insufficient levels of organic matter and other undesirable physicochemical and microbiological properties (e.g., [589, 590]). The chromatic characteristics of mine spoils or worn mine soil can provide valuable insights into their weathering chronology, chemical attributes, and physical composition. The presence of bright red and brown hues in spoils and soils typically signifies a certain level of oxidation and leaching of the substance. The aforementioned materials exhibit characteristics such as lower pH levels, reduced salt content, diminished fertility, decreased pyrite content, and increased vulnerability to physical weathering in comparison to darker-colored materials. The presence of gray hues in rocks, spoils, and soils typically signifies reduced levels of oxidation and leaching, resulting in better pH values and fertility levels within these materials. Rocks, spoils, and mine soils that exhibit a dark gray or black color are known to possess substantial quantities of organic matter and tend to exhibit acidity, as reported by Sheoran et al. [453].

The available evidence suggests that the spontaneous process of natural colonization of dark-gray spoils, as well as other similar mining sites, can be highly effective in establishing mature and functional ecosystems within a span of 100 years [453; 591; 592]. The process of rehabilitating mine wasteland typically necessitates human intervention, also known as aided regeneration, in order to meet restoration goals within an acceptable time frame [20, 593].

# 5.3 Effects of mining on soil biological properties

## 5.3.1 Soil microorganisms

Ghose [818] reported that soil microbes have a significant impact on aggregate stabilization, which is crucial for preserving appropriate structural conditions for cultivation and promoting porosity for crop growth. The activity of soil microbes decreases when there is disruption in the soil layers, and it exhibits a sluggish rate of independent recovery.

Soil microbes encompass a variety of bacterial species that play an active role in the decomposition of plant materials. Additionally, there are fungal species that engage in symbiotic relationships with numerous plants, aiding in the uptake of nitrogen and phosphorus in return for carbon. According to Williamson and Johnson [514], the microorganisms generate polysaccharides that enhance soil aggregation and exert a beneficial influence on plant growth. According to [594], locations characterized by a thriving population of soil microorganisms have consistent soil aggregation, while areas with reduced microbial activity tend to have compacted soil and inadequate aggregation. According to [595], the presence of microbial activity diminishes as the depth and duration of mining activities increase, resulting in the continued accumulation of topsoil. The quantity of adenosine triphosphatase, which serves as an indicator of microbial activity, experiences a significant decrease to minimal levels within a short span of a few months. According to [596], the microbial response to glucose has a decelerated pattern throughout all depths, indicating a decline in metabolic rates over time.

## 5.3.2 Bacteria

Bacteria are of significant importance in the process of organic material degradation, particularly during the initial phases characterized by elevated moisture levels. During the advanced phases of decomposition, fungi typically exhibit dominance. Rhizobia are a type of unicellular bacteria that belong to the family Rhizobiaceae. They establish a mutually beneficial association, known as symbiosis, with legume plants. The bacteria in question have the ability to extract nitrogen from the atmosphere, a form that is not readily used by plants, and subsequently transform it into ammonia ($NH_4^+$), a nitrogen compound that plants can effectively utilize [597].

Both free-living and symbiotic plant growth-promoting rhizobacteria have the potential to enhance plant growth through various mechanisms. These include the provision of bioavailable phosphorus for plant uptake, nitrogen fixation for plant utilization, sequestration of trace elements such as iron through siderophores, production of plant hormones like auxins, cytokinins, and gibberellins, as well as the reduction of plant ethylene levels [598; 599].

The relocation of soil layers and their subsequent stockpiling result in the displacement of bacteria residing in the first upper layers, ultimately leading to their settlement at the base of the pile beneath densely packed soil. During the initial year, a surge of activity takes place in the just-formed top layer when bacteria become exposed to atmospheric oxygen. According to Williamson and Johnson [514], the bacterial numbers at the surface show minimal alteration after a storage period of 2 years. However, it is noteworthy that less than half of the initial populations endure at depths above 50 cm.

## 5.3.3 Mycorrhizal fungi

They are a group of symbiotic organisms that form mutualistic associations with the roots of most plants. Arbuscular mycorrhiza (AM) fungi are widely distributed soil microorganisms found in a variety of environments and climates. According to Gould et al. [826], the mycorrhizal fungus's hyphae network is disrupted when soils are originally disturbed and stored. The importance of mycorrhizal connections for the survival and growth of plants, as well as the uptake of nutrients such as phosphorus and nitrogen, has been extensively demonstrated in the literature, particularly in the context of phosphorus-deficient abandoned soils [615].

One significant genus of AM is *Glomus*, which has the ability to establish symbiotic relationships with a diverse range of host plants, such as sunflower [600]. The combined application of *Trichoderma koningii* and AM fungus resulted in enhanced plant development of *Eucalyptus globulus* in the presence of heavy metal contamination, as demonstrated in studies conducted by [601, 602].

According to [603], there is a little reduction in the possibility of viable mycorrhizal inoculum during the initial 2 years of storage. The viability of mycorrhiza in stored soils exhibits a significant decline, perhaps reaching values as low as one-tenth of those observed in undisturbed soil [973]. Miller et al. [925] reported that the viability of mycorrhizal fungi is influenced by soil water potential, which is considered a crucial element. In the context of soil water potential, it has been observed that mycorrhizal propagules exhibit enhanced survival rates throughout prolonged storage periods under conditions of lower soil water potential (indicative of drier soil). Conversely, when soil water potential is above the threshold of −2 MPa, the duration of storage becomes a more critical factor in determining the viability of mycorrhizal propagules. In arid regions, the presence of deep stocks may not pose a significant risk to the survival of mycorrhizal propagules.

## 5.3.4 Soil microbial activities

Microbial activity is inhibited significantly in heavy metal-contaminated soils [468]. Kandeler et al. [233], found that microbial biomass in an As-contaminated soil near the mine was lower than that far away from the mine. [605] reported that a higher content of As above the threshold could decrease soil microbial biomass significantly.

Soil enzymes play an important role in the process of organic matter decomposition and nutrient cycling. Previous studies have shown that As in the soil reduces the activities and effectiveness of soil enzymes [e.g., 168, 468]. [606], for instance, found that activities of almost all enzymes in the soil studied were significantly reduced by 10–50 times with increased concentration of heavy metals including As.

Gao et al. [168], assessed the effects of toxic elements combined pollution on soil enzyme activities and microbial community structure. The study found that soil microbial populations were significantly lower under polluted sites than under control treatments, with soil bacteria decreasing the most in population size than the other soil microbes such as fungi and actinomycetes. Further, elevated heavy metal concentrations and toxicant levels differently impacted soil enzyme activities, with inhibition of phosphatase, urease, and dehydrogenase activity.

Prach and Pysek [607] assessed the As mobility in mine tailings and its impact on soil enzyme activity in South Korea. The study found that soil enzyme activities were remarkably affected with the presence of high contents of As. More particularly, the abundance and activities of soil enzymes negatively correlated with the water-soluble As fraction in the contaminated soils. Thus, the soil enzyme activities were mainly affected by the As water-soluble fraction than by the other fractions. In the same study, the treatments that decreased the As water-soluble fraction enhanced the soil enzyme activities.

## 5.4 Conclusions

In conclusion, mining activities have a significant impact on soil characteristics. The pH levels of soil in mines can be affected by weathering and oxidation, leading to variations in acidity. This can affect the movement and accessibility of trace elements. Mine spoils often suffer from impaired soil fertility, with deficits in macronutrients. The presence of metallic micronutrients can hinder plant growth. The physical features of mine soils, such as the presence of rocks and texture, can influence water and nutrient retention capacity. Additionally, mining activities can impact soil aggregation and moisture content. Overall, mining activities have a profound influence on soil properties, which can have long-lasting effects on ecosystems and agricultural productivity.

Albert Kobina Mensah

# Chapter 6
# Topsoil and its management during stockpiling

**Abstract:** The process of mineral mining often leads to the removal of vegetation and depletion of nutrients from the topsoil, which is crucial for plant growth. Bulldozers and heavy machinery are used in mining to extract minerals, resulting in the depletion of soil fertility and productivity. Stockpiling topsoil also negatively affects the soil quality and beneficial soil microbes. Preserving and directly returning topsoil to mining sites are more effective methods to retain soil fertility compared to stockpiling. Deep-ripping and fertilizer application can enhance the physical environment and replenish depleted nutrients. Directly returning topsoil mitigates redundant processing, reduces the need for clearing additional land, and maintains indigenous plant species richness. Managing and conserving topsoil are crucial for successful land reclamation and revegetation in mining regions.

## 6.1 Introduction

The majority of vital nutrients for plants are often found in the uppermost layer of soil, known as the topsoil, in the soil profile. The process of mineral mining results in the removal of vegetation, which consequently leads to the inevitable depletion of certain plant nutrients from the affected area. Furthermore, the mining procedure involves the utilization of bulldozers and other heavy machines to remove the uppermost layer of soil, which is subsequently extracted in order to obtain the desired mineral. The shallow depth of the topsoil, approximately 20 cm, renders it crucial for providing essential nutrients to plants. However, the utilization of bulldozers and other heavy equipment in the process of scraping results in the depletion of soil fertility and productivity. Consequently, this practice exposes the subsurface, which is unsuitable for cultivating crops.

As illustrated by [735] study, the soil organic carbon (SOC) concentration in the Prestea/Bogoso mined area of Golden Star Resources was measured at 0.14%. This value falls below the recognized threshold for soil fertility, suggesting a disturbance in ecosystem functioning and the depletion of the litter layer as a result of mineral mining activi-

**Albert Kobina Mensah**, Soil Research Institute, Council for Scientific and Industrial Research, Kumasi, Ghana, e-mail: albert.mensah@rub.de, albertkobinamensah@gmail.com, ORCID: https://orcid.org/0000-0001-5952-3357

https://doi.org/10.1515/9783111662046-006

ties. Organic matter values are classified as low when they fall below 4%, medium when they range from 4% to 8%, and high when they exceed 8%.[1]

The availability of topsoil is a crucial determinant in the attainment of a prosperous and successful restoration of degraded and decommissioned mining sites. The topsoil is of utmost significance, particularly in its role in promoting the growth and establishment of indigenous species [608]. Topsoil is employed to cover insufficient substrate and improve the growing conditions for plants [453]. For example, in the agricultural approach to mine soil reclamation, the mine-contaminated site is regraded, the landform reconstruction is done, and the topsoil is used to spread on the surface before the planting of the species is done. This scenario makes topsoil an indispensable resource in mine land restoration engineering.

According to Mensah et al. [324], the presence of mounds formed by the buildup of topsoil during mineral extraction has been observed to affect the biological, chemical, and physical properties of the soil. For instance, Amegbey [721] reported that the practice of stockpiling has a negative effect on the quality of soil resources. Furthermore, the stockpiles are subjected to anaerobic conditions, resulting in the demise of many plant propagules and a significant decline in the populations of beneficial soil microbes.

Amegbey [737] provided empirical evidence demonstrating that the fresh soil obtained from the sand mine located in Eneabba, Western Australia, exhibited a significantly higher seed count, ranging from five to ten times greater in comparison to soil that had been stored for a period of 3 years. These reports corroborate those made by Ghose [178] during a study on topsoil management in mining in India. In his study, the population of soil living organisms and their activities declined as the age of the stockpiling period increased (see Fig. 6.1).

The accessibility of topsoil is restricted and is typically not preserved in many cases [453]. Moreover, in areas characterized by a tropical climate, where the predominant portion (90%) of yearly precipitation is received within a condensed 3-month time frame known as the rainy season, the preservation of topsoil and the sustenance of soil integrity pose continuous problems for water quality and the food chain. The act of recycling topsoil is not frequently done; instead, it is more common to acquire topsoil from adjacent regions for the purpose of restoring degraded mined-out areas [453].

Harris et al. [611] reported an observed increase in the population of anaerobic bacteria and a simultaneous decrease in aerobic bacteria at a depth of roughly 1 m within the stockpile. The insufficient oxygenation within the stockpile hinders the process of nitrification, leading to the accumulation of ammonia in the anaerobic zones.

---

1 University of Connecticut, College of Agriculture and Natural Resources, Cooperative Extension System (2003). Interpretation of Soil Test Results. Available: www.soiltest.uconn.edu/factsheets/InterpretationResults_new.pdf. Accessed on 26 Nov. 23.

Following the process of soil removal and subsequent reinstatement from the stockpile, there is a rapid reestablishment of the aerobic microbial population, often above the standard level, as observed by Williamson and Johnson in [1025]. Furthermore, it is suggested that nitrification be conducted at levels beyond the standard rates [453]. If a soil that has been restored has a high concentration of ammonia, it is anticipated that the subsequent production of nitrate will surpass the usual values. Consequently, there exists a substantial probability of nitrogen loss to the adjacent environment via mechanisms such as leaching and denitrification [610]. The process of nitrate leaking into watercourses presents a notable hazard to both the aquatic ecosystem and the accessibility of potable water [609] in relation to the induction of eutrophication.

Eutrophication is a phenomenon characterized by the transportation of chemicals, such as nitrates, sulfates, and phosphates, by runoff into water bodies, leading to an increase in nutrient concentrations within these aquatic ecosystems. This phenomenon results in a notable augmentation of plant growth within aquatic ecosystems, hence exerting a significant impact on the overall water depth and volume available. Furthermore, it should be noted that the release of nitrogen from soil in the form of gaseous nitrogen or nitrous oxides has the potential to contribute to the deterioration of the ozone layer [611; 612].

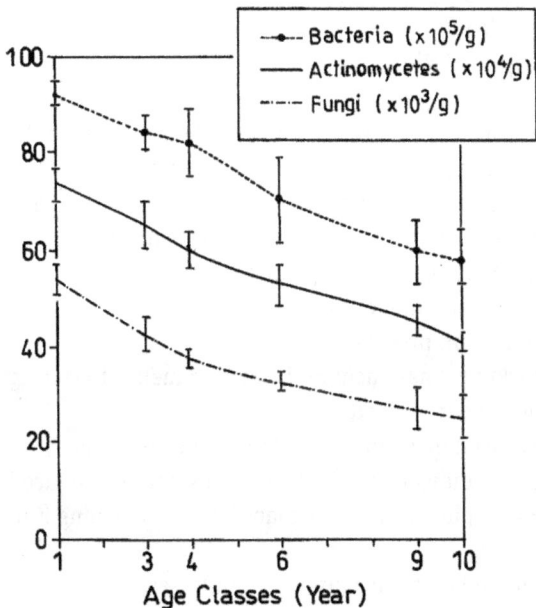

**Fig. 6.1:** Effects of stockpiling topsoil on soil microbial populations in mine land reclamation. Source: Ghose [178].

## 6.2 "Direct return" of topsoil in mine land restoration

There may have been a significant period of time between the initial removal of topsoil and the subsequent reapplication of the same soil to the reclaimed area. Therefore, if stockpiled soil is not stored adequately, its qualities will gradually degrade, leading to a state of biological non-productivity [179]. Amegbey [737] suggests that it is advisable to replenish the topsoil in areas where landform reconstruction has been fully accomplished, a practice sometimes referred to as "**direct return**." Fig. 6.2 Topsoil stockpiling can eventually lead to reduced nutrient cycling and lower availability of nutrients, which can potentially impair the entire revegetation and reclamation success [359].

Thus, during direct return, the topsoil that was removed at the initiation and operation stages of the mines' project life cycle is saved and maintained for eventual use during the revegetation time. This is something that can be done in order to bring the land back to its natural state. Consequently, in direct return, the soil that was stripped from the land prior to mineral mining is relocated directly to an area where the landform redesign is complete and revegetation is ready to be commenced on the land. These ready-made soils and lands are, as a result, prepared to accept the soil.

In certain mining operations, it is common practice to perform deep-ripping (subsoiling) on the mine floor either before or after the application of soil on the newly formed landscape. This process is carried out to disrupt the compacted materials resulting from the characteristics of the regolith materials and the movement of heavy machinery [613].

The objective is to manipulate a straight or winged tine within the subsoil with the purpose of enhancing the physical environment to facilitate root growth and penetration in the reconstructed substrate or soil. Fertilizers, for example, those of P and N sources, are then applied to the soil to reintroduce or replenish the depleted nutrients.

Direct return offers various advantages in comparison to the practice of stockpiling and conserving topsoil for further rehabilitation purposes:
- It mitigates the occurrence of redundant processing.
- The necessity of establishing stockpiles may include the requirement of clearing additional land. Direct return aids in curtailing this.
- Stockpiling has a significant negative impact on the quality of soil resources.
- The density and species richness of indigenous plants are substantially reduced when an area is restored using stockpiled soil rather than directly returning it to the site.
- It avoids excessive handling of the soil that can cause nutrient losses.
- It avoids the introduction of alien species and aids in maintaining native species.
- Direct return is relatively cheaper. Here, the topsoil is readily available, and one does not need to buy topsoil for reclamation.

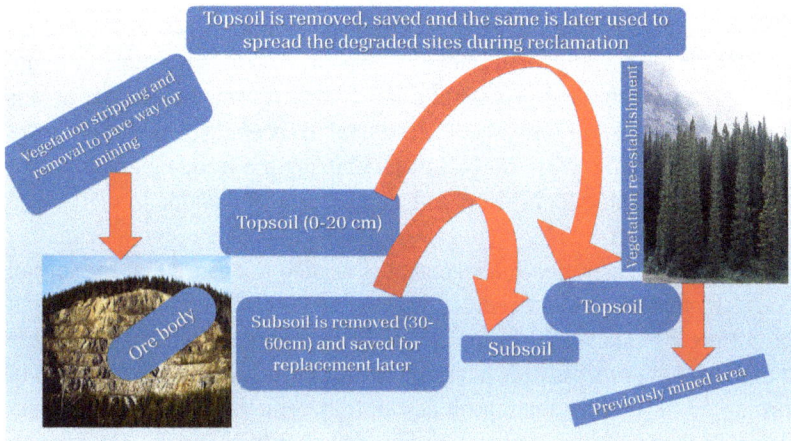

**Fig. 6.2:** Direct return model for soil handling in mine site restoration schemes. Ideally, the topsoil and subsoil are placed directly onto the newly prepared landscape. If necessary, the soils are stockpiled only for a few hours or days at most. This is the most effective manner of retaining soil fertility. Source: Author's own construct.

## 6.3 Managing and conserving topsoil for successful reclamation/revegetation

The inclusion of topsoil is a crucial element in the process of land reclamation within mining regions [614]. If the topsoil is not extracted individually at the outset for the purpose of subsequent replacement within the same region, it can result in significant harm. Preserving topsoil for future utilization is crucial in safeguarding the medium against pollution and erosion, hence enhancing its quality and productivity when later employed for reclamation [1066]. In this regard, some reports (e.g., [615]) suggested that the use of careful handling and storage techniques can effectively safeguard the physical and chemical properties of topsoil both during its storage period and subsequent redistribution onto designated regions. The application of site-specific modern technologies in monitoring and implementing these measures can effectively mitigate deterioration and create favorable conditions for plant growth [178].

### 6.3.1 Removing and storing topsoil

It is recommended to remove topsoil as a first step after clearing vegetation in disturbed regions and before beginning any surface disturbance activities like blasting, drilling, mining, or any other activity. The stripping must be done in seasons when the soil is dry. This technique significantly reduces the risk of compaction and soil structure damage by minimizing spreading and remolding. Extended periods of rainfall are not conducive to stripping the soil. The process of clearing topsoil from a particular area is frequently achieved by means of scrapers. It is essential to carefully plan the routing of scrapers during this operation with the goal of minimizing the frequency of machine travel over the soil in order to prevent soil compaction and damage to the soil structure.

Furthermore, careful operation management is needed to ensure that the topsoil and subsoil are removed to the desired depths. It is advised to extract the soils and keep them in different containers. It is thought to be unfavorable to inter-mine these soils during the stripping process. It is advised that before any disturbance occurs in the areas earmarked for mining, the topsoil be carefully removed and placed aside in a separate layer. It is advised to extract a 15 cm layer that includes the "A" horizon plus any consolidated material (if the total amount available is less than 15 cm) when the topsoil depth is less than 15 cm. After that, this extracted mixture needs to be divided up and spread out uniformly as surface soil. It is recommended to separate and replace the B horizon with subsoil, together with a portion of the C horizon that has appropriate qualities for root growth [178]. The process of stockpiling topsoil in mounds is done for two main reasons, as argued by [178]:

i.   when it is not possible to redistribute the topsoil on the regraded mined area, and
ii.  when it has to be protected against wind and water erosion, compaction, and any further activities that may affect its productivity during its reuse for mined land restoration projects.

If there is a need to stockpile topsoil for its later reuse, it is advisable to store it by following these recommendations:

–   Minimize the duration of storing stockpiles. The extended duration of storage periods has a detrimental impact on topsoil productivity due to the elimination of anaerobic organisms that are unable to withstand high temperatures. Furthermore, extended periods of storage can lead to the depletion of vital nutrients in plants, rendering them susceptible to erosion by water and wind, which can wash away the topmost nutrient-rich layers.
–   Provide the stockpile with a maximum and broad surface area. The top of the stockpiles should be flat enough to support crop growth and control erosion by water.
–   Keep the stockpiles at a low height, preferably below 2 m (maximum height of 5 m), and with a maximum slope of 18.5° [178]. Generally, coarser soils should have higher heights (e.g., 5 m is recommended for sandy soils), and finer soils should be kept at lower heights (e.g., 1 m for clayey soils).

- It is recommended to implement revegetation measures on the stockpiles to safeguard the soil against erosion, discourage weed growth, and maintain the presence of beneficial soil microorganisms.
- In order to retain biological activity in the topsoil, the local provenance plant species are sometimes sown on the stockpiles to develop a green cover.
- It is crucial to select stockpile locations that will not be disturbed by future mining activities, as excessive handling can have detrimental effects on the soil structure.

## 6.3.2 Redistribution of topsoil in the area to be reclaimed

In order to maintain productivity, preserve natural habitats, and conserve soil and water resources, numerous authors have recommended the practice of overlaying the spoil overburden with topsoil of varying depths. The thickness of the topsoil has an impact on several soil properties, such as water-holding capacity, nutrient-supplying capability, buffering capacity, and plant root depth. The uniform redistribution of topsoil is essential to ensure its compatibility with the needs of the species intended for the restoration of the mined area to its pre-mined potential. This involves considering the location and other physical characteristics of the topsoil.

## 6.4 Conclusions

The following conclusions are drawn about the chapter:
a) The majority of vital nutrients for plants are often found in the topsoil, which is the uppermost layer of soil.
b) The stockpiling of topsoil also has negative effects on soil quality and the populations of beneficial soil microbes.
c) The availability of topsoil is crucial for the restoration of degraded mining sites, but it is often not preserved and is instead acquired from adjacent regions.
d) Direct return of topsoil, where it is saved and maintained for later use, is a more effective method of retaining soil fertility compared to stockpiling.
e) Deep-ripping and the application of fertilizers can enhance the physical environment and replenish depleted nutrients in the soil.
f) Managing and conserving topsoil are crucial for successful land reclamation and revegetation in mining regions.

Albert Kobina Mensah

# Chapter 7
# Effects of mining on the accumulation and pollution with potentially toxic elements

**Abstract:** This chapter discusses the presence and impact of potentially toxic elements (PTEs) in the environment, particularly in relation to mining activities. It highlights the sources of PTEs, such as natural processes and human activities, and the potential health risks associated with their contamination of soil, water, and air. It also explores the specific effects of PTEs such as arsenic, cadmium, mercury, lead, antimony, selenium, and thallium on human health, including the development of various cancers and neurological disorders. The importance of implementing strategies to mitigate PTE contamination and protect human, plant, and animal health is emphasized.

## 7.1 Natural sources

The most common types of potentially toxic elements (PTEs) that have been discovered to pollute groundwater and aquifers worldwide are those that are found naturally in the environment [460]. According to Escobar et al. [148], they could originate from hydrological, geological, or soil-forming biogeochemical processes. The lithology of the source rock materials, volcanic activity, weathering history, transport, sorption, biological activity, and precipitation all influence the composition of soil PTE during normal soil-forming conditions [230].

For instance, the concentration of arsenic in sedimentary rocks can range from 1.7 to 400 mg/kg, while in igneous rocks, it can range from 1.3 to 3.0 mg/kg [148]. High concentrations of As in the environment can also be caused by volcanic eruptions, other natural processes such as weathering and rock deterioration, and biological activity [81].

Natural sources can also originate from atmospheric emissions or from desorption and dissolution of minerals high in arsenic [76]. Matschullat [302] stated that volcanoes release 17,150 tons of As into the atmosphere, oceans release 27 tons, and burning wood, oil, and naturally occurring forest fires release 125–3,345 tons. As is widely distributed as particles in the atmosphere [455]. Arsenic in the atmosphere is also caused by wind mobilization and deposition, marine aerosols, and industrial exhausts. Burning fossil fuels or using smelters (a process known as dry deposition) releases particles that land

**Albert Kobina Mensah,** Soil Research Institute, Council for Scientific and Industrial Research, Kumasi, Ghana, e-mail: albert.mensah@rub.de, albertkobinamensah@gmail.com,
ORCID: https://orcid.org/0000-0001-5952-3357

https://doi.org/10.1515/9783111662046-007

on the ground. The part of atmospheric As that dissolves in rain is also known as wet deposition [253]. Arsines that naturally escape from swampy fields and marshy soil are another natural source of As.

## 7.2 Anthropogenic sources

Arsenic concentrations of only a few nanograms per cubic meter are found in areas free of and less affected by human activity [81]. Bissel and Frimmel [81] state that the naturally occurring concentration of As in soils varies from 0.01 to 100 mg/kg. However, the highest amounts of As deposition in sediments and soils are caused by industrial pollutants [253]. For example, [616] reported that As concentrations in mining streams, which are primarily derived from acid mine drainage (AMD), ranged from 5 µg/L to 72 mg/L in seven countries in southeast Asia, Africa, and Latin America.

Burning fossil fuels in homes and power plants is another human-caused source [81]. By volatilizing $As_4O_6$, which condenses in the flue system, burning coal releases As into the atmosphere [302]. Thermal power plant fly ash has the potential to contaminate soil [81].

Furthermore, as noted by Goldberg [181] the application of arsenical fungicides, herbicides, and insecticides in the wood industry and agriculture may be considered human sources. As residues can reach levels of up to 2 g/kg in soils that have been exposed to long-term applications of arsenical insecticides, herbicides, and pesticides [e.g., 413, 519]. Lead-acid batteries are made mostly from As, while gallium arsenide, a semiconductor used in computers and other electrical devices, is made from very small amounts of extremely pure As metal [355]. These could be other human-driven causes of As pollution for the ground, surface, and soil waters.

## 7.3 Soil contamination and pollution with PTEs from gold mining

High PTE concentrations in water, soil, and air can be released as a result of mining and processing As-bearing minerals [132]. The extraction of As-rich sulfide ores releases contaminated waste in addition to As being released naturally through weathering and oxidation processes. This waste is dumped as tailing heaps, which can cause an environment-wide release of As and other hazardous elements. These also have the unintended consequence of contaminating several environmental media, including the atmosphere, surface runoff, precipitation, groundwater, and soil. As a result, contamination from gold mining can originate from either anthropogenic sources (from improper handling and management of mining wastes) or from geogenic sources (from natural oxidation of the FeAsS parent material, the mining ore) [96, 132, 216, 300, 319, 321, 393, 467].

The case of PTE pollution from anthropogenic and natural sources results from gold mining, along with the paths of contamination and their effects on the environment, society, and human health. As previously indicated, the primary cause of natural contamination is the oxidation of FeAsS, the ore used in gold mining. Other sulfides such as pyrite ($FeS_2$), which will produce acidity and release As during oxidation, may also be present in these tailings [121, 122]. According to Posada-Ayala et al. [393], mine tailings – which include sulfide minerals like arsenopyrite (FeAsS) – are a particularly rich source of As in this regard.

In other cases, As-rich tailing spoils could be carried by the wind or the air and end up in the surrounding area, endangering both human health and the environment. Furthermore, As might be activated by precipitation or during floods [333]. Such polluted areas are also capable of introducing and dispersing As over a large area through volcanic eruptions. Moreover, AMD causes As to be produced naturally through gold mining.

When reactive sulfide minerals, primarily pyrite and arsenopyrite, are exposed to air, oxygen, water, and microbiological activity, AMD occurs spontaneously [162]. In conclusion, the primary sulfide minerals found in gold mining sites and spoils, FeAsS or FeS, naturally oxidize to cause AMD. In addition to air (oxygen), rain drop temperature and water saturation levels, microbial activity, and the extent of metal sulfide exposure are physical, biological, and chemical elements that affect AMD formation. One of the biggest environmental issues that the mining industry faces is AMD, often known as acid rock drainage (ARD) waters. Equations (7.1) and (7.2) [132] provide an illustration of this:

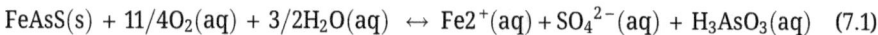

$$FeAsS(s) + 11/4O_2(aq) + 3/2H_2O(aq) \leftrightarrow Fe2^+(aq) + SO_4^{2-}(aq) + H_3AsO_3(aq) \quad (7.1)$$

$$H_3AsO_3 + 1/2O_2(aq) \leftrightarrow H_2AsO_4^- + H^+ \quad (7.2)$$

Anthropogenic sources of PTE contamination resulting from gold mining activities encompass various factors, such as the abandonment of mining spoils, the collapse of mine tailings, the leaching of effluents from mine tailings, the spillage from tailing dams, the spillage from mine sites, the surface runoff from mine sites, the stockpiling of mine wastes, the processing of mine waste, and the inadequate handling of hazardous materials associated with mining operations.

## 7.4 Plant and human health effects of potentially toxic elements

Arsine gas (AsH$_3$), for instance, is reported as the most toxic compound [253, 446], and the fatal dose is 250 mg/m$^3$ at an exposure time of 30 min [455]. As earlier explained, the toxicity of various forms of As strongly depends on their oxidative states and chemical structures [348]. The inorganic forms in soil are accumulated and transported through the food chain; they become toxic and in turn tend to affect various life forms [455].

Short-term exposure to high concentrations, even at a single dose, may be fatal to humans, whereas long-term exposure to low concentrations may induce the development of different diseases or disorders [937]. High concentrations of PTEs in abandoned or active mining spoils are a serious public health issue that could threaten both human health and the environment's integrity, which may indeed cause pollutant release and soil and water contamination [320, 333]. These contaminated areas serve as major point sources of pollution and PTE reservoirs.

Further, these PTEs present possible health hazards to those who live close to mining zones because they can spread to neighboring plants, food crops, soil, surface water, and groundwater. The distribution and availability of these elements in the environment are influenced by the geology of the area (as PTEs may be lithogenically enriched in primary minerals), as well as unsustainable mining practices that release toxic elements into environmental media. In this regard, the release of PTEs previously embedded in the crystals of metal-rich rock minerals during weathering may enrich significant soil PTEs. Consequently, when PTEs become part of the soil, they may be carried by wind or water and are eventually deposited with suspended particles settling on the ground through a process called dry or wet deposition, thus causing enrichment of extended neighboring areas beyond the immediate mine waste deposition.

The effects of contaminants on human health are often linked to (i) the specificity of the contaminants and (ii) human exposure (pathways and duration) to them (Naidu and Biswas, 2023). Artisanal gold mining, for example, has led to mercury pollution, river sedimentation, erosion, and unreclaimed fields [323]. Strategies should be implemented to restore the land or find alternative uses for affected areas. The goal is to mitigate risks to human, plant, and animal health and restore soil functionality [515]. This allows for the safe utilization of mining spoil.

## 7.4.1 Effects of PTEs on plant health

Arsenic is not an essential element, so it does not have any "expected" concentration in plants [333]. However, plants growing in uncontaminated soil contain As <3.6 mg/kg [172], while Kabata-Pendias [230] reported that the value ranges between 0 and 1.5 mg/kg. There are also other cases where terrestrial plants may accumulate As by root uptake from soil or by absorption of airborne As deposited on the leaves [455, 468]. Toxicity symptoms include inhibition of seed germination, decrease in plant height, depressed tillering, reduction in root growth and some necrosis, tissue death, decrease in shoot growth, lower fruit and grain yield, and sometimes lead to death. Additionally, discolored and stunted roots, withered and yellow leaves, and reductions in chlorophyll and protein contents and in photosynthetic capacity are reported [e.g., 455, 468].

## 7.4.2 Effects of PTEs on human health

Globally, elements in soils developed from the underlying rocks play an integral role in the health of the exposed people [239]. Medical geology has linked several diseases to environmental pollution and PTE poisoning. In this scenario, PTEs build up in the environment to dangerous levels, which lead to several noncommunicable diseases.

The International Agency for Research on Cancer (IARC) has classified arsenic as a Class-I human carcinogen [217]. Acute and chronic exposure to arsenic can cause numerous human health problems [89]. These include dermal, respiratory, cardiovascular, gastrointestinal, hematological, hepatic, renal, neurological, developmental, reproductive, immunological, genotoxic, mutagenic, and carcinogenic effects (such as liver cancer) [468].

High arsenic in food crops grown in mining-contaminated areas of Ghana can cause adverse effects on human health [e.g., 363]. Chronic exposure to As causes many clinical manifestations, where cutaneous lesions are the highest reported [468]. Among these manifestations are melanosis (hyperpigmentation), keratosis, and leukomelanosis (hypopigmentation) [e.g., 455, 468, see Fig. 7.1]. Arsenic is also a well-known carcinogen, causing skin, lung, bladder, liver, and kidney cancers [217, 242].

*Mycobacterium ulcerans*, the bacterium responsible for Buruli ulcer, thrives in As-contaminated areas [617]. Interestingly, As has been found to pose a health threat to women and children in gold mining areas, with median As hazard index values ranging from 3,000 to over 65,000 [321]. The oxidation of As-bearing minerals, which usually occurs in mine spoils, provides a medium for this ulcer-causing bacterium, and more than 2,000 cases of this ulcer have occurred in mining districts in Ghana (Duker et al., 2004).

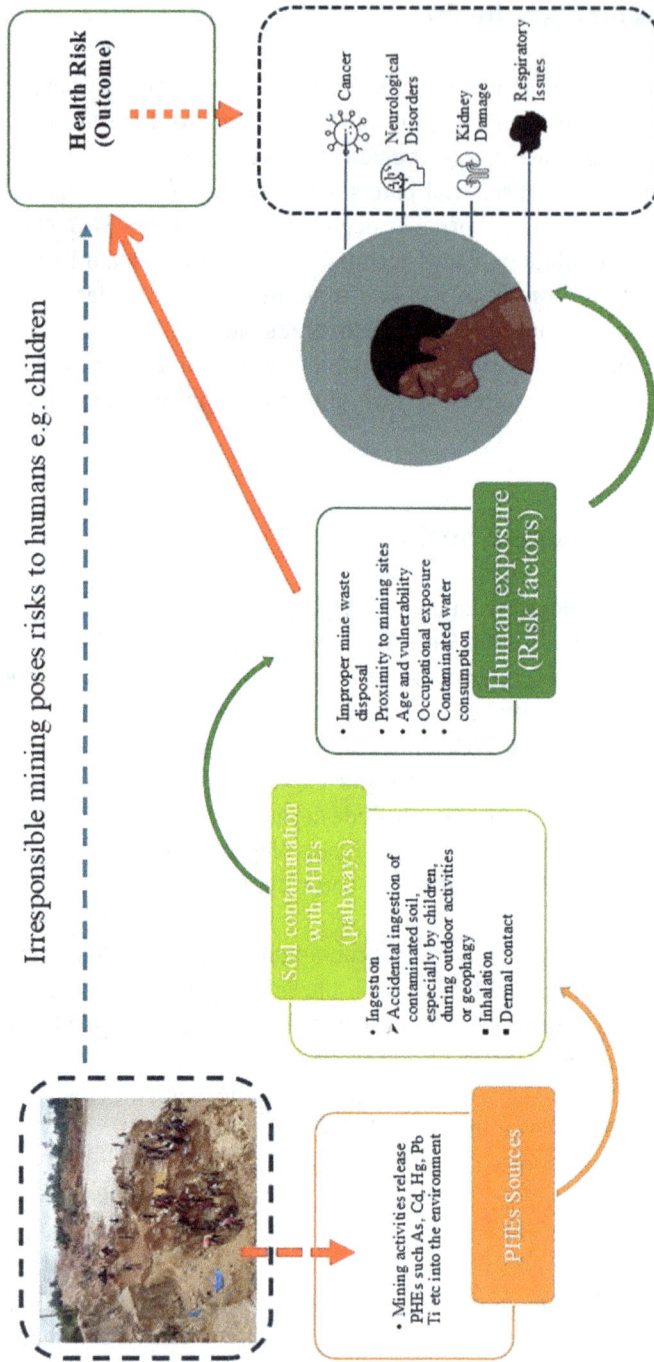

**Fig. 7.1:** Exposure to potentially harmful elements and the associated human health impacts.

Bortey-Sam et al. [89] studied the human health risk from metalloids via consumption of food animals near gold mines in Ghana. The study detected As contents ranging from 52% to 100% in the gizzard, liver, kidney, and muscle of livestock in this mining area. This was attributed to the nature of the gold-bearing ore (e.g., pyrites and arsenopyrites); processing of the ore resulting in the production of arsenic trioxide ($As_2O_3$) distributed throughout the area; accumulation in surface soil and water; and contamination of the food chain in the mining surroundings.

Reports from Ouedraogo [373] also indicated that two deaths linked to high-As drinking water occurred in this area. Further, a study in Bangladesh revealed cattle poisoning with As with symptoms of restlessness, diarrhea, instability in walking, convulsions, panting, and salivation. Such symptoms were followed by death within the next 12–36 h [141].

More specifically, PTEs including Cd, Hg, Pb, Sb, Se, and Ti are found to have no vital role in plant and human nutrition and are hence considered detrimental to human health when present in high concentrations above specified limits [239, 515]. For example, the three major heavy metals – Cd, Hg, and Pb – are not recognized to have any vital biological functions in humans; hence, they are referred to as the "big three heavy metal poisons." Cadmium accumulation in the body affects various vital enzymes and impairs the proper functioning of kidneys [515].

Excessive copper levels can cause detrimental effects by interfering with cellular processes known to play a role in Parkinson's disease [80]. Wilson disease is another factor, a condition of Cu metabolism that is hereditary and causes Cu buildup and toxicity in the liver and brain. The adenosine triphosphatase copper transporter gene

**Fig. 7.2:** Children using abandoned As-contaminated mine tailings as a playground in southwestern Ghana.

(ATP7B), which encodes a protein that transports copper from hepatocytes into the bile, is the cause of this condition [392].

Beckers and Rinklebe [1062] reported that Hg(II) and other inorganic forms of Hg in soil are linked to complex health problems due to their capacity to enter the food chain and bioaccumulate even at low exposure levels. Additionally, Hg is linked to birth defects in newborn babies, many neurological diseases, and cognitive dysfunctions in school children, and many such cases have been reported in mining communities [e.g., 56, 87, 117, 353]. This situation is dire in mining areas where mine lands are abandoned and serve as play ground for children (See Fig. 7.2).

# 7.5 Potentially toxic elements in soil

## 7.5.1 Variation of PTE impacts across ecosystems and land use types

The environmental and health impacts of PTEs vary across different ecosystems and land use types. For example, agricultural lands are particularly vulnerable to PTE contamination due to the potential for PTEs to enter the food chain through crops [1063]. In contrast, forested ecosystems may experience reduced biodiversity and altered nutrient cycling as a result of PTE contamination, but the direct human health risks may be lower unless these areas are used for foraging or water supply ([858]; Adriano et al., 2005).

Urban areas, where soils are often contaminated by a mix of industrial pollutants and traffic emissions, may pose significant health risks due to the proximity of human populations to contaminated sites (Wei & Yang, 2010). In these areas, exposure to PTEs can occur through multiple pathways, including soil dust, urban gardening, and direct contact with contaminated soils [25, 710].

Wetlands and aquatic ecosystems are particularly sensitive to PTE contamination due to the mobility of these elements in water. PTEs can be transported over long distances in water, leading to widespread contamination of aquatic habitats and the bioaccumulation of toxic elements in fish and other aquatic organisms [942]. This not only affects wildlife but also poses significant risks to human populations that rely on these ecosystems for food and water ([1065, 1066]).

## 7.5.2 Managing PTEs in reclaimed soils

The management of PTEs in reclaimed soils is a critical challenge for ensuring the long-term sustainability of these lands. Strategies such as phytoremediation, the use of soil amendments to immobilize PTEs, and regular monitoring of soil and plant tissue are essential for mitigating the risks associated with PTEs [85, 486]. However, the persistence of these elements in the environment and the complexity of their behavior in soils mean that ongoing efforts are needed to manage their impacts effectively [9, 935].

# 7.6 Behavior and mobility of potentially toxic elements (PTEs) in reclaimed soils

## 7.6.1 Behavior of PTEs in reclaimed soils

The behavior of PTEs in reclaimed soils is a complex process influenced by various chemical, physical, and biological factors. These elements can exist in multiple forms within the soil, ranging from free ions in soil solution to complex compounds bound to soil particles [230]. Understanding the behavior of PTEs is crucial for assessing their potential impact on the environment and human health [304, 1067].

PTEs can exist in different chemical forms or species in soils, including free metal ions, complexes with organic and inorganic ligands, precipitates, or adsorbed species on soil particles [1067]. The speciation of PTEs determines their mobility, bioavailability, and toxicity [230]. For instance, heavy metals like lead (Pb) and cadmium (Cd) are more toxic and mobile in their ionic forms, whereas they are less bioavailable when bound to organic matter or precipitated as insoluble salts [25].

## 7.6.2 Persistence of PTEs in reclaimed soils

PTEs can persist in reclaimed soils for extended periods, posing long-term environmental risks. The persistence of these elements depends on their chemical form, soil conditions, and the effectiveness of reclamation practices [9]. Certain PTEs, like lead and cadmium, are known to have long residence times in soils due to their low mobility under specific conditions, which can lead to prolonged environmental hazards [25, 230].

### 7.6.2.1 Longevity of PTE contamination

Many PTEs, particularly heavy metals, are highly persistent in the environment due to their nonbiodegradable nature. Unlike organic pollutants that can be broken down by microorganisms, PTEs do not degrade and can remain in the soil for decades or even centuries [9]. The persistence of PTEs is particularly concerning in reclaimed soils, where residual contamination from mining activities can continue to pose risks to soil health, water quality, and human health long after the cessation of mining [25, 230].

### 7.6.2.2 Environmental risks of persistent PTEs

The persistence of PTEs in reclaimed soils can lead to chronic exposure for plants, animals, and humans. This long-term exposure can result in bioaccumulation and biomagnification, where PTEs concentrate in living organisms and move up the food chain, increasing their toxicity at higher trophic levels [1004]. Persistent PTEs can also lead to ongoing soil degradation, reducing the effectiveness of reclamation efforts and limiting the potential for sustainable land use [25, 230].

## 7.7 Conclusions

In conclusion, PTEs such as arsenic, cadmium, mercury, lead, antimony, selenium, and thallium can occur naturally in the environment but are also released through human activities, particularly in mining and industrial processes. These PTEs can contaminate soil, water, and air, posing significant health risks to humans and other organisms. Exposure to high concentrations of PTEs can lead to acute toxicity and even death, while chronic exposure to low concentrations can result in the development of diseases and disorders. The presence of PTEs in mining areas is a major concern, as it can lead to the contamination of surrounding vegetation, crops, soil, surface water, and groundwater. Strategies for land restoration and alternative land use are necessary to mitigate these risks and protect human, plant, and animal health. Arsenic, in particular, is a well-known carcinogen and can cause various types of cancer. Other PTEs such as cadmium, mercury, and lead also have detrimental effects on human health, affecting crucial enzymes, kidney function, and neurological processes. Overall, the presence of PTEs in the environment poses significant health risks and requires effective management and remediation strategies.

Albert Kobina Mensah*, Emmanuel Amoakwah,
and Ephraim Sekyi-Annan

# Chapter 8
# The power of plants in cleaning and stabilising potentially toxic elements in mine-contaminated soils

**Abstract:** This study provides a current review of the impact and presence of potentially toxic elements (PTEs) in the environment as they pertain to mining operations. The study emphasizes the origins of PTEs, including both natural and anthropogenic processes, as well as the possible health hazards that may result from their release into the atmosphere, water, and soil. The study also presents the impacts of PTEs like arsenic, cadmium, mercury, lead, antimony, selenium, and thallium on human health, including the onset of diverse tumors and neurological disorders. We emphasize the importance of implementing strategies to mitigate PTE contamination to protect the health of humans, plants, and animals. Additionally, we examine and explain in detail the significance of soil biogeochemical or governing factors, which affect the mobilization of potentially toxic elements in mine-contaminated soils. Furthermore, these parameters assist the scientist or the project manager in determining which remediation and mitigation strategies are most suitable for the contaminated site. In particular, it elucidates how soil pH affects the toxicity and behavior of metal and non-metal species. It explains that lower pH levels enhance the mobility of cationic trace elements, while higher pH levels increase the mobility of anionic species. This review thoroughly elaborates on the significance of soil redox chemistry in relation to the mobility and solubility of trace elements, particularly in the presence of inundation. Furthermore, this study examines the effect of redox conditions on PTE transport in flooded soils and aquatic environments. Ultimately, we argue compelling justifications for eco-friendly alternatives, revegetation, phytocleaning of PTEs, and restoration of contaminated mine sites.

**Keywords:** Soil redox chemistry, sesquioxides, phytoremediation, potentially toxic elements, arsenopyrites, soil biogeochemistry

---

**\*Corresponding author: Albert Kobina Mensah**, Soil Research Institute, Council for Scientific and Industrial Research, Kumasi, Ghana, e-mail: albert.mensah@rub.de, albertkobinamensah@gmail.com, ORCID: https://orcid.org/0000-0001-5952-3357
**Emmanuel Amoakwah, Ephraim Sekyi-Annan**, Soil Research Institute, Council for Scientific and Industrial Research, Kumasi, Ghana

https://doi.org/10.1515/9783111662046-008

# 8.1 Introduction

Mining has profound impacts on the ecosystem, leading to soil and water quality deterioration due to heavy metal accumulation. This negative impact extends to the aquatic habitats of various fish species. The activity also impacts the volume of water supply in mining towns, resulting in an increase in water treatment costs. This, in turn, has a negative impact on the economic well-being of the people who live in these regions and rely on water resources. Mensah et al. [324] and Mensah and Tuokuu [323] have highlighted various shortcomings that contribute to the enduring impacts of mining in local communities, such as insufficient community engagement, burdensome registration procedures for small-scale mining enterprises, ineffective mining policies, and lax regulatory enforcement.

Certain elements, including arsenic, cadmium, mercury, lead, antimony, selenium, and thallium, have the potential to be harmful to living organisms. The environment naturally contains these elements, but human activities, particularly in mining and industrial processes, also release them into the surroundings. These particular toxic elements have the potential to pollute soil, water, and air, presenting considerable health hazards to both humans and other living beings. High concentrations of potentially toxic elements (PTEs) can be extremely harmful, causing acute toxicity and even death. On the other hand, long-term exposure to low concentrations of these substances can contribute to the development of various diseases and disorders. The presence of PTEs in mining areas is a significant concern due to the potential contamination it poses to nearby vegetation, crops, soil, surface water, and groundwater.

It is crucial to implement strategies for land restoration and alternative land use in order to effectively mitigate these risks and safeguard the health of humans, plants, and animals. People widely recognize arsenic as a potent carcinogen capable of inducing a range of cancerous conditions. Other PTEs, such as cadmium, mercury, and lead, can have harmful impacts on human health, disrupting important enzymes, kidney function, and neurological processes. Overall, the presence of PTEs in the environment poses significant health risks and necessitates efficient management and remediation approaches.

Ensuring the mining industry's long-term sustainability necessitates the gradual restoration and rehabilitation of polluted and damaged mining sites during both the active phase and the end of the mine's project life cycle. In addition, it is crucial to incorporate sustainable practices to address the environmental damage caused by mineral extraction, rehabilitate the mine sites, and ensure the safe reuse of the land. Implementing such measures is crucial to safeguard the ecosystem and promote the welfare of both humans and animals in mining regions. There is a pressing requirement for thorough remediation studies to investigate the potential risks associated with polluted, abandoned gold mining spoils [332].

So far, numerous physical and chemical methods have been used or proposed to restore or recover the condition of polluted and deteriorated mine sites. The scientific literature continues to debate the use of physical and chemical stabilization techniques in the restoration of mine-affected lands. These procedures can either impede or enhance the movement of PTEs in contaminated sites [376, 1068]. Using a scientific approach, one can address the issue by filling excavated holes, reshaping the terrain, and ultimately replacing the contaminated soil with clean, uncontaminated topsoil.

Furthermore, the physical approach employs various methods to ensure that hazardous compounds do not spread into the surrounding ecosystems and mining communities. These methods include the construction of barriers, caps, and liners, which effectively contain the compounds [515]. The approach has also involved chemical treatments with oxides, like iron oxides, to keep contaminants stable, make them less soluble, and reduce their harmful effects on people and the environment [251, 376, 509].

Mensah et al. [332] conducted a study that found applying compost and manure to a polluted mine site in Ghana significantly increased the bioavailable As content. In their study, they observed that the use of compost resulted in a 332% increase, whereas the use of manure resulted in a 315% increase in the bioavailable As. Hussain et al. [1069] conducted a study that revealed a noticeable rise in soil organic carbon content following the application of cow dung manure in a field with arsenic contamination. The content or presence of dissolved organic carbon (DOC) led to a decrease in As's availability in a Pakistani paddy field. Alternatively, changes in soil pH and redox conditions could cause the breakdown of organic carbon and metal oxides used in chemical treatments. When PTEs adhere to or absorb their surfaces, they could potentially release these substances, causing groundwater pollution and food chain contamination [150, 833, 268]. Furthermore, Yang et al. [520] observed that using biochar in a flood-affected region facilitated the solubility of As, consequently resulting in increased mobilization and release of the contaminant.

Thus, using these physicochemical methods to clean up large areas of heavily polluted land might require a lot of money and knowledge about how to carefully watch, measure, and evaluate all the time. Moreover, the potential to use these methods to rehabilitate extensive areas of contaminated and degraded mining sites may not align with the standards of socio-environmental and economic sustainability for the mining company, host mining communities, and the surrounding ecosystems. Therefore, the effective application of these physicochemical techniques of soil remediation to restore extensive regions of heavily contaminated land may necessitate substantial financial resources and expertise in meticulous monitoring, continuous measurement, and assessment.

Utilizing phytoremediation through revegetation, the mining industry has the potential to achieve long-term sustainability for the host mining community, its environment, and its economy. This method, which involves using plants to restore polluted and damaged mine sites, offers a cost-effective solution. For instance, we could employ this approach to address the issue of heavy metal contamination in gold mine soils [360, 964]. Revegetation on mined land could lessen the harmful effects of PTEs on human health by limiting their movement, stopping metals from entering water sources, and keeping the environment clean and free of metal contamination. In addition, revegetation can help with carbon sequestration, control wind and water erosion, ensure water quality, reduce runoff from polluted mining waste, maintain and enhance soil fertility, and preserve soil moisture.

In Ghana, revegetation is an emerging field of environmental engineering that demands further research and exploration to enhance its practicality. Furthermore, it is crucial to thoroughly document and evaluate the advantages and disadvantages of different case studies and experiments conducted both in the field and in a greenhouse, as well as the practical applications in rehabilitating mine-damaged sites by various mining companies in Ghana. This approach not only amplifies success stories and valuable insights but also tackles the challenges of restoring and revitalizing numerous polluted and degraded mine sites. In order to achieve this objective, we thoroughly:

i.   provide the most up-to-date review and explanation of the factors that may affect the mobilization and bioavailability of PTEs in mine-affected soils;
ii.  present a scientific argument for using revegetation as a remediation method for mine sites that contain PTEs.

## 8.2 Soil variables regulating bio-availability of toxic elements in contaminated soils influencing bioavailability and mobilization of potentially toxic elements

### 8.2.1 Impact of soil pH

Soil pH is crucial to how metal and non-metal species bind to and migrate through soil, as well as their oxidation state and toxicity. When the soil is acidic, $H^+$ ions move metal cations out of the soil components' cation exchange capacity (CEC). This frees the metal cations from sesquioxides and variable-charged clays. Metals are more accessible in soil solutions at low pH because soil organic matter retains metallic substances less. At low pH (below 5.5), numerous cationic PTEs, such as Cd, Cu, Hg, Ni, Pb, and Zn, become more soluble and mobile in soil. Thus, lower pH enhances cationic trace element mobility and availability, while higher pH increases anionic species mobility and availability [36]. Higher pH decreases silicate secondary mineral electronegativity, making anionic PTEs more available. The oxides' positive charge and capacity to bind and stick decrease with high pH. They adhere less to anionic PTEs than cationic ones [1070].

Many studies show that pH controls PTE solubility and availability. Mamindy-Pajany et al. [292] found the best way to stabilize As in polluted sediments by looking at how As(V) interacted with mineral adsorbents that contained Fe, like hematite and goethite, and how that interacted with pH. Beiyuan et al. [69] found that pH affects how mobile and plant-available As is in polluted soil. At high pH and low redox potential, As is more mobile. Additionally, Al Abed et al. [18] batch leaching studies at different pH levels demonstrated a strong pH dependence on As and Fe leaching. This experiment looked at how pH affects As availability and solubility, as well as adsorption and mobility. Violante et al. [500] found that decreasing mineral positive charge affects soil sorption capacity as pH rises. Thus, soil sorption of anionic PTE species decreases with increasing pH [500] and increases with decreasing pH.

Catalano et al. [634], Martin et al. [906], and Violante et al. [1015] describe anions' sorption capability using "inner-sphere sorption and outer-sphere sorption" phenomena. Specifically sorbed ions or molecules replace $OH^-$ or $OH_2$ groups on variable-charge minerals. These steps facilitate the binding of OH-group protons at low pH, a process known as "inner-sphere sorption" [1015]. Arsenic(V) and As(III) produce inner-sphere complexes [500]. As(V) can form distinct surface complexes on inorganic soil components depending on pH and surface coverage, according to [943]. In many pH ranges, As(V) is sorbed more than As(III). In contrast, Violante et al. [500] found that in a basic pH medium, ferrihydrite sorbed arsenite more than arsenate.

The study by Mamindy-Pajany et al. [292] showed how pH affects the adsorption of As species by examining the adsorbent's ability to ionize and surface charge phenomena. Additionally, negatively charged arsenate species ($H_2AsO_4^-$, $HAsO_4^{2-}$, and $AsO_4^{3-}$) are more common when the pH is higher. The surface charges of adsorbents can be positive, negative, or zero depending on the pH. This is due to how protons move between the solution and the mineral surface. Iron oxides have a zero charge point at a pH value or range known as the pH at the point of zero charge ($pH_{ZPC}$). This value or range is where the mineral surface charge of iron oxides is zero, that is, the surface charge is neutral. Mamindy-Pajany et al. [292] reported that a positive surface charge, such as when pH falls below $pH_{ZPC}$, promotes arsenate adsorption. In this situation, the surface of adsorbents undergoes protonation, resulting in an increase in positive charges on the surface.

## 8.2.2 Impact of soil redox chemistry

In sediments and soils, water saturation changes chemical and biological characteristics, affects microbial populations, and governs soil activities. Flooding causes soils to go from aerobic to anaerobic, causing reducing and oxidizing reactions [133]. Under flooding conditions, the redox potential (EH) controls anions and cationic PTE solubility. This can happen directly because the soil moves when it rains or dries, or indirectly because the EH changes pH, DOC, and the chemical behavior of Al, Fe, Mn, and

S. A drop in EH during flooding and rainy regimes can increase pH, which boosts As mobility, and vice versa [e.g., 354, 406].

Redox potential can also reduce metals, like $Al^{3+}$ to $Al^{2+}$, $Fe^{3+}$ to $Fe^{2+}$, and $Mn^{3+}$ to $Mn^{2+}$. This can liberate bound arsenic or mercury from reduced Al/Fe/Mn oxides. In long-term dry conditions, oxidizing circumstances produce the opposite effect [431, 432]. Redox indirectly affects PTE mobilization by altering the contents of the soil organic matter. As a metal and non-metal transporter, the surface of soil carbon may sequester and hold PTEs and eventually govern their mobility and bioavailability. Numerous positive ions on carbon's surface act as fertile sorption grounds for PTEs, decreasing or boosting their mobility. Under redox conditions, organic matter affects PTE mobility in numerous ways. Mensah et al. [323] came up with these suggestions about how C or OM content affects PTE mobility:

i)   Reductive dissolution of organic matter (OM) frees bound PTE
ii)  Changes in pH during OM desorption free-held PTE
iii) Production of dissolved organic matter due to microbial biomass reduction in soil affects PTE release
iv)  Release of Mn- and Fe-oxyhydroxides that are bound to organic matter from their bonds

By altering and regulating the mobilization of potentially mobile PTE fractions, the redox potential could influence or govern PTEs' environmental impact. For instance, these PTEs can become mobile due to their weak bonding or weak adsorption on mineral surfaces, as well as their binding to amorphous/low-crystalline iron oxides and crystalline iron oxides. Environmental variables like pH and redox could release these components. The following paragraphs explain how these occur.

Reducing conditions may liberate Fe oxide fractions attached to the soil colloid, potentially leading to environmental pollution and food chain accumulation [268]. Flooding and EH may affect plant uptake and transport of critical elements and pollutants [1071–1072]. Redox conditions alter the oxidation states of PTEs, such as As and Hg. This impacts their mobility and toxicity in watery and wet soils [1073]. As(III) is more mobile than As(V). Soil-reducing agents that convert As(V) to As(III) may facilitate these. For instance, organic matter availability or purposely flooded fields could hasten the alteration of As(V) to As(III) and consequently enhance As availability. Inorganic soil components strongly retain As(V), causing microbial oxidation and As immobilization. Well-drained areas contain arsenic(V) and As(III), as well as As-0 and arsine ($H_2As$) [85].

In environments with very little oxygen, such as flooded or waterlogged ecosystems, previous research has shown that sulfide minerals (specifically CdS) can form due to the reduction of sulfate compounds. As per the study conducted by Palansooriya et al. [376], this phenomenon is observed in paddy soils with lower levels of Cd, where the solubility of Cd decreases as a result of enhanced synthesis of CdS. This phenomenon highlights the significance of sulfur in effectively immobilizing Cd, particularly in paddy rice fields.

## 8.2.3 Effects of metal oxyhydroxides (Al/Fe/Mn)/sesquioxides

Metal oxides (Al, Fe, and Mn) are necessary for PTE to travel through soils and water [85, 143, 250]. Sesquioxides are the standard term for metal oxides. Sesquioxides are oxides of an element (or radical) with a 2:3 element-oxygen atom count. Phosphorus(III) and aluminum ($Al_2O_3$) oxides are sesquioxides. Many sesquioxides, such as $Al_2O_3$, $La_2O_3$, and $Fe_2O_3$, contain a metal in the +3 oxidation state and an oxide ion $O_2$. The soil contains iron and aluminum sesquioxides. Alkali metal sesquioxides are unique as they include both peroxide $O_2$ and superoxide $O_2$ ions. For example, rubidium sesquioxide $Rb_4O_6$ is $(Rb^+)_4(O_2)(O_2)$. Other metalloid and non-metal sesquioxides that do not belong are phosphorus(III) oxide ($P_4O_6$), dinitrogen trioxide ($N_2O_3$), and boron trioxide ($B_2O_3$).

Due to their amphoteric nature and large, active surfaces, metal oxides can bind and remove soil pollutants through specific sorption, co-precipitation, and inner- and outer-sphere complexes [85, 250]. Iron hydrous oxide surfaces may help retain PTE. For example, it can adsorb on surfaces that are positively charged, and vice versa. Due to their huge surface areas, metal oxides such as ferric, manganese, aluminum, titanium, magnesium, and cerium oxides are promising PTE adsorbents [1074–1075]. Mench et al. [1076] found reduced water-extractable As concentration and plant tissue uptake when they added iron oxides to As-contaminated garden soils. Hartley and Lepp [1077] tested four Fe-bearing additives for As reduction in canal dredging, coal fly ash deposits, and low-level alkali waste. Goethite significantly decreased the As content in plant shoots compared to iron grit, iron(II) sulfate + lime, and iron(III) sulfate + lime.

Protonation and deprotonation of metal oxyhydroxides, or sesquioxides, are dynamic mechanisms that can regulate PTE behavior in contaminated watersheds, floodplains, wetlands, and degraded mining sites. These mechanisms emerge in high pH and low redox potential conditions. For metal oxyhydroxides, deprotonation occurs in higher pH environments, whereas protonation occurs in acidic environments. High pH causes surface hydrolysis, which increases hydroxyl ions. Increased hydroxyl ions lower the surface sorption capacity. Under the aforementioned conditions, cationic PTEs such as lead, cadmium, copper, zinc, and others have a higher sorption capacity. Sesquioxides generally reduce the accessibility of toxic metals, reducing their ability to contaminate the ecosystem and food chain. For instance, their surfaces get de-protonated during high pH or alkaline conditions and consequently exacerbate the mobility and accessibility of anionic PTEs, such as arsenic and selenium. This phenomenon causes environmental degradation and food chain pollution.

When exposed to reduced pH, sesquioxide surfaces like aluminum, iron, and manganese gain protons. These effects increase metal oxides' anionic PTE adsorption but decrease their cationic PTE adsorption. Therefore, protonation and deprotonation are critical to regulating PTE absorption in food crops as well as regulating PTE mobilization into surface and groundwater.

## 8.2.4 Impact of soil organic matter

Carbon in the soil is responsible for transporting PTEs; hence, soil carbon is frequently considered to be a carrier and sequester of PTEs. It affects As (im)mobilization by forming carbonates, carbon-As complexes, functional groups on carbon surfaces, and positive and negative surface ions. These occur via its indirect effects on medium pH, and the supply and activation of microbes needed for As reduction and oxidation. A lower pH reduces organic matter's negative surface charge, increasing its As adsorption capability, and vice versa. Carbonates in calcareous soils or sediments buffer pH fluctuations [133]. Thus, producing carbonates from soluble organic carbon, such as during soil liming, may increase negative charges and As solubility and mobilization.

By feeding bacteria, organic materials indirectly impact As transport and migration. Additionally, C can stimulate microbial activity, which may catalyze redox reactions with electron acceptors such as oxygen and iron [133]. Mobility and availability in contaminated mining spoils are greatly affected by these processes. Wastewater can increase soil C and nutrient levels. Redox effects on PTE mobilization are discussed in Section 8.1.3.

Organic matter decomposition generates humic acids, significantly affecting PTE availability. Native soil humic compounds affect PTE availability. The soil solution cannot release PTEs because high-molecular-weight humic acids firmly bind them, preventing environmental pollution and food chain contamination. However, low-molecular-weight humic acids can chelate PTEs and prevent their adsorption onto solid surfaces, increasing their mobility. Thus, low-molecular-weight humic acids are extremely active in soil and improve PTE binding. The equation shows that humic acids strongly depend on pH:

$$R - COOH \longleftrightarrow R - COO^- + H^+ \tag{8.1}$$

where R is the carbon chain of organic matter, charge acquisition by organic matter occurs due to ionization, where $H^+$ is dissociated from or onto the active functional group [1078], and the charge generated is highly influenced by soil pH. At neutral to high soil pH, the negatively charged deprotonated form ($R-COO^-$) is abundant and tends to adsorb cationic PTEs.

Heavy metals have the unique ability to form organometallic complexes with the functional groups of organic matter due to their soft acid nature. The primary active functional groups include carboxyl, alcohol, and phenol. The presence of metals is heavily influenced by the ability of functional groups to bind with them. When dissolved organic carbon is present, it can reduce the adsorption of metals onto soil constituents. This happens because the dissolved organic carbon competes with the metal ions for binding sites and can also form organometallic complexes. Additionally, the dissolved organic carbon may be preferentially adsorbed onto the solid phase. This phenomenon is particularly noticeable and impactful in soil with a near-neutral pH. This is because, at higher pH levels, there is a greater abundance of free metal ions.

When organic matter is added to the soil, it boosts the soil's cation exchange capacity (CEC), allowing it to effectively hold onto important cationic PTEs. As previously stated, this ability is heavily influenced by soil pH. When the pH is between 6 and 8, humic acids undergo a transformation that results in an increased surface area. This transformation also leads to the highest level of retention for PTEs, as noted by [962]. At this pH range, the affinity of humic acids for PTEs intensifies, thereby exerting control and influence over the mobility of metals. It is important to note that the stability of metal-ligand complexes decreases as soil pH decreases, which supports the role of R-COO- in the complexation of metals.

Soil organic matter consists of decomposed plant and animal remains, as well as microorganisms. In addition, soil organic matter (SOM) contains humus or humic compounds that collaborate to restrict the movement of PTEs in polluted mine sites. Ion exchange, complexation, and adsorption are the main ways in which organic matter retains PTEs. In addition, the surface of the soil organic matter contains phenol, carboxyl, carboxylate, and amino functional groups, which serve as binding sites for PTEs [376].

## 8.2.5 Impact of soil cation exchange capacity

Understanding the cation exchange capacity of soil involves grasping its ability to swap ions on exchangeable sites with ions of equal or greater charge. This phenomenon is a result of changes in pH on the colloidal surface of the soil or the replacement of one ion or charged particle with another, which occurs due to differences in their ionic sizes. Aluminum has the ability to displace calcium from the exchange sites because of its larger positive charge. In certain situations, magnesium and calcium can sometimes replace each other on the exchange site due to their similar ionic sizes. Therefore, soil ion binding capacity is determined by CEC [376].

The presence of positive or negative charges on the soil colloid impacts their ability to regulate the transport of potentially harmful elements. Soil colloids play a crucial role in facilitating soil reactions at the exchange site. It is worth noting that soils with a higher cation exchange capacity possess a greater capacity to retain cationic PTEs. Therefore, soils that are primarily composed of negative ions will effectively restrict the mobility of positive or cationic PTEs. The movement of PTEs and CEC in soils is affected by the soil type. Palansooriya et al. [376] reported that certain soil components, such as clay minerals, metal oxides, and organic matter, have the ability to enhance CEC and create a larger surface area for the sorption of PTEs, thereby restricting their mobilization and availability.

## 8.2.6 Impact of particle size distribution

Understanding the movement of PTEs is closely tied to the composition of soil particles, which is shaped by the varying proportions of clay, silt, and sand. Soils that contain a greater proportion of clay tend to have higher concentrations of both positive and negative ions. These ions play a crucial role in regulating the release and dispersal of harmful elements in areas such as polluted watersheds, floodplains, agricultural fields, and mining sites. When examining the size distribution of the soil, it becomes apparent that the clay fractions contain significantly higher concentrations of sesquioxides, including Al, Mn, and Fe, in comparison to the coarse fractions like sand particles.

In a study conducted by Palansooriya et al. [376], it was found that soils rich in clay minerals have a greater capacity to adsorb substances. This is due to their permanent or variable charges, which enable them to retain higher levels of PTEs. In addition, the large surface area of clay particles provides ample space for the adsorption of both positively and negatively charged PTEs. Furthermore, the clay fractions are mainly composed of sulfides and have a larger surface area. In the study conducted by Mensah et al. [321], it was found that the presence of sulfides plays a crucial role in regulating the release of arsenic. The research focused on the effects of arsenic pollution in mine spoils located in the south-western region of Ghana and its potential impact on human health.

## 8.2.7 Impact of soil electrical conductivity (soil salinity)

Soil electrical conductivity (EC) is a chemical characteristic that affects the movement of PTEs in contaminated soils. Soil EC values that are higher suggest an uptick in salinity and an increase in pH. Soil salinity is often associated with an increase in the exchangeable cations found in the soil, such as calcium, potassium, magnesium, and sodium. These soil nutrients can potentially compete with other elements for adsorption sites, as shown in a study by [1079]. This could lead to greater movement of PTEs in soil that has a high salt concentration. When cadmium and other PTEs come into contact with chlorides in saline soil, they can cause changes in their mobility, as observed by [1080].

Liming the soil with the intention of lowering the acidity of the soil and raising the pH can, as a consequence, lead to an increase in the amount of salt that is present in the soil, which can eventually lead to the mobilization and availability of anionic PTEs. Therefore, in salt-affected soils, the expected high EC associated with such sites, when polluted with PTEs, could release anionic toxic elements and enhance their sensitivity to food chain contamination. As a result, it is recommended that the process of liming acidic contaminated sites be carried out with caution. This is due to the fact that the increase in soil pH that is brought about by the presence of exchangeable or basic cations has the potential to exacerbate the problem of pollution in the riparian environment and have a detrimental effect on the health of surrounding inhabitants.

## 8.2.8 Impact of aging of the mine spoil

Contaminated sites and their neglect over time play a significant role in the build-up, accessibility, and eventual release of PTEs. Younger soils often contain a greater amount of potentially mobile components of PTEs or a larger quantity of mobile PTEs in comparison to older soils. When it comes to older soils, the proportions of PTEs that are either unavailable or residual generally take precedence. Therefore, the age of mine sites, whether they are currently in operation or have been abandoned, can impact the extent to which metals or metalloids are released into the surrounding environment. This can lead to contamination of the food chain and potential health effects for animals and humans.

More precisely, it is expected that the percentage of PTE found in the mining sites or contaminated mine spoils and tailings will have a higher concentration of PTE. It is expected that abandoned mine waste or residue that has not been reclaimed for a long time will gradually move PTEs from the accessible or available portion into the inaccessible areas. According to Antoniadis et al. [36], the PTE can become permanently trapped in interlayer soil sites over time. This trapping can be caused by cations like K that are fixed within the lattice or by the presence of Al polymers. Also, in aged-mining sites, the process of soil pedogenesis gradually moves the elements from the accessible portion, causing them to become enclosed or immobilized by the crystalline structure. As a result, the elements will slowly change from primary minerals to secondary minerals.

As stated by Shaheen and Rinklebe [431], the age of soil plays a role in the development of pedogenesis, potentially leading to an accumulation of crystalline Fe-oxides and hydroxides within the soil. Furthermore, the Feo to Fed ratio, commonly referred to as the activity ratio, can offer insights into the soil's age and reveal a gradual transition toward the crystallization of Fe oxides as time progresses [445]. Therefore, one could argue that older contaminated sites are mainly characterized by secondary minerals of elements. Due to their age or current activity, older mine spoil or tailings may contain a higher concentration of PTEs in forms that are not readily available, known as the residual fraction, compared to younger or currently active mining tailings or spoils. This claim is backed up by Mensah et al. [321], who observed that secondary forms of arsenic, such as scorodite, were widespread in abandoned mine-tailings.

On the other hand, primary forms were more prevalent in the active tailings. Mensah et al. [321] studied arsenic contamination, the possibility of mobilization, and the resulting impact on human health in south-western Ghana. They provided additional explanation, stating that the presence of scorodite in the mine tailings indicates that this secondary form of arsenic formed as a result of exposure to oxygen at the site. This occurred as a result of the oxidation and weathering of the primary mineral at the contaminated mine sites, specifically arsenopyrite, as mentioned in the studies by DeSisto et al. [121]. In a study conducted by Drewniak and Sklodowska [132], it was demonstrated that scorodite is the primary secondary As mineral found in mine-waste heaps and industrial deposits. The site has been neglected and left barren for many years without any efforts to reclaim or provide protective measures, such as vegetation cover. Von der Heyden [1018] argued that remedial action at these sites should prioritize the mitigation of scorodite rather than arsenopyrites based on these observations.

### 8.2.9 Impact of anions (chlorides, sulfate, phosphate, and nitrate)

Understanding the behavior of anions can have a significant impact on how PTEs are released from mining spoil and their potential effects on the environment. Their understanding of the chemical properties also impacts the selection of soil remediation methods to improve the condition of polluted and degraded mine sites. In a study on soil remediation incubation, Mensah et al. [332] reported a correlation between elevated chloride levels and the application of manure and NPK fertilizer to a mine tailing site heavily contaminated with arsenic. They provided an explanation for how chloride can displace As-anions from sorption sites, especially in soils amended with manure and inorganic fertilizer. This displacement can potentially increase the uptake, availability, and release of As.

Previous reports have indicated that the presence of chloride can impact the mobilization of As from the soil. This can occur through competition with As(V) or As(III) for adsorption sites or by altering the electrostatic charge on soil minerals [354]. In contaminated mine sites, the presence of anions can lead to the mobilization of PTEs and limit the availability of positive PTEs. For example, when anions like carbonates, chloride, phosphates, sulfates, and nitrate are present on the soil exchangeable sites, they can displace radicals or ions that are of equal or smaller ionic size. In addition, the production of negative charges on the soil colloid is increased, which in turn enhances their ability to absorb cationic PTEs.

## 8.3 Plant techniques for PTE remediation

This method involves the use of plants to facilitate the removal of metal and metalloid contaminants from the soil matrix [333, 465, 1067]. According to [868], phytoremediation, a green approach using plants to remediate toxic compounds, is a cost-effective, socially acceptable, and environmentally friendly technology for soil and groundwater clean-up. Reeves et al. [1067] reported that hyperaccumulator plants have focused on the possibility of the removal of large amounts of elements from the soil, supported by the application of conventional fertilizers. They added that the rate of biomass production and the concentration of the desired element that can be achieved in harvestable plant matter are critical factors to be considered when selecting species for phytoremediation. Other factors to consider in selecting suitable species for mine land remediation are provided by other authors [e.g., 38, 318, 335].

It is thus recommended that hyperaccumulation plants are those plant species that are capable of accumulating 1,000 mg As/kg in the dry matter of any aboveground tissue (e.g., Baker et al., [289, 738]). Recently, Mensah et al. [333] observed that *Chromolaena odorata*, an indigenous plant near an abandoned mining spoil in Ghana, could offer potential for cleaning As from mining sites. In that study, *Chromolaena odorata* had an As translocation factor of 4.7, further implying its ability to accumulate As from the mining soils. Translocation factor, bioconcentration factor, and bioaccumulation factor are other indices used to appraise the phytoremediation potential of plant species.

## 8.4 Revegetation in mitigating soil pollution with potentially toxic elements

Research conducted by Mensah et al. [321] highlights the significance of mine wastes as contributors to soil contamination. These wastes serve as point sources for PTEs such as Al, As, Cd, Cr, Cu, Fe, Mn, Ni, Pb, Ti, V, and Zn. Heavy metals can be transferred to nearby plants, food crops, soil, surface water, and plants due to changes in the soil biogeochemistry and environmental conditions. Understanding the impact of metal pollution in soils and sediments is crucial for safeguarding the health of both people and ecosystems. It can have far-reaching consequences, such as disrupting aquatic life, compromising food safety, and compromising the quality of water sources.

In different scenarios, children who regularly play on soil that is contaminated with metals or in abandoned mine sites face the risk of consuming metals [58, 440]. In addition, as mentioned by Mensah et al. [321], communities might come into contact with unfenced abandoned mine spoil sites and inhale polluted dust or consume contaminated mine wastes. In regions where gold mining is prevalent, the ingestion of soil PTEs remains a major route of exposure that impacts human health. Various studies have highlighted the significance of this pathway [e.g., 41, 58, 321]. It is possible for this to occur unintentionally through the ingestion of dust, dirt, and soil, or intentionally through the practice of geophagy, which involves purposefully consuming soil or earth materials such as clay [1081]. In Latin America and Africa, geophagy is a common practice among pregnant women [96, 571]. It is widely recognized that metal ores, tailing deposits, soil, and dust in gold mining sites pose significant health risks to both humans and animals [96]. For instance, both short-term and prolonged exposure to As can lead to various health problems in humans, such as skin damage, cancer symptoms, and circulatory system abnormalities [35, 86].

In addition, Wilson disease is caused by an accumulation of copper in the body due to consuming contaminated food and water. This condition affects multiple organ systems, including the liver and the brain [1062]. Also, exposure to and accumulation of certain elements in essential body organs may be associated with a range of health issues, including reproductive, developmental, hepatic, hematological, and immuno-logical health defects. These issues can include abortions, infertility, birth defects, and malformations [727]. Nevertheless, the detrimental impact on ecological integrity and human health caused by poor handling of mine wastes [73], runoff from stockpiles [324], and the neglect of untreated mine spoils [321] often surpasses the advantages linked to gold mining in Ghana, such as job opportunities and government revenues. These activities contribute to the increased release and migration of harmful substan-ces into the environment, which can have negative effects on people's livelihoods and health.

Addressing metal contamination in gold mine spoils has traditionally focused on soil washing and flushing, replacing contaminated soil with clean soil, and chemically stabilizing the soil using oxides to decrease metal solubility. In certain situations, methods have been employed to reduce the movement and spread of detrimental sub-stances. These techniques involve the implementation of various methods such as built barriers, caps, and liners [515]. These techniques can be quite expensive for many mining corporations and can be challenging to implement. In addition, PTEs do not break down like organic chemicals do. Therefore, the clean-up process typically involves either removing them or implementing stabilization and immobilization techniques to minimize their absorption, surface runoff, or leaching.

Phytoremediation, the use of native plants, is proposed as a practical and cost-effective solution to tackle heavy metal contamination in gold mine soils [360, 964]. When choosing plants for phytoremediation on contaminated sites, recent studies [e.g., 38, 384, 1082–1083] have recommended using native species that are well-suited to the specific location being studied. In addition, native and indigenous species are preferred over foreign ones because they are better adapted to the local environment and can seamlessly integrate into the functioning ecosystem [335]. As per [1063], phy-toremediation is an emerging area of research that necessitates additional compre-hensive studies to gain a complete understanding of its potential applications. For in-stance, Mensah et al. [333] have conducted research on the ability of certain native plants to remediate mine-contaminated spoils. These plants include *Pueraria mon-tana*, *Alchornea cordifolia*, *Lantana camara*, and *Pityrogramma calomelanos*-fern, as mentioned in studies by Bansah and Addo [57], Issaka and Ashraf [863], Liu et al. [35], and Petelka et al. [384]. Native plant species have the potential to clean up these PTEs.

*Chromolaena odorata* and fern are effective for phytoextraction, with different transfer factors for various elements.

Based on the higher transfer factors for these elements, it appears that *Chromolaena odorata* would be a more favorable option compared to the fern. In addition, *Alchornea cordifolia* and *Pueraria montana* have been found to be effective in the phytoremediation of Cu, Ti, and Zn, as their TFs are above 1. Perhaps *Lantana camara* could be a potential option for Cu and Zn. Thus, local plants like *Chromolaena odorata* and *Pityrogramma calomelanos*-fern have the ability to extract elements such as As, Cu, Ti, and Zn from the contaminated mine soils and farms. In a study conducted by Liu et al. [35] on Cd-contaminated soils in China, they focused on *Lantana camara* and its potential for Cd remediation. The researchers found that this plant demonstrated remarkable Cd tolerance, with bioaccumulation and translocation values exceeding 1.

Moreover, *Pteris vittata* L. demonstrated significant potential for phytoextraction of Pb compared to the other eight native plants studied by [1063] in their research on cleaning up metal-contaminated mine tailings in Malaysia. Therefore, through the reduction of food chain contamination, limiting mobility, minimizing metal transfer into surface and groundwater, and ensuring a metal-contamination-free environment, phytoremediation mitigates the adverse impacts of PTEs on human health. Further, it can aid in carbon sequestration, control wind and water erosion, ensure water quality, reduce discharge from contaminated mining waste, enhance soil fertility, and conserve soil moisture.

In a study conducted by Mensah et al. [327], it was demonstrated that the addition of manure and iron oxide led to the phytostabilization of several elements, including Co, Mn, Hg, Mo, Ni, and Zn, in ryegrass. This was supported by the observation of bioconcentration factor (BCF) values that exceeded 1. The study findings showed that the use of manure and iron oxide together greatly enhanced the effectiveness of ryegrass phytostabilization. Thus, it was deduced that the combined use of manure and oxides can be applied to mine tailings to improve the growth of ryegrass and reduce the release of PTEs into the surrounding environment.

In another study by Mensah et al. [332], a 60-day pot experiment was conducted to explore the effectiveness of ryegrass (*Lolium perenne*) in phytocleaning when combined with various soil organic and inorganic amendments. It was discovered that ryegrass has a higher concentration of As in its roots compared to its shoots, making it potentially more effective for phytostabilization. These findings were reinforced by BCF values above 1 and lower BAC and TF values, which were below 1.

# 8.5 Identifying appropriate phytoremediation technology

## 8.5.1 Direct and assisted phytoremediation

There are two main categories of phytoremediation: direct and indirect. Indirect, or plant-assisted, remediation involves the collaborative interaction between plants and a soil improvement method to address the presence of pollutants in contaminated soil.

## 8.5.2 Indirect or assisted phytoremediation

Improving soil quality involves using different organic or inorganic supplements to enhance the soil. These amendments may consist of various substances, including biochar, compost, manure, iron oxide, inorganic fertilizers, sewage sludge, red mud, and other similar materials. These adjustments can be used individually or in combination with each other. Using pot experiments can help in evaluating and selecting the best revegetation scheme for restoring degraded mining sites. As per Tordoff et al. [483], it is recommended to initially test large-scale revegetation schemes through small-scale glasshouse pot trials and subsequently conduct extensive field trial programs. By identifying the major limitations to plant growth, one can choose and plan the appropriate remedial treatment [1084].

Pot experiments in soil science offer a controlled and versatile platform to investigate a range of research questions related to soil properties, plant growth, nutrient management, and environmental impacts. The findings from these experiments become crucial for the progress of sustainable agriculture practices and environmental management strategies. For instance, in a pot experiment conducted by Addai et al. [6], the following significant improvements were observed in the soil quality when different soil amendments (e.g., rice husk biochar, poultry litter compost, and NPK) were applied to a contaminated mine soil, either separately or in combination:

a) Improvements in the soil pH values resulted in higher soil pH levels, which in turn reduced soil acidity.
b) Increased organic matter content and greater availability of phosphorus.
c) Increased soil CEC for calcium, potassium, magnesium, and sodium ions.
d) Decreased concentrations of soil exchangeable acidity.
e) Enriched essential nutrient content in terms of N, P, and K leads to a significant 42% increase in the nitrogen content.
f) Increased uptake and utilization of P and K by plants, with a remarkable 128% and 101% increase, respectively.

Further, Addai et al. [6] observed that the use of biochar, poultry litter compost, and their combinations led to a significant decrease in Hg uptake (measured in mg/kg) in the pot. In fact, the reduction reached up to 49% compared to the control. One crucial aspect that affects the effectiveness of phytoremediation is how well plants can rapidly absorb significant amounts of metal. Hyperaccumulators have the remarkable ability to accumulate significant amounts of metal in their tissue, regardless of the metal concentration in the soil, as long as the specific metal is present. This property sets them apart from moderate accumulators currently employed in phytoextraction, where the amount of absorbed metal is determined by the concentration in the soil.

Understanding the availability of metals in the soil solution is crucial for determining how efficiently plant roots can absorb them, regardless of the overall metal content in the soil. Some researchers suggest using different chemicals to increase the amount of metal available for plant uptake in order to improve the speed and quantity of metal removal by plants. Various acidifying agents, fertilizer salts, and chelating materials have been suggested for this purpose ([757, 758, 779, 889]). These chemicals enhance the availability of metal in the soil solution by either releasing or displacing metal from the solid phase of the soil or by increasing the solubility of precipitated metal species.

Research in this field has had some degree of success, but there are concerns about the potential consequences of releasing significant amounts of toxic metals into soil water. The effectiveness of phytoextraction relies heavily on the presence of heavy metals in the soil solution, which helps plants absorb them more efficiently. Various amendments have been suggested to improve the accessibility of heavy metals for successful phytoextraction research.

## 8.5.3 Direct or unaided phytoremediation

On the other hand, direct phytoremediation involves plants absorbing and breaking down contaminants within their tissues. This method involves the direct planting of vegetation on the damaged soil without the use of soil amendments. Usually, this type of natural rejuvenation involves using plant species that are well-suited to the local climate and soil conditions in the specific geographic area. For a thorough understanding of species selection and potential survival in degraded mining spoil, it is recommended that the project officer or environmentalist observe the immediate vicinity, such as the abandoned mine-tailings. Examining the species found on the outskirts of a damaged ecosystem or mine spoil can give us valuable information about their ability to survive in the challenging conditions of a degraded or nutrient-poor mining site.

## 8.6 Plant elemental uptake, phytoremediation potential, and identifying plant species for phytoremediation

Understanding plant element uptake and phytoremediation potential requires evaluating how specific plant species can absorb and store contaminants from their surroundings, aiming to clean up polluted environments. This process allows for a better understanding of the plants that are successful in removing contaminants from the soil and reducing the movement of contaminants. It can be used to inform and guide strategies for phytoremediation. For a thorough assessment of suitable plant species for phytoremediation projects, it is essential to consider the effectiveness of metal accumulation.

Plants can be classified based on the level of metal accumulation they show and the specific organ where the accumulation takes place. When studying metal accumulation in plants, it is important to analyze the levels of trace elements in different parts of the plant, such as shoots, roots, stems, fruits, seeds, and other plant components. When evaluating a plant species' capacity to handle and store metals in its tissues, concentration values are compared to both normal and phytotoxic levels. Plant accumulation levels are compared to established threshold criteria to assess their potential for element accumulation.

These calculations are applicable to phytoremediation experimental trials and projects conducted either in the greenhouse or in the field. These methods can also be used to analyze and measure the bio-accessibility and toxicity of metals in food crops cultivated in polluted regions. To assess the phytoremediation capacity of the selected plant species, two main parameters or indices need to be taken into account:
i.   Soil-to-plant transfer factor
ii.  Root-to-shoot plant transfer factor

### 8.6.1 Soil-to-plant transfer factors

The BCF and bio-accumulation coefficient (BAC) are utilized for evaluating the accumulation of PTE in plant roots and shoots. BCF calculates the ratio of element concentration in plant roots to soil content using eq. (8.5), while BAC calculates the ratio of element concentration in plant shoots to soil content using eq. (8.4). These factors evaluate the capacity of a plant species to eliminate PTE from contaminated soil. Plants with BCF/BAC values below 1 are not suitable for phytoextraction, as indicated by studies conducted by Fitz and Wenzel [160], Yoon et al. [528], and Shaheen and Rinklebe [434]:

$$\text{BAC} = \frac{\text{PTE content in shoot}}{\text{Total PTE content in soil}} \qquad (8.4)$$

$$\text{BCF} = \frac{\text{PTE content in root}}{\text{Total PTE content in soil}} \qquad (8.5)$$

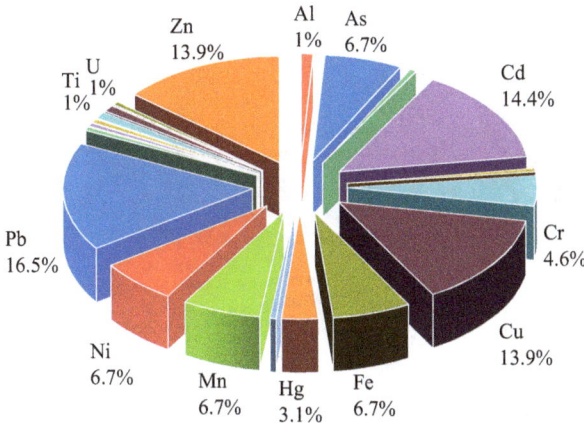

**Fig. 8.1:** Twenty classes of PTEs in mine-contaminated sites targeted for remediation by sixty-four plant species. The pie chart was produced from data compiled from publications between 2014 and 2024.

### 8.6.2 Root-to-shoot plant transfer factor

Understanding the translocation factor (TF) is essential in evaluating the phytoremediation capabilities of different plant species. It calculates the movement of elements from the roots to the above-ground parts, such as the shoot, stem, or leaf (eq. 8.6). An ideal TF of 1 is advised for classifying a plant species that shows promise in phytoremediation. The TF is crucial in determining the classification of plant species for phytoremediation, as it describes the unique characteristics of hyper-accumulating species:

$$\text{TF} = \frac{\text{PTE content in shoot}}{\text{PTE content in soil}} \qquad (8.6)$$

# 8.7 Phytoremediation mechanisms by plant species

Phytoremediation is a broad term that encompasses various mechanisms by which plants naturally remove contaminants from polluted environments. Plants have the ability to break down organic pollutants and remove and stabilize metals at contami-

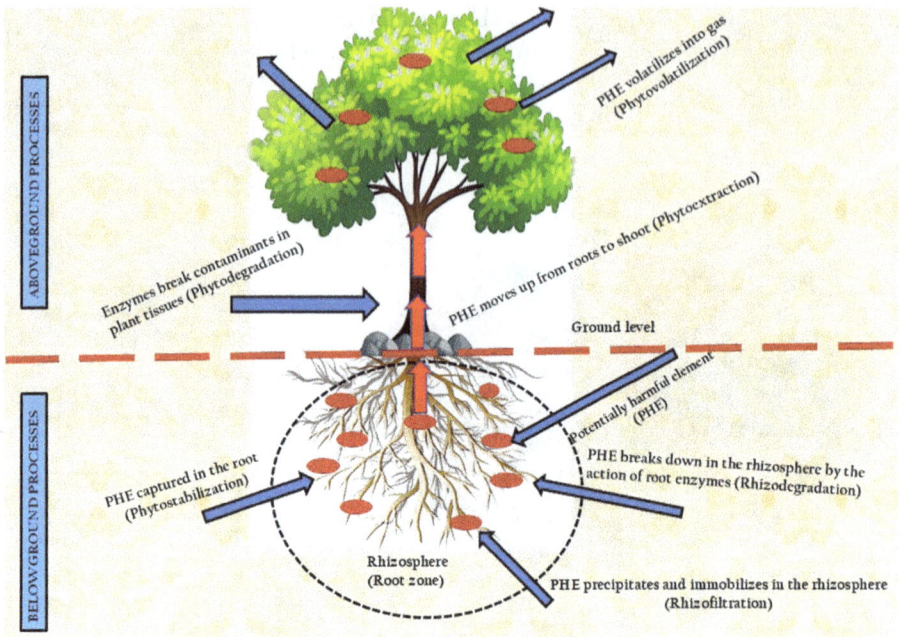

**Fig. 8.2:** An overview of different mechanisms of action for phytoremediating metal-polluted environments.

nated sites. There are several ways to accomplish this, as outlined below. The methods used to remediate sites contaminated with heavy metals differ slightly from those used for sites polluted with organic contaminants. Phytoremediation encompasses a wide range of techniques and applications, each with its own unique mechanism for addressing metal pollution in the environment (Fig. 8.1). Plants have varying abilities to eliminate, trap, or break down pollutants, and the specific pollutants they can target also vary. Figure 8.2 shows an overview of different mechanisms of action for phytoremediating metal-polluted environments.

Phytoextraction is a process in which plant roots absorb pollutants and store them in their aboveground tissues [445]. Phytoextraction is a commonly used technique in remediation practices that involves the removal and disposal of contaminated plants or the extraction of valuable pollutants for commercial purposes, as explained by McIntyre [308]. Based on the research conducted by Pathak et al. [382], it has been noted that more than 500 plant species have the capacity to accumulate high levels of heavy metals in their above-ground structures.

Phytodegradation is a process that involves the use of plants and their enzymes to break down organic contaminants, usually occurring within the plant tissues [308]. These substances are highly effective for chemical compounds that have the ability to move within plants, like herbicides [389]. Phytostabilization is a remediation technique that utilizes specific plant species to prevent the movement of contaminants, especially heavy metals, and prevent their dispersion, such as leaching into groundwater or entering the food chain. This is accomplished by reducing their bioavailability [445]. Phytovolatilization is the process by which plants absorb pollutants, like mercury (Hg) and selenium (Se), convert them into volatile forms, and then release them into the environment as gases [24]. Phytostimulation, also known as rhizodegradation, is a fascinating process where plants enhance soil microbial activity in the root zone to aid in the breakdown of organic pollutants, like petroleum hydrocarbons [389]. Plants release sugars and acids that act as catalysts for microbial activity, initiating the biodegradation of organic pollutants.

However, there are certain limitations to phytoremediation, including the long duration it requires and the need for careful agronomic maintenance. Gerhardt et al. [174] reported that the use of phytoremediation for inorganic contaminants can lead to a significant delay in the re-utilization of the land compared to traditional methods, resulting in a time burden. This leads to the locking of capital, which requires careful consideration when conducting cost analysis. In addition, the effectiveness of the technique is limited to pollution levels that are categorized as low to moderate.

## 8.8 Conclusions and future research

Mining has substantial environmental and social consequences, such as soil degradation, water quality problems, and negative impacts on fish habitats. In order to achieve long-term sustainability, it is crucial for the mining industry to prioritize the restoration and rehabilitation of polluted sites at every stage of the project. It's important to understand the significant impact that metal oxides like Al, Fe, and Mn have on the movement of PTEs in soil. This is because of their expansive surface areas.

Understanding the impact of soil organic matter on the mobilization of PTEs involves considering various factors, such as carbonates and organometallic complexes. Green remediation can effectively tackle heavy metal contamination problems, minimize the spread of metals in the food chain, prevent metal transfer into surface and groundwater, and ensure a clean environment free from metal contamination. Revegetation can also aid with carbon sequestration, control wind and water erosion, and preserve soil fertility.

The selection of the right plant species should be done with great care, taking into consideration the root depth and contact area needs of the plant species. Trees are typically utilized in the sense that they have a better penetration potential due to the presence of their tap root system, which is able to anchor into deeper layers of the soil profile beyond the topsoil that is visible on the surface. When it comes to grasses, they are most commonly used as cover crops, which help to stabilize the surface soil, reduce the risk of erosion, and conserve both soil and moisture.

**Tab. 8.1:** Various plant species tested for the remediation of PTEs in mine-contaminated sites.

| Metal targeted for cleaning | Plant species | References |
|---|---|---|
| Cu | *Azolla pinnata* and *Lemna minor* | Sandhya Bharti et al. [75] |
| | *Dittrichia viscosa, Cistus salviifolius*, and *Euphorbia pithyusa* subsp. *cupanii* | MN Jimenez et al. [224] |
| | *Solanum viarum* Dunal sp. | Thays França Afonso et al. [13] |
| | *Eleocharis acicularis* | Nguyen Thi Hoang Ha et al. [209] |
| | *Eucalyptus grandis* and *Ailanthus altissima* | Hakime Abbaslou and Somayeh Bakhtiari [1] |
| | *Jatropha curcas* L. | Paloma Alvarez-Mateos et al. [27] |
| | *Ricinus communis* L. | Alejandro Ruiz Olivares et al. [368] |
| | *Typha latifolia* and *Chrysopogon zizanioides* | Alexander Kofi Anning and Ruth Akoto [32] |
| | *Brassica juncea* L. | J. C. Mendoza-Hernandez et al. [312] |
| | *Chrysopogon zizanioides* L. | Ritesh Banerjee et al. [55] |
| | *Bidens pilosa* and *Plantago lanceolata* | Robson Andreazza et al. [31] |
| | *Brassica juncea* | Luiz AB Novo [358] |
| | *Lolium multiflorum* | Violeta Mugica-Alvarez et al. [343] |
| | *Helianthus annuus* L. | Robson Andreazza et al. [31] |
| | *Erica Australia* and *Erica andevalensis* | M.M Abreu et al. [3] |
| | *Chrysopogon zizanioides* | Carmen Vargas et al. [499] |
| | *Typha angustifolia* L. | Lo Vun Yen and Kartini Saibeh [525] |
| | *Eucalyptus globulus* | Taoufik El Rasafi et al. [142] |
| | *Epilobium fragilis, Carthamus oxyacantha*, and *Verbascum speciosum* | Maleyeri et al. [291] |
| | *Miconia zamorensis, Axonopus compressus*, and *Erato polymnioides* | I Chamba-Eras et al. [108] |
| | *Prosopis tamarugo, Schinus molle*, and *Atriplex nummularia* | Lam et al. 2017 |
| | *Chrysopogon zizanioides* | Meenu Gautam and Madhoolika Agrawal [171] |
| | *Dittrichia viscosa* | A. Buscaroli et al. [97] |
| | *Alchornea cordifolia* | Mensah et al. [333] |

**Tab. 8.1** (continued)

| Metal targeted for cleaning | Plant species | References |
|---|---|---|
| | *Brassica* sp. | Antoniades et al. [36] |
| | *Chromolaena odorata* | Mensah et al. [333] |
| | *Fern* | Mensah et al. [333] |
| | *Pueraria montana* | Mensah et al. [333] |
| | *Zea mays* | Antoniades et al. [37] |
| | *Hwentia* | Bansah and Addo [57] |
| | *Musa sapientum* | Bansah and Addo [57] |
| | *Leucaena leucocephala* | Bansah and Addo [57] |
| | *Terminalia superba* | Bansah and Addo [57] |
| | *Ryegrass* | Mensah et al. [332] |
| Cd | *Azolla pinnata* and *Lemna minor* | Sandhya Bharti et al. [75] |
| | *Paspalum conjugatum* and *Pinus massoniana* | Lin Zhang et al. [533] |
| | *Eucalyptus grandis* and *Ailanthus altissima* | Hakime Abbaslou and Somayeh Bakhtiari [1] |
| | *Jatropha curcas* L. | Paloma Alvarez-Mateos et al. [27] |
| | *Ricinus communis* L. | Alejandro Ruiz Olivares et al. [312] |
| | *Brassica juncea* L. | J. C. Mendoza-Hernandez et al. [368] |
| | *Aloe burgersfortensis* and *A. castanea* | Joao Marcelo-Silva et al. [295] |
| | *Clidemia sericea* D. Don | Elvia Valeria Durante-Yanez et al. [134] |
| | *Typha angustifolia* L. | Lo Vun Yen and Kartini Saibeh [525] |
| | *Atriplex hortensis* | S. Sai Kachout et al. [415] |
| | *Pteris ensiformis, Boehmeria nivea, Aster prorerus, Hydrocotyle sibthorpioides,* and *Eleusine indica* | Pan et al. [378] |
| | *Clidemia sericea* D. Don | EV Durante-Yanez et al. [134] |
| | *Brassica napus* | Ammaiyappan Selvam and Jonathan Woon-Chung Wong [426] |
| | *Brassica napus, Brassica parachinensis,* and *Zea mays* | Ammaiyappan Selvam and Jonathan Woon-Chung Wong [427] |
| | *Epilobium fragilis, Carthamus oxyacantha,* and *Verbascum speciosum* | Maleyeri et al. [291] |

**Tab. 8.1** (continued)

| Metal targeted for cleaning | Plant species | References |
|---|---|---|
| | *Inula viscosa, Euphorbia dendroides, Poa annua, A. donax, Cistus salviifolius,* and *Helichrysum italicum* | M. Barbafieri et al. [60] |
| | *Pelargonium roseum* | Majid Mahdieh et al. [290] |
| | *Helianthus annuus* L. | Khalid A. Alaboudi et al. [19] |
| | *Canna indica* L. | V. Subhashini and V.V.S. Swamy [469] |
| | *Atriplex halimus* | J. A. Acosta et al. [4] |
| | *Brassica juncea* | Sunayana Goswami and Suchismita Das [183] |
| | *Miconia zamorensis, Axonopus compressus,* and *Erato polymnioides* | I Chamba-Eras et al. [108] |
| | *Prosopis tamarugo, Schinus molle,* and *Atriplex nummularia* | Elisabeth J. LA, et al. (2017) |
| | *Chrysopogon zizanioides* | Meenu Gautam and Madhoolika Agrawal [171] |
| | *Dittrichia viscosa* | A. Buscaroli et al. [97] |
| | *Brassica* sp. | Antoniades et al. [36] |
| | *Zea mays* | Antoniades et al. [37] |
| | *Hwentia* | Bansah and Addo [57] |
| | *Musa sapientum* | Bansah and Addo [57] |
| | *Leucaena leucocephala* | Bansah and Addo [57] |
| | *Terminalia superba* | Bansah and Addo [57] |
| | *Ryegrass* | Mensah et al. [332] |
| As | *Jatropha curcas* L. | Paloma Alvarez-Mateos et al. [27] |
| | *Acacia mangium* | Ruhan A. Rosli et al. [411] |
| | *Typha latifolia* and *Chrysopogon zizanioides* | Alexander Kofi Anning and Ruth Akoto [32] |
| | *Brassica juncea* L. | J. C. Mendoza-Hernandez et al. [312] |
| | *Lemna gibba* L. | Martin Mkandawire et al. [2005] |
| | *Erica australis* and *Erica andevalensis* | M.M Abreu et al. [3] |
| | *Brassica juncea* | Byong-Gu Ko et al. [247] |

**Tab. 8.1** (continued)

| Metal targeted for cleaning | Plant species | References |
|---|---|---|
| | *Pteris ensiformis, Boehmeria nivea, Aster prorerus, Hydrocotyle sibthorpioides*, and *Eleusine indica* | Pan et al. [378] |
| | *Epilobium fragilis, Carthamus oxyacantha,* and *Verbascum speciosum* | Maleyeri et al. [291] |
| | *Pteris vittata* L. | Simone Cantamessa et al. [99] |
| | *Alchornea cordifolia* | Mensah et al. [333] |
| | *Brassica* sp. | Antoniades et al. [36] |
| | *Chromolaena odorata* | Mensah et al. [333] |
| | *Fern* | Mensah et al. [333] |
| | *Pueraria montana* | Mensah et al. [333] |
| | *Ryegrass* | Mensah et al. [332] |
| Ni | *Azolla pinnata* and *Lemna minor* *Solanum viarum* Dunal sp. | Sandhya Bharti et al. [75] Thays França Afonso et al. [13] |
| | *Eleocharis acicularis* | Nguyen Thi Hoang Ha et al. [209] |
| | *Jatropha curcas* L. | Paloma Alvarez-Mateos et al. [27] |
| | *Brassica juncea* L. | J. C. Mendoza-Hernandez et al. [312] |
| | *Aloe burgersfortensis* and *A. castanea* | Joao Marcelo-Silva et al. [295] |
| | *Pongamia pinnata* | Yu et al. [1064] |
| | *Typha angustifolia* L. | Lo Vun Yen and Kartini Saibeh [525] |
| | *Pelargonium roseum* | Majid Mahdieh et al. [290] |
| | *Canna indica* L. | V. Subhashini and V.V.S. Swamy [469] |
| | *Chrysopogon zizanioides* | Meenu Gautam and Madhoolika Agrawal [171] |
| | *Dittrichia viscosa* | A. Buscaroli et al. [97] |
| | *Brassica* sp. | Antoniades et al. [36] |
| | *Zea mays* | Antoniades et al. [37] |
| Zn | *Azolla pinnata* and *Lemna minor* | Sandhya Bharti et al. [75] |
| | *Dittrichia viscosa, Cistus salviifolius*, and *Euphorbia pithyusa* subsp. *cupanii* | MN Jimenez et al. [224] |
| | *Solanum viarum* Dunal sp. | Thays França Afonso et al. [13] |

**Tab. 8.1** (continued)

| Metal targeted for cleaning | Plant species | References |
|---|---|---|
| | *Eleocharis acicularis* | Nguyen Thi Hoang Ha et al. [209] |
| | *Eucalyptus grandis* and *Ailanthus altissima* | Hakime Abbaslou and Somayeh Bakhtiari [1] |
| | *Jatropha curcas* L. | Paloma Alvarez-Mateos et al. [27] |
| | *Ricinus communis* L. | Alejandro Ruiz Olivares et al. [312] |
| | *Typha latifolia* and *Chrysopogon zizanioides* | Alexander Kofi Anning and Ruth Akoto [32] |
| | *Chrysopogon zizanioides* L. | Ritesh Banerjee et al. [55] |
| | *Brassica juncea* | Luiz AB Novo [358] |
| | *Lolium multiflorum* | Violeta Mugica-Alvarez et al. [343] |
| | *Erica australis* and *Erica andevalensis* | M.M Abreu et al. [3] |
| | *Chrysopogon zizanioides* | Carmen Vargas et al. [499] |
| | *Typha angustifolia* L. | Lo Vun Yen and Kartini Saibeh [525] |
| | *Eucalyptus globulus* | Taoufik El Rasafi et al. [142] |
| | *Atriplex hortensis* | S. Sai Kachout et al. [415] |
| | *Pteris ensiformis, Boehmeria nivea, Aster prorerus, Hydrocotyle sibthorpioides*, and *Eleusine indica* | Pan et al. [378] |
| | *Avena sativa, Hordeum vulgare*, and *Brassica juncea* | Stephen D. Ebbs and Leon V. Kochian [139] |
| | *Inula viscosa, Euphorbia dendroides, Poa annua, A. donax, Cistus salviifolius*, and *Helichrysum italicum* | M. Barbieri et al. [60] |
| | *Canna indica* L. | V. Subhashini and V.V.S. Swamy [469] |
| | *Bidens triplinervia* and *Senecio* sp. | Jaume Bech et al. [65] |
| | *Atriplex halimus* | J. A. Acosta et al. [4] |
| | *Miconia zamorensis, Axonopus compressus*, and *Erato polymnioides* | I Chamba-Eras et al. [108] |
| | *Prosopis tamarugo, Schinus molle*, and *Atriplex nummularia* | Elisabeth J. LA, et al. (2017) |
| | *Chrysopogon zizanioides* | Meenu Gautam and Madhoolika Agrawal [171] |

**Tab. 8.1** (continued)

| Metal targeted for cleaning | Plant species | References |
|---|---|---|
| | *Dittrichia viscosa* | A. Buscaroli et al. [97] |
| | *Cytisus scoparius* | |
| Pb | *Azolla pinnata* and *Lemna minor* | Sandhya Bharti et al. [75] |
| | *Paspalum conjugatum* and *Pinus massoniana* | Lin Zhang et al. [533] |
| | *Artemisia capillaris, Taraxacum mongolicum, Medicago sativa*, and *Plantago asiatica* | Nan Lu et al. [284] |
| | *Dittrichia viscosa, Cistus salviifolius*, and *Euphorbia pithyusa* subsp. *cupanii* | MN Jimenez et al. [224] |
| | *Solanum viarum* Dunal sp. | Thays Franca Afonso et al. [13] |
| | *Eleocharis acicularis* | Nguyen Thi Hoang Ha et al. [209] |
| | *Eucalyptus grandis* and *Ailanthus altissima* | Hakime Abbaslou and Somayeh Bakhtiari [1] |
| | *Jatropha curcas* L. | Paloma Alvarez-Mateos et al. [27] |
| | *Ricinus communis* L. | Alejandro Ruiz Olivares et al. [368] |
| | *Typha latifolia* and *Chrysopogon zizanioides* | Alexander Kofi Anning and Ruth Akoto [32] |
| | *Brassica juncea* L. | J. C. Mendoza-Hernandez et al. [312] |
| | *Lolium multiflorum* | Violeta Mugica-Alvarez et al. [343] |
| | *Clidemia sericea* D. Don | Elvia Valeria Durante-Yanez et al. [134] |
| | *Erica Australia* and *Erica andevalensis* | M.M Abreu et al. [3] |
| | *Typha angustifolia* L. | Lo Vun Yen and Kartini Saibeh [525] |
| | *Eucalyptus globulus* | Taoufik El Rasafi et al. [142] |
| | *Atriplex hortensis* | S. Sai Kachout et al. [415] |
| | *Pteris ensiformis, Boehmeria nivea, Aster prorerus, Hydrocotyle sibthorpioides*, and *Eleusine indica* | Pan et al. [378] |
| | *Clidemia sericea* D. Don | EV Durante-Yanez et al. [134] |
| | *Inula viscosa, Euphorbia dendroides, Poa annua, A. donax, Cistus salviifolius*, and *Helichrysum italicum* | M. Barbieri et al. [60] |

**Tab. 8.1** (continued)

| Metal targeted for cleaning | Plant species | References |
|---|---|---|
| | *Pelargonium roseum* | Majid Mahdieh et al. [290] |
| | *Helianthus annuus* L. | Khalid A. Alaboudi et al. [19] |
| | *Canna indica* L. | V. Subhashini and V.V.S. Swamy [469] |
| | *Bidens triplinervia* and *Senecio* sp. | Jaume Bech et al. [65] |
| | *Atriplex halimus* | J. A. Acosta et al. [4] |
| | *Miconia zamorensis, Axonopus compressus*, and *Erato polymnioides* | I Chamba-Eras et al. [108] |
| | *Prosopis tamarugo, Schinus molle*, and *Atriplex nummularia* | Elisabeth J. LA, et al. (2017) |
| | *Chrysopogon zizanioides* | Meenu Gautam and Madhoolika Agrawal [171] |
| | *Dittrichia viscosa* | A. Buscaroli et al. [97] |
| | *Alchornea cordifolia* | Mensah et al. [333] |
| | *Brassica* sp. | Antoniades et al. [36] |
| | *Chromolaena odorata* | Mensah et al. [333] |
| | *Fern* | Mensah et al. [333] |
| | *Pueraria montana* | Mensah et al. [333] |
| | *Hwentia* | Bansah and Addo [57] |
| | *Musa sapientum* | Bansah and Addo [57] |
| | *Terminalia superba* | Bansah and Addo [57] |
| Mn | *Azolla pinnata* and *Lemna minor* | Sandhya Bharti et al. [75] |
| | *Solanum viarum* Dunal sp. | Thays Franca Afonso et al. [13] |
| | *Eleocharis acicularis* | Nguyen Thi Hoang Ha et al. [209] |
| | *Eucalyptus grandis* and *Ailanthus altissima* | Hakime Abbaslou and Somayeh Bakhtiari [1] |
| | *Ricinus communis* L. | Alejandro Ruiz Olivares et al. [368] |
| | *Brassica juncea* L. | J. C. Mendoza-Hernandez et al. [312] |
| | *Chrysopogon zizanioides* L. | Ritesh Banerjee et al. [55] |
| | *Lolium multiflorum* | Violeta Mugica-Alvarez et al. [343] |
| | *Aloe burgersfortensis* and *A. castanea* | Joao Marcelo-Silva et al. [295] |

**Tab. 8.1** (continued)

| Metal targeted for cleaning | Plant species | References |
|---|---|---|
| | *Prosopis tamarugo, Schinus molle*, and *Atriplex nummularia* | Elisabeth J. LA, et al. (2017) |
| | *Chrysopogon zizanioides* | Meenu Gautam and Madhoolika Agrawal [171] |
| | *Zea mays* | Antoniades et al. [36] |
| | *Hwentia* | Bansah and Addo [57] |
| | *Musa sapientum* | Bansah and Addo [57] |
| | *Leucaena leucocephala* | Bansah and Addo [57] |
| | *Terminalia superba* | Bansah and Addo [57] |
| Cr | *Azolla pinnata* and *Lemna minor* | Sandhya Bharti et al. [75] |
| | *Solanum viarum* Dunal sp. | Thays Franca Afonso et al. [13] |
| | *Eleocharis acicularis* | Nguyen Thi Hoang Ha et al. [209] |
| | *Jatropha curcas* L. | Paloma Alvarez-Mateos et al. [27] |
| | *Brassica juncea* L. | J. C. Mendoza-Hernandez et al. [312] |
| | *Chrysopogon zizanioides* L. | Ritesh Banerjee et al. [55] |
| | *Typha angustifolia* L. | Lo Vun Yen and Kartini Saibeh [525] |
| | *Canna indica* L. | V. Subhashini and V.V.S. Swamy [469] |
| | *Chrysopogon zizanioides* | Meenu Gautam and Madhoolika Agrawal [171] |
| Fe | *Azolla pinnata* and *Lemna minor* | Sandhya Bharti et al. [75] |
| | *Eleocharis acicularis* | Nguyen Thi Hoang Ha et al. [209] |
| | *Eucalyptus grandis* and *Ailanthus altissima* | Hakime Abbaslou and Somayeh Bakhtiari [1] |
| | *Jatropha curcas* L. | Paloma Alvarez-Mateos et al. [27] |
| | *Brassica juncea* L. | J. C. Mendoza-Hernandez et al. [312] |
| | *Chrysopogon zizanioides* L. | Ritesh Banerjee et al. [55] |
| | *Pongamia pinnata* | Xiumei Yu et al. [1064] |
| | *Typha angustifolia* L. | Lo Vun Yen and Kartini Saibeh [525] |
| | *Eucalyptus globulus* | Taoufik El Rasafi et al. [142] |
| | *Epilobium fragilis, Carthamus oxyacantha*, and *Verbascum speciosum* | Maleyeri et al., [291] |

**Tab. 8.1** (continued)

| Metal targeted for cleaning | Plant species | References |
|---|---|---|
| | *Prosopis tamarugo, Schinus molle*, and *Atriplex nummularia* | Elisabeth J. LA, et al. [2017] |
| | *Chrysopogon zizanioides* | Meenu Gautam and Madhoolika Agrawal [171] |
| | *Dittrichia viscosa* | A. Buscaroli et al. [97] |
| Hg | *Jatropha curcas* L. | Paloma Alvarez-Mateos et al. [27] |
| | *Typha latifolia* and *Chrysopogon zizanioides* | Alexander Kofi Anning and Ruth Akoto [32] |
| | *Clidemia sericea* D. Don | Elvia Valeria Durante-Yanez et al. [134] |
| | *Jatropha curcas* L. | Jose Marrugo-Negrete et al. [298] |
| | *Clidemia sericea* D. Don | EV Durante-Yanez et al. [134] |
| | *Brassica* sp. | Antoniades et al. [36] |
| Ti | *Pongamia pinnata* | Xiumei Yu et al. [1064] |
| | *Alchornea cordifolia* | Mensah et al. [333] |
| | *Chromolaena odorata* | Mensah et al. [333] |
| | *Fern* | Mensah et al. [333] |
| | *Pueraria montana* | Mensah et al. [333] |
| Al | *Zea mays* | Antoniades et al. [37] |
| | *Brassica juncea* L. | J. C. Mendoza-Hernandez et al. [312] |
| U | *Nicotiana tabacum* L. | Mirjana D. Stojanovic et al. [466] |
| | *Lemna gibba* L. | Martin Mkandawire et al. [2005] |
| Sr | *Euphorbia macroclada, Verbascum cheiranthifolium*, and*Astragalus gummifer* | Ahmet Sasmaz and Merve Sasmaz [418] |
| V | *Pongamia pinnata* | Xiumei Yu et al. [1064] |
| Si | *Eleocharis acicularis* | Nguyen Thi Hoang Ha et al. [209] |
| Co | *Aloe burgersfortensis* and *A. castanea* | Joao Marcelo-Silva et al. [295] |
| Sn | *Jatropha curcas* L. | Paloma Alvarez-Mateos et al. [27] |
| Ba | *Solanum viarum* Dunal sp. | Thays Franca Afonso et al. [13] |
| Mn | *Chrysopogon zizanioides* | Meenu Gautam and Madhoolika Agrawal [171] |

In the end, it is essential to successfully implement suitable agricultural practices, such as crop management, harvesting, processing, and disposal, in order to prevent the potential re-release of toxins into the surrounding environment. In the future, it will be necessary to conduct additional research in the form of laboratory investigations, incubation studies, experiments conducted in greenhouses or pots, field experimental trials, systematic reviews, and scoping reviews in order to widen the scope of revegetation in Ghana.

Albert Kobina Mensah
# Chapter 9
# Achieving mining sector sustainability

**Abstract:** Sustainable development focuses on human well-being, economic advancement, social connections, and environmental sustainability. It involves the development of renewable substitutes for nonrenewable resources. In the mining sector, Ghana should adopt strategic approaches to achieve sustainability, including water management, climate change, biodiversity conservation, mine closure strategies, sustainable metal mining practices, and tailings management. Regulation 81 of LI 2182 mandates reclamation and restoration of degraded mined sites, but there is inadequate information and documentation about these processes. A systematic review on post-mining landscape restoration in Africa has shown no progress, hindering the development of national restoration programs. Compliance with mine closure regulations and reclamation efforts is increasing among mining companies.

## 9.1 Introduction

Sustainable development promotes a comprehensive perspective on human well-being, encompassing economic advancement, robust social connections, and environmental sustainability, in order to contribute toward global preservation efforts. The Brundtland Commission, in its 1987 report titled "Our Common Future," provides a definition of sustainable development as the form of development that effectively fulfills the requirements of the current generation while ensuring that the capacity of future generations to fulfill their own needs remains intact.

The definition of sustainable development, as established collaboratively by the World Wide Fund for Nature (WWF), the World Conservation Union (IUCN), and the United Nations Environment Programme (UNEP), entails the provision of a satisfactory standard of human existence that is achieved by operating within the ecological carrying capacity of the supporting ecosystems.

According to the Development Assistance Committee (DAC) of the Organization for Economic Cooperation and Development (OECD), sustainable development involves achieving a harmonious equilibrium between economic, social, and environmental goals. This is accomplished by integrating these objectives through policies and practices that mutually reinforce each other, while also acknowledging the need

**Albert Kobina Mensah,** Soil Research Institute, Council for Scientific and Industrial Research, Kumasi, Ghana, e-mail: albert.mensah@rub.de, albertkobinamensah@gmail.com,
ORCID: https://orcid.org/0000-0001-5952-3357

https://doi.org/10.1515/9783111662046-009

for trade-offs. Sustainability refers to the capacity to maintain a specified behavior over an extended period of time without depleting essential resources or causing significant harm to the environment or society.

The theory of the three pillars of sustainability posits that in order to achieve comprehensive sustainability, all the three pillars must exhibit sustainability. The three primary dimensions of sustainable development encompass the environment (referred to as the planet), the economic aspect (often associated with profits), and the social component (pertaining to humans). When all three pillars exhibit robustness, individuals reside within a societal framework wherein the prevailing standard of living is characterized by superior quality. The region has a pristine and salubrious environment, accompanied by a commendable degree of economic prosperity and a thriving level of social satisfaction.[1]

*Social sustainability* refers to the capacity of a social entity, such as a nation, household, or institution, to operate at a specified level of social welfare and cohesion in a sustainable manner over an extended period of time. The concept of social sustainability lacks a clear and universally accepted definition in terms of its practical and implementable aspects. Therefore, it can be argued that this particular pillar is the least robust due to the lack of consensus over the correct orientation. The measurement can be quantified using the gross national happiness index. Issues such as warfare, chronic poverty, pervasive unfairness, and inadequate education rates might be regarded as manifestations of an unsustainable societal framework.[2]

*Economic sustainability* is achieved when a political entity, such as a nation, maintains a percentage of its people below a predetermined minimum standard of living threshold that is deemed desirable. It is imperative that the percentage be maintained at a significantly low level, perhaps 5% or below, as those falling below this threshold experience considerable pain, manifesting either in bodily ailments stemming from compromised health or psychological distress. Economic sustainability refers to the capacity of an economy to maintain a specified level of economic production in a continuous and enduring manner.[3]

*Environmental sustainability* refers to the ongoing ability to maintain a balance between the rate at which renewable resources are harvested, pollution is generated, and non-renewable resources are depleted, without causing long-term harm or depletion. If a certain entity or system lacks the ability to be perpetuated indefinitely, it can be concluded that it is not sustainable.[4] The most crucial among the three pillars is

---

1 The three pillars of sustainability. http://www.thwink.org/sustain/glossary/ThreePillarsOfSustainability.htm. Accessed on 27 November 2023.
2 http://www.thwink.org/sustain/glossary/SocialSustainability.htm. Accessed on 27 November 2023.
3 http://www.thwink.org/sustain/glossary/EconomicSustainability.htm. Accessed on 27 November 2023.
4 http://www.thwink.org/sustain/glossary/EnvironmentalSustainability.htm.

this. In his seminal work published in 1990, Herman Daly put forth a set of criteria that must be satisfied in order for a system to achieve environmental sustainability. These criteria are as follows:

1. In the case of renewable resources, the rate at which these resources are harvested should not surpass the rate at which they naturally regenerate. This concept is commonly referred to as sustainable yield.
2. With regard to pollution, it is imperative that the rates at which waste is generated from various projects do not exceed the environment's capacity to assimilate and process such waste. This principle is known as sustainable waste disposal.
3. In the context of nonrenewable resources, the depletion of these resources should necessitate the simultaneous development of renewable substitutes that can adequately fulfill the functions previously served by the nonrenewable resources.

## Historical developments toward sustainability

The traditional goal of public policy was progress, which was seen as a continuous movement towards an ideal future. However, the Club of Rome's 1972 report "The Limits to Growth" challenged this notion. The report argued that resources are finite, and growth dependent on them cannot be endless.

Two key reports, the World Conservation Strategy and Our Common Future, provided the answer to sustainable development; thus, the concept of sustainable development was born. The World Conservation Strategy defined conservation as managing human use of the biosphere to yield the greatest sustainable benefit to present generations, while maintaining its potential to meet the needs and aspirations of future generations. Development was defined as the modification of the biosphere and the application of human, financial, living, and non-living resources to satisfy human needs and improve the quality of human life.

For development to be sustainable, it must consider social and ecological factors as well as economic ones, including the living and non-living resource base and the long-term and short-term advantages and disadvantages of alternative actions. The Brundtland Commission report advocated for a new era of economic growth based on policies that sustain and expand the environmental resource base, which is essential to relieve poverty in the developing world.

Sustainable development meets the needs of the present without compromising the ability of future generations to meet their own. The Brundtland Report acknowledged the limitations imposed by the present state of technology and social organization on environmental resources and the biosphere's ability to absorb human activities. However, we can manage and improve technology and social organization to pave the way for a new era of economic growth.

In conclusion, sustainable development must rest on political will, considering population size and growth in harmony with the changing productive potential of the ecosystem.

## The concept of environmental sustainability: history and the progress made till date

The concept of environmental sustainability has evolved over time, with the goal of public policy traditionally being progress. People viewed progress as a constant journey towards an ideal future, rooted in economic principles rather than moral or philosophical beliefs. However, in 1972, the Limits to Growth Report challenged this notion, arguing that resources are finite and growth dependent on them cannot be endless.

Two key reports, the World Conservation Strategy and Our Common Future, provided the answer to sustainable development; thus, the concept of sustainable development was born. The World Conservation Strategy defined conservation in human terms as "the management of human use of the biosphere so that it may yield the greatest sustainable benefit to present generations while maintaining its potential to meet the needs and aspirations of future generations." Development was defined as "the modification of the biosphere and the application of human, financial, living, and non-living resources to satisfy human needs and improve the quality of human life."

For development to be sustainable, it must take into account social and ecological factors as well as economic ones: the living and non-living resource base and the long-term and short-term advantages and disadvantages of alternative actions.

In 1972, the United Nations Conference on Human and Environment in Stockholm resulted in the adoption of 7 declarations and 26 principles aimed at promoting human action and development towards the environment and the sustainability of the ecosystem.

According to Declaration number 4, underdevelopment is the primary cause of most environmental problems in developing countries, where millions of people live far below the minimum standards necessary for a decent human existence. Therefore, developing countries must direct their efforts to development, bearing in mind their priorities and the need to safeguard and improve the environment. Conversely, industrialised countries should strive to narrow the gap with developing countries, given that environmental issues typically stem from industrialisation and technological advancement.

In conclusion, the concept of environmental sustainability has evolved over time, with the goal of promoting sustainable development and addressing the challenges faced by both developed and developing nations.

The United Nations Conference on the Human Environment (1972) and the World Commission on Environment and Development (WCD) both emphasize the importance of maintaining and restoring the earth's capacity to produce vital renewable resources. While Principle 4 emphasizes the role of major groups in achieving sustainable development, Principle 3 emphasizes the need to maintain and improve sustainable development.

The Brundtland Commission report advocates for sustainable growth, which meets the needs of the present without compromising the ability of future generations to meet their own. However, the concept of sustainable development does imply limits imposed by the present state of technology and social organization on environmental resources and the biosphere's ability to absorb the effects of human activity. We can manage and improve technology and social organization to pave the way for a new era of economic growth.

Agenda 21 is an international blueprint that outlines actions that governments, international organizations, industries, and communities can take to achieve sustainability in the 21st century. It defines sustainable development using 27 principles, with the objective of alleviating poverty, hunger, sickness, and illiteracy worldwide while halting the deterioration of ecosystems that sustain life.

Agenda 21 consists of four main sections: social and economic dimensions; conservation and management of resources for development; strengthening the roles of major groups; and means of implementation. The first section examines the underlying human factors and problems of development, trade, integrated decision-making, and the role of major groups in achieving sustainable development. The second section explores the necessary resources to mobilize for sustainable futures, such as finance, technology, education, institutional and legal structures, data and information, and the development of national capacity in relevant disciplines.

In conclusion, the Rio Declaration on Environment and Development provides a framework for achieving sustainable development, emphasizing the importance of maintaining and improving the earth's capacity to produce vital renewable resources.

Johannesburg, South Africa hosted the World Summit on Sustainable Development (WSSD) from 26 August to 4 September 2002. It focused on poverty eradication, changing unsustainable consumption and production patterns, and protecting and managing the natural resource base of economic and social development. The summit also emphasized the importance of globalization and its potential to improve living standards for all.

The United Nations Conference on Sustainable Development (UNCSD) took place in Rio de Janeiro, Brazil, on 20-22 June 2012, with the main objective of renewing political commitment towards realising the Rio agenda. As such, it is commonly referred to by other experts as Rio plus 20. The conference resulted in a focused political outcome document with clear and practical measures for implementing sustainable development.

Member States decided to launch a process to develop a set of Sustainable Development Goals (SDGs), which will build upon the Millennium Development Goals and converge with the post-2015 development agenda. The project's specific objectives included preparing a substantive contribution to the UNCSD debate in 2012, constructing a coherent vision for sustainable development in the 21st century, synthesizing analytical and applied policy work regarding menus of policy options for sustainable, green growth, and promoting sustained, inclusive, and sustainable economic growth.

In 2015, over 150 world leaders met in New York to adopt an ambitious new sustainable development agenda, Agenda 2030. The agenda consists of a declaration, 17 Sustainable Development Goals, 169 targets, a section on means of implementation, renewed global partnership, and a framework for review and follow-up.

The Sustainable Development Goals include ending poverty, ensuring healthy lives and well-being for all, ensuring inclusive and equitable quality education, achieving gender equality, ensuring availability and sustainable management of water and sanitation, ensuring access to affordable, reliable, sustainable, and modern energy, promoting sustained, inclusive and sustainable economic growth, building resilient infrastructure, reducing inequality, making cities and human settlements inclusive, safe, resilient, and sustainable, ensuring sustainable consumption and production patterns, taking urgent action to combat climate change, conserving and sustainably using oceans, seas, and marine resources, protecting terrestrial ecosystems, promoting peaceful and inclusive societies for sustainable development, strengthening the means of implementation, and revitalizing the global partnership for sustainable development.

Sustainable development is a holistic approach to human well-being that includes economic progress, strong social bonds, and environmental sustainability. It was defined by the Brundtland Commission in 1987 as meeting present needs without compromising future generations' ability to meet their own needs. The World-Wide Fund for Nature, the World Conservation Union, and the United Nations Environment Program jointly define sustainable development as providing the quality of human life while living within the carrying capacity of supporting ecosystems.

## 9.2 Attaining sustainability in the mining sector

Mineral deposits possess the characteristic of being nonrenewable. How is the concept of sustainability applied to nonrenewable mineral resources? Mineral resources exhibit a nonrenewable characteristic, and their present exploitation and consumption would lead to their ultimate depletion or exhaustion, thus adversely affecting the welfare of future generations. The upcoming generation is expected to face a scarcity of resources as a result of their depletion. The concept of the three pillars of sustainability asserts that in order to attain holistic sustainability, it is imperative to maintain the three pillars comprising economic (profit), environmental (planet), and social (people) components.

According to Mikesell [336], the attainment of sustainability in mineral reserves can be accomplished by annually conserving and reinvesting an equivalent sum to the present value of the annual net revenue derived from the trade of mining commodities. Placer Dome, a multinational gold mining corporation, defines sustainability as the conscientious approach to the planning, establishment, execution, and cessation of mining activities, with due consideration and responsiveness to the societal, environmental, and economic requirements of both current and future generations within the local communities and nations in which it conducts its operations.

According to Gibson [820], the acceptability of mining is contingent upon the effective remediation of its immediate negative impacts and the provision of socioeconomic benefits that serve as a means to facilitate a transition toward a more sustainable future for the local population. According to [1085], it is emphasized that for mining to be regarded as a contributor to sustainability, it is necessary to attain a net positive impact on the environment and human well-being.

There exist substantial variations in regulatory frameworks across different regions of the world; hence, adhering to regulations does not automatically guarantee the implementation of effective environmental practices. Hilson and Murck [205], contend that developing countries are characterized by a nascent stage of environmental legislation and inadequate enforcement systems. Therefore, a mining operation that complies with or exceeds the legal standards may not always be promoting environmental improvement or sustainable development.

According to [619], a sustainable mine can be defined as a mining operation that effectively addresses the needs of both current and future generations. This entails taking into account the negative impacts on the environment, economy, and society, and ensuring that the costs associated with these impacts are borne by the mining company itself. In the context of developing countries, achieving sustainability often necessitates surpassing the minimum legal requirements.

According to the International Council on Mining and Metals (ICMM), sustainable growth within the mining industry encompasses investments that are characterized by financial profitability, technical appropriateness, environmental soundness, and social responsibility. It is assumed that these areas are of equal importance; hence, the emphasis on one area usually leads to a crisis across the entire area of mining activity [620].

## The AGENDA 21 and sustainable development in 21st Century

The AGENDA 21 document encompasses all necessary elements for establishing and achieving sustainable development in the 21st century and beyond. The document delineates compelling principles that can act as a catalyst for attaining intergenerational and intragenerational fairness within the economy, society, and our environment.

For example, its primary concept emphasises that human beings are the focal point of all developmental projects or initiatives. In this context, it becomes imperative for policymakers to contemplate the ramifications of their actions, not only to

provide sustenance and maintain livelihoods but also to assess the environmental repercussions of any developmental initiatives. In general, any activity that creates
profits for firms and corporations without implementing steps to safeguard the environment and the social fabric of the community in which they operate cannot be
deemed sustainable.

Principle number 10 underscores the necessity for public engagement and collaboration with those directly impacted by proposed developments. Such consultation
should extend beyond merely informing the community about proposed developments on their lands; it should actively engage them, solicit their ideas, and incorporate their inputs and proposals throughout all phases of the project life cycle, i.e.,
from the initiation to the closure.

Principle 20 mandates the involvement of women in all developmental initiatives.
Principle number 15 mandates the initial conduct of an environmental impact assessment, implementation of strategies to enhance positive project outcomes, mitigate adverse effects, and reassess any consequences inadequately addressed in environmental management plans.

Principle number 22 advocates for the incorporation and acknowledgement of indigenous knowledge in addressing environmental sustainability. Principle 22 of
AGENDA 21, a document that serves as a blueprint for sustainable development in the
21st century, asserts that indigenous peoples and their communities, along with other
local communities, play a crucial role in environmental management and development due to their knowledge and traditional practices. States must acknowledge and
actively promote their identity, culture, and interests, facilitating their meaningful involvement in the pursuit of sustainable development.

## Example: Application of AGENDA 21 principle number 22 (i.e., integration of indigenous knowledge) in achieving sustainable development in Ghana

Prior to the twentieth century, the management of watersheds in numerous Ghanaian
communities was predominantly governed by religiously rooted restrictions, including the implementation of taboos and the preservation of sacred groves, aimed at preventing encroachment on these vital resources. These strategies fostered a profound
respect for soil and water resources.

We regarded our rivers as divine entities, and therefore, any act of destruction
towards these vital resources could invoke or attract retribution from the deities.
These punishments may affect the entire community or society.

The consequences may include an erratic water supply, the desiccation of rivers,
a decrease in soil productivity and fertility, extended periods of dryness leading to
unpredictable rainfall, a reduction in crop yields, famine, or a decline in fish catches.
As a result, individuals were apprehensive about the repercussions linked to violating

these norms, taboos, and customs. Ultimately, our waters remained pristine, our soil retained its fertility, and productivity was sustained; our forests were conserved, and our crops produced bountiful harvests.

For instance, Ofosu-Mensah [1084] delved into the intricate details to elucidate the resource-curse hypothesis, employing Akyem Abuakwa, a prominent mining community in Ghana, as a case study. He articulates:

*During the precolonial period, the Abuakwa people held the River Birem in high esteem, viewing it as a fundamental source of their prosperity and nourishment. The discovery of gold within the Birem River further elevated their veneration, likening it to that of a divine entity. In his 1904 exposition, linguist Kwabena Pro elucidated the reverence the people of Osino held for the Birem, regarding it as a sacred river. He articulated, "we do not eat from that river because the river is our god and when it asked us whether we liked gold or fish we said gold, and so they [the Osino people] left the fish in the river."*

Currently, the pollution of rivers by galamsey operators reflects a troubling disregard for the detrimental environmental impacts of their actions. Regrettably, traditional governance of watersheds has encountered numerous obstacles in the evaluation and assessment of environmental harm, the enforcement of regulations to hold unlawful land users accountable, and the integration of land users' rights with regulatory and management policies. The emergence of Christianity, modernity, Western education, and urbanisation, coupled with the aspiration to harness the nation's resources, has diminished the efficacy and reverence for traditional constraints aimed at environmental protection.

Therefore, it is essential to integrate the traditional knowledge of the community with contemporary scientific understanding. When the knowledge of indigenous peoples is overlooked, potential for progress remains significantly hindered. Consequently, it is essential to utilise and enhance the indigenous knowledge possessed by the community in areas such as cultivation, climate change mitigation, land management, soil fertility management, water resource conservation, and biodiversity conservation, among others, for their benefit.

## 9.3 Have mining corporations operating in Africa made sufficient progress in terms of promoting sustainability within their mining communities?

The use of the "precautionary principle" is considered a fundamental component of sustainable growth in the mining industry. The aforementioned principle, as stated in the 1992 Intergovernmental Agreement on the Environment, asserts that when there is a possibility of causing substantial or irreversible damage to the environment, the lack of absolute scientific certainty should not be used as a reason to postpone the adoption of actions intended to prevent environmental deterioration (Agenda 21, 1992). This specific principle is classified as the 15th among a set of 227 principles that are designed to promote the achievement of sustainable development in the twenty-first century. The concepts mentioned are further elaborated in a document referred to as Agenda 21.

The persistent question is around the methods by which a country can achieve sustainability in regard to the exploitation of a finite resource. Considering the finite nature of mineral resources, such as gold, it is crucial to explore strategic approaches that Ghana might adopt to achieve sustainability in the utilization of its nonrenewable mineral resources. To optimize the benefits obtained by mining communities through resource exploitation, what strategies might be employed? Mining communities primarily encounter the environmental and social consequences that arise from mineral explorations.

One of the three requirements delineated by Herman Daly [119], in his arguments advocating for a sustainable society posits that the exhaustion of nonrenewable resources should include the simultaneous advancement of renewable alternatives that possess comparable worth. The ICMM asserts that sustainable growth in the mining sector entails investments that exhibit financial profitability, technical appropriateness, environmental soundness, and social responsibility.[5] Figure 9.1 captures the three fundamental pillars that serve as exemplars of sustainability within the mining industry.

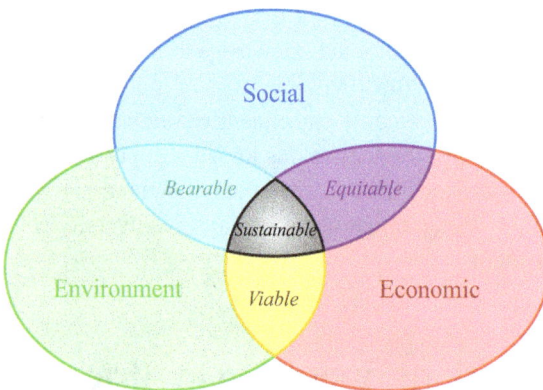

**Fig. 9.1:** The theory of the three pillars of sustainability posits that in order to achieve comprehensive sustainability, all the three pillars must exhibit sustainability.

---

5 http://www.icmm.com/en-gb/environment. Accessed on 27 November 2023.

A sustainable mine can be characterized as a mining operation that prioritizes the protection of the environment, engages with the local population, and ensures economic viability, while also aiming to maximize income from mineral extraction. Therefore, it is crucial to employ suitable strategies to address the environmental deterioration caused by mining activities, both while the mine is operational and during its closing period. It is vital to exert endeavors in order to alleviate the detrimental environmental effects associated with mineral extraction, while simultaneously optimizing its advantageous outcomes. The main focus of this book will be the examination of efforts to achieve environmental sustainability in the mining sector. The focus of our study will be directed toward the revegetation stage of land reclamation, with a particular emphasis on the restoration of regions that have undergone degradation due to mining operations.

## 9.4 International institutional frameworks governing sustainable development in the mining sector

### 9.4.1 International Council on Mining and Metals (ICMM)

The ICMM is a globally recognized organization that is committed to promoting safety, equity, and sustainability within the mining and metals sector. The initiative aims to enhance environmental and social performance by fostering collaboration among 27 mining and metals firms, as well as more than 30 regional and commodity associations.

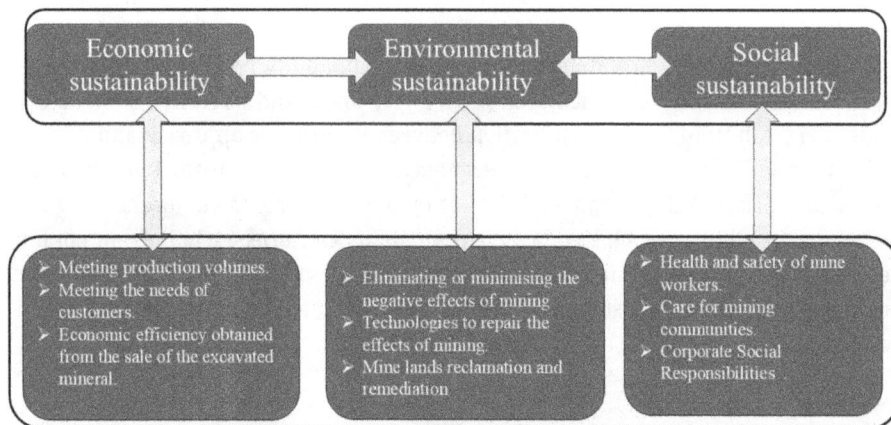

**Fig. 9.2:** The three pillars of mining sector sustainability. Adapted and modified after Dubinski (2013).

They function as catalysts for driving change, thus augmenting the societal impact of the mining industry. The areas of focus encompassed in this study are water management, climate change, biodiversity conservation, mine closure strategies, sustainable metal mining practices, and tailings management.[9]

### 9.4.2 Mining, Minerals and Sustainable Development Project (MMS-DP)

Initiated in 2002, MMS-DP (Mining, Minerals and Sustainable Development Project) was a research project aimed at exploring how the mining and minerals sector could contribute to the global transition to sustainable development. Prior to the World Summit on Sustainable Development (WSSD) in 2002, the World Business Council for Sustainable Development contracted the International Institute for Environment and Development to undertake a 2-year independent process of research and consultation – MMS-DP. Its four broad objectives were:

i.   To assess the global mining and minerals sector in terms of transition to sustainable development
ii.  To identify how the services provided through the minerals supply chain can be delivered in ways that support sustainable development
iii. To propose key elements for improving the minerals system
iv.  To build platforms for analysis and engagement for ongoing communication and networking among all stakeholders in the sector.[6]

### 9.4.3 Global Reporting Initiative (GRI), 1997

A sustainability report refers to a publication issued by a firm or organization that provides information regarding the economic, environmental, and social consequences resulting from its routine operations. The sustainability report serves as a platform for the company to articulate its core principles and governance structure, while also establishing a clear connection between its strategic approach and its dedication to fostering a sustainable global economy. The Global Reporting Initiative (GRI) assists enterprises and governmental bodies on a global scale in comprehending and conveying their influence on significant sustainability concerns, including but not limited to climate change, human rights, governance, and social welfare. This facilitates the implementation of measures that provide positive outcomes in terms of social, environmental, and economic aspects, benefiting all stakeholders involved.[7]

---

6  http://pubs.iied.org/pdfs/G00910.pdf. Accessed on 27 November 2023.
7  https://www.globalreporting.org/information/sustainability-reporting/Pages/default.aspx

## 9.5 Institutional frameworks governing sustainable development in the mining sector in Ghana

The key legislations regulating mining activities in Ghana include the 1992 Constitution of Ghana, the Minerals and Mining Act, 2006 (Act 703) as amended by the Minerals and Mining (Amendment) Act, 2015 (Act 900), the Minerals and Mining (Amendment) Act, 2019 (Act 995), as well as the Minerals Commission Act, 1993 (Act 450).

The Minerals and Mining Policy of Ghana stipulates clearly that the government's objective is "to achieve a socially acceptable balance, within the environmental regulatory framework, between mining and the physical and human environment and to ensure that internationally accepted standards of health, mine safety, and environmental protection are observed by all participants in the mining sector."

Regulation 81 of LI 2182 of the Minerals and Mining Regulations of Ghana makes it a mandatory legal requirement for miners to embark on reclamation and restoration of degraded mined sites. The said Regulation stipulates that any land used for mineral exploration is rehabilitated and as far as possible returned to the condition in which it was prior to the mining operations [621]. In addition, it is a priority of the government of Ghana to ensure redressing land degradation through integrated landscape management with several projects being initiated and implemented, including the multisectoral mining integrated project.

Restoration of mined areas is an important step after mining to ensure that degraded lands are returned to conditions suitable for productive use. The main aim of restoring mined lands is to reestablish vegetation cover, restore biodiversity, stabilize the soil and water conditions, and thus restore the provision of ecosystem goods and services [50]. The mining regulations, discussed above, stipulate procedures for ensuring that mining activities are done in ways that minimize environmental impacts, and as such, all mining activities are required to follow the guidelines. A major drawback of galamsey is the fact that degraded mined lands cannot be easily restored, usually requiring long periods of time to naturally restore such lands to productive status.

It is estimated that the land area under surface mining in Ghana may have more than doubled within the last decade due to the proliferation of both galamsey and main mining operations [51, 462, 622]. However, there is inadequate information and documentation about restoration and reclamation of mine sites in the country. The reclamation and restoration successes, as well as basic information on these reclaimed sites, remain generally unavailable, a situation which hampers knowledge transfer.

Furthermore, a paper by Festin et al. [158] on the restoration of post-mining landscapes in Africa demonstrated that despite the long history of mining in Africa, no systematic review has summarized advances in restoration research and practices after mining disturbances. This hinders the successful development of national restoration programs that inform policy and contribute positively to sustainable development in the country and within the region. Compliance and adherence to mine clo-sure regulations have been somewhat positive among mining companies in the coun-try, and restoration or reclamation efforts are thought to be gradually rising.

### 9.5.1 Reclamation bond

This document pertains to a reclamation security agreement between the Environ-mental Protection Agency of Ghana (EPA-Ghana) and mining corporations. Its purpose is to outline the procedures and strategies for reclaiming lands that have been im-pacted by the activities of these companies. In this instance, the company makes a deposit into an account that accrues interest. The account is held jointly by the com-pany and EPA, as stated by [623]. The specified sum serves as a kind of assurance and may be allocated toward the restoration of environmentally damaged mining sites in the event that the firm fails to fulfill this obligation throughout the decommissioning stage of the mining project's life cycle.

This action aligns with the regulations set forth in the EPA Act 494 of 1994 and LI 1652 of 1999. These regulations mandate that mining companies must acquire an envi-ronmental permit by submitting a positive environmental impact assessment of their operations. Additionally, these companies are obligated to provide a reclamation secu-rity bond to the Environmental Protection Agency. This bond serves as a guarantee and can be utilized to restore the environmentally damaged mining sites in the event of the company's failure to do so.

## 9.6 Conclusions

The following conclusions are drawn about the chapter:
a)  Sustainable development promotes a comprehensive perspective on human well-being, encompassing economic advancement, robust social connections, and envi-ronmental sustainability in order to contribute toward global preservation efforts.
b)  The three primary dimensions of sustainable development encompass the envi-ronment (referred to as the planet), the economic aspect (often associated with profits), and the social component (pertaining to humans).
c)  In the context of nonrenewable resources, the depletion of these resources should necessitate the simultaneous development of renewable substitutes that can ade-quately fulfill the functions previously served by the nonrenewable resources.
d)  The use of the "precautionary principle" is considered a fundamental component of sustainable growth in the mining industry.

e) Considering the finite nature of mineral resources, such as gold, it is crucial to explore strategic approaches that Ghana might adopt to achieve sustainability in the utilization of its nonrenewable mineral resources.

f) Regulation 81 of LI 2182 of the Minerals and Mining Regulations of Ghana makes it a mandatory legal requirement for miners to embark on reclamation and restoration of degraded mined sites.

g) Restoration of mined areas is an important step after mining to ensure that degraded lands are returned to conditions suitable for productive use.

h) Compliance and adherence to mine closure regulations have been somewhat positive among mining companies in the country, and restoration or reclamation efforts are thought to be gradually rising.

i) Additionally, these companies are obligated to provide a reclamation security bond to the Environmental Protection Agency.

Albert Kobina Mensah

# Chapter 10
# Rehabilitation and restoration of degraded mined sites and soils

**Abstract:** In this chapter, many terminologies that are important to mine land restoration are defined. Clarification is provided regarding the definitions of soil degradation, restoration, remediation, reclamation, rehabilitation, and revegetation. But in this book and in this chapter, all these phrases are used interchangeably. Furthermore, I talked about the many methods for revegetating degraded mine fields and provided examples and diagrams to support my points. Particular strategies for recovering mine lands are described including agricultural, adaptive, and ameliorative strategies as well as mixed strategies and different chemical, physical, and biological techniques. It is advised to use the mixed revegetation strategy, which works well when using a single methodology or approach is impractical or does not produce the required results.

## 10.1 Definition of terms and concepts (see Fig. 10.1)

### 10.1.1 Restoration

It is the broader concept of restoring degraded mine lands and making the soil or land usable again. Restoring the degraded mining sites to their pristine condition is the ultimate goal, but that is almost an impossibility. Therefore, the pertinent definition of restoration refers to the process of returning anything to its previous state or position or to a condition that is unimpaired or flawless. The act of restoration involves returning something to its original condition or to a state of robust health and vitality. The connotation encompasses both the restoration to an initial state and the attainment of a state that is characterized by perfection and soundness [see 89].

### 10.1.2 Remediation

Remediation refers to the process of addressing or resolving a problem or deficiency. The term "to remedy" can be defined as the act of rectifying or making something

**Albert Kobina Mensah,** Soil Research Institute, Council for Scientific and Industrial Research, Kumasi, Ghana, e-mail: albert.mensah@rub.de, albertkobinamensah@gmail.com, ORCID: 0000-0001-5952-3357

https://doi.org/10.1515/9783111662046-010

good. The primary focus lies on the procedural aspects rather than the final outcome achieved [see 89]. The word "remediation" comes from the Latin word *remedium* which means to restore and to clean ([779, 1086]); this has been further described by Tonelli et al. [481] in their work on the history of phytoremediation. Here, strategies are taken to either remove the pollutants or contaminant from the environment via mobilization techniques or restrict and reduce their mobility to prevent them from being available to cause harm via immobilization techniques. As a result, these procedures effectively bind the pollutant, thereby preventing any potential harm to living creatures. The measures also have the objective of minimizing their absorption into food crops, hence safeguarding human health.

## 10.1.3 Reclamation

Reclamation is the process of rendering land suitable for agricultural purposes. The term "reclaim" is defined as the act of restoring something to its appropriate condition. The connotation does not pertain to reverting back to an initial condition, but rather to achieving a state that is practical or beneficial. Hence, substitution emerges as a viable alternative. The term can also imply "to replace." Here, it could mean to provide substitute in place of the original or to provide an equivalent to take the place of the previous state or the original condition. Reclamation is seen as a favorable and essential measure aimed at restoring mined regions to an environmentally acceptable state, whether for the purpose of reinstating the previous land use or facilitating a new utilization. One approach to managing lands is to prioritize their conservation and protection. Another perspective, as suggested by Bastida (2010i), is to maximize the economic value of these resources.

## 10.1.4 Rehabilitation

Rehabilitation can be described as the process of restoring an entity to its previous condition or status. This phenomenon bears resemblance to the concept of restoration; however, it does not necessarily imply the attainment of perfection. In colloquial parlance, the term "rehabilitated" typically implies that the subject in question has not attained the same level of originality or soundness as it would have if it had undergone restoration [see 89]. Certain ecological functions that were impaired during extraction of the mineral are repaired: for example, controlling soil erosion around a tailings dam, taking measures to improve soil quality of the degraded ecosystem, and using plant-based strategies to clean up toxic elements such as arsenic from As-contaminated mining spoil. Here, the emphasis is on the action in progress or the process of restoring, but not the final goal of achieving the needed previous level.

Thus, rehabilitation refers to the systematic restoration of the environmental damages caused by mining activities. Rehabilitation in mine land restoration projects often consists of two primary phases: the design of land forms and the reconstruction of a stable land surface, followed by the process of revegetation or the establishment of an alternative land use on the reconstructed land form [737]. In this book, I will focus on the revegetation phase of rehabilitation and its impacts in restoring soil quality of a degraded mining site as well its remediation potential for potentially toxic elements (PTEs) in contaminated-mined soils. Please see Fig. 10.1.

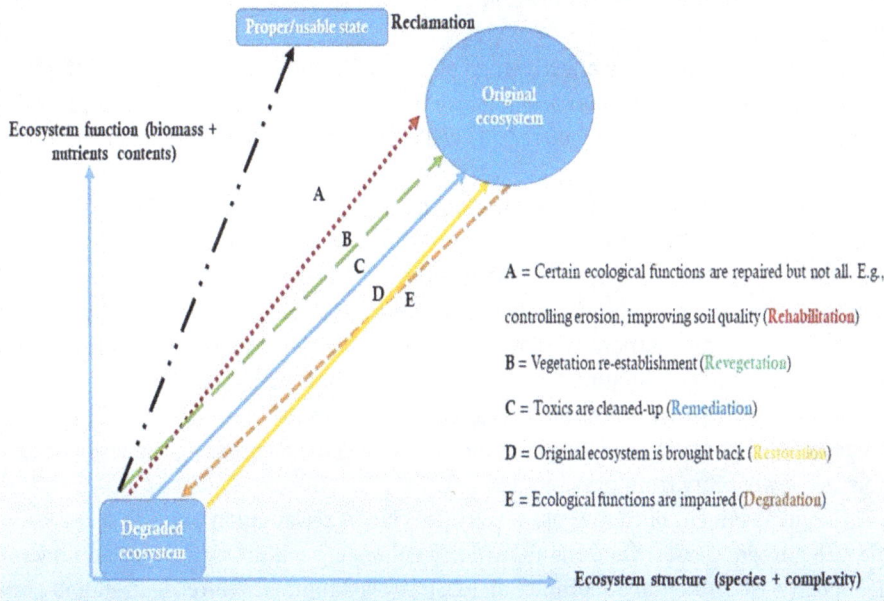

Fig. 10.1: Different options in degraded mined restoration. Source: Adapted and modified after Bradshaw [89].

## 10.1.5 Revegetation of degraded mined soils

According to [759], the conventional approach to remediating degraded soils involves implementing extended periods of forest fallowing in a rotational manner. The composition and density of the soil have a direct impact on the long-term stability of the recovered plant community [453]. Soil serves as the fundamental basis for this process. In order to restore land to its original state and maintain a self-sustaining ecosystem, reclamation strategies should encompass various aspects such as soil structure, soil fertility, microbial populations, topsoil management, and nutrient cycling [453]. The reclamation and revegetation of abandoned mined areas are frequently constrained by several physical and chemical characteristics found in the soil. These characteristics encompass, but are not restricted to low pH levels, elevated metal and metalloid concentrations (including metal salts), insufficient nutrient levels (e.g., N, P, and K), and inadequate or absent soil structure.

Mining activities can lead to the degradation or loss of soil structure and functions, resulting in soil toxicity, limited nutrient availability, and unfavorable soil texture. The form and function of soil, while constituting only a component of a larger ecosystem, serve as a microcosm that reflects the characteristics and dynamics of the entire ecosystem. According to [762], if these issues are not addressed, the reestablishment of vegetation and the restoration of ecosystem function may prove to be challenging or unattainable.

In order to enhance the effectiveness of rehabilitation and reforestation efforts, it is imperative to conduct studies and analyses on the biological and physiological traits of regenerated or newly planted trees as well as the underlying mechanisms that impact their productivity [880].

In laboratory settings, mulches have been observed to enhance soil moisture conditions. However, it is important to note that achieving successful revegetation in real-world environments necessitates the elimination of salts from the root zone, as highlighted by Grigg et al. [831]. The presence of high salinity is a significant constraint for the establishment of plants, as it hampers the process of seedling emergence by delaying germination [750]. Consequently, this leads to an extended period during which the growth medium has to retain moisture [842]. Additionally, it hinders the growth of plants by inducing osmotic stress, hence exacerbating the issue of low moisture availability from precipitation [831].

The management of topsoil plays a crucial role in reclamation plans as it serves to mitigate nitrogen losses while simultaneously enhancing soil nutrient levels and microbial activity. According to Sheoran et al. [453], revegetation is well-recognized and considered to be an effective method for mitigating erosion and safeguarding soil quality during the reclamation process. Ecological restoration and mine reclamation have emerged as integral components of the sustainable development agenda in numerous nations. Sheoran et al. [453], add that effective planning and environmental management practices can significantly mitigate the environmental consequences associated with mining activities, hence contributing to the preservation of ecological variety.

When endeavoring to rehabilitate a native ecosystem, the initial revegetation attempt is unlikely to provide vegetation that is identical to the original. The primary objective of the initial revegetation effort is to develop the fundamental elements necessary for a self-sustaining system, ensuring that the ideal vegetation complex is achieved through successional processes [see 453].

The optimal period for initiating vegetation growth is contingent upon the temporal pattern and consistency of precipitation throughout the seasons. It is imperative to ensure that all necessary preparation tasks are carried out prior to the period when seeds are most likely to encounter the requisite conditions for germination and viabil-ity, namely consistent rainfall and appropriate temperatures [453].

Coppin et al. [776] provide a definition of revegetation as the systematic undertak-ing of vegetation establishment and subsequent maintenance, which is carried out as an integral component of reclamation, rehabilitation, or restoration efforts (see Fig. 10.1). One example of land reclamation is the filling of surface mines, followed by the recon-toring of the land and the establishment of a vegetative cover that serves to safeguard the soil [948].

The primary emphasis of mine restoration endeavors has been directed toward nitrogen-fixing species belonging to the legume family as well as various grasses, herbs, and trees. According to Sheoran et al. [453], plants that possess the ability to tolerate metals have proven to be efficacious in soils that are both acidic and contain high concentrations of heavy metals. Wong [1026] argues that the restoration of plant cover on overburden dumps can effectively achieve several objectives including stabi-lization, pollution control, visual enhancement, and mitigation of risks to human pop-ulations.

Phytoextraction, which involves the use of green plants for remediation pur-poses, shows promise as a feasible approach for the decontamination of extensive soil areas. Furthermore, it has been suggested that this method could potentially serve as a viable alternative to existing soil clean-up techniques [757, 799, 857, 1022].

The presence of vegetation has a crucial function in safeguarding the soil surface from erosion and facilitating the formation of fine particles [483, 772]. The degrada-tion process can be reversed through the stabilization of soils, achieved by the growth of vast root systems. After being established, plants have the ability to enhance soil organic matter (SOM), reduce soil bulk density, regulate soil pH, and facilitate the up-ward movement and accumulation of mineral nutrients in an accessible state.

The root systems of these organisms enable them to function as scavengers of nutrients that are not easily accessible. The nutrients are accumulated by plants and subsequently redeposited onto the soil surface in the form of organic matter. This organic matter facilitates the availability of nutrients via microbial decomposition [773, 906].

## 10.2 Techniques employed for the restoration of vegetation on mined soils

Williams and Bellitto [1024] propose that the practice of revegetation can be broadly classified into three primary approaches: agricultural, ameliorative, and adaptive. The majority of contemporary revegetation initiatives, particularly those that are more challenging in nature, employ a hybrid approach that combines both ameliorative and adaptive strategies.

## 10.2.1 Agricultural approach

The agricultural approach refers to a methodology or perspective that is employed in the field of agriculture. According to Williams and Bellitto [1024], the agriculture strategy, while somewhat antiquated, has historically been the most commonly employed method for revegetation and continues to be mandated under the Surface Mining Control and Reclamation Act of 1977 (SMCRA) regulations for coal mining projects. *This approach involves the replacement of topsoil, the application of fertilizers, and the planting of indigenous species to restore vegetation on the disturbed sites.*

Unfortunately, the practice of salvaging topsoil was not commonly observed during the majority of historical mining operations, resulting in its limited availability for utilization. Given the inherent mineralization of soils in the vicinity of mine sites, the transportation of appropriate soils often necessitates extensive distances to be covered, resulting in economic inefficiency. Hence, the utilization of ameliorative and adaptive methods, which involve the direct revegetation of mined fields without topsoil, is frequently considered favorable.

While these technologies have been developed and employed in Europe since the 1960s, it is widely considered that the initial effective implementation of these methods on the North American continent occurred at the California Gulch Superfund site, situated near Leadville, Colorado, in 1992.

An example of the implementation of the agricultural approach can be observed in the reclamation strategy undertaken by the AngloGold Ashanti Company Limited at Obuasi in Ghana. This strategy aimed to restore and rehabilitate the mined lands owned by the company as outlined by Tetteh et al. [478].

## 10.2.2 Ameliorative approach

The ameliorative approach refers to a method or strategy aimed at improving or mitigating a certain situation or condition. It involves manipulating the soil by amending and improving its physical, chemical, and biological properties before establishing the vegetation. The addition of the soil amendments aids to create conditions that make it easy for the established plants to obtain nutrients easily. For example, biochar can be added to the soil to improve the soil pH by liming its high acidity to be in the alkaline range so that heavy metals such as copper, zinc, cadmium, and other cations become immobile. Such action reduces the transport of the metals into food crops and in the process curtails food chain contamination.

When compared to the untreated arsenic (As)-contaminated mine soil, Mensah et al. [332] found that the addition of iron oxide doses reduced the readily bioavailable As by 93%; while compost, manure, inorganic fertilizers, and biochar increased it by 106–332%, 24–315%, 19–398%, and 28–47%, respectively, with a significantly higher impact for the 5% doses. The study thus recommended a combination of methods such as biochar in combination with compost/and or manure, or biochar in mixture with iron oxide. Such strategy encourages vegetation reestablishment on the degraded mining sites to aid restore ecological system functions and consequently reduce As availability to the surroundings.

Williams and Bellitto [1024] in their works demonstrated that the ameliorative technique involves the chemical modification of soils in order to rectify any identified issues with the spoil or tailing. The methodology can be employed to increase the pH level as well as decrease the solubility and accessibility of heavy metals to plants. Furthermore, the ameliorative technique can be employed to mitigate the likelihood of continued acid formation as well as reduce both total and leachable metal concentrations.

The approach involves determining an appropriate combination and proportion of inexpensive waste products, along with commonly used reclamation additives including organic matter, lime, biochar, iron oxides, red mud, manure, compost, sewage sludge, and phosphate fertilizers, among others, in order to achieve or amend the necessary soil physicochemical properties that can facilitate or support the growth and reestablishment of introduced vegetation.

The amelioration combination can comprise specific major nutrients that have been identified as lacking in the soil. For instance, using a pot experiment, Addai et al. [6] investigated the effects of biochar, poultry litter compost, and inorganic fertilizer on the uptake of PTEs and soil quality improvement of a decommissioned mine tailings in Ghana. The results show that biochar, poultry litter compost, and NPK significantly increased soil pH, organic matter, available P, and cation exchange capacity (CEC), but reduced exchangeable acidity. Furthermore, tailings amended with NPK increased N uptake by 42%, while biochar and compost increased P and K uptake by 128% and 101%, respectively. The PTE uptake was reduced by 49% in the pot. However, biochar, compost, and inorganic fertilizer enhanced Cd, Cr, and Pb uptake, potentially posing lettuce food chain contamination issues. The study therefore concluded that inorganic NPK fertilizer and rice husk biochar can improve soil quality of the mine tailing for safe food production.

Empirical evidence has demonstrated that in the absence of any modifications, mining spoils and tailings present significant challenges for the establishment and growth of vegetation. This can persist as an issue even after an extended period of 8–10 years of surface exposure [830]. For instance, a PTE-like arsenic can shift from available fractions to the residual unavailable fractions in the soil mineral as the age of the spoil increases.

In this regard, Mensah et al. [321] found that there was substantially greater proportion of arsenic in the residual fraction in an abandoned mine spoils compared to active spoils in southwestern Ghana. Additionally, in the same study, there were relatively higher proportions of arsenic bound to the amorphous iron oxide fractions in the active mining spoils compared to abandoned bulk spoils. They attributed these differences in arsenic proportions between the two mining spoils to age differences between both spoils. The abandoned mining spoil was older than the active mining spoil. Shaheen et al. [432] explained that amorphous components dominate the Fe oxide content of weakly formed and younger soils. They added that increasing soil age and the process of soil pedogenesis may shift PTE contents toward increasing fractions of crystalline Fe-oxides and hydroxides. These explanations may account for why it may be difficult to recover or restore the original ecosystem functions as the mining spoils are left abandoned for longer periods.

Furthermore, arsenic which is mostly the dominant PTE in mining tailings may convert from their primary forms into secondary forms as the land is abandoned for longer periods. For instance, Mensah et al. [323, 316] reported that scorodite was the dominant form of arsenic species in a mine tailing in southwestern Ghana. Similarly, Drewniak and Sklodowska [132] found that scorodite is the most dominant secondary As mineral in mine-waste heaps and industrial deposits. Scorodite is generally considered a secondary product of the natural oxidation and weathering of arsenopyrite, a primary form of arsenic in mine tailings [e.g., 320, 323].

The study conducted by [831] investigated the efficacy of sawdust or straw mulch additives in mitigating the negative characteristics of mine spoils and enhancing the effectiveness of revegetation endeavors. The application of mulch in laboratory experiments resulted in enhanced infiltration rates, higher soil moisture retention capacity, and decreased strength of the surface crust.

In the agricultural domain, it was shown that the incorporation of mulches to a depth of 0.15 m, with application rates of at least 20 t/ha of straw or 80 t/ha of sawdust, was necessary to effectively counteract the upward movement of salts through capillary action during drying periods. Additionally, this practice was found to promote the growth of vegetation cover to a desirable level. Mulching is one of the major soil and water conservation measures applied for conserving soil moisture and modifying soil physical and chemical environment.

Further, the utilization of organic mulch amendments has the potential to enhance the effectiveness of revegetation efforts on saline-sodic bentonite spoils. This finding is supported by the studies conducted by [747]. Additionally, [957] found promising initial outcomes in terms of plant establishment when employing a straw mulch treatment. According to [831], the utilization of organic mulches could potentially serve as a viable alternative amendment for the purpose of revegetating saline-sodic spoils in central Queensland.

In South Central Uganda, Kakaire et al. [232] investigated the effect of mulching on soil hydrophysical properties in Kibaale sub-catchment. Samples were obtained between 0 and 20 cm depths and under 0, 5, 10, and 15 cm mulch thickness levels. The application of mulch resulted in a notable enhancement of various soil hydrophysical parameters that were examined including bulk density, saturated hydraulic conductivity ($K_{sat}$), field capacity, wilting point, porosity, soil organic matter (SOM), and mean weight diameter (MWD) over a span of two seasons. The highest level of improvement was observed with a mulch thickness of 10 cm, whereas the hydraulic conductivity ($K_{sat}$), porosity, and soil organic matter (SOM) exhibited significant variations ($P < 0.05$) in response to different mulch thicknesses. The study suggested the use of a mulch layer of 10 cm.

The utilization of reclaimed topsoil at a depth of 0.3 m can also present a viable approach for the restoration of vegetation in mine-degraded spoils [830]. Nevertheless, the availability of appropriate topsoil reserves may be limited, necessitating the use of other amendments. The inclusion of gypsum in laboratory settings has also been found to have several beneficial effects including the reduction of dispersion, enhancement of hydraulic conductivity, and mitigation of crust strength during the drying process [749].

However, the effectiveness of gypsum application in promoting vegetation establishment and growth in field conditions has not been substantiated by previous studies [797, 958]. According to the study conducted by [957], the introduction of gypsum was seen to enhance the electrical conductivity of the treated substance, thereby intensifying the challenges associated with elevated saline levels in relation to plant development and viability.

Grigg et al. [831] conducted a field trial at the Goonyella Riverside open-cut coal mine in central Queensland, Australia, specifically on an out-of-pit dump. The analysis revealed that the spoil exhibits typical physical and chemical attributes including a relatively high clay content. The clay types present in the spoil exhibit reactivity in the presence of sodium, which, in conjunction with magnesium, predominantly influences the soil's exchange capacity. Additionally, the spoil exhibits high salinity levels, primarily attributed to an abundance of chloride. Furthermore, the spoil has low organic matter content and is deficient in major plant nutrients.

The presence of mine spoils and tailings creates a highly inhospitable environment for the development and survival of plants. Without any modification, the process of natural plant colonization is characterized by limited occurrence and the prevalence of species that are capable of tolerating high levels of salt [766].

The addition of several soil amendments including sawdust, wood residues, sewage sludge, and animal manures has been found to enhance soil productivity. These amendments encourage microbial activity, which in turn contributes to the provision of essential nutrients (N and P) and organic carbon to the soil [453]. Such strategy of ameliorating the contaminated mine spoil with soil amendments in order to create some conducive soil environments for establishment of vegetation can be referred to as **assisted reclamation or assisted revegetation**. This reduces the length of time it may take for the vegetation to establish for revegetation success.

### 10.2.3 Adaptive approach

This approach refers to a method or strategy that is flexible and responsive to changing circumstances or conditions. According to Williams and Bellitto [139], the adaptive strategy entails the recognition, definition, and establishment of plant species that exhibit ecotypic differentiation, indicating their ability to adapt to and tolerate the specific conditions of a given site. The utilization of an in vitro plant tolerance testing approach enables the efficient and economical screening of a substantial quantity of plants to assess their tolerance toward the particular environmental conditions present at the designated location. This approach entails the application of tissue culture methodologies and the modification of growth media by chemical adjustments in order to replicate the precise environmental parameters of the site. The germination and initial root growth of a plant serve as reliable indicators of its reaction to the prevailing environmental circumstances at a given site. Previously, it was not recognized that certain varieties of these specific species exhibited tolerance toward the regular site conditions.

Furthermore, the reclamation process at historical mine sites is frequently hindered by various challenges including but not limited to acidic soil conditions, elevated levels of metal contaminants, and the inherent susceptibility of the soil to erosion. Moreover, the restoration efforts are further complicated by the presence of steep slopes and the adverse effects of exposure due to the hilly terrain and high altitude. The coexistence of unfavorable soil physical and chemical variables alongside challenging exposure settings can render natural recovery infeasible and human rehabilitation arduous within disturbed regions. It is not uncommon to find regions that have experienced disturbance and have afterward remained devoid of vegetation for a period exceeding 100 years. The previous endeavors to restore vegetation in these regions have frequently yielded outcomes that were deemed unsatisfactory by the relevant stakeholders [1024].

Another practical illustration of the application of an adaptive strategy to land revegetation may be found in the reclamation endeavors undertaken at Haller Park, located within the Bamburi Cement Mines near Mombasa, Kenya. The process of land reclamation was initiated in 1971 with the initial implementation of planting 26 distinct tree species within open quarries. After a duration of 6 months, a mere three species managed to persist. The species observed in the study were *Casuarina equisetifolia* and *Conocarpus lancifolius.* The two plant species under discussion are the banana tree and the coconut palm. The taxonomic designation *Casuarina* sp. refers to an unspecified species within the *Casuarina* genus.

The detected pioneer species exhibited superior characteristics, as it demonstrated the ability to withstand saline water despite its adaptation to arid environments. Additionally, this species possesses the capacity to fix atmospheric nitrogen inside its root system. Furthermore, it is an evergreen tree that consistently sheds and regenerates foliage. Last, it displays rapid growth, achieving a height of 2 m within a span of 6 months. Equisetifolia is a species of evergreen tree in the family Casuarina-ceae. It is also known as the *Casuarina* tree or the "Whistling Pine." Equisetifolia, originally native to Australia, has become a prevalent tree species along the East African coast.

The leaves of *Casuarina* trees exhibit a notable abundance of tannins. This hinders the breakdown process by microorganisms. To address the issue at hand, the introduction of the millipede species *Epibolus pulchripes* was implemented. This particular species had the capability to effectively decompose the needle leaves of the *Casuarina* plant, thereby facilitating the production of the needed humus within the system. For over two decades, humus has been generated partially by this method. Due to the process of revegetation, the initial area of 2 km$^2$ that was undergoing rehabilitation experienced colonization by insects and several other living forms. The systematic implementation of indigenous coastal vegetation commenced in 1989.

By the turn of the millennium, the park had witnessed the introduction of over 300 indigenous plant species, the establishment of 30 mammalian species, and the habitation of 180 avian species. Certain creatures were introduced into the environment under the designation of "orphans." Some individuals sought refuge, while others were intentionally brought in [988].

## 10.2.4 Using the combined approach

This strategy entails the integration of two or three previously examined methodologies: either the amalgamation of agriculture with an ameliorative approach; the combination of an agricultural approach with an adaptive approach; the integration of an ameliorative approach with an adaptive approach; or the incorporation of an agricultural approach with both the ameliorative and adaptive approaches. This strategy proves to be advantageous in situations where the implementation of a single methodology or approach fails to yield the desired outcome or is not feasible.

According to the study conducted by Tordoff et al. [483] the proposed methodology could entail the utilization of a protective layer imbued with cultivars that exhibit tolerance toward metals in reclaiming mine contaminated sites. This strategy may offer some benefits that surpass those of individual conventional methods. The reason for this choice may be that the cost of using covering material in large quantities is prohibitively expensive, and the direct implementation of tolerating material also carries a significant risk.

## 10.3 Conclusions

In conclusion, this chapter provides a comprehensive understanding of the key terms related to mine land restoration. It clarifies the definitions of soil degradation, restoration, remediation, reclamation, rehabilitation, and revegetation, although these terms are used interchangeably throughout the book. The chapter also includes diagrams and examples to illustrate different approaches to revegetation of degraded mine lands. It further outlines specific methodologies for reclaiming mine lands, including agricultural, adaptive, and ameliorative approaches, as well as combined approaches and various chemical, physical, and biological processes. It is recommended to employ a combined revegetation strategy when a singular approach is not feasible or fails to achieve desired outcomes.

Reginald Tang Guuroh and Albert Kobina Mensah

# Chapter 11
# Case studies in mine land revegetation and remediation employing various approaches

**Abstract:** This chapter examines the status and progress of restoration efforts in mined sites in Ghana, with a focus on large-scale licensed formal mine companies. In addition the chapter also summarizes restoration experiments and trials from unlicensed artisanal small scale mining. We also provide an overview of the regulatory framework governing the mining sector sustainability in Ghana and emphasize the importance of restoring degraded lands to ensure productive use. We highlight the reclamation efforts and strategies implemented by these companies including the use of indigenous and exotic tree species, topsoil amendment, erosion control measures, and planting of various species. Specific examples of restoration activities are provided such as the reclamation of gold mining sites in Ahafo by Newmont Ghana, the restoration work at the Iduapriem mine site by AngloGold Ashanti Ltd, and the rehabilitation activities at the Damang and Tarkwa mines by Gold Fields Ghana Ltd. While acknowledging the efforts of these companies, the chapter also emphasizes the need for more information and documentation on restoration and reclamation efforts in Ghana. It calls for knowledge transfer and the development of national restoration programs to contribute to sustainable development in the country.

## 11.0 Introduction

Globally, the mining industry is expanding tremendously due to the increasing global demand for precious metals including gold, diamond, manganese and bauxite. In many parts of West Africa, including Ghana, the rising incidences of unlicensed small-scale surface mining (galamsey) and the adoption of unsustainable mineral exploration practices as well as inappropriate mine closure mechanisms and poor regulation or weak enforcement of mining regulations have resulted in increased levels of potentially toxic element (PTE) contamination and associated risks to soils and the environment [803, 949].

**Reginald Tang Guuroh,** Forestry Research Institute of Ghana-Soil Research Institute, Council for Scientific and Industrial Research, Fumesua-Kumasi, Ghana
**Albert Kobina Mensah,** Soil Research Institute, Council for Scientific and Industrial Research, Academy Post Office, PMB, Kwadaso-Kumasi, Ghana

https://doi.org/10.1515/9783111662046-011

Restoration of mined areas is an important step after mining to ensure that degraded lands are returned to conditions suitable for productive use. The main aim of restoring mined lands is to re-establish vegetation cover, restore biodiversity, stabilize/ reduce contamination of the soil and water conditions, and thus restore the provision of ecosystem goods and services [734]. The mining regulations of Ghana (e.g., the 1992 Constitution of Ghana, the Minerals and Mining Act, 2006 (Act 703) as amended by the Minerals and Mining (Amendment) Act, 2015 (Act 900) and the Minerals and Mining (Amendment Act, 2019 (Act 995) as well as the Minerals Commission Act, 1993 (Act 450)) stipulate procedures for ensuring that mining activities are done in ways that minimize environmental impacts and as such mining activities are required to follow the guidelines. The Environmental Protection Agency (EPA) has outlined strategies for rehabilitating and consequently ensuring biodiversity conservation of mined out sites by mining companies in Ghana [799]. Large-scale licensed mining companies in Ghana are mandated by law to commit to sustainable mining practices and such companies have implemented various initiatives including incorporating biodiversity management and reclamation of mine spoils within their operational procedures as recommended by the International Council on Mining and Minerals (ICMM) and the EPA. In the case of galamsey, the actors do not usually adhere to any regulations making reclamation efforts more challenging.

It is estimated that the land area under surface mining in Ghana may have more than doubled within the last decade due to proliferation in both galamsey and main mining operations (Awotwi et al., [737]; Forkuor et al., [809]). However, there is inadequate information and documentation about restoration and reclamation of mine sites in the country. The reclamation and restoration successes as well as basic information on these reclaimed sites remain generally unavailable; a situation which hampers knowledge transfer. Furthermore, a paper by [807] on restoration of post-mining landscape in Africa demonstrated that despite the long history of mining in Africa, no systematic review has summarized advances in restoration research and practices after mining disturbance. This hinders the successful development of national restoration programs that inform policy and contribute positively to sustainable development in the country and within the region. Compliance and adherence to mine closure regulations have been somewhat positive among mining companies in the country and restoration or reclamation efforts are thought to be gradually rising.

This chapter examines the status and progress of restoration efforts in mined sites in Ghana. It documents reclamation activities, including experimental trials, carried out by both large-scale licensed formal mine companies and those done by researchers and other organizations on abandoned galamsey sites. We highlight the reclamation efforts and strategies implemented by these companies including the use of indigenous and exotic tree species, topsoil amendment, erosion control measures, and planting of various species. There is the need for capacity building, knowledge transfer and the development of national restoration programs to contribute to sustainable development in the country.

# 11.1 Restoration efforts conducted by large-scale licensed formal mining firms in Ghana

## 11.1.1 Newmont Ghana Limited, Ghana, West Africa

Newmont, a prominent gold producer in Ghana, maintains its position as the largest in the Ghana through its operations in two distinct areas: Newmont Ahafo and Newmont Akyem. The Newmont Ahafo gold mining complex is situated inside the Ahafo region, encompassing a total land area of 2,000 ha. The Akyem gold mine, which commenced operations in 2013, is an open-pit mining facility that is now projected to conclude its operations by 2024. Newmont has undertaken reclamation efforts on its gold mining sites in Ahafo, encompassing a cumulative area of 70 ha, as of the year 2015, in alignment with its sustainability plan. According to the company's assessment, the state of reclamation enables the utilization of the area for agricultural purposes, along with other potential applications. Furthermore, during the time frame spanning from 2015 to 2016, the company documented that a total of 56 ha had undergone closure and reclamation processes.

## 11.1.2 AngloGold Ashanti Ltd

AngloGold Ashanti is a prominent mining corporation in Ghana, boasting two active sites known as Obuasi and Iduapriem. The Iduapriem mine is situated around 10 km to the South West of Tarkwa inside the Western region of Ghana. Festin et al. [158] and Tetteh et al. [477] documented that AngloGold Ashanti implemented a comprehensive approach to restoration of the Iduapriem mine site in close proximity to Tarkwa, Ghana, encompassing physical, chemical, and biological processes. The mining firm possessed a total area of 110 ha as a concession and commenced its mining operations in the year 1991. Restoration efforts have been undertaken on historical sites with respective ages of 2, 5, 9, and 11 years.

The organization employed a restoration technique that encompassed a series of steps: (1) the implementation of earthwork and slope-battering techniques is recommended to achieve a visually harmonious integration of the landscape; (2) the application of oxide material is suggested to consolidate the soil and improve its stability; (3) it is advised to amend the topsoil and utilize manufactured technosols, such as poultry droppings and cow manure, along with fertilizers, in conjunction with the oxide material; (4) to mitigate runoff and erosion, the establishment of crest drains is proposed; (5) additionally, the cultivation of cover crops such as *Puereria phaseoloides* (Roxb.) Benth. and *Centrosema pubescens* Benth. is recommended to further enhance erosion control; (6) the subsequent step involves the planting of seedlings of *Acacia mangium* Willd., *Gliricidia sepium* (Jacq.) Kunth ex Walp., *Leucaena leucocephala* (Lam.) de Wit, and *Senna siamea* (Lam.) Irwin and Barneby, followed by regular weeding, pruning, and application of fertilizers. Several other plant species have been effectively utilized in various applications.

These include *Colocasia esculenta*, often known as cocoyam, *Terminalia superba*, also referred to as Ofram, *Pityrogramma calomelanos*, commonly known as Fern, *Xylopia aethiopica*, known as Hwentia, *Musa sapientum*, commonly referred to as banana, *Chromolaena odorata*, also known as Acheampong weed, and *Theobroma cacao*, commonly known as cocoa [57].

According to the 2022 sustainability report of AngloGold Ashanti, the incorporation of postclosure economic concerns is a fundamental aspect of their mine closure strategy. In Obuasi, there are two regions that have undergone reclamation efforts. The first area was originally occupied by a mining shaft and processing plant, while the second area was the site of reprocessed tailings dams.

## 11.1.3 Gold fields Ghana Ltd

According to the company's official website, rehabilitation efforts have been conducted at the East Tailings and Storage Facility, which is no longer in operation, located at the Damang mine. According to the study, a total area of more than 100 ha within the decommissioned East Tailings Storage Facility at Damang mine underwent rehabilitation in the year 2019. This rehabilitation process involved the cultivation of many types of trees including oil palm, coconut, mango, and cashew. Furthermore, there have been reports of rehabilitation efforts at the Tarkwa mine, specifically targeting inactive waste rock piles. These efforts involve the implementation of nitrogen-fixing trees, indigenous flora, and economically valuable trees such as rubber, teak, and oil palm.

In the year 2020, Abosso Goldfields Ltd (Damang mine) organized an Open House event aimed at engaging community people and key stakeholders. During this event, the company focused on presenting and discussing its "concurrent rehabilitation strategy" as well as various methodologies employed for revegetation and maintenance procedures. Concurrent rehabilitation refers to the practice when a firm undertakes the rehabilitation of affected areas concurrently with continuing mining operations rather than deferring such efforts until mining activities have ceased entirely. According to the report, the mine has successfully restored 506 ha of formerly damaged land, transforming it into thriving vegetation and fertile agriculture. Additionally, the mine intends to allocate a budget of up to US$20 million for the reclamation of an additional 1,472 ha of degraded land.

## 11.1.4 Chirano Gold Mines Ltd

The study encountered difficulties in locating sufficient reclamation information pertaining to Chirano Gold Mines Ltd from both published scientific literature and other web sources. In cases where certain information was obtained, it was notably limited in scope and lacked specific specifics regarding the precise acts that were carried out. The Technical Report for the year 2022 released by the company highlighted the closure and rehabilitation of a tailing storage facility spanning 45 ha within the mine. The rehabilitation process, initiated in 2018, involved the introduction of both indigenous and exotic species. However, the report did not furnish additional information regarding the specifics of the rehabilitation efforts. According to the research, the vegetation within the repaired site has exhibited robust growth, signifying favorable conditions. Furthermore, the area has emerged as an exemplary example for the restoration of tailing storage facilities.

## 11.2 Restoration in unlicensed artisanal small-scale mining sector (galamsay)

In unlicensed artisanal mining, operators usually do not follow laid down mining procedures and regulations aimed at ensuring sustainable mining, and as a result, the impact of mining is much higher than in licensed mine areas. In the case of unlicensed mining, two options become available for restoration, that is, passive and active approaches.

In the passive approach, the causes of degradation are reduced or completed stopped but the area is allowed to recover through natural processes without any active human intervention. This is arguably the most commonly applied strategy in unlicensed mine sites across Ghana (Fig. 11.1).

In the active approach, human manipulation or intervention is applied to accelerate the reclamation of degraded mine lands. Some examples of such active restoration in unlicensed mine areas are presented below.

### 11.2.1 Student and Youth Travel Organization

The Student and Youth Travel Organization aims to provide a complete range of cross-cultural programs, support services to youth, volunteers, students, educators, and institutions, gap year organizations, and provide assistance to the needy in education, health, and many other sectors. As part of the organization's activities, it undertook a project on reclamation of degraded mine areas in some parts of Ghana. The objective of the project was to reclaim at least 5,000 ha of degraded mine land in 20 communities located within the Fanteakwa District and East Akyem Municipality of the Eastern region by the year 2020. Planting was done using indigenous trees but also including vegetables and other food crops together with the involvement of farmers.

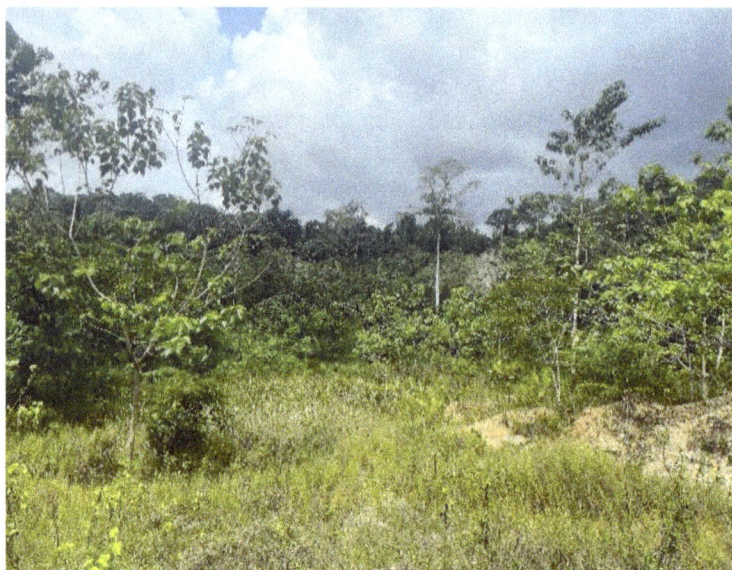

**Fig. 11.1:** Abandoned mine areas left to recover through natural rejuvenation.

The initiative uses various approaches including participatory stakeholder action (local communities, local leadership, relevant government agencies, educational institutions as well as tourists) and involvement of tourists/youth exchange volunteer programs to promote intercultural exchange and learning in the affected communities. The organization also reported that they had previously successfully reclaimed 40.5 ha in Kazigu in Upper East region.

## 11.2.2 The Ghana land restoration and small-scale mining project

The Government of Ghana has recently initiated the Ghana Landscape Restoration and Small-Scale Mining Project aimed at reclaiming or restoring various degraded mine lands. The project is being implemented with the involvement of the Environmental Protection Agency (EPA) and Ministry of Lands and Natural Resources (MLNR). The project is funded by the International Development Association.

The objective of the project is to strengthen integrated natural resource management and increase benefits to communities in targeted savannah and cocoa forest landscapes. The project is making significant progress including the provision of inputs for sustainable land and water management practices, forest management planning and restoration, formation of the Community Resource Management Area (CREMA), wildlife monitoring, trainings on online license application, and identification of pilot sites for rehabilitation of mined out areas. Procurement and environmental and social due diligence for several planned activities are also progressing well. In respect of reclamation of abandoned mine areas, the project targets to achieve 2,000 ha by 2027.

### 11.2.3 Mine land reclamation using oil palm by Solidaridad

Solidaridad Ghana, a nongovernmental organization, has reported on its website that it is helping to reclaim degraded mine sites in the Central region of Ghana with oil palm through its National Initiative for Sustainable Climate-Smart Oil Palm Smallholders (NIS-COPS) Program. NISCOPS is a 4-year program funded by the Dutch Ministry of Foreign Affairs designed to implement climate-smart initiatives in the oil palm sector in Ghana, Nigeria, Indonesia, and Malaysia (Fig. 11.2). The NGO reported oil palm as a resilient crop which happens to be one of the most suitable crops for reclamation of degraded mine lands.

**Fig. 11.2:** Oil palm planted on mine site. Picture credit: Solidaridad Ghana.

### 11.2.4 Mine land reclamation by tropenbos

Tropenbos Ghana, a nongovernmental organization, has undertaken an experimental trial reclamation activity at Manso Yawkrom and Asarekrom both in the Ashante region and one at Agyareago in the Asante Akim municipality also in the Ahanti region.[1] The activities are under the "Securing Food and Ecosystem Services in

---

1 https://gna.org.gh/2022/03/tropenbos-ghana-engages-community-leaders-on-reclamation-of-mined-sites/

Mining Plagued Regions in Ghana Project." Under this initiative, the project has established 3.88-ha experimental rehabilitation plot at Manso Yawkrom, 2.4-ha plot at Asarekrom, and another 2.9-ha plot at Agyareago.

Some of the activities the project engaged in ahead of tree planting on mine sites
are supporting communities to cover mine pits, leveling the sites, filling up sites with
topsoil, and finally the actual planting of selected tree species. Some of the species
used in reclamation are *Acacia magium*, *Senna siamea*, *Khaya anthoteca* (*Mahogany*
spp.), *Terminalia superba* (Ofram), *Terminalia ivorensis* (Emere), and *Tetrapleura tet-
raptera* (Prekese).

# 11.3 Various research trials involving plant species growth and phytoremediation trials on mine lands

## 11.3.1 Mine land reclamation research by CSIR-Forestry Research Institute of Ghana

The CSIR-Forestry Research Institute of Ghana conducted pilot studies on reclamation
of degraded mine sites between 2017 and 2019 in the Bibiani-Ahwianso-Bekwai district
and the Atewa range. These studies focused on testing the survival and growth potential of selected tree species on unamended mine soils.

In the Bibiani-Ahwianso-Bekwai district where the study was conducted at Nka-
tieso, four tree species were tested: *Terminalia superba*, *Nauclea diderrichii*, *Senna siamea*, and *Gmelina arborea*. The results showed that both seedling survival and growth
rates were highest for *Gmelina arborea*. *Terminalia superba* recorded the lowest
growth in the mined site while *Terminalia superba* and *Nauclea diderrichii* (both indigenous species) were the worst performers in terms of survival in the mined site.

## 11.3.2 Other experimental trials

Another experiment by Mensah et al. [334] assessed the phytoremediation potential of
four native plans namely *Alchornea cordifolia*, *Chromolaena odorata*, *Lantana camara*, *Pitryogramma calomelanos*, and *Pueraria montana*. The results showed that
there was a high mobility of copper and zinc into various plant parts. *Chromolaena
odorata* was found to have a good phytoremediation ability for copper while all the
five species were good for absorbing zinc into both shoots and roots. The study further reported that *Chromolaena odorata* is a good species for phytoextraction of As,
Cu, Ti, and Zn in contaminated mine soils.

In general, a report by the International Growth Centre costing reclamation of galamsey operations in eleven selected Metropolitan District Assemblies (MDAs) in the former Western region of Ghana (i.e., covers today's Western and Western North regions) estimated that a total amount of GH 987,707,164.53 is required for returning lands affected by galamsey to states close to their original status. In the 11 MDAs alone, the report estimated 1,845 abandoned galamsey sites with an estimated 5,532 active operating sites that will soon become abandoned.

### Case study 1 [332]

**Title**: Compost and manure addition enhanced arsenic phytostabilization potential of ryegrass grown in a contaminated gold mine spoil

**Type of experiment:** Greenhouse pot experiments

**Study background and objectives:** Gold mine spoils are threats to environmental health and thus raise health concerns for humans and animals. A pot experiment was conducted to study arsenic (As) phytostabilization potential of ryegrass (*Lolium perenne*) using compost, iron oxide, and manure as soil amendments. The study was conducted on an abandoned mine spoil with mean total As of about 5,104 mg/kg.

**Materials and methods:** Soil amendments were each applied at the rate of 5% (w/w) to the As-contaminated soil. The amendments were applied separately or in combination in equal proportion. Total, readily available (extracted with $(NH_4)_2SO_4$) and specific-sorbed/exchangeable As (extracted with $(NH_4)H_2PO_4$) and plant As contents were determined after harvest. Bioconcentration factor (BCF) and bioaccumulation concentration (BAC) were used to determine the plant soil-to-root and -shoot transfer coefficients, respectively. Additionally, the translocation factor (TF) was used to obtain the root-to-shoot transfer efficiency.

**Results and discussion:** Manure addition increased the soil total As by 41%, readily available As by 243%, and specific-sorbed/exchangeable As by 38%. Manure addition is enhanced root uptake by 134%, whilst its combination with compost improved uptake by 101%. Consequently, application of manure alone and in combination with compost resulted in BCF above 1. Arsenic phytostabilization potential of ryegrass was influenced by soil readily available and specific-sorbed/exchangeable As, total As, $Cl^-$, $NO_3^-$, $SO_4^{2-}$, EC, and P.

**Conclusions:** Compost and manure addition improve As phytostabilization potential of ryegrass when applied to a contaminated gold mine spoil.

### Case study 2 ([624], 2023)

**Title:** Rehabilitation plan for coal pit revegetation area East Kalimantan

**Type of experiment:** Field experiment

**Study background and objectives:** The reclamation of the degraded mining region, particularly the disposal area, is imperative in order to safeguard the potential for its future utilization. The rehabilitation of open pit mining sites should prioritize the restoration of the area's production capacity and ecological stability, thereby ensuring a strong contribution to the ecosystem. Its objective was to establish a production forest capable of providing long-term environmental support. The aim of this study was to develop a reclamation plan for a mine pit by conducting physical and chemical characterization of the site, evaluating the effectiveness of the drainage system, assessing soil fertility and potential soil contamination, and identifying indigenous crop and forest plant species with rapid growth capabilities. The primary emphasis lies in the analysis of soil conditions impacted by the issue of acid mine drainage.

**Materials and methods:** The exercise encompassed an area of approximately 0.5 ha within an open-pit coal mine, undertaken through collaborative efforts between Korea and Indonesia. Soil characterisation was done to evaluate the spread of heavy metals and their mobility in soil and surface water. For this, soil samples were collected from various demarcated sampling points. The site was divided into five zones, with representative samples collected from five points within the target area. The samples were then prepared by air drying and sieving through a 2 mm sieve. They were later saved for laboratory analyses. The soil characterisation was done for the soil physico-chemical properties and their heavy metal contents. The physico-chemical parameters included the soil texture, colour, pH, total nitrogen and nitrate, total organic carbon, moisture, CEC, P and Si availability, Ca-Mg-K oxides, and heavy metal contents.

**Results and discussion:** The findings indicated that the restoration target site in the mining site had a significant incline and was characterized by acidic soil, with a pH range of 4.2–4.6. Additionally, the site displayed low levels of exchangeable cations and limited availability of phosphate. The levels of heavy metals detected did not surpass the established contamination threshold, therefore obviating the need for additional processing measures to mitigate the potential leaching of heavy metals. The regular surveillance of the drainage system holds significant importance as it directly impacts the frequency of watering and the management of leaching.

**Conclusion:** The heavy metal pollution witnessed at the site was basically a problem of sulphide oxidation. This necessitates constant monitoring of water and soil. The problem of acid mine drainage is a major issue in mine land reclamation, especially those that have to do with sulphide mineral dominance. The sulphide undergoes a reaction with oxygen and consequently releases heavy metals into the soil solution. Such reaction also reduces the soil pH and consequently raises the soil acidity of the contaminated and abandoned mining sites. Despite low metal content and low mobility, difficulties in revegetation were noticed in some areas. Continuous monitoring and GIS analysis are essential for successful reclamation plans. Understanding soil conditions can help prioritise mitigation efforts and ensure proper control over environmental parameters at each mining operation step.

**Case study 3** [625]
**Type of experiment:** Field experiment
**Title:** Postmining reclamation of manganese waste dump

**Study background and objectives:** The purpose of this study was to show how mining-affected areas might benefit impacted people in the long run through quantifiable gains in social justice, economic growth, and environmental restoration. The cleanup of a significant waste dump (Pit B waste dump), which makes up 8.1% of the 175 km$^2$ concession that has been disturbed during the concession's 100 years of operation, at Ghana Managanese Company (GMC) shows the mine's dedication to maintaining environmental sustainability, enhancing the socioeconomic ties with the neighborhood, and lowering liability upon closure.

**Materials and methods:** The Pit B waste dump was reclaimed using 406,800 m$^3$ of topsoil and plantations of palm trees intercropped with cassava, pineapple, and plantain. Laboratory research was conducted to assess the commercial viability of processing waste rock dumps for resale as building aggregate, road construction material, railway ballast, or forestation. The local community was involved in the reclamation planning and building process, with the Municipal Assembly and all interested parties participating to prevent disputes and promote project acceptability. Stakeholders included the Environmental Protection Agency (EPA), Minerals Commission, Tarkwa Nsuaem Municipal Assembly, chiefs, elders, and citizens. Soil testing was conducted before planting, with GPS positions determined to ensure uniform planting distances. Agronomic techniques like weed removal and dead branch cutting were used to ensure healthy growth. Two months after planting, 250 g of 15:15:15 NPK fertilizer was applied to each plant, and 500 sacks containing cow dung and fowl droppings were dispersed as natural manure.

**Results and discussion:** The mined soils faced challenges in vegetation maintenance, but reclamation of waste dumps like GMC offered economic and livelihood support. Soil enrichment methods stabilized vegetation, improved visuals, and bioaccumulated metals in plants. Reclaimed sites generated jobs and income, and community involvement was crucial for successful reclamation projects.

**Conclusions:** The study discussed the reclaimed 20-ha waste dump used for 32 years for manganese disposal. The waste was converted into arable land using sound environmental and engineering principles, ensuring sustainability and supporting catchment communities' livelihoods. The reclaimed land showed lower heavy metal contents in cassava and plantain compared to undisturbed farming lands. The study emphasized the importance of community involvement in the reclamation process for successful project execution.

### Case study 4 [359]
**Type of experiment:** Pot experiment
**Title:** Effect of topsoil stockpiling on soil properties and organic amendments on tree growth during gold mine reclamation in Ghana
**Study background and objectives:** Mine land reclamation involves restoring mined land to premining conditions or altering it for another productive use. However, most disturbed mine sites in Ghana have not been reclaimed due to poor management

practices such as poor slope stabilization and lack of topsoil application. Topsoil, a valuable resource, is crucial for successful reclamation as it contains building blocks for plant growth, essential microbes, and labile plant nutrients. Newmont Ghana Gold Limited (NGGL) salvages all topsoil and stockpiles it for future reclamation. However, topsoil stockpiling can decrease soil organic matter (SOM) content, leading to adverse effects on the physical, chemical, and biological properties of the soil resource. Organic amendment (OA) is a valuable source of SOM that can improve soil physical, chemical, and biological properties following disturbances like mining. Using OA with stockpiled topsoil (TSP) is a vital management practice to improve the supply of nutrients and soil physical and chemical characteristics. This study investigates the effect of topsoil stockpiling on soil properties and OA on tree growth at NGGL. The hypothesis is that topsoil stockpiling will cause significant adverse impacts on soil properties, while amendment of TSP with organic materials like composted sewage sludge (CSS) and poultry layer manure (PLM) will significantly promote the survival and growth of planted trees at the site.

**Materials and methods:** A field trial was conducted on a waste rock dump in the Amoma Pit, where salvaged topsoil and subsoil were covered with biological geotextiles to control erosion and seeded with deep-rooted perennial plants in 2009. Soil samples were collected and analyzed for pH, nutrients, SOM, electrical conductivity, and texture. Three experimental plots were divided into three treatments: no amendment, PLM, and CSS. Six-month-old potted seedlings were planted. The study aimed to assess the impact of organic amendments on tree growth and survival. Tree survival after planting was recorded and tree growth parameters (diameter and height) were measured using appropriate equipment. Statistical analyses were performed using IBM SPSS Statistics Version 25.0. The study tested two hypotheses: the effect of topsoil stockpiling on soil properties and the effect of OA treatment on tree growth. The results were compared with baseline soil data before mining operations.

**Results:** Statistical analysis revealed that topsoil stockpiling did not have any significant adverse impact on the measured soil properties compared with the reference plot. One-way ANOVA combined with LSD and Duncan post hoc tests ($\alpha$ = 0.05) also indicated no significant influence of organic amendments on tree growth. Competition from herbaceous plants due to ineffective weed control was observed to be the main driving factor hindering survival and growth of planted trees.

**Conclusions:** This study reveals that topsoil salvaging and stockpiling did not significantly impact soil properties during NGGL's mine operation at the Amoma Pit. This is due to the proper TSM strategy implemented by NGGL, which generated optimum soil conditions. The company's revegetation approach, seeding cover crops in the first year and postponing tree species planting until the second year, coupled with inadequate weed control, influenced tree survival and growth. Therefore, planting forest tree species concurrently with erosion and sediment control ground cover species, coupled with adequate weed control, is perceived to influence tree survival and growth.

**Case study 5** (Martínez-Martínez et al. cited in FAO, 2018)[2]
**Type of experiment:** Field
**Title:** Aided phytostabilization: an effective remediation technique for tailings in SE Spain
**Study background and objectives:** Mining was established in Sierra Minera de Cartagena–La Unión (Murcia, Spain) for a period over 2,500 years. The extraction of mineral sulfides (mostly ZnS and PbS) resulted in the production of significant quantities of tailings. The tailing ponds were left abandoned after the operation ceased in 1991 and were a significant cause for concern because to the high concentration of poisonous metal(loid)s and posed risks to the surrounding communities. Tailings from Zn/Pb mining exhibited reduced fertility, declined organic matter, and high acidity, which made it difficult to establish native plants. This study assessed the efficacy of assisted phytostabilization on a tailings pond 30 months later, focusing on the transfer of metal(loid) elements to plant species.
**Materials and methods:** The investigation was conducted at the Santa Antonieta tailings pond, situated in the Cartagena–La Unión mining zone. The pond covered an area of 1.4 ha. The utilization of marble waste as a carbonates source served the purpose of neutralizing acidity, immobilizing metals, and enhancing soil structure. Pig slurry, along with the solid waste (manure), were utilized as a means of introducing organic material and essential nutrients into the soil to promote its development and facilitate the growth of vegetation.

The following species were planted in 2012: *Atriplex halimus* L., *Cistus albidus* L., *Helichrysum stoechas* (L.) *Moench., Hyparrhenia hirta* (L.) *Stapf., Lavandula dentata* L., *Lygeum spartum* (L.) *Kunth., Rosmarinus officinalis* L., *Phagnalon saxatile* (L.) *Cass, Piptatherum miliaceum, Cynodon dactylon, Limonium caesium, Sonchus tenerrimus,* and *Atriplex halimus.*

**Results:** The use of marble waste, pig slurry, and manure effectively facilitated the restoration of an acidic tailings pond through the process of assisted phytostabilization. The approach resulted in an increase in soil pH, cation exchange capacity (CEC), total organic carbon (TOC), and nutrient content. It also enhanced soil structure and significantly reduced the mobility of metals, particularly cadmium (Cd), lead (Pb), and zinc (Zn), by 90–99%.
**Conclusion:** *Lygeum spartum* and *Piptatherum miliaceum* showed efficacy in phytostabilizing Pb, Zn, and As by efficiently accumulating significant metal concentrations in their roots while minimizing transfer to the aboveground parts.

---

2 This case study was taken from Rodríguez-Eugenio, N., McLaughlin, M. and Pennock, D. 2018. Soil Pollution: a hidden reality. Rome, FAO. 142 pp.

## 11.4 Lessons from successful long-term reclamation projects

Long-term reclamation projects that have achieved sustainable outcomes offer valuable lessons that can be applied to improve future practices. These lessons highlight the importance of adopting a holistic and adaptive approach to land reclamation, taking into account the specific conditions of each site and the need for ongoing monitoring and management.

### 11.4.1 Importance of initial soil preparation

One of the key lessons from successful long-term reclamation projects is the importance of thorough initial soil preparation. This includes the reconstruction of the soil profile, the application of soil amendments, and the establishment of a suitable drainage system. Proper soil preparation creates a foundation for successful vegetation establishment and long-term soil health. For example, in the oil sands reclamation projects in Alberta, Canada, careful reconstruction of the soil profile, including the placement of suitable topsoil and subsoil layers, has been critical to the successful restoration of these lands.

### 11.4.2 Selection of appropriate plant species

The selection of appropriate plant species is another critical factor in the success of long-term reclamation efforts. Successful projects have often used a combination of fast-growing pioneer species to stabilize the soil and slow-growing native species that contribute to the development of a diverse and resilient plant community. In many cases, the use of native species has been shown to be particularly effective in promoting long-term ecosystem recovery, as these plants are better adapted to local conditions and can support a wider range of wildlife.

### 11.4.3 Ongoing monitoring and adaptive management

Long-term success in land reclamation is often achieved through ongoing monitoring and adaptive management. Monitoring allows for the early detection of issues such as soil erosion, PTE mobilization, or vegetation failure, enabling timely interventions. Adaptive management, which involves adjusting management practices based on monitoring results, has been a key component of successful projects. For example, in some reclaimed mining areas in Germany, adaptive management practices, such as adjusting soil pH through lime applications or reseeding with different plant species, have been used to address emerging challenges and ensure the long-term sustainability of the site.

### 11.4.4 Community involvement and long-term commitment

Successful long-term reclamation projects often involve the active participation of local communities and a long-term commitment from stakeholders. Community involvement ensures that reclamation efforts are aligned with local needs and values, increasing the likelihood of project success. Moreover, long-term commitment from mining companies, government agencies, and other stakeholders is essential for providing the resources and support needed to maintain and manage reclaimed lands over time.

## 11.5 Applying lessons to improve future reclamation practices

The lessons learned from successful long-term reclamation projects can be applied to improve future practices, particularly in areas where reclamation efforts have faced challenges. These lessons emphasize the need for a comprehensive and site-specific approach to reclamation, one that considers the full range of environmental, social, and economic factors [115].

### 11.5.1 Emphasizing soil health and ecosystem function

Future reclamation practices should place a greater emphasis on restoring soil health and ecosystem function rather than merely achieving surface stabilization. This approach includes adopting practices that enhance soil fertility, promote biodiversity, and support the long-term sustainability of reclaimed land [261]. Techniques such as agroforestry, the use of cover crops, and the integration of ecological principles into reclamation planning can contribute to more resilient and sustainable outcomes [626].

### 11.5.2 Enhancing remediation of PTEs

Given the persistence and risks associated with PTEs in reclaimed soils, future reclamation efforts should incorporate more effective remediation strategies. This approach may include the use of advanced soil amendments, such as biochar or engineered nanoparticles, that can immobilize PTEs more effectively as well as the continued development of phytoremediation techniques that target specific contaminants [68]. Additionally, ongoing research into the behavior of PTEs in reclaimed soils is essential for refining remediation practices and reducing long-term environmental risks [627].

### 11.5.3 Strengthening monitoring and adaptive management

To ensure the long-term success of reclamation efforts, future projects should adopt robust monitoring programs and adaptive management frameworks. These programs should track key indicators of soil health, vegetation success, and PTE mobility, with the flexibility to adjust management practices as needed [628]. The integration of remote sensing technologies and data-driven decision-making tools can enhance the effectiveness of monitoring and management efforts [629].

### 11.5.4 Fostering collaborative and inclusive reclamation efforts

Finally, future reclamation practices should foster collaboration among all stakeholders including local communities, government agencies, industry, and researchers. Inclusive reclamation efforts that engage communities in decision-making and management processes are more likely to achieve sustainable outcomes [503]. Additionally, building long-term partnerships between industry and environmental organizations can provide the resources and expertise needed to address complex reclamation challenges [630].

In summary, long-term studies of land reclamation efforts have revealed important insights into the sustainability of reclaimed lands, highlighting both successes and ongoing challenges. By applying the lessons learned from successful projects, future reclamation efforts can be improved, leading to more resilient and sustainable ecosystems in areas affected by mining and other industrial activities. A commitment to soil health, effective PTE remediation, adaptive management, and community involvement will be keys to achieving these outcomes [631, 762].

## 11.6 Conclusion

This chapter examined the status and progress of restoration efforts in mined sites in Ghana, focusing on large-scale licensed formal mining companies and small-scale unlicensed artisanal miners. It also provided an overview of the regulatory framework governing the mining sector sustainability in Ghana and emphasized the importance of restoring degraded lands to ensure productive use. Additionally, we highlighted the reclamation efforts and strategies implemented by these companies including the use of indigenous and exotic tree species, topsoil amendment, erosion control measures, and planting of various crops. The chapter also drew the following summaries and recommendations to ensure a sustainable mineral mining sector in Ghana:

– documented some research trials involving plant species growth and phytoremediation on mine lands, showing the potential of some native plants to accumulate and stabilize heavy metals in contaminated soils;
– concluded that the restoration of mined lands in Ghana faces many challenges such as policy gaps, logistical constraints, financial limitations, poor education, and stakeholder neglect;
– recommended ensuring collaboration among the stakeholders, including traditional leaders, in decision-making, establishing local offices, consolidating regulatory bodies, implementing capacity building initiatives, and providing geological data to support artisanal and small-scale mining; and
– called for more research to explore plant species suitable for reclamation, to analyze soil quality and heavy metal residues at reclaimed sites, and to evaluate the social, economic, and environmental benefits of revegetation.

Albert Kobina Mensah

# Chapter 12
# Plant species used in revegetation and their corresponding impacts

**Abstract:** A strategic plan for plant succession is necessary for rehabilitating areas impacted by mining activities. The plan should include the integration of rapidly growing grass species, leguminous plants, and various grasses and shrubs to establish ground cover, stabilize the surface, and maintain continuous vegetative coverage. The selection of suitable plant species for remediation and revegetation should prioritize their capacity to accumulate and transport pollutants, detoxify or immobilize pollutants, generate substantial biomass, withstand toxic pollutant levels, and exhibit rapid growth and a short life cycle. A variety of plant life including grasses, legumes, shrubs, and trees is suggested for long-term remediation efforts. Trees are often chosen for large-scale land decontamination and revegetation because they have the capacity to produce biomass, reach deep into the soil, aid in nutrient cycling, and access nutrients and contaminants in lower soil layers. Caution is advised when utilizing specific tree species, like *Eucalyptus*, due to potential adverse impacts on plant species and biodiversity. Genetically modified plants for phytoremediation and monoculture are subjects of contention.

## 12.1 Introduction

The process of rehabilitating areas affected by mining activities, known as mine spoils, necessitates the development of a strategic plan for plant succession. This strategy aims to establish a diverse plant community that effectively covers the surface area and enhances soil fertility [341]. Ideally, the recommended sequence should involve the incorporation of fast-growing grass species to promptly establish ground cover and stabilize the surface. Additionally, leguminous plants should be included to facilitate nitrogen fixation, while other grasses and shrubs should be incorporated to ensure a sustained and enduring vegetative cover [341]. Furthermore, it is worth considering both nodulating and non-nodulating legume species throughout the revegetation process.

**Albert Kobina Mensah,** Soil Research Institute, Council for Scientific and Industrial Research, Kumasi, Ghana, e-mail: albert.mensah@rub.de, albertkobinamensah@gmail.com, ORCID: https://orcid.org/0000-0001-5952-3357

https://doi.org/10.1515/9783111662046-012

## 12.2 Identifying appropriate plant species for remediation and revegetation

Certain characteristics are necessary for plant species that are used in the remediation of pollutants in the environment. These characteristics include the capacity to accumulate and translocate pollutants, the ability to detoxify or immobilize pollutants, the capability to produce a high biomass, and tolerance to potentially toxic concentrations of pollutants as well as rapid growth and a short life cycle [33, 306, 458]. Cappa and Pilon-Smits [100] assert that herbaceous plants are highly favored as hyperaccumulating species due to their quick development, substantial above-ground biomass, and broad root systems. In the context of sustained remediation programs, Guerra Sierra et al. [192], Pandey and Souza-Alonso [380], and [632]. advocate for the utilization of a diverse assemblage of vegetation comprising grasses, legumes, shrubs, and trees. These authors highlight the advantageous characteristics of grasses such as their rapid growth, while emphasizing the importance of incorporating specialized species that fulfill distinct functions at different stages of the remediation process.

According to Pandey et al. [381], plant species that are nonpalatable, perennial, and possess economic value are considered more favorable for sustainable remediation in comparison to invasive plant species. In order to address various types of contaminants, a comprehensive process involving prospecting, research, and development is necessary to choose appropriate plant species for remediation purposes [311, 457]. Trees, grasses, and shrubs, particularly those of the leguminous kind, are commonly suggested for the purpose of revegetation and restoration of mining sites that have undergone degradation and contamination.

In addition to the identification of candidate plant species in their native environments, the field of plant breeding, which encompasses biotechnological methods, shows potential for the advancement of more effective phytoremediators. This can be achieved through the transfer of genes responsible for pollutant hyperaccumulation from low-biomass wild species to higher-biomass cultivated species or through the utilization of advanced genetic engineering techniques [280]. A significant portion of this is accomplished through crossover, a recognized breeding method that involves gene transfer through natural means rather than technical enhancement.

Nevertheless, the utilization of genetically modified commodities continues to be a subject of contention on a global scale, as several nations maintain divergent perspectives toward their implementation and utilization inside their respective territories [see 633, 2012; 634, 347]. The apparent efficacy of genetically altered products for phytoremediation in laboratory settings raises important considerations regarding their acceptance in field applications.

The topic of discussion also revolves around the use of monoculture and mixed plantation practices or employing the use of either native or introduced species.

## 12.3 The utilization of trees

The utilization of trees is occasionally contemplated as a means of rehabilitating depleted mining soils. In addition, it has been observed that trees possess the ability to enhance soil fertility [735]. In many instances, spoil bank regions are afforested with trees to serve the dual purpose of generating revenue and conceal the visual impact of these unattractive banks [635].

Trees are commonly favored for extensive land decontamination and revegetation endeavors for several reasons. Firstly, they exhibit high efficiency in generating biomass, thereby contributing a greater amount of organic matter to the soil. Secondly, their ability to penetrate deeper into the soil profile surpasses that of herbaceous species. Thirdly, trees provide plant roots and exudates that actively contribute to the cycling of nutrients. Additionally, they have the potential to mitigate soil compaction. Lastly, trees possess the capability to access nutrients and contaminants present in deeper horizons.

According to Young [1054], it is recommended that tree species utilized in revegetation and soil decontamination programs possess specific characteristics. These characteristics include the ability to thrive in nutrient-deficient soils, a high rate of nitrogen fixation, leaf litter with a well-balanced and nutrient-rich composition, low levels of lignin and polyphenols per unit volume, and the ability to either rapidly decompose litter for nutrient release or slowly decay litter for soil coverage purposes.

Additionally, it is important that these tree species do not contain hazardous chemicals in their root remnants or litter, exhibit a strong tap root system, demonstrate a high rate of biomass generation from leaves, and possess a dense network of fine roots capable of forming mycorrhizal associations. For example, the use of monoculture *Eucalyptus* species in revegetation especially in dry environments should be exercised with caution as they exhibit allelopathic effects on other plant species and also dry up soil moisture.

Allelopathy is a phenomenon that involves the advantageous or detrimental impacts exerted by one plant on another plant, encompassing both cultivated and invasive species. These effects arise from the discharge of bioactive compounds, referred to as allelochemicals, which are released through various mechanisms such as leaching, root exudation, volatilization, residue decomposition, and other related processes. This phenomenon occurs within both natural and agricultural environments [636]. Allelochemicals that exhibit negative allelopathic effects play a significant role in the defense mechanism of plants against herbivory, which refers to the consumption of plants by animals as their major source of sustenance [1087]. Certain plant species, including *Tectona grandis*, *Eucalyptus* sp., and the neem tree, have been seen to demonstrate allelopathic properties.

The presence of *Eucalyptus* spp. plantations has been observed to have an impact on the diversity of forest species as a result of its allelopathic effects [637, 1088]. In this regard, Mensah [335] observed that there were a sparse and reduced vegetation beneath a forest canopy in *Eucalyptus* plantation in a sub-catchment in Kenya, which suggested a decline in biodiversity under the forest. He further observed that there were a few plants growing with the *Eucalyptus* plantation, but were often found in the periphery of the plantation, around 2–3 m away from the plantation. The predominant vegetation observed consisted of *Cynodon dactylon* grass and *Acacia mearnsii*, commonly known as the "wattle" tree.

In contrast, the natural forest exhibited a correlation with a diverse array of plant species in the undergrowth such as *C. dactylon*, *Podocarpus* sp., *A. mearnsii*, *Croton macrostachyus*, and many legume species. This encounter suggests that there is a higher level of biodiversity, leading to a greater diversity of litter falling. This might potentially result in a more balanced functioning of the ecosystem, as there would be an increased accumulation of soil organic matter and a potentially faster rate of decomposition, leading to the formation of humus.

In their study, Dutta and Agrawal [136] conducted research to examine the impact of tree plantations on soil characteristics and microbial activity in coalmine spoil land. They observed that the elevated levels of total nitrogen, when compared to the fresh mine, were attributed to the accumulation of organic matter in the soil through root systems and the leaching of nitrogen from the herbaceous vegetation present in the plots.

Additionally, it was shown that throughout the rainy season, the deceased microbial population serves as an extra substrate, thereby significantly enhancing the process of mineralization. The increased levels of mineral nitrogen observed during the rainy season could potentially be attributed to the presence of readily decomposable substances such as glucose, sucrose, amino acids, and amides [136]. On the other hand, the decline in N mineralization can potentially be attributed to a decreased microbial biomass, leading to the immobilization of nutrients [136].

The practice of reclamation forestry involves the establishment of forests on land that has been eroded or degraded, and it has been shown to effectively enhance soil fertility [529]. Additionally, the author asserts that the utilization of shifting cultivation serves as an exemplification of the forest's ability to replenish the fertility that is depleted during the planting period.

Trees exhibit a high level of efficiency in generating biomass, contributing a greater amount of organic matter to the soil, both in the aerial and subterranean portions, compared to other plant species. The roots of these plants have a larger vertical extent in the raw mine stones compared to the grass, and with some assistance, they are able to penetrate into the less densely packed spoil strata located beneath the clay "cap" [638]. The normal root depth for herbaceous species is 50 cm, whereas for trees it is 3 m. However, certain phreatophytes that access groundwater have been observed to reach depths of 15 m or more, particularly in arid conditions [639].

Trees have the potential to enhance soil quality through various mechanisms. These include the maintenance or augmentation of soil organic matter, biological nitrogen fixation, absorption of nutrients from the lower layers of the soil, as well as the roots of understory herbaceous plants, increased water infiltration and retention, mitigation of nutrient loss through erosion and leaching, improvement of soil physical characteristics, and reduction of soil acidity while enhancing biological activity. Furthermore, trees play a crucial role in the creation of new self-sustaining top soils. The decomposition of plant litter and the release of root exudates contribute to the nutrient cycling process in soil [640–641, 771, 921].

In addition, the implementation of tree-based rehabilitation strategies has the potential to mitigate compaction tendencies in mine wastes. According to [759], if these newly developed soils exhibit enhanced drainage properties, there will be a decrease in the amount of water retained at the soil surface, hence reducing the likelihood of soil erosion.

Additionally, a viable hypothesis suggests that trees, as a whole, exhibit greater efficiency in the uptake of nutrients produced by weathering compared to herbaceous plants. Rock weathering is a process that leads to the release of essential elements such as potassium, phosphorus, bases, and micronutrients. This release is particularly prominent in the B/C and C soil strata, which are frequently accessed by tree roots [529].

## 12.4 Using grasses and shrubs

They are commonly regarded as pioneer crops due to their ability to establish early vegetation. Grasses exhibit both beneficial and detrimental impacts on mining sites. The utilization of soil stabilizers is often necessary for the purpose of stabilizing soils; nevertheless, it is important to note that their application may potentially result in competition with the process of woody regeneration. Grasses, namely those of the C4 variety, have the capacity to exhibit enhanced resilience in the face of drought, limited soil nutrient availability, and various other climatic stressors. The root systems of grasses exhibit a fibrous nature, which serves to impede the process of soil erosion. Additionally, these root systems include soil-forming capabilities, leading to the gradual development of an organic soil layer.

Consequently, grasses cover the soil surface and play a crucial role in stabilizing soil, conserving soil moisture, and potentially engaging in competitive interactions with undesirable weed species. According to previous studies conducted by [642], [643] and [644], it is crucial for the initial cover to facilitate the growth of various self-sustaining plant communities.

Grasses are thus commonly recognized as pioneering crops that facilitate early vegetation establishment. Cover crops are known to provide assistance in soil stabilization and have a diverse range of benefits. When considering the selection of grasses for soil remediation purposes, it is important to prioritize certain traits. These features include the ability to tolerate metal pollution, a reasonably fast growth rate, and efficient coverage of the soil surface. Alvarenga et al. [26] state that grasses possess the ability to minimize the build-up of toxins in their aboveground parts, which helps prevent contamination of the food chain. Consequently, grasses are frequently employed in the process of phytostabilization. In the context of gold mine tailings remediation, the utilization of grasses may present certain limits despite their ability to acquire a higher proportion of pollutants in their root systems.

In a greenhouse study conducted by Mensah et al. [324], ryegrass was utilized to investigate the effects of arsenic treatment in a gold mine tailing located in southwestern Ghana. The researchers reached the conclusion that, in order to effectively remove As contaminants with ryegrass, it would be necessary to extract the roots of the plant during the harvesting process. Nevertheless, the translocation of the contamination into the root system of the ryegrass is limited to the surface level of the mine spoil or tailing due to the lateral nature of the roots, which prevents them from accessing deeper layers of the mine tailings. Hence, it may be argued that grasses are more aptly suited for safeguarding the soil surface of the contaminated field against erosion. Moreover, these materials could potentially contribute to the containment of the metal or metalloid, reducing their mobility and mitigating nonpoint source pollution in the surrounding ecosystem.

## 12.5 Mixed species versus single species

Monoculture refers to the agricultural practice of cultivating a single crop species over a large area (See Fig. 12.1). The concept of mixed species refers to the coexistence or interaction of different species within a given ecological community (Fig. 12,2).

In marginal habitats with limited or no maintenance, the necessity of including a diverse range of plant species arises due to the inherent unpredictability of one species' performance [341]. Monocultures exhibit various disadvantages including increased susceptibility to pests and diseases, limited fulfillment of multiple use and conservation roles, a lack of a balanced ecosystem, and the inability to provide a more stable system in the presence of environmental variability and directional change compared to mixed cultures. In contrast, mixed forestry systems may exhibit more stability due to their enhanced capacity to safely retain carbon over extended periods. In brief, mixed systems have the potential to provide greater contributions toward the mitigation of climate change compared to monocultures.

According to Morgan [341], it is recommended that the species composition consist of a combination of grasses, forbs, and woody species including both bushes and trees. However, there may be instances such as some types of gully reclamation or along pipe line corridors, where special considerations render such a mixture undesirable.

For example, Young [529] argues that data and induction indicate that the direct impacts of tree canopies in terms of providing cover and preventing soil loss are comparatively weaker than those of ground cover or cover crops in the context of erosion control. Therefore, the most efficient method for directly preventing soil erosion is through the implementation of a surface litter cover, which includes crop leftovers, tree trimming, or a combination of both.

Wulder et al. [645] reported that after conducting a 10-year observation, the combined litter production of a mixed plantation consisting of *Cunninghamia lanceolata* and *Michelia macclurei* in a 1:1 proportion was found to be 43% greater than that of a pure *C. lanceolata* plantation. According to Parrotta's [955] findings, it was noted that litter generation in mixed plantations tended to be higher compared to monospecific *Eucalyptus* plantings in Puerto Rico. In a study conducted by [646], it was shown that the annual litter quantity of 55-year-old *Pinus massoniana* and M. mixed forest of Macclurei exhibited an 11.2% increase in productivity compared to a stand of *P. massoniana* of the same age. The composition of species plays a significant role in determining litter generation within a given climate range [647, 648].

Gregorowius et al. [649] conducted a comprehensive analysis of the significance of species mixes in enhancing plantation productivity. Nevertheless, there is a lack of comprehensive understanding of the impact of broadleaved trees in mixed plantations on the quantity and distribution of litter deposition.

According to the findings of [650], the inclusion of *A. mearnsii* in conjunction with *Eucalyptus globules* resulted in enhanced levels and rates of nitrogen (N) and phosphorus (P) cycling through the process of aboveground litter fall as compared to the use of *Eucalyptus* alone. Previous studies have indicated that the levels of nitrogen (N) and phosphorus (P) recycled through the process of litter fall were greater in ecosystems containing a combination of nitrogen-fixing trees and non-nitrogen-fixing trees compared to ecosystems consisting only of non-nitrogen-fixing tree species [651, 652, 653, 654].

## 12.6 Indigenous/native/local species versus exotic/ introduced/foreign species

The role of alien or native species in reclamation necessitates meticulous deliberation, as the introduction of newly exotic species may potentially lead to their pest-like behavior in other contexts. Hence, it is imperative to exercise caution when selecting candidate species for revegetation in order to prevent their potential transformation into hazardous weeds within the context of local to regional floristic dynamics. In the context of artificial introduction, it is crucial to prioritize the selection of species that exhibit strong adaptability to the specific characteristics of the local ecosystem. The

**Fig. 12.1:** A former mining site restored using single species with *Acasia mangium*.

**Fig. 12.2:** Mixed species consisting of shrubs and grasses under tree plantation on a reclaimed mining site in Ghana. The tree species comprises *Terminalia superba*, *Terminalia ivoriensis*, *Acasia mangium*, and *Tectona grandis*. Shrubs consisted of *Chromolaena odorata* and *Centrosema pubescens*.

adoption of indigenous species is often favored over exotic species due to their higher likelihood of integration into a fully operational ecosystem and their adaptation to local climatic conditions [655, 656].

Whenever feasible, it is advisable to prioritize the selection of indigenous species. According to Morgan [341], conducting an examination of the surrounding areas can provide valuable insights into the species that are most likely to endure and flourish. Singh et al. [995] also observed that indigenous leguminous species exhibit more pronounced enhancements in soil fertility indices as compared to indigenous nonleguminous species. According to Sheoran et al. [453], it has been observed that native legumes exhibit greater efficacy in elucidating disparities in soil characteristics compared to foreign legumes within a limited time frame.

The consideration of introduced or exotic species should not be dismissed, particularly in cases where the local environment has experienced a decline that surpasses that of neighboring areas or if the population of native species is limited. The revegetation plan should incorporate provisions that facilitate the natural process of plant succession. In numerous instances, the primary goal is to introduce pioneer species with the purpose of promptly providing vegetation cover and enhancing the soil, thus enabling the subsequent establishment of native species as the pioneer plants diminish [341].

Nitrogen is a prominent limiting component in mine waste, necessitating constant application of N-fertilizer to sustain the robust growth and long-term survival of plants [657, 658]. One potential alternate strategy could involve the introduction of leguminous plants and other species that possess the ability to fix atmospheric N. Species that are capable of fixing N have a significant impact on the fertility of soil due to their ability to produce litter that is rich in nutrients and easily decomposable as well as their role in the turnover of fine roots and nodules. The process of mineralization of N-rich litter from these species facilitates significant transfer to companion species and subsequent cycling, hence facilitating the establishment of a self-sustaining ecosystem [659].

In summary, the employing native plant species in phytoremediation is suggested as a financially viable solution for environmental management, specifically for addressing the issue of heavy metal pollution in soils resulting from gold mining activities [360, 660]. In the given context, several researchers [e.g., 38, 661, 662, 2024; 384] have recently proposed the utilization of locally adapted native species as a prudent approach in the selection of candidate plants for phytoremediation of contaminated sites. Likewise, it is generally desired to prioritize the inclusion of native and indigenous species over foreign species due to their potential to seamlessly integrate into the existing local ecosystem and exhibit superior adaptability to the local environment [335].

For instance, Mensah et al. [333] investigated the phytoremediation potential of five different native naturally growing plant species in the surrounding of the abandoned mine spoil in southwestern Ghana: *Chromolaena odorata*, *Pityrogramma calomelanos-fern*, *Alchornea cordifolia*, *Lantana camara*, and *Pueraria montana*. The five native plant species were tested for As, Cu, Ti, and Zn, with all having the highest zinc concentrations above 50 mg/kg. *L. camara* had the most Zn, followed by *Chromolaena odorata*, *Pueraria montana*, *Alchornea cordifolia*, and fern. Copper was the second most abundant, followed by As and Ti. This study was unique in examining these plant species for Ti remediation in mine spoil soils. The species are shown in Fig. 12.3, and the concentrations of potentially toxic elements in the plant parts are displayed in Fig. 12.4.

**Fig. 12.3:** Plant species growing near an abandoned mining spoil in southwestern Ghana. Source: Mensah et al. [333].

## 12.7 Factors to consider in selecting plant species for revegetation

According to Morgan [341], the selection of species for revegetation purposes should be based on several key parameters. These parameters include the ability of the species to exhibit rapid growth, resilience against diseases and pests, capacity to outcompete fewer desirable species, adaptability to the local soil composition, and adaptability to the prevailing climatic conditions in the area.

Young [1054] added that when selecting tree species for revegetation programs aimed at improving or restoring soil fertility, certain qualities should be considered. These qualities include the ability to thrive in poor soil conditions, a high rate of nitrogen fixation, a balanced nutrient content in foliage litter, low levels of lignin and polyphenols, and the ability to either rapidly decay litter for nutrient release or decay at a moderate rate for soil cover purposes.

Additionally, the selected species should not contain toxic substances in their litter or root residues, possess a well-developed tap root system, exhibit a high rate of

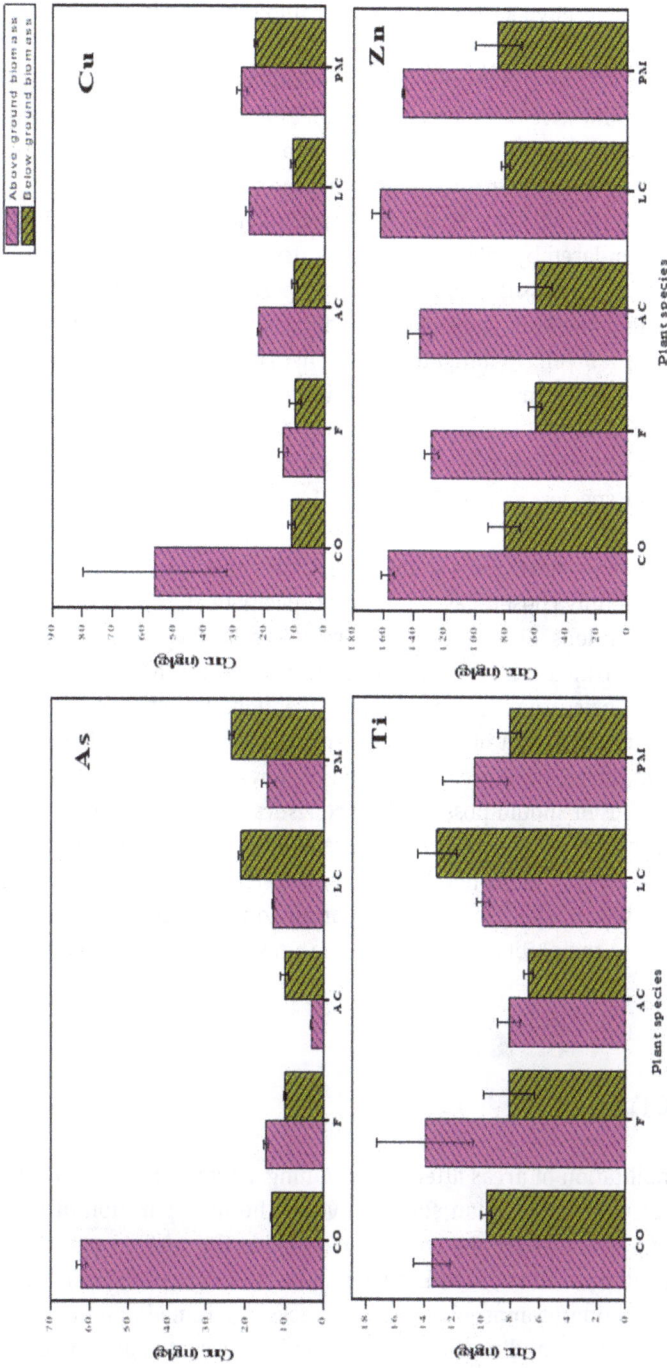

**Fig. 12.4:** Concentration of As, Cu, Ti, and Zn (mg/kg) in the five native plant species growing near the abandoned mining spoil. CO, *Chromolaena odorata*; F, fern (*Pityrogramma calomelanos*); AC, *Alchornea cordifolia*; LC, *Lantana camara*; PM, *Pueraria montana*. Source: Mensah et al. [333].

leafy biomass production, and have a dense network of fine roots capable of forming mycorrhizal associations.

The choice of species will be influenced by the specific soil and climatic conditions in the area, with the objective of establishing a consistent distribution of vegetation rather than a patchy arrangement. This approach is recommended to prevent the accumulation of runoff and localized erosion. Achieving the optimal condition can be challenging in cases where the soils exhibit limited water retention capacity, toxicity, or when plant growth is hindered by cold temperatures, dryness, or exposure.

The selection of plants that possess advantageous properties is a critical consideration [341]. To illustrate this, the Mine Lands Reclamation Unit in Nigeria's Jos Plateau opted for *Eucalyptus* sp. as a means to mitigate the expenses associated with the reclamation of mine spoil mounds resulting from tin mining activities. The utilization of certain crops for fuel and pole timber has led to notable reductions in the soil's base status and pH, as documented by Alexander [22]. This had negative implications for the growth of other plant species, as highlighted by Morgan [341] and as explained earlier.

The process of restoring eroded ecosystems necessitates the utilization of plant species that have been carefully chosen based on their capacity to endure, regenerate, or reproduce in harsh conditions. These conditions are determined by factors such as the composition of the deposited material, the exposed position on the surface, and the ability to stabilize the structure of the soil [288]. Conventional practice in the field of revegetation involves the selection of drought-tolerant and rapidly growing plant species or forage that can thrive on soils with limited nutrient availability.

The plants that are chosen should possess characteristics that facilitate their successful establishment, rapid growth as well as the development of dense canopies and root systems. Sheoran et al. [453], indicated that acidity is a significant determinant in inhibiting vegetation growth in specific regions. In order to thrive in such environments, plants must possess the ability to withstand the presence of metal pollutants [101, 663].

## 12.8 Conclusions

In conclusion, the rehabilitation of areas affected by mining activities requires a strategic plan for plant succession. This plan should involve the incorporation of fast-growing grass species to establish ground cover and stabilize the surface as well as leguminous plants for nitrogen fixation and other grasses and shrubs for sustained vegetative cover. The selection of appropriate plant species for remediation and revegetation should consider their ability to accumulate and translocate pollutants, detoxify or immobilize pollutants, produce high biomass, tolerate toxic concentrations of pollutants, and have rapid growth and a short life cycle. The use of diverse vegetation comprising grasses, legumes, shrubs, and trees is recommended for sustained remediation programs.

Trees, in particular, are favored for extensive land decontamination and revegetation due to their ability to generate biomass, penetrate deeper into the soil, contribute to nutrient cycling, and access nutrients and contaminants in deeper horizons. However, caution should be exercised when using certain tree species, such as *Eucalyptus*, as they may exhibit allelopathic effects on other plant species and impact biodiversity. The utilization of genetically modified plants for phytoremediation is a topic of debate, and the use of monoculture or mixed plantation practices should be considered. Overall, the selection of appropriate plant species and the use of trees can effectively enhance soil fertility and contribute to the restoration of degraded mining sites.

Albert Kobina Mensah

# Chapter 13
# Concurrent rehabilitation/revegetation

**Abstract:** Progressive land restoration should be the focal point of a comprehensive environmental management program that incorporates the reclamation of mining sites. By adopting this methodology, financial gains are maximized, environmental and residential harm is minimized, and resources are utilized in an efficient fashion. Effective rehabilitation necessitates the identification of mining impacts, the safeguarding of affected regions, and the strategic planning and evaluation of operations.

## 13.1 Introduction

The reclamation of mining sites should not be seen just as an activity to be undertaken at or shortly before mine closure [448]. Instead, it should be incorporated into a comprehensive program for efficient environmental management across all stages of resource development, encompassing exploration, construction, operation, and closure. The achievement of successful reclamation and revegetation is facilitated by meticulous consideration of various factors.

Plans for restoration and the reclamation efforts should be taken ahead and incorporated into the environmental impacts assessment and the environmental management plans at every stage of the mine's project life cycle. Sheoran et al. [448] reported that these stages include initial planning, clearance of the site, removal and storage of soil, replacement of soil, selection of appropriate species, re-establishment of vegetation, and maintenance of these places for future use.

Thus the emergence of progressive land restoration as a concept can be attributed to the acknowledgment of the adverse environmental and social consequences associated with conventional mining methods. Throughout history, mining operations have frequently resulted in extensive landscape degradation, typically due to a lack of regard for the potential long-term repercussions. The development and acceptance of progressive restoration schemes have been driven by a combination of factors including heightened awareness of environmental challenges, mounting regulatory restrictions, and a trend toward sustainable development practices.

**Albert Kobina Mensah,** Soil Research Institute, Council for Scientific and Industrial Research, Kumasi, Ghana, e-mail: albert.mensah@rub.de, albertkobinamensah@gmail.com, ORCID: https://orcid.org/0000-0001-5952-3357

https://doi.org/10.1515/9783111662046-013

As the mining sector progressively prioritized environmental and social responsibility, corporations began incorporating restoration planning as an integral component of their mining operations. The recognition that ecosystems have a higher likelihood of effective recovery when restoration efforts are implemented during active mining rather than postcompletion has played a significant role in the development of the concurrent or simultaneous or progressive land restoration strategy.

The aforementioned transition exemplifies a larger pattern observed in the resource extraction sectors, wherein there is a growing inclination toward adopting sustainable and socially responsible approaches. This notion embodies an adaptive approach to attaining sustainability in the mining sector, wherein reclamation activities are carried out concurrently or in parallel with the operational phase of the mine project life cycle.

In contrast to the conventional approach of implementing reclamation and restoration measures during the closure phase of mining projects, the present approach entails restoring the land progressively during the mining process. These endeavors contribute to the mitigation of the ecological, agricultural, and residential impacts resulting from mineral extraction activities, hence facilitating a reduction in the expenses associated with land reclamation.

The use of this approach guarantees the efficient utilization of all available resources and mitigates the potential for ecosystem functional loss or resource depletion throughout the operational phase and following the extraction of mineral resources. According to Chugh [113], the primary objective of concurrent mining and rehabilitation is to optimize the financial returns for the mining sector throughout and after the extraction of natural resources. Concurrent rehabilitation (CR) can also be referred to by various words including progressive rehabilitation, simultaneous rehabilitation, zonal rehabilitation, or "pipe solution" mining field rehabilitation.

Additionally, Chugh [113] stated that the objectives of concurrent mining and reclamation are to reduce or improve the following: (1) the utilization of farmland for agricultural purposes throughout and after the complete mining operation, (2) the availability of surface and groundwater resources for agriculture and economic growth, and (3) the revenue generated from the utilization of all available resources.

In contemporary mining practices, the process of reclamation is typically undertaken in mine regions following the conclusion of mineral extraction activities across expansive territories. The practice described here is commonly referred to as traditional reclamation, also known as TR or an "end-of-pipe solution." It involves doing land grading, soil placement, and agricultural production mostly after the land has achieved stability following mining activities [664–665].

# Brief conceptual framework guiding concurrent reclamation (Reclamation integrated into the mine's project life cycle)

Reclamation is a fundamental aspect of a mine's life cycle. Contemporary mining relies on reclamation to preserve the integrity of the environment and the lifestyles affected by this land use. Open-pit mining necessitates the use of numerous reclamation techniques, in addition to ore extraction and the management of substantial waste rock.

In mine reclamation planning In the Canadian North document prepared for the Canadian Arctic Resources Committee, Brian Bowman and Doug Baker [1068], recounted that reclamation techniques are should often times be considered as an integral part of the mineral extraction process. Ideally, the implementation of reclamation procedures should align with mine development and occur gradually throughout the mine's lifespan.

This minimizes final closing expenses and the handling of substantial quantities of material during the operation or active and eventually at the closure stages of the mines. Table 13.1 summarises the need for inclusion of the concurrent reclamation action into the mine's project life cycle (i.e., from the initiation or pre-mining planning through monitoring and evaluation till the decommissioning or closure stage of the mine).

**Tab. 13.1:** Impacts (positive and negative) of mining projects on the environment and the community.

| Negative impacts | Positive impacts |
|---|---|
| Pre-construction/planning/initiation phase:<br>1. Plan of removal of vegetation<br>2. Plan of evacuation of locals<br>3. Plan of contamination of water bodies<br>4. Plan of soil erosion<br>5. Plan of loss of livelihood<br>6. Development of mine pits may breed mosquitoes<br>7. Development of mine pits may pose danger to livestock and humans<br>8. There will be loss of agricultural productive lands<br>9. Development and construction of roads may affect biodiversity and lead to loss of habitats<br>10. Conflicts and community grievances | 1. Employment to the community<br>2. Initiation of investments in the local community<br>3. Training and education of the indigenes<br>4. Building of recreational sites such as parks for children |

**Tab. 13.1** (continued)

| Negative impacts | Positive impacts |
|---|---|
| Construction phase: | 1. Employment to the local community |
| 1. Soil erosion | 2. Sharing of cleared vegetation with the |
| 2. Removal of vegetation | community. E.g. firewood |
| 3. Loss of people's livelihoods | 3. Increased commercial activities in the mining |
| 4. Depletion of soil fertility | community |
| 5. Loss of arable fields | 4. Capacity building of mine workers |
| 6. Wastes generation | 5. Boost in local economic activities |
| 7. Loss of natural habitats | |
| 8. Destruction of heritage sites | |
| 9. Destruction of sites with religious significance | |
| 10. Human right abuses | |
| 11. Conflicts between the mines and the community | |
| 12. Disruption in social norms and values | |
| 13. Potential increase in social vices | |
| 14. Fire | |
| 15. Surface and groundwater pollution | |
| 16. Alteration of drainage and surface hydrology | |
| 17. Sanitation issues | |
| Operation phase: | 1. Investments in social amenities such as |
| 1. Spillage of tailings | schools, roads, hospitals, recreational centres, |
| 2. Water contamination of surface and | etc. |
| groundwater | 2. Education and training of community residents |
| 3. Soil contamination | 3. Employment of locals |
| 4. Soil erosion | 4. Residents will be assisted in their efforts to |
| 5. Conflicts | improve their former living standards and |
| 6. Accidents and fire | income earning capacities |
| 7. Social impacts such as high cost of living and | 5. Investment in environmental programs |
| goods and services | 6. Resettlers will be integrated socially and |
| 8. Wastes generation | economically into host communities |
| 9. Sanitation issues | 7. Business support to local people |
| 10. Stockpiling of mine wastes which affect soil | 8. Boost to the local economy |
| fertility | 9. Payment of royalties to the local assembly |
| 11. Loss of arable lands | 10. Generation of revenues |
| 12. Air pollution | |
| 13. Creation of mine pits may pose danger to human and livestock | |
| 14. Noise pollution from mine plants | |
| 15. Loss of aesthetic and recreational sites | |

**Tab. 13.1** (continued)

| Negative impacts | Positive impacts |
|---|---|
| Closure/decommissioning phase: | 1. Employment creation as locals could be engaged in reclamation works |
| 1. Abandonment of pits may pose danger to humans and livestock | 2. Restored mine fields could be put under agriculture, forestry and tourism |
| 2. Abandoned pits may breed mosquitoes and affect the health of residents | 3. Revenue generation |
| 3. Abandoned mine spoils may contain potentially toxic elements that may affect the health of women and children | 4. Recreational parks to the mining community |
| 4. Abandoned mine spoils may lead to loss of fertile and productive fields | |
| 5. Abandoned mine spoils may deprive the residents and communities of their livelihoods | |
| 6. Abandoned mine spoils may cause surface and groundwater pollution | |
| 7. Loss of playing grounds for children | |
| 8. Air pollution | |
| 9. Public safety may be affected | |

# 1 Pre-mining planning

This phase constitutes a component of the project commencement stage within the life cycle of mining projects. Two major objectives are to conduct the site assessment and gather baseline data. The preliminary evaluation/assessment may encompass the environmental and social ramifications of the project, ensuring that measures are devised to mitigate and actions implemented to establish the necessary edifices prior to the project's initiation.

The preliminary evaluations may also encompass a health risk assessment of the proposed project. Socio-economic appraisals of the projects have also been initiated. This will inform the construction or expansion of mining facilities, the design and construction of tailings dams, and the demarcation of areas for waste stockpile storage. It will also encompass the measures to be implemented to mitigate environmental consequences, enhance positive effects, and review any planned impacts that can be managed.

Baseline data can be gathered for future comparisons and to evaluate and monitor progress made towards the rehabilitation of mining sites during and after reclamation. Examples of these baseline data encompass data collection on the physico-chemical parameters of the soil and its biological attributes. The substantial mineral composition of the soil, the soil fauna and flora, and the heavy metals are all documented.

## 2 Active mining operations

The concurrent rehabilitation approach entails reclaiming the land simultaneously throughout the mine project life cycle. Thus, it may be proposed that key actions here include concurrently taking actions aimed at reclaiming the land whilst the mining or the mineral extraction process is still in the operational phase or still active. In addition to this, topsoil management and taking actions aimed at vegetation reestablishment at the degraded site are paramount.

Specifically, measures are employed to maintain the function and the productivity of the soil. One of the major hindrances in mine land reclamation projects is access to and availability of clean topsoil for spreading during the regreening of the degraded sites. As reiterated in previous chapters, thus it may be required for the environmental department section of the mining company to employ steps aimed at preserving the topsoil that was stripped and removed during the operation phase of the mine project life cycle.

This is done in order that the stripped topsoil can be reused during the reclamation of the site. Saving topsoil that was collected insitu at the mine sites is beneficial in that such action aids in facilitating reclamation and revegetation success. In this scenario, the native soil adapts to the local climates better, aids in curtailing the introduction of foreign and alien species, and curtails the spread and colonisation of resistant and stubborn pests and diseases into the native ecological system.

## 3 Monitoring and Evaluation

This is done to ensure that entities and organisations regulated by the regulatory institutions are complying with the actions stipulated in the EMP and that they are taking steps to achieve a cleaner production. Thus, monitoring and evaluation activities include, among other things, the following:

a. Inspection of surface (open pits) and/or underground mine workings;
b. Inspection and maintenance of mine waste facilities, including the effectiveness of any cover and/or seepage capture systems; and
c. Mechanisms for contingency and response planning and implementation.
d. Water quality monitoring, including assessing the extent of contamination and pollution, and the release of effluents, and ensuring there is no damage to aquatic and terrestrial resources
e. Compliance with environmental safety and standards
f. Community engagements as to regular updates and contact with host mining communities
g. Tailings dam management

In Ghana, monitoring and evaluation may also include environmental auditing, which includes environmental performance, rating, and disclosure, which comes in the form of sustainability reporting. This includes the use of EPA's ANKOBEN rules and system. The rating concept provides the guidelines for the use of colours as indicators of performance levels.

AKOBEN employs a colour scheme to convey the message to the general public in simple terms. The system rates the sustainability compliance of companies using numerical cut-off points, cards, and colour coding. the colours range for each category of the five colour codes. They range from gold to red, with gold indicating the best performance and red indicating the worst performance.

AKOBEN's methodology is classified into three groups: rating criteria, rating concept, and rating rules [1069]. The rating criteria consist of seven parameters (Fig. 13.1):
a.   Legal issues;
b.   Hazardous waste management;
c.   Toxic and non-toxic releases;
d.   Monitoring and reporting;
e.   Environmental best practices;
f.   Community complaints; and
g.   Corporate social responsibility (CSR)

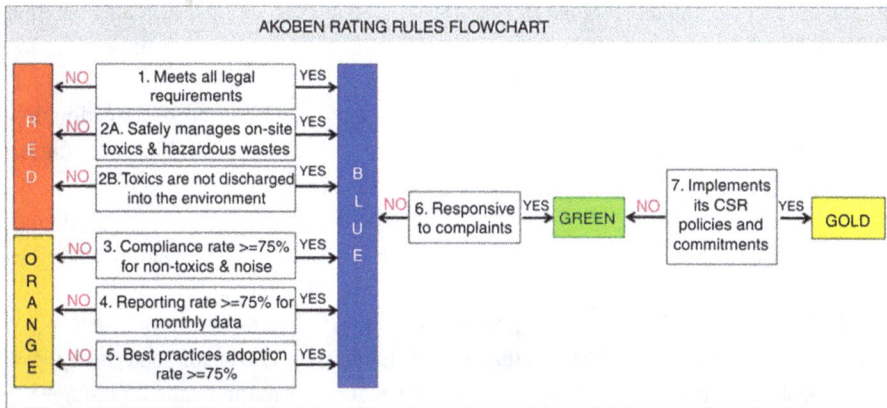

**Fig. 13.1:** Ghana's AKOBEN Rating system in evaluating environmental performance, rating and disclosure.

The software system has been designed in such a way that there is zero tolerance for noncompliance with legal and hazardous waste management. Thus, even if a company complies 99 percent with both legal and hazardous waste management requirements, the system would still rate the company "red." Final rate awarded to an industry is based on the industry's worst performance.

Three main ways of generating the data used for rating are monthly reports submitted to the EPA by industries, routine site inspections, and annual site audits. Report cards are generated for industries to inform them about their performance against the indicators. On this report card, the overall rate given to a company is shown in addition to the actual percentage obtained. After evaluating and analysing the final results, the ratings are released to the public primarily through the press and the internet.

## 4 Post-mining stage where the final assessment and completion of reclamation activities

During the closing phase of mining, the residual effects of abandoned mine land on the community's air, water, and soil must be considered. Furthermore, steps must be implemented to mitigate or restrict the dissemination of point and non-point source pollution issues. If the mining area is assessed to potentially endanger human health and the environment in the short or long term, remediation activities are prioritised [969]. Remediation entails decontaminating the area, particularly the soil, by physical, chemical, and biological means [160].

Thereafter, an evaluation is conducted to determine the feasibility of restoring the landscape to its original condition. The restoration process entails the extensive enhancement of damaged regions to reinstate the ecological equilibrium and original biodiversity of the ecosystem, both functionally and structurally, as it previously existed. Restoration of the degraded sites to its former glory is almost an impossibility. It may take several years to restore the ecological integrity to its pristine condition, i.e., just as the way it was before. When ecosystem restoration is unfeasible, reclamation is considered.

Reclamation is typically good for abandoned sites when restoring the original environment is unfeasible. In this instance, the ecosystem can be transformed into alternative land uses that can benefit the riparian or host mining communities. In this regard, [969] highlighted that there are instances of mines and quarries where the restoration techniques are inapplicable. They further added that such scenario can lead to an inevitable transition from traditional land uses to new, stable, permanent, and advantageous land uses, ultimately facilitating the entire rehabilitation of the area. Here, the new land uses, while unrelated to the prior ones, enhance the area's environmental, social, and economic conditions. The post-mine land can thus, be developed into alternative uses other than the prior land use. These options should be ones that are implemented based on effective consultations with all relevant stakeholders and such actions should benefit the host mining communities. [308] indicated that at the closure of the mine, the land can be redeveloped for agriculture, forestry, lakes or pools, intensive recreational land use, non-intensive recreational land use, conservation, and pit backfilling. [308] further reiterated that the objective of post-mining land use is to achieve economic and sustainable outcomes that fulfil human needs while safeguarding life and the environment.

Pagouni et al. [969] proposed two possibilities for a sustainable mineral industry that includes mine closure plans: repurposing and co-purposing. Repurposing might involve transforming the local economy and mitigating the effects of mine closure by incorporating existing infrastructure into the region's growth. Finally, the coexistence of similar activities remains a possibility. Co-purposing refers to the introduction of a new activity inside the mining and quarry sector that does not compete with mining operations or land management activities, resulting in a smooth, favourable transfer with developmental opportunities.

## 13.2 Ensuring success of concurrent rehabilitation

For an effective implementation and success of CR in the field, Chugh [113] outlined four main elements that will aid to amplify the benefits and minimize the negative impacts (also summarized in Fig. 13.2):

1. *Identifying the mining area's impact on the surface and subsurface* is crucial for developing effective CR strategies. This involves assessing the extent of damage and influence on variables like forest land, farmland, aquifer, or coal seam. Reliable data and operational software are needed to document and analyze these factors.
2. *Identifying activities and resources requiring protection within the impacted regions* includes extracting agricultural produce, demolishing structures, relocating residents, or extracting subsoil or loam. It also helps identify areas where topsoil and/or subsoil may be stored before use.
3. *Strategic planning and evaluation of mining and CR operations* in designated regions is necessary to safeguard resources, reduce costs, and optimize the objective function. This may involve iterative pursuit of design elements to achieve the intended function.
4. *Evaluating the impact of CR implementation on specified goals* provides feedback on the effectiveness of planned CR activities and may necessitate iterative evaluation of alternative reclamation strategies, mining plans, and community and regional development initiatives.

```
┌─────────────────────────┐
│  Identify mining area/s │ ◄──┐
│        influence        │    │
└─────────────────────────┘    │
         │                     │
         ▼                     │
┌─────────────────────────┐    │
│ Identify activities/    │ ◄──┤
│ resources to protect    │    │
│ within the affected area│    │
└─────────────────────────┘    │
         │               ▲     │
         ▼               │     │
┌─────────────────────────┐    │
│  Planning and analysis  │ ◄──┤
│  of mining and CMR      │    │
│  activities             │    │
└─────────────────────────┘    │
         │               ▲     │
         ▼               │     │
┌─────────────────────────┐    │
│ Assess CMR implementation│───┘
└─────────────────────────┘
```

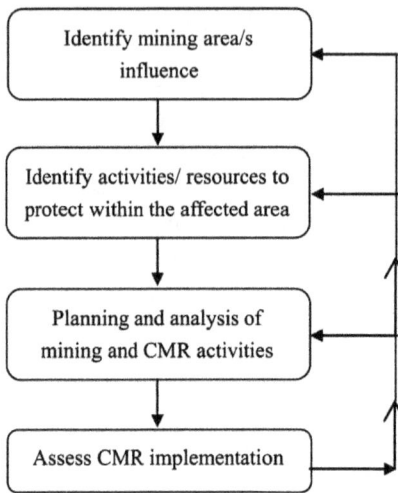

**Fig. 13.2:** Elements of critical scientific planning for concurrent or progressive mine land rehabilitation. Source: Chugh [113].

**Tab. 13.2:** Measures that should be taken at each phase of the mine's project lifecycle to achieve mining sector sustainability. This can be considered as part of concurrent reclamation of mining projects.

| Preconstruction/ planning/initiation phase | Construction phase | Operation phase | Closure/ decommissioning phase |
|---|---|---|---|
| Submission of EIS and EAP | Measures for mitigation/ reduce/avoid/manage hazardous materials e.g. padding, fencing of tailing and other contaminated sites, laying of storage pipes | Submission of EMP and EAP | Reclamation |
| Resettlement plans for residual impacts | Health and safety of the mine workers/mine community | Community engagement and conflict resolution | Dismantle and removal of storage, transmission pipes |
| Compensation of affected people | Accident management | Development of alternative livelihood programs e.g. roads, schools, water infrastructure, training and education, hospitals, training centres for the community, recreation parks for children, etc. | Recreational sites |

**Tab. 13.2** (continued)

| Preconstruction/ planning/initiation phase | Construction phase | Operation phase | Closure/ decommissioning phase |
|---|---|---|---|
| Public hearing/ community engagements | Fire management | Concurrent rehabilitation- reclaim as you mine | Tourism |
| Plans for closure of the undertaking | Waste generation measures | Stockpiling of mine wastes/ spoils and measures to protect | |
| Measures for mitigation -e.g. tailings management, waste management, spillage pond, etc. | Measures to address erosion – planting grasses | Tailings management – planting grasses around mine tailings, erosion management | |
| Reclamation plans | Compensation measures | Monitoring and evaluation | |
| Conflict management plans and other community engagement techniques | Community engagement, addressing complaints, resolving conflicts, etc. | Fire/accidents management | |
| | Dusts, air and noise pollution management. | Health and safety of mine workers | |
| | Construction of tailing dam, spillage (seepage) pond, and its management. | Corporate social responsibilities | |
| | Designing of storage sites for stockpiled soils. | Environmental performance reporting, rating and disclosure | |
| | Vegetation clearing, removal and management – share with community for firewood, making biochar, compost, etc. | | |
| | Protection of sacred places e.g. grave yards, national parks and forests, ancient temples, etc. | | |

# 13.3 Conclusions and summaries

In conclusion, the reclamation of mining sites should be integrated into a comprehensive program for efficient environmental management throughout all stages of resource development. The concept of progressive land restoration has emerged as a response to the adverse environmental and social consequences of conventional mining methods. Progressive restoration involves restoring the land concurrently with the mining process rather than waiting until after completion.

This approach optimizes financial returns for the mining sector, reduces ecological and residential impacts, and ensures the efficient utilization of resources. To ensure the success of CR, it is important to identify the impact of mining on the surface and subsurface, protect activities and resources within impacted regions, strategically plan and evaluate mining and restoration operations, and evaluate the impact of restoration implementation on specified goals.

## Summaries of key points for concurrent reclamation

### A. Environmental impacts of mining:
1) Mining for minerals, regardless of how large, medium, or small they are, have negative effects on the ecological system
2) Detrimental effects particularly from the surface or pit, causes a great deal of damage to the land, the fauna and flora of the soil
3) Human health implications on host mining communities arise from the environmental degradation caused by mineral extraction.

### B. Regulatory framework:
4) All undertakings or projects that are anticipated to have negative influence on the environment are required to undergo an environmental impact assessment (EIA)
5) The EIA includes environmental management plans (EMP) developed by the corporations involved.
6) Plans for reclamation are included in the EMP, which are typically implemented at the closing phase of the mine project life cycle.
7) The goal is to achieve sustainability in the governance of mineral resources.

### C. Reclamation plans:
8) Traditionally, degraded lands and soils are supposed to be restored to their previous level.
9) If restoration is not possible, the property could be put into alternative uses; e.g., the land can be repurposed.

10) The repurposing of the land can be done for agriculture, recreation, tourism, and other activities such as play grounds, swimming pools, forestry, fish farms, and other similar activities.

## D. Concurrent reclamation:

11) In concurrent rehabilitation, mining corporations are encouraged to undertake reclamation activities during the operational phase of the mining cycle.
12) This unconventional strategy indicates that reclamation operations are implemented to restore ecological integrity during the active phase of the mine.
13) Concurrent rehabilitation mandates that actions are undertaken to restore the site from the project initiation phase through to the operational, implementation, monitoring, evaluation, and decommissioning phases of the mine.
14) In effect, the mining sites are concurrently reclaimed throughout the mining process, as opposed to the conventional method where reclamation occurs post-mine closure.

## E. An operational example:

15) For example, a company extracts resources from zone A, they initiate and implement reclamation measures at A, while the company proceeds to mine in zone B.

Albert Kobina Mensah, Bernd Marschner, Sabry M. Shaheen, and Joerg Rinklebe

# Chapter 14
# Arsenic in a highly contaminated gold mine spoil in Ghana: mobilization and potential of soil amendments to reduce the water-soluble arsenic content and improve soil quality

**Abstract:** Gold mine spoils pose potential threats to environmental resources and consequently raise health concerns for humans. We sequentially extracted arsenic from soil collected from an abandoned mining site for geochemical fractions (water-soluble (FI); specific-sorbed/exchangeable (FII); poorly (FIII)- and well-crystalline (IV) Fe oxide; and residual (FV)). We also employed a preliminary soil incubation study to investigate the potential of soil amendments for amelioration of the water-soluble arsenic contents and soil quality improvement. Compost, iron oxide, manure, and rice husk biochar were each applied at the rates of 0.5%, 2%, and 5% (w/w) to the soil; and

***

**Albert Kobina Mensah**, Department of Soil Science and Soil Ecology, Institute of Geography, Ruhr-Universitaet Bochum, Universitaet Strasse 150, Bochum 44801, Germany; Department of Arid Land Agriculture, Faculty of Meteorology, Environment, and Arid Land Agriculture, King Abdulaziz University, Jeddah 21589, Saudi Arabia; Department of Environment, Energy and Geoinformatics, University of Sejong, 98 Gunja-Dong, Guangjin-Gu, Seoul, Republic of Korea, e-mail: albert.mensah@rub.de
**Bernd Marschner**, Department of Soil Science and Soil Ecology, Institute of Geography, Ruhr-Universitaet Bochum, Universitaet Strasse 150, Bochum 44801, Germany; Department of Arid Land Agriculture, Faculty of Meteorology, Environment, and Arid Land Agriculture, King Abdulaziz University, Jeddah 21589, Saudi Arabia; Department of Environment, Energy and Geoinformatics, University of Sejong, 98 Gunja-Dong, Guangjin-Gu, Seoul, Republic of Korea, e-mail: bernd.marschner@rub.de
**Sabry M. Shaheen**, Laboratory of Soil and Groundwater Management, Institute of Foundation Engineering, Water and Waste Management, School of Architecture and Civil Engineering, University of Wuppertal, Pauluskirchstraße 7, 42285 Wuppertal, Germany; Department of Soil and Water Sciences, Faculty of Agriculture, University of Kafrelsheikh, 33516 Kafr El-Sheikh, Egypt; Department of Arid Land Agriculture, Faculty of Meteorology, Environment, and Arid Land Agriculture, King Abdulaziz University, Jeddah 21589, Saudi Arabia; Department of Environment, Energy and Geoinformatics, University of Sejong, 98 Gunja-Dong, Guangjin-Gu, Seoul, Republic of Korea, e-mail: shaheen@uni-wuppertal.de, smshaheen@agr.kfs.edu.eg
**Joerg Rinklebe**, Laboratory of Soil and Groundwater Management, Institute of Foundation Engineering, Water and Waste Management, School of Architecture and Civil Engineering, University of Wuppertal, Pauluskirchstraße 7, 42285 Wuppertal, Germany; Department of Arid Land Agriculture, Faculty of Meteorology, Environment, and Arid Land Agriculture, King Abdulaziz University, Jeddah 21589, Saudi Arabia; Department of Environment, Energy and Geoinformatics, University of Sejong, 98 Gunja-Dong, Guangjin-Gu, Seoul, Republic of Korea, e-mail: rinklebe@uni-wuppertal.de

https://doi.org/10.1515/9783111662046-014

NPK (+S) (15:15:15 + 11) fertilizer was applied at 5, 0.2, and 0.1 g/kg. We found that the sequential extraction experiment revealed higher content of arsenic mainly associated with FIII (49%). Results from the incubation experiments indicate that 5% iron oxide reduced the water-soluble content by 93.4%. Further, 5% compost reduced phytoavailable arsenic by 32.5%. Compost, manure, and biochar at 5% improved the mine spoil total C and N, whereas lone treatment with compost and manure at 5% also improved the soil exchangeable $K^+$, $Mg^{2+}$, and $Na^+$. But application of manure, compost, and iron oxide at 5% provided sorption sites for P, reducing available P from 118.5 mg/kg in the control to 60.3 (by −49%), 12.6 (by −89%), and 7.1 (by −94%) mg/kg, respectively. We thus conclude that 5% iron oxide may very effectively reduce migration and associated environmental health risk of water-soluble As.

**Keywords:** Arsenic geochemical fractions, gold mine spoil, risk assessment, soil quality, soil remediation

# 14.1 Introduction

The persistent toxicity of arsenic (As) in soil and water is recognized as a significant environmental health risk in numerous countries including Ghana [321, 376]. The global average concentration of arsenic in uncontaminated soils is documented at 6.8 mg/kg [230]. However, increased levels can be attributed to the natural weathering processes of arsenic-rich parent materials as well as anthropogenic influences including gold mining, various industrial operations, and the accumulation of contaminated sediments [214]. These sources can mobilize arsenic into surface or subsurface waters, potentially resulting in toxic impacts on biotic communities [292]. In Ghana, these issues have been intensified by inadequate management of mining tailings, accumulation of mining wastes, failure of tailings structures, and the neglect of mine tailings. The implications of these actions present significant environmental health concerns for the surrounding ecosystem, subsequently leading to potential health risks for local residents [73, 86].

The total content method for evaluating arsenic risk to human and environmental health may lead to an overestimation of potential risks, while neglecting the actual ecological toxicity effects, mobility, and bioavailability of the element [199, 283, 500]. The fraction of water-soluble arsenic content is likely the most critical aspect associated with environmental risks and has demonstrated a positive correlation with arsenic concentrations observed in field conditions [500]. Consequently, this fraction of arsenic represents a significant concern, as its occurrence in the soil can lead to substantial ecological and environmental hazards, resulting in the contamination of both surface and groundwater sources [500]. According to Karak et al. [234], the water-soluble fraction represents the most mobile and toxic component of arsenic in soil and aquatic environments, readily forming outer-sphere complexes on mineral surfaces [763].

A variety of organic and inorganic amendments have been utilized for the remediation of contaminated mine soils. These include manures, compost [437], biosolids, sawdust, wood ash, lime [436], biochar, charcoal [407], zeolites, coal fly ash [436], and metal oxides/hydroxides such as Fe, Al, and Mn [214, 509]. Metal oxides are extensively utilized for the remediation of arsenic in contaminated soils, attributed to their robust sorption and immobilization capabilities [e.g., 251, 376]. According to the findings of omárek et al. [251], the introduction of iron oxides, such as iron grit, effectively reduces the mobile and bioavailable forms of arsenic, thereby mitigating their potential environmental hazards and limiting absorption by soil organisms, crops, and humans. The benefits of iron oxides include their enduring stability, minimal environmental impact, reduced occupational hazards, high efficacy, cost-effectiveness, natural abundance, and their role in promoting plant growth.

Therefore, the application of iron oxides to an abandoned arsenic-contaminated mine spoils could serve as a viable approach to mitigate potential human health hazards linked to arsenic exposure. Compost and manure have the potential to immobilize arsenic from soil through mechanisms such as adsorption and/or complexation with particulate organic matter, as documented by Gadepalle et al. [173] and Hartley et al. [199]. The application of biochar can enhance soil fertility and facilitate the re-establishment of vegetation on sites by introducing organic matter into the soil [326]. These amendments have the capacity to rectify soil pH issues at contaminated sites, supply essential nutrients to enhance or restore the fertility of soils affected by mining activities, and reduce the phytoavailability of harmful elements to crop plants. The selection of remediation strategies is significantly influenced by the characteristics of the soil, with soil pH being particularly crucial, as it can impact the solubility and mobility of arsenic within the adjacent ecosystem [214].

To date, limited scientific investigations exist regarding the effectiveness of soil amendments, such as iron oxides, in reducing the water-soluble content of arsenic and enhancing the quality of the soil in the abandoned gold mining site. We hypothesize that soil amendments could serve as a viable approach to mitigate the water-soluble fraction of arsenic present in the gold mine spoil. Further, we assume that the introduction of soil amendments to mine-contaminated spoil could enhance the soil quality and facilitate the re-establishment of vegetation. Consequently, this is a preliminary study to (i) assess whether in situ application of oil amendments to an arsenic-contaminated mining site could effectively reduce the water-soluble and phytoavailable fractions of arsenic and (ii) evaluate if the application of soil amendments can promote the enhancement of soil quality.

## 14.2 Materials and methods

### 14.2.1 Site characteristics and total element contents

Soil samples were obtained from arsenic-contaminated gold mine tailings site in the western region of Ghana. The site's texture is predominantly sandy with 62% sand, 36% silt, and 3% clay. The soil exhibits a neutral pH of 7, total carbon content of 1%, total organic carbon concentration of 5 mg/L, total aluminum level of 3,390 mg/kg, total calcium concentration of 10,893 mg/kg, and total manganese content of 357 mg/kg. The site exhibits elevated concentrations of As and Fe at 1,807 mg/kg and 19,348 mg/kg, respectively (Tab. 14.1). Additional information regarding the site, sample collection and preparation, climatic conditions, and physicochemical qualities is documented in Mensah et al. [321].

**Tab. 14.1:** Basic soil physicochemical properties and element total contents in the mine spoil.

| y | Unit | Value | SD |
|---|------|-------|----|
| pH | – | 7 | 0 |
| EC | µS/cm | 632 | 1 |
| TOC | mg/L | 5 | 1 |
| TC | | 1 | 0 |
| TN | | 0 | 0 |
| Sand | % | 62 | 15 |
| Silt | | 36 | 14 |
| Clay | | 3 | 1 |
| As | | 1,807 | 0 |
| Al | | 3,390 | 1 |
| Ald | | 233 | |
| Alo | | 150 | |
| Ca | | 10,893 | 1 |
| Fe | | 19,348 | 1 |
| Fed | mg/kg | 4,919 | |
| Feo | | 5,110 | |
| K | | 388 | 0 |
| Mg | | 6,174 | 1 |
| Mn | | 357 | 0 |
| Mnd | | 73 | |
| Mno | | 105 | |
| P | | 305 | 0 |

## 14.2.2 Soil amendment characterization

The soil amendments utilized in the study included rice husk biochar, iron oxides, NPK fertilizer (15:15:15), compost, and cow dung manure. Compost was derived from partially decomposed dung of small animals (rabbits, rats, and guinea pigs), poultry manure, domestic organic waste, soil, and plant litter. The manure was cow dung sourced from a livestock farm. Iron oxide was purchased from the Amazon marketplace in Germany and primarily consisted of iron and oxides. The amendments were subjected to oven-drying at 80 °C, subsequently crushed, homogenized, and filtered through a sieve with a size of less than 0.63 mm. Subsequently, they were analyzed for pH, electrical conductivity (EC), total carbon and nitrogen, total elements, and heavy metals, similar to the soil analysis.

The total carbon concentration in the amendments was usually elevated, with the highest levels seen in biochar (42%), followed by manure (40%) and compost (24%). The total nitrogen concentration was 1% for biochar, 2% for compost, and 3% for manure. The iron concentrations were greatest in compost (15,120 mg/kg), followed by manure (570 mg/kg) and charcoal (313 mg/kg). Composts exhibited the highest concentrations of Al, Mn, and P at 5,888 mg/kg, 1,864 mg/kg, and 27,360 mg/kg, respectively (Tab. 14.2). All soil additions had heavy metal concentrations below detection limits, except for compost, which contained As content at 3 mg/kg.

**Tab. 14.2:** Basic physicochemical properties and element total contents in the mine spoil.

| Parameter | Unit | Biochar | Compost | Manure |
|---|---|---|---|---|
| pH | | 7 | 7 | 8 |
| EC | µS/cm | 122 | 1,946 | 207 |
| TC | % | 42 | 24 | 40 |
| TN | | 1 | 2 | 2 |
| C/N | | 59 | 11 | 16 |
| Al | mg/kg | 304 | 5,888 | 722 |
| As | | b.d.l | 3 | b.d.l |
| Ca | | 1,344 | 36,880 | 30,720 |
| Fe | | 313 | 15,120 | 570 |
| K | | 6,488 | 30,400 | 14,920 |
| Mg | | 847 | 11,432 | 5,920 |
| Mn | | 359 | 1,864 | 174 |
| Na | | 1,157 | 3,752 | 5,112 |
| P | | 1,277 | 27,360 | 7,408 |

below detection limit

## 14.3 Arsenic geochemical fractions

Arsenic is predominantly found in the amorphous iron oxide fraction (FIII; 49% of total As) and the residual fraction (FV; 38.5% of total As). The sequence included the exchangeable fraction (FII; 10.7%), the water-soluble fraction (FI; 1.2%), and subsequently, the high-crystalline Fe oxide fraction (FIV; 0.5%) (Fig. 14.1). The findings suggest that arsenic in the polluted mining site was closely associated with amorphous iron oxide and residual fractions, as also reported by Mensah et al. [321]. Water-soluble arsenic (FI), the most harmful and bioavailable portion to the environment, is estimated to vary from 5 to 13 mg/kg in mining and heavily contaminated soils [170, 473].

The mobile percentage of arsenic ($\sum$FI – FIV) in the mining spoil soil is 1,183.6 mg/kg, constituting 61.5% of the total. Arsenic in the potentially mobile fraction may be solubilized due to the reductive dissolution of iron-bearing minerals [e.g., 438]. Consequently, this arsenic component may be present in diverse reducing and oxidizing circumstances, thereby presenting toxicological and environmental risks. The mobile percentage of arsenic (FI + FII) was 229 mg/kg, constituting 12% of the total arsenic concentration. Elevated arsenic mobility from arsenic-contaminated gold mine tailings has been documented in Ghana by Mensah et al. [321], in China by Tang et al. [473], and in Spain by García-Sánchez et al. [170].

## 14.4 Experimental design/incubation study

The study utilized an incubation study and laboratory experiments. Five distinct treatments were administered, each at three varying levels, with three replications for each. Compost, iron oxide, manure, and rice husk biochar were individually administered at concentrations of 0.5%, 2%, and 5% (w/w) to 300 g of soil in a jar; additionally, NPK(+S) (15:15:15 + 11) fertilizer was added at rates of 5, 0.2, and 0.1 g/kg (Tab. 14.3). Water was introduced to achieve a soil water content of 70% field capacity, and the setups, including the control, were maintained in a dark room at 20 °C, organized in a completely randomized design for 28 days.

## 14.5 Chemical analyses

### 14.5.1 Analyses of soil properties

Soil samples for analysis were taken at 1- and 28-day intervals. They were analyzed for their gravimetric moisture content, pH, EC, dissolved organic carbon (DOC), anions ($F^-$, $Cl^-$, $NO_3^{3-}$, $PO_4^{3-}$, and $SO_4^{2-}$), water-soluble arsenic, and phytoavailable arsenic. To measure pH, EC, DOC, and the anions, 10 g of moist soil was weighed into a 100-mL PE bottle, followed by the addition of 50 mL of deionized water. The mixture was shaken for 2 h at ambient temperature and subsequently filtered using 0.45-μm filter paper. The filtrates were analyzed for pH, EC, DOC, and anions. Furthermore, samples of the soils post-28-day incubation were collected, preserved, and oven-dried at 60 °C; they were subsequently analyzed for exchangeable cations, cation exchange capacity (CEC), total carbon, nitrogen, and accessible phosphorus.

The available phosphorus in the soil was determined using the P-Bray method, soil moisture content was measured via the gravimetric water content method, exchangeable cations ($Ca^{2+}$, $Mg^{2+}$, $K^+$, and $Na^+$) were determined using the hexamine cobalt chloride method, and cation exchange capacity was estimated by summing all exchangeable cations. Soil pH was measured using a pH meter (Sentix 41, WTW GmbH, Weilheim, Germany), EC was determined with an EC meter (TetraCon 325, WTW GmbH, Weilheim, Germany), DOC, total carbon, and nitrogen were analyzed using an elemental analyzer (Vario MAX cube, Elementar Analysensysteme GmbH, Hanau, Germany), and anions were analyzed using ion chromatography (Metrohm, 881 Compact IC Pro).

### 14.5.2 Determination of water-soluble As, phytoavailable As, and mobile As

For the assessment of water-soluble As and phytoavailable As in the incubated samples, 2 g of the obtained wet samples was utilized. The contents were determined in accordance with Wenzel et al. [513]. To extract water-soluble arsenic, 25 mL of 0.05 M $(NH_4)_2SO_4$ solution was introduced to the moist soil, shaken for 4 h at 20 °C, centrifuged at 4,000 rpm for 15 min, and subsequently filtered using 0.45-μm filter paper. For phytoavailable arsenic, the same process was employed, but extraction was conducted using 25 mL of 0.05 M $(NH_4)H_2PO_4$, with shaking for 16 h. The As concentrations were quantified from the filtrates using ICP. The mobile arsenic was determined by summing the water-soluble arsenic and phytoavailable arsenic.

## 14.6 Data treatment and statistical analyses

We conducted a one-way ANOVA to compare the means of various treatment rates (control, biochar, compost, manure, iron oxide, and NPK fertilizer). Prior to running the ANOVA, we performed a normality assessment of the data utilizing the Shapiro-Wilk test. In these tests, variables with $P > 0.05$ were deemed normally distributed, whereas those with $P < 0.05$ were classified as not normally distributed. We conducted multiple range tests utilizing Tukey's honestly significant difference test to compare treatment averages at $P < 0.05$. Furthermore, we conducted a Pearson moment correlation matrix analysis among water-soluble As, phytoavailable As, mobile As, and various potential As-controlling soil factors including moisture content, pH, EC, DOC, anions ($F^-$, $Cl^-$, $NO_3^-$, $PO_4^{3-}$, and $SO_4^{2-}$), available P, and exchangeable cations. Statistical analyses were conducted utilizing IBM SPSS Statistics 25 (NY, USA), and figures illustrating means of measured values were generated with OriginPro 9.1 b215 (Origin-Lab Corporation, Northampton, USA) software.

## 14.7 Results and discussion

### 14.7.1 Changes in pH, electrical conductivity (EC), and soluble anion contents ($Cl^-$, $NO_3^-$, and $SO_4^{2-}$)

The alterations in soil physicochemical parameters after treatments with soil amendments during 24-h and 28-day incubations are presented in Tabs. 14.3a and 14.3b. During the 24-h interval, the soil pH fluctuated from 6.1 to 7.6, with the maximum pH recorded in the 5% iron oxide treatment and the minimum in the 5% compost treatment. The 5% iron oxide treatment increased the pH of the control by 13%, whereas the 5% compost addition reduced the pH by 8.6%; however, these changes were not statistically significant (Tab. 14.3a). During the 28-day incubation period, the addition of compost, manure, and biochar at 5% significantly ($P < 0.05$) increased the soil pH in the control from 6.8 to 7.9, 7.8, and 7.5, respectively (Tab. 14.3b). The increase in pH following the addition of iron oxides, compost, manure, and biochar may be attributed to the presence of positive charges. Danila et al. [120] reported an increase in soil pH of arsenic-contaminated mining soil treated with 2% iron oxides in Sweden.

EC ranged between 1,207.3 and 2,275.7 µS/cm throughout the entire incubation period. The application of NPK fertilizer at 5 g/kg and 5% manure additions considerably ($P < 0.05$) increased the EC value in the control from 1,577.3 to 2,275.7 µS/cm and 2,101.0 µS/cm, respectively, by the conclusion of the 28-day incubation period. In the same time frame, 5% iron oxide decreased the EC from 1,577.3 to 1,266.7 µS/cm (Tab. 14.3b).

The concentration of chloride in the control soil rose with the incremental addition of compost, manure, and NPK fertilizer. Throughout the 24-h incubation period, 5% compost markedly elevated chloride levels in the control soil from 31.8 to 175.1 mg/kg, while 5% manure augmented it to 286.9 mg/kg, and 5 g/kg NPK fertilizer further enhanced it to 425.5 mg/kg. Similarly, throughout the 28-day incubation period, the addition of 5% compost markedly elevated chloride level from 26.8 to 168.3 mg/kg, further to 266.5 mg/kg with the incorporation of 5% manure, and to 441.7 mg/kg with the application of 5 g/kg NPK fertilizer.

The most significant increase in chloride concentration in our samples was observed with 5 g/kg NPK fertilizer. Nonetheless, 2% iron oxide markedly diminished chloride levels by 79%, whilst 5% iron oxide decreased it by 69%. Moreover, biochar at concentrations of 0.5% and 5% dramatically lowered chloride levels by 75% and

**Tab. 14.3a:** Changes in soil physicochemical parameters (moisture, pH, EC, DOC, and anions) within 24-h incubation period following treatments with soil amendments.

| Treatment | pH | SD | EC (μS/cm) | SD | TOC (mg/L) | SD | Moisture (%) | SD | Chloride (mg/kg) | SD | Nitrate (mg/kg) | SD | Sulfate (mg/kg) | SD |
|---|---|---|---|---|---|---|---|---|---|---|---|---|---|---|
| Control | 7.0abc | 0.8 | 1,612.7abcd | 51.9 | 11.2a | 6.8 | 23.3a | 0.8 | 31.8ab | 12.1 | 9.6ab | 0.8 | 5,573.6 cd | 949.0 |
| C0.5 | 5.9a | 0.5 | 1,686.0bcd | 60.8 | 16.7a | 3.2 | 23.9a | 1.8 | 26.1b | 7.5 | 21.91c | 3.6 | 4,922.9abcd | 218.5 |
| C2 | 6.6abc | 0.1 | 1,756.0cde | 47.3 | 20.7a | 1.1 | 23.0a | 0.4 | 67.9b | 2.2 | 55.38d | 3.0 | 4,707.6abcd | 52.5 |
| C5 | 6.1ab | 1.1 | 1,839.3de | 55.8 | 44.69b | 3.2 | 24.2a | 0.4 | 175.07d | 7.9 | 126.77e | 3.2 | 4,214.43a | 305.7 |
| F0.5 | 7.4bc | 0.0 | 1,704.0bcd | 9.0 | 5.2a | 0.3 | 21.8a | 1.4 | 37.7a | 58.2 | 10.5ab | 1.8 | 5,119.4abcd | 131.5 |
| F2 | 7.5c | 0.0 | 1,589.0abc | 6.6 | 8.0a | 0.8 | 23.0a | 1.3 | 8.9a | 4.4 | 9.5ab | 0.9 | 4,651.2abc | 56.6 |
| F5 | 7.6c | 0.0 | 1,697.7bcd | 34.0 | 10.7a | 0.3 | 23.8a | 2.0 | 8.7a | 3.2 | 8.9ab | 1.3 | 5,013.8abcd | 189.5 |
| M0.5 | 7.5c | 0.4 | 1,642.0abcd | 14.7 | 17.2 | 5.5 | 11.5a | 1.1 | 40.2ab | 4.3 | 8.6ab | 1.6 | 5,315.2bcd | 231.5 |
| M2 | 7.0abc | 0.8 | 1,784.3cde | 37.0 | 53.75b | 23.0 | 23.5a | 0.3 | 123.06c | 8.4 | 10.9ab | 3.0 | 4,915.7abcd | 22.2 |
| M5 | 6.9abc | 0.2 | 1,952.33e | 44.7 | 114.4c | 5.4 | 25.5a | 0.7 | 286.87e | 6.5 | 14.4b | 3.6 | 4,722.4bcd | 165.8 |
| N0.1 | 7.4bc | 0.0 | 1,522.0ab | 70.5 | 3.8a | 1.0 | 27.4a | 3.7 | 16.5a | 7.4 | 8.2ab | 0.9 | 4,427.39ab | 335.1 |
| N0.2 | 7.4bc | 0.0 | 1,580.7abc | 166.4 | 5.1a | 0.4 | 25.2a | 2.6 | 31.7ab | 8.2 | 10.2 | 0.6 | 4,725.9abcd | 545.0 |
| N5 | 7.5c | 0.2 | 2,410f | 10.0 | 8.7a | 2.4 | 22.6a | 6.0 | 425.53 f | 10.2 | 9.2ab | 1.1 | 5,709.9d | 180.6 |
| B0.5 | 7.5c | 0.1 | 1,426.3a | 196.8 | 3.9a | 1.6 | 22.7a | 1.1 | 7.4a | 2.2 | 6.5a | 0.3 | 4,964.8abcd | 312.5 |
| B2 | 7.2abc | 0.4 | 1,557.7abc | 51.5 | 11.3a | 8.0 | 22.1a | 1.3 | 16.6a | 4.7 | 6.9a | 1.3 | 4,843.3abcd | 149.9 |
| B5 | 7.3abc | 0.4 | 1,716.0bcd | 37.4 | 4.0a | 1.5 | 22.3a | 1.1 | 10.6a | 2.7 | 4.8a | 0.8 | 4,708.9abcd | 156.4 |

C0.5, C2, and C5 = soil treatment with compost at 0.5%, 2%, and 5%, respectively; F0.5, F2, and F5 = iron oxide at 0.5%, 2%, and 5%, respectively; M0.5, M2, and M5 = manure at 0.5%, 2%, and 5%, respectively; N0.1, N0.2, and N5 = NPK fertilizer at 0.1, 0.2, and 5 g/kg, respectively; and B0.5, B2, and B5 = rice husk biochar at 0.5%, 2%, and 5%, respectively. Means with different letters differ significantly among treatments at $P < 0.05$ and same letters indicate no significant differences among means according to the Tukey's honestly significant different test.

**Tab. 14.3b:** Changes in soil physicochemical parameters (moisture, pH, EC, DOC, and anions) at the end of 28-day incubation period following treatments with soil amendments.

| Treatment | pH | SD | EC (µS/cm) | SD | TOC (mg/l) | SD | Moisture (%) | SD | Cl⁻ (mg/kg) | SD | NO₃⁻ (mg/kg) | SD | SO₄²⁻ (mg/kg) | SD |
|---|---|---|---|---|---|---|---|---|---|---|---|---|---|---|
| Control | 6.8a | 0.3 | 1,577.3e | 35.9 | 11.5b | 1.7 | 21.9a | 1.5 | 26.8de | 1.1 | 7.8a | 1.9 | 5,217.9a | 140.6 |
| C0.5 | 7.6bc | 0.1 | 1,844.0fg | 73.7 | 6.4ab | 1.9 | 18.9a | 7.2 | 24.5bcde | 7.7 | 18.1b | 3.8 | 5,057.6a | 212.4 |
| C2 | 7.6bc | 0.1 | 1,863.0fgh | 74.1 | 11.2ab | 0.8 | 23.0a | 1.9 | 74.7f | 2.4 | 55.0c | 2.1 | 4,646.2a | 314.9 |
| C5 | 7.9c | 0.1 | 1,981.3fg | 41.9 | 25.5c | 2.9 | 21.5a | 0.5 | 168.3g | 0.7 | 6.1a | 0.0 | 4,577.8a | 42.6 |
| F0.5 | 7.3ab | 0.0 | 1,245.3ab | 138.5 | 1.5a | 2.4 | 33.0a | 1.4 | 8.9abcd | 6.0 | 9.0a | 1.4 | 5,023.5a | 695.9 |
| F2 | 7.4bc | 0.1 | 1,317.3abcd | 53.2 | 6.7ab | 3.5 | 36.0a | 4.5 | 5.6a | 0.9 | 8.1a | 0.6 | 4,795.0a | 205.0 |
| F5 | 7.4bc | 0.1 | 1,266.7abc | 61.2 | 6.0ab | 2.5 | 27.3a | 11.8 | 8.1abc | 2.1 | 8.3a | 0.3 | 4,869.4a | 330.7 |
| M0.5 | 7.6bc | 0.0 | 1,711.7ef | 55.2 | 8.3ab | 1.1 | 22.9a | 0.2 | 36.6e | 0.9 | 7.9a | 3.2 | 5,454.0a | 98.1 |
| M2 | 7.8bc | 0.2 | 1,895.3fgh | 96.6 | 36.7c | 11.5 | 23.5a | 0.2 | 118.0f | 5.1 | 6.0a | 1.7 | 3,608.0a | 2,670.6 |
| M5 | 7.8bc | 0.0 | 2,101.0hi | 197.5 | 104.2d | 5.0 | 22.7a | 0.7 | 266.5h | 8.7 | 8.9a | 3.0 | 4,883.0a | 157.2 |
| N0.1 | 7.3ab | 0.1 | 1,238.3ab | 60.1 | 4.5ab | 2.3 | 29.7a | 15.2 | 13.8abcd | 6.1 | 7.9a | 1.9 | 4,393.9a | 251.0 |
| N0.2 | 7.4bc | 0.1 | 1,207.3a | 56.1 | 5.7ab | 3.0 | 32.4a | 9.2 | 25.7cde | 6.4 | 6.9a | 0.8 | 4,186.9a | 134.7 |
| N5 | 7.6bc | 0.4 | 2,275.7i | 107.9 | 6.5ab | 2.9 | 21.9a | 1.7 | 441.7j | 16.3 | 7.9a | 1.0 | 5,691.2a | 281.3 |
| B0.5 | 7.3ab | 0.5 | 1,538.0de | 11.8 | 4.0a | 0.0 | 22.5a | 0.5 | 6.6ab | 3.6 | 7.2a | 4.0 | 4,994.4a | 76.3 |
| B2 | 7.3ab | 0.1 | 1,484.0bcde | 28.0 | 4.4ab | 0.0 | 22.2a | 0.9 | 10.5abcd | 1.9 | 4.5a | 0.9 | 4,608.1a | 250.5 |
| B5 | 7.5bc | 0.1 | 1,515.0cde | 44.5 | 2.2a | 0.4 | 23.2a | 0.4 | 8.5abc | 1.0 | 3.0a | 0.3 | 4,879.7a | 145.3 |

C0.5, C2, and C5 = soil treatment with compost at 0.5%, 2%, and 5%, respectively; F0.5, F2, and F5 = iron oxide at 0.5%, 2%, and 5%, respectively; M0.5, M2, and M5 = manure at 0.5%, 2%, and 5%, respectively; N0.1, N0.2, and N5 = NPK fertilizer at 0.1, 0.2, and 5 g/kg, respectively; and B0.5, B2, and B5 = rice husk biochar at 0.5%, 2%, and 5%, respectively. Means with different letters differ significantly among treatments at $P < 0.05$ and same letters indicate no significant differences among means according to the Tukey's honestly significant different test.

68.3%, respectively. Significant alterations in nitrate concentrations were seen alone with the addition of compost. Specifically, 0.5%, 2%, and 5% compost increased the nitrate concentration in the control from 9.6 to 22 mg/kg, 55.4 mg/kg, and 126.8 mg/kg, respectively. No treatments resulted in notable alterations in the concentration of sulfates in our soils.

## 14.7.2 Changes in soil exchangeable cations and CEC

At the conclusion of the 28-day incubation period, the incorporation of soil amendments did not significantly influence the soil's exchangeable $Ca^{2+}$ or the overall cation exchange capacity (CEC). Nonetheless, the only application of compost and manure at 5% enhanced the soil's exchangeable $K^+$, $Mg^{2+}$, and $Na^+$ levels. The enhancement in soil exchangeable cations following the incorporation of compost and manure may be attributed to the comparatively elevated concentrations of the cations present in the compost and manure utilized in the study (Tab. 14.2). The elevated levels of the cations in soils amended with compost and manure may be ascribed to their release into soil exchange sites from functional groups, such as carboxylic and phenolic acids, present on the surfaces of stabilized organic matter [277]. Mensah and Frimpong [326] observed that the maximum compost application rate of 2% enhanced the exchangeable magnesium, potassium, and sodium contents in degraded soils in Ghana.

Elevated contents of exchangeable cations in soil have environmental ramifications for regulating soil pH, preserving nutrient retention (thereby mitigating the loss of cations and other nutrients), and influencing the transport and mobilization of arsenic in polluted mining soils. Compost application is said to have a liming impact owing to its abundance of alkaline cations including Ca, Mg, and K, which are released from organic matter through elevated mineralization rates [e.g., 159, 326]. Thus, elevated pH may enhance As solubility and mobilization, as elucidated in Section 14.3.2. The presence of the basic cations may increase the soil's ion binding capacity, hence enhancing its sorption ability [376], which could subsequently bind arsenic to their surfaces and curtail its availability.

## 14.7.3 Changes in soil C, N, and P availability

The addition of 5% compost and manure considerably influenced DOC in our experiment. Composting at 5% over a 24-h period notably elevated the DOC in the control from 11.2 to 44.7 mg/L, whereas 5% manure raised the DOC to 114.4 mg/L. Comparable observations were recorded during the 28-day incubation period (Tabs. 14.3a, b). However, 0.5% iron oxides considerably reduced DOC by 86.6%, whilst 5% biochar decreased DOC by 80%. Consequently, elevated doses of compost and manure increased the DOC levels, whereas iron oxides and biochar diminished these levels in our soils.

The elevated DOC in the manure-treated soils may result from the availability and re-lease of carbon content through mineralization.

At the conclusion of the 28-day incubation period, compost, manure, and biochar enhanced the total carbon content of the mining spoil in conjunction with increased amendment application rates, see Fig. 14.4. The carbon content in the control in-creased by 58% with the addition of 5% compost, 115.7% with 5% manure, and 99% with 5% biochar treatment. Consequently, the most significant increases in carbon content in the mining spoil were observed with elevated applications of manure and biochar, which can be ascribed to the higher carbon concentrations in the biochar (42%) and manure (40%) utilized in the studies.

In a comparable incubation investigation, Frimpong et al. [164] found that carbon concentrations were elevated in soils amended with cow dung manure and biochar compared to the control group. Comparable results were observed regarding the total nitrogen content in the soil, where increased applications of compost, manure, and biochar resulted in a notable enhancement of nitrogen levels in the mine spoil. The incorporation of compost, manure, and biochar at 5% each resulted in percentage in-creases in the nitrogen content of mine debris by 227%, 344%, and 59%, respectively (Tab. 14.3). Furthermore, increased biochar applications resulted in an enhancement of the spoil C/N ratio (+25%), but greater quantities of compost, manure, and NPK fer-tilizer caused a notable reduction in the C/N ratios.

The compost and manure exhibited comparatively elevated intrinsic nitrogen concentrations, suggesting a greater rate of nitrogen mineralization than that of bio-char. This elucidates the comparatively elevated C:N ratio in biochar-amended soils, suggesting a potential reduction in the rates of breakdown and mineralization. The conclusion of these findings is that nutrients in compost and manure-enhanced soils may be readily lost through leaching, infiltration, and runoff, whereas those in bio-char amended soils may be retained and stored for an extended duration or many seasons for crop utilization. Lehmann et al. [267] indicated that compost is readily de-gradable, but biochar possesses an aromatic structure and a refractory character, ren-dering it very resistant to breakdown.

Duku et al. [791] indicated that the application of biochar to soils reduces ammo-nium loss by leaching and $NH_3$ volatilization. Moreover, De Gryze et al. [785] indicated that biochar mitigates nutrient loss in soils and promotes nutrient recycling, hence yielding beneficial effects on crop production over time through gradual release into the soil. The elevated carbon content in the manure/biochar-treated mine soils sug-gests that the application of biochar and/or manure may promote carbon accumula-tion and sequestration in these soils.

The available phosphorus in the soil of our study varied from 7.1 to 123.8 mg/kg, with the minimum recorded in 5% iron oxide and the maximum in 5% biochar as seen in Fig. 14.5. The availability of phosphorus in the soil diminished proportionally with increased applications of compost, iron oxides, and manure. Consequently, ma-nure, compost, and iron oxides may have supplied sorption sites for phosphorus. As a result, the concentration of mine spoil accessible phosphorus decreased from 118.5 mg/kg in the control to 60.3 mg/kg (−49%), 12.6 mg/kg (−89%), and 7.1 mg/kg (−94%) with the addition of 5% manure, compost, and iron oxides, respectively. The observed decreases can be ascribed to the elevated levels of Al, Fe, and Mn present in the compost and manure utilized in the study, whereas the significant reduction attributed to FeO is likely a result of its substantial Fe concentration.

Elevated levels of Al, Fe, and Mn can sequester P at soil exchange locations, thereby diminishing its availability. An (2019) found that the utilization of Fe-based sorbent reduced the available phosphorus content in soil, resulting in phosphorus deficit for plant growth. Furthermore, [1040] indicated that elevated iron concentrations in a wetland diminished phosphorus mobilization into neighboring rivers and mitigated potential eutrophication through sorption onto precipitating Fe(III) oxyhydroxides in northeastern Germany.

Moreover, available P may be complexed or precipitated on the surfaces of compost and manure-treated mine ruin soils due to the presence of positively charged ions such as calcium, potassium, magnesium, and sodium. We conclude that the decrease in soil available phosphorus in the mining soil resulting from the application of compost and manure may indicate that these organic amendments are ineffective in facilitating the complexation of cations, such as iron and aluminum, which are known to cause phosphorus fixation in soils. Diogo et al. [787] observed an increase in soil available phosphorus in a mining soil in western Rwanda, attributable to the complexation of iron and aluminum cations by manure, which is commonly associated with phosphorus fixation in tropical soils.

## 14.7.4 Water-soluble and phytoavailable As as affected by addition of biochar, compost, iron oxide, manure, and inorganic fertilizer

### 14.7.4.1 Water-soluble As

The water-soluble arsenic concentration in the control soil (20.5 mg/kg) is thrice greater than the standard total concentration of 6.8 mg/kg for the global soil average [230]. Overall, water-soluble arsenic decreased as the amount of iron oxides increased during the incubation period. For example, 5% iron oxide decreased water-soluble arsenic from 20.5 to 1.4 mg/kg (i.e., −93.4%) over the 28-day incubation period (Tab. 14.3), while arsenic solubility diminished by 89% within 24 h. The mining spoil soil already possesses a significant concentration of total Fe (19,343 mg/kg, Tab. 14.1).

The geochemical fractionation data (Fig. 14.1) demonstrated that a significant proportion of arsenic was associated with amorphous iron oxides (45%), while 0.5% constituted the water-soluble fraction. Consequently, we posit that further treatment of the mining debris with iron oxides has diminished the already minimal water-soluble amount. Danila et al. [120] and Komárek et al. [251] indicated that elevated Fe concentrations in mine soils correspondingly enhance the immobilization efficiency of As. The adsorption and binding of arsenic by iron oxide can be ascribed to the abundance of positive charges on the surfaces of iron oxides. Positive charges may predominate under acidic conditions, as elucidated in Section 14.3.2.

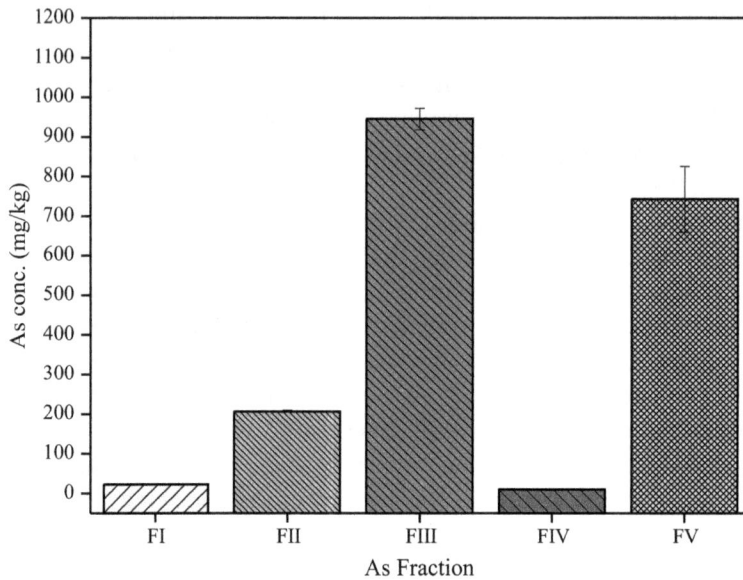

**Fig. 14.1:** Distribution of As geochemical fractions in the mine spoil in mg/kg. Fr. I = non-specifically sorbed As (water-soluble), Fr. II = specifically sorbed-As, Fr. III = amorphous iron oxide fraction, Fr. IV = high-crystalline Fe oxide fraction, FV = residual fraction.

The stability of FeO in arsenic immobilization is significantly influenced by the pH and redox conditions of the environment. Under reductive circumstances and elevated pH, arsenic bound to the surfaces of iron oxide may be released and become accessible [146, 145, 165]. Furthermore, iron oxyhydroxides in soil may gradually convert into more crystalline Fe oxides (e.g., hematite), which possess a reduced specific surface area [120]. This may result in the desorption of previously adsorbed arsenic and enhance its leaching over time [301].

Conversely, As solubility rose correspondingly with the addition of compost, manure, and NPK fertilizer throughout both the 24-h and 28-day incubation periods. For example, water-soluble arsenic elevated to 101.0 mg/kg (+317%) with 5% compost and to 101.1 mg/kg with 5 g/kg NPK (+277%) (Tab. 14.3). The 5% increase in water-soluble arsenic due to compost, manure, and NPK fertilizer may be attributed to the elevated salinity and chloride levels resulting from the incorporation of these amendments into the soil.

The elevated EC values and chloride concentrations in soils treated with compost, manure, and NPK fertilizer further substantiate this observation (Tabs. 14.2a and b). The elevated salinity and chloride levels may facilitate the desorption of complexed arsenic, resulting in its release. Furthermore, there were highly significant positive correlations between water-soluble arsenic, EC, and chloride during the 28-day incubation period (Tab. 14.4 and Tab. 14.5). Consequently, the solubility of water-soluble arsenic in our study may be significantly influenced by salinity and chloride levels, with elevated salinity and chloride concentrations in the mine spoil potentially leading to increased arsenic solubility, and conversely.

Furthermore, the elevation of water-soluble arsenic (Fig. 14.2) may be associated with the elevated carbon content in the compost and manure utilized in our study. Hartley et al. [199] documented elevated levels of arsenic in compost-amended soils at a brownfield site in the UK, attributing the increase to the production of arsenic-DOC complexes on the surfaces of the compost. Our current analysis revealed a substantial positive correlation between water-soluble arsenic and DOC ($r$ = +0.53, $P$ < 0.01; Tab. 14.5). Consequently, we posit that water-soluble arsenic content may have formed complexes with DOC in the soils modified with compost and manure, thereby enhancing arsenic availability.

The incorporation of 5% biochar markedly enhanced arsenic solubility by 76% over 24 h and by 46% over 28 days. The enhancement in arsenic solubility due to biochar may be attributed to deprotonation [69], resulting from the pH elevation of the mine debris following biochar application. Biochar elevates soil pH and exhibits liming effects, as documented by Mensah and Frimpong [326]. This can be ascribed to the existence of basic cations in the ash generated during the biochar production process [471]. Thus, a rise in pH may enhance arsenic solubility from the mine debris throughout the 28-day incubation period, as corroborated by other studies [e.g., 69]. This is further evidenced by the substantial positive connection ($r$ = +0.64, $P$ < 0.01) between pH and water-soluble arsenic during the 28-day duration.

### 14.7.4.2 Phytoavailable As

The application of 5% compost diminished the phytoavailable arsenic in mine debris from 265.6 to 179.4 mg/kg (−36%) (Tab. 14.3 and Fig. 14.3). The decrease in phytoavailable arsenic can be ascribed to the elevated levels of aluminum, iron, and manganese in the compost amendments, as indicated in Tab. 14.2. The elevated levels of Al, Fe, and Mn may have offered sorption sites for the bioavailable arsenic in the soil, thereby facilitating its reduction for potential uptake. Furthermore, the compost comprises organic components and possesses a comparatively high carbon content of 24%. Elevated soil carbon levels can serve as reservoirs of plant-available arsenic, thereby diminishing its absorption. This is further evidenced by the substantial negative correlation ($r$ = −0.6, $P$ < 0.05; Tab. 14.5) between total soil carbon and plant-available arsenic. Consequently, an increase in carbon content correlates with a decrease in the availability of arsenic for

**Tab. 14.4:** Percentage increase and decrease in mine spoil soil water-soluble As, phytoavailable As, and soil quality parameters at the end of the 28-day incubation following addition with soil amendments.

| Treatment | Water-soluble As | Phytoavailable As | Mobile As | K (mmolc/kg) | Mg (mmolc/kg) | Na (mmolc/kg) | N% | C% | C/N | Avail P (mg/kg) |
|---|---|---|---|---|---|---|---|---|---|---|
| C0.5 | - 112 | 12 | 4 | - 511 | - 20 | - 121 | - 19 | - 4 | 13 | 34 |
| C2 | - 328 | 24 | 3 | - 1,795 | - 47 | - 296 | - 92 | - 20 | 38 | 64 |
| C5 | - 317 | 36 | 15 | - 4,285 | - 122 | - 625 | - 227 | - 58 | 52 | 89 |
| F0.5 | 56 | 4 | 7 | - 2 | - 25 | 39 | - 3 | - 4 | - 1 | 42 |
| F2 | 84 | 8 | 12 | - 212 | - 12 | - 25 | - 13 | - 5 | 7 | 80 |
| F5 | 93 | 7 | 13 | 0 | - 19 | 12 | - 29 | - 14 | 11 | 94 |
| M0.5 | - 23 | - 9 | - 10 | - 173 | - 23 | - 105 | - 42 | - 12 | 22 | 1 |
| M2 | - 166 | 14 | 3 | - 928 | - 42 | - 497 | - 132 | - 37 | 41 | 38 |
| M5 | - 312 | 12 | - 7 | - 2,099 | - 68 | - 1,082 | - 344 | - 115 | 52 | 49 |
| N0.1 | 10 | 1 | 1 | - 53 | 6 | 19 | 6 | 4 | - 3 | 23 |
| N0.2 | - 9 | 8 | 7 | - 89 | - 2 | 33 | 3 | 3 | 1 | 22 |
| N5 | - 277 | 11 | - 6 | - 2,060 | - 46 | - 22 | - 65 | - 1 | 39 | 21 |
| B0.5 | - 150 | 0 | - 9 | - 43 | - 10 | 47 | - 7 | - 10 | - 3 | - 5 |
| B2 | - 27 | 0 | - 2 | - 167 | 5 | 46 | - 26 | - 51 | - 19 | 6 |
| B5 | - 46 | 14 | 11 | - 402 | 10 | 54 | - 59 | - 99 | - 25 | - 4 |

C0.5, C2, and C5 = soil treatment with compost at 0.5%, 2%, and 5%, respectively; F0.5, F2, and F5 = iron oxide at 0.5%, 2%, and 5%, respectively; M0.5, M2, and M5 = manure at 0.5%, 2%, and 5%, respectively; N0.1, N0.2, and N5 = NPK fertilizer at 0.1, 0.2, and 5 g/kg, respectively; and B0.5, B2, and B5 = rice husk biochar at 0.5%, 2%, and 5%, respectively. (-) means percentage increase as compared to the control; (+) means percentage decrease as compared to the control.

**Tab. 14.5:** Correlation matrix between water-soluble As, phytoavailable As, mobile As, and relevant As-governing soil chemical factors at the end of 28-day incubation period.

| | pH | TOC | Chloride | Nitrate | Sulfate | Water-soluble As | Phytoavailable As | Mobile As | EC | ExCa | ExK | ExMg | ExNa | TC | Avail P | N |
|---|---|---|---|---|---|---|---|---|---|---|---|---|---|---|---|---|
| TOC | NS | 1 | 0.52* | NS | NS | 0.5* | NS | NS | NS | Ns | 0.55* | 0.63** | 0.97** | 0.68** | NS | 16 |
| Chloride | NS | 0.52* | 1 | NS | NS | 0.78** | -0.55* | NS | 0.93** | NS | 0.68** | 0.62* | 0.54* | NS | NS | 16 |
| Nitrate | -0.7** | NS | NS | 1 | -0.51* | 0.62* | -0.7** | NS | NS | -0.55* | 0.84** | 0.81** | NS | NS | -0.54* | 16 |
| Sulfate | ns | NS | NS | -0.51* | 1 | NS | ns | NS | NS | NS | NS | NS | NS | NS | NS | 16 |
| Water-soluble As | -0.58* | 0.5* | 0.78** | 0.62* | NS | 1 | -0.8** | NS | 0.74** | -0.51* | 0.9** | 0.78** | 0.63** | NS | NS | 16 |
| Phytoavailable As | 0.54* | NS | -0.55* | -0.7** | NS | -0.8** | 1 | 0.62* | -0.52* | 0.57* | -0.83** | -0.73** | -0.6* | -0.6* | NS | 16 |
| Mobile As | Ns | NS | NS | Ns | NS | NS | 0.62* | 1 | NS | NS | NS | NS | NS | -0.55* | NS | 16 |
| EC | NS | NS | 0.93** | NS | NS | 0.74** | -0.52* | NS | 1 | NS | 0.61* | 0.54* | NS | NS | NS | 16 |
| CEC | NS | NS | NS | NS | NS | NS | NS | NS | NS | NS | NS | 0.67** | 0.52* | NS | -0.61* | 16 |
| C | NS | 0.68** | NS | NS | NS | NS | -0.6* | -0.55* | NS | NS | NS | NS | 0.64** | 1 | NS | 16 |

NS: nonsignificant; *indicates significant relationships at $P < 0.05$; **indicates significant relationships at $P < 0.01$; $N$ = number of observations; TOC = total organic carbon; EC = electrical conductivity; ExCa = exchangeable calcium; ExK = exchangeable potassium; ExMg = exchangeable magnesium; ExNa = exchangeable sodium; CEC = cation exchange capacity; TC = soil total carbon; avail P = soil available P; $n$ = number of observations.

plant absorption and vice versa. Karczewska et al. [245] indicated that amending soil with organic materials, such as compost, effectively reduced the concentration of plant-available arsenic in the shoots of ryegrass. Furthermore, Gadepalle et al. [173] discovered less plant-available arsenic in soils amended with 5% compost.

None of the treatments substantially influenced alterations in the phytoavailable arsenic in the mining spoil with the exception of 2% compost and manure, 5 g/kg NPK, and 5% biochar at the 28-day incubation period (Tab. 14.3). During the 28-day incubation period, 5 g/kg NPK and 5% biochar marginally diminished As phytoavailability (Fig. 14.3) by 11% and 14%, respectively. The decrease in phytoavailable arsenic with the addition of NPK is presumed to result from its displacement by the phosphorus provided by the inorganic fertilizer. As functions as a phosphorus analog and is absorbed by plants through a phosphorus transporter mechanism [310, 387].

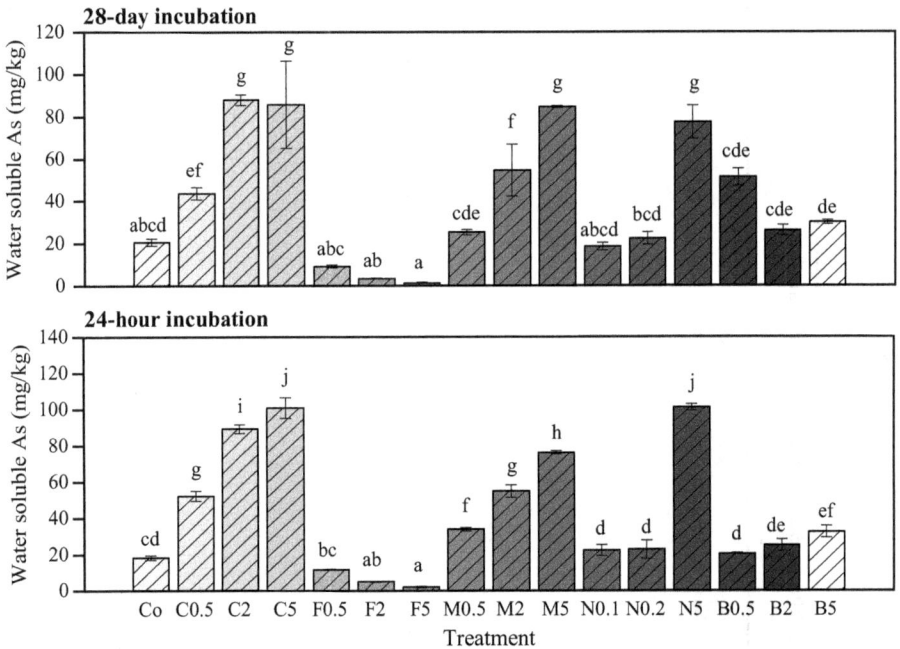

**Fig. 14.2:** Changes in the mine spoil water-soluble As contents during the 1-day and 28-day incubation period following treatments with biochar, compost, iron oxide, manure and inorganic fertiliser. Co = control; C0.5, C2, C5 = soil treatment with compost at 0.5%, 2% and 5%, respectively; F0.5, F2, F5 = iron oxide at 0.5%, 2% and 5%, respectively; M0.5, M2, M5 = manure at 0.5%, 2% and 5%, respectively; N0.1, N0.2, N5 = NPK fertiliser at 0.1g/kg, 0.2g/kg and 5 g/kg, respectively; and B0.5, B2, B5 = rice husk biochar at 0.5%, 2% and 5%, respectively. Charts represent means of three replicates and error bars represent their standard deviations. Means with different letters differ significantly among treatments at P < 0.05 and same letters indicate no significant differences among means according to the Tukey's Honestly Significant Different test.

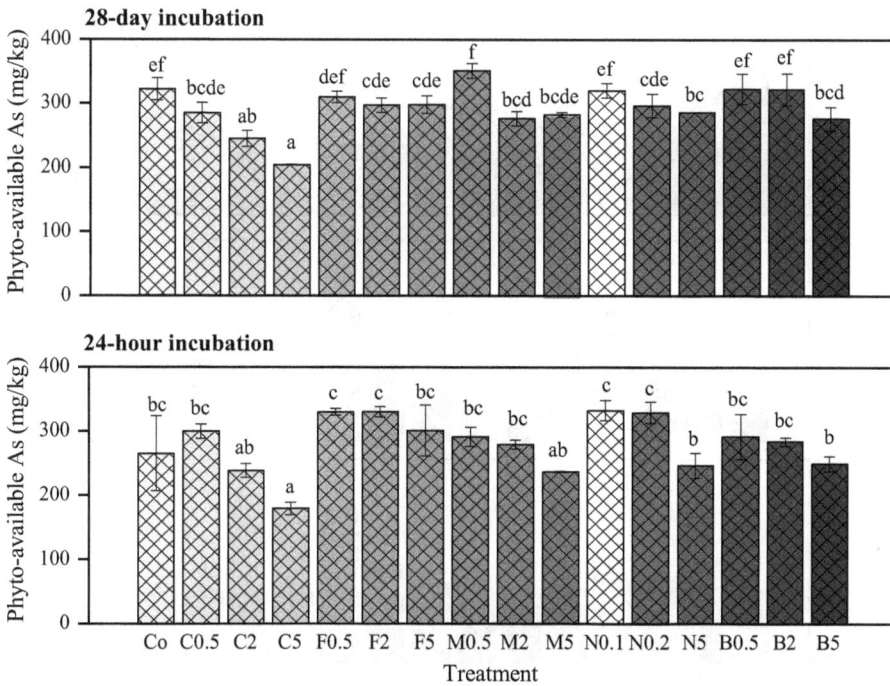

**Fig. 14.3:** Changes in the mine spoil phyto-available As contents during the 1-day and 28-day incubation period following treatments with biochar, compost, iron oxide, manure and inorganic fertiliser. Co = control; C0.5, C2, C5 = soil treatment with compost at 0.5%, 2% and 5%, respectively; F0.5, F2, F5 = iron oxide at 0.5%, 2% and 5%, respectively; M0.5, M2, M5 = manure at 0.5%, 2% and 5%, respectively; N0.1, N0.2, N5 = NPK fertiliser at 0.1g/kg, 0.2g/kg and 5 g/kg, respectively; and B0.5, B2, B5 = rice husk biochar at 0.5%, 2% and 5%, respectively. Charts represent means of three replicates and error bars represent their standard deviations. Means with different letters differ significantly among treatments at $P < 0.05$ and same letters indicate no significant differences among means according to the Tukey's Honestly Significant Different test.

In this context, we can hypothesize that phosphorus may have been absorbed by displacing the phytoavailable arsenic from the soil solution. This is corroborated by Beesley et al. [68], who observed that fertilization with phosphorus (P) inhibited arsenic (As) uptake by plants and promoted phosphorus availability, as arsenic and phosphorus compete for binding sites in soils, resulting in less soil phytoavailable arsenic for plant absorption. Furthermore, it has been shown that phosphorus deficit in soil solution might augment arsenic uptake, whilst elevated phosphorus levels may impede arsenic absorption [386].

The decrease in plant-available arsenic due to biochar and manure may have resulted from their relatively high carbon content and the presence of cations such as aluminum, iron, and manganese on the surfaces of these materials (Tab. 14.2). Beesley et al. [68] observed that the incorporation of biochar into contaminated mining soil diminished

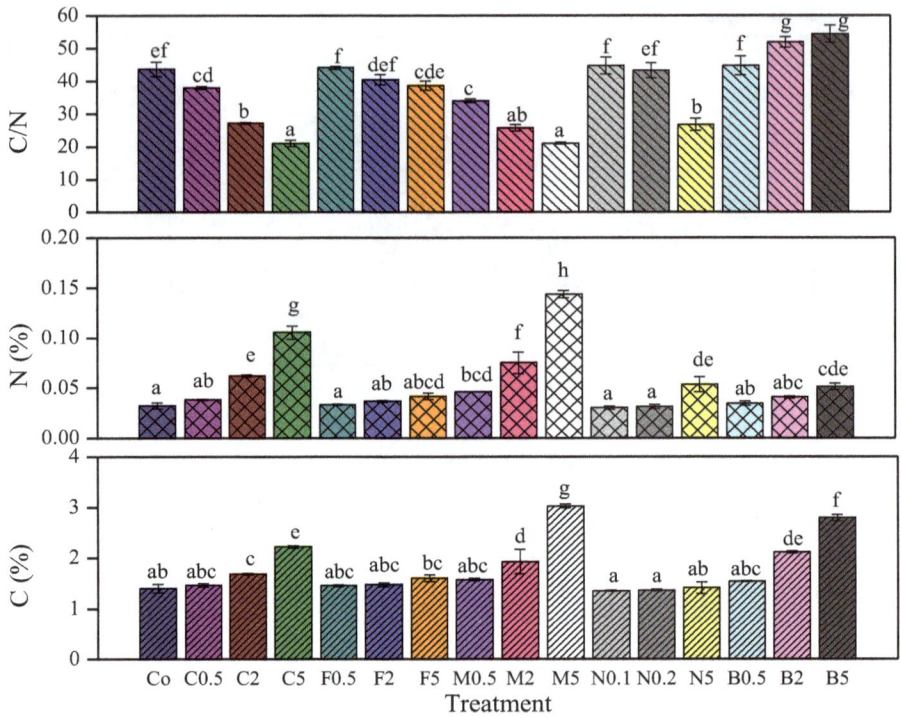

**Fig. 14.4:** Changes in the mine spoil soil percentage carbon, nitrogen content and C-N ratio at the end of the incubation period following treatments with biochar, compost, iron oxide, manure and inorganic fertiliser. Co = control; C0.5, C2, C5 = soil treatment with compost at 0.5%, 2% and 5%, respectively; F0.5, F2, F5 = iron oxide at 0.5%, 2% and 5%, respectively; M0.5, M2, M5 = manure at 0.5%, 2% and 5%, respectively; N0.1, N0.2, N5 = NPK fertiliser at 0.1g/kg, 0.2g/kg and 5 g/kg, respectively; and B0.5, B2, B5 = rice husk biochar at 0.5%, 2% and 5%, respectively. Means with different letters differ significantly among treatments at $P < 0.05$ and same letters indicate no significant differences among means according to the Tukey's Honestly Significant Different test.

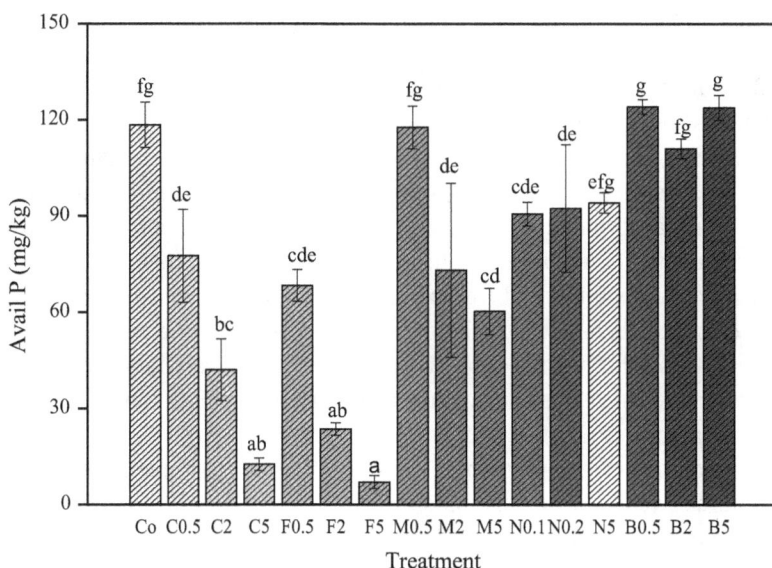

**Fig. 14.5:** Changes in the mine spoil soil available phosphorus at the end of the incubation period following treatments with biochar, compost, iron oxide, manure and inorganic fertiliser. Co = control; C0.5, C2, C5 = soil treatment with compost at 0.5%, 2% and 5%, respectively; F0.5, F2, F5 = iron oxide at 0.5%, 2% and 5%, respectively; M0.5, M2, M5 = manure at 0.5%, 2% and 5%, respectively; N0.1, N0.2, N5 = NPK fertiliser at 0.1g/kg, 0.2g/kg and 5 g/kg, respectively; and B0.5, B2, B5 = rice husk biochar at 0.5%, 2% and 5%, respectively. Means with different letters differ significantly among treatments at $P < 0.05$ and same letters indicate no significant differences among means according to the Tukey's Honestly Significant Different test.

the phytoavailable arsenic level, hence reducing the absorption and toxicity-transfer risk to tomato plants. Biochar may have shown some efficacy in reducing phytoavailable arsenic, although caution is warranted in its application to arsenic-contaminated soils. This is due to the potential elevation of soil pH linked to the liming effects of biochar [315] and the generation of soluble carbon [745], which may simultaneously result in arsenic mobilization. As previously mentioned in Section 14.3.2, elevated soil pH may enhance the release of arsenic, hence amplifying its ecotoxicological impacts. Beesley et al. [68] proposed that the amalgamation of biochar with iron oxides could be advantageous, as iron oxides may mitigate arsenic phytotoxicity.

## 14.8 Conclusion

In an abandoned mine dump in Ghana, we evaluated arsenic (As) mobilization and conducted preliminary incubation study to investigate the efficacy of soil amendments in reducing available As. The sequential extraction experiment demonstrated an elevated concentration of arsenic predominantly linked to weakly crystalline and amorphous iron oxide components, suggesting a significant potential for mobilization and ecotoxicological danger of arsenic under varying conditions. The use of 5% iron oxide to the polluted mine spoil may effectively mitigate arsenic migration into surface and groundwater, thereby lessening related environmental and human health risks. Moreover, treatment with 5% compost may have the capacity to lessen the available arsenic for plant absorption. This suggests the potential of compost to mitigate probable food chain contamination linked to arsenic in the mining region. Nonetheless, the use of manure, compost, and iron oxide may create sorption sites for phosphorus and decrease its availability for plant uptake. Furthermore, the application of compost, manure, and biochar to mine spoil may enhance the total carbon and nitrogen levels, along with the exchangeable potassium, magnesium, and sodium contents in the soil. These initial findings establish a foundation and success criteria to facilitate the integrated use of iron oxides, compost, and manure for reducing arsenic mobilization and improving the soil quality of arsenic-contaminated mining soil. We propose a comprehensive field application and an extended study to appraise the enduring effectiveness of the treatments in mitigating arsenic mobilization into the surrounding ecosystem and enhancing soil quality.

Jewel Andoh and Albert Kobina Mensah

# Chapter 15
# Significance of revegetation of degraded mining sites

**Abstract:** Mining activities can lead to soil degradation, but restoring mined lands through afforestation and revegetation can improve soil quality and fertility. Enhancing soil quality is crucial in regions lacking natural vegetation, and organic matter input from trees and mining spoils can enhance soil structure and nutrient availability. Revegetation also plays a crucial role in reducing soil acidity and erosion. Afforestation is effective in reclaiming salty and alkaline soils, while phytoremediation using native plants can help mitigate metal pollution. Overall, restoration efforts can provide significant benefits to ecosystems, society, and the environment.

## 15.1 Introduction

The revegetation of degraded and contaminated mining sites is integral to broader land restoration initiatives. It typically involves the restoration of flora on previously damaged mining spoil and tailings. This is often conducted following the landform rebuilding of the mining site. The concept of revegetation of mining areas contaminated with heavy metals and/or degraded in soil quality and productivity can contribute to achieving sustainable mineral resource governance. Mining corporations must undertake sustainable initiatives to remediate the environmental damage inflicted on the land during and after gold extraction. This section examines how revegetation contributes to cleaner production and eventually a sustainable mineral mining sector in Ghana.

## 15.2 Improving and replenishing soil quality

Mining leads to soil degradation through compaction, soil erosion, and leaching [335], affecting agricultural productivity. Therefore, restoring the mined lands will improve the quality by improving plant available nutrients, increasing organic matter content

**Jewel Andoh,** Forestry Research Institute of Ghana, Council for Scientific and Industrial Research, Fumesua-Kumasi, Ghana
**Albert Kobina Mensah,** Soil Research Institute, Council for Scientific and Industrial Research, Academy Post Office, PMB, Kwadaso-Kumasi, Ghana

https://doi.org/10.1515/9783111662046-015

and microbial activity. Restoring mined lands could control erosion and minimize offsite drainage. Morgan [341] revealed that restoration through afforestation can reduce runoff and soil loss in gullied areas averagely by about 73% and 83%, respectively. Also, Dutta and Agrawal [136] noted that plantation establishment can improve the soil texture of mined lands. The established plantations increase soil pH since the organic matter input modifies the pH of the soil.

## 15.3 Enhancing and reinstating soil fertility

The regeneration and preservation of fertility are particularly crucial in regions lacking the tropical forest vegetation required for expedited natural restoration [732]. In addition to mitigating the direct impact of heavy rainfall on soil and impeding the flow of runoff, trees and their debris also contribute to the prevention of erosion on exposed soil surfaces, as documented by [761]. Furthermore, the decomposition of forest biomass yields organic matter, which enhances soil fertility, structure, and various hydrophysical properties, as noted by Anane-Sakyi [722, 761], and Ingram [218]. According to [738], the contribution of herbs, shrubs, and trees to the soil in the form of organic litter during fallow periods accelerates the restoration of natural fertility to the fallow area.

According to [712], it was also noted that in undisturbed fallow, there is a movement of nutrients from the soil to the mat layer, then to the vegetation, and then back to the soil and mat layer through the process of litter fall. The mat layer mechanism additionally functions as a mechanism for nutrient preservation, as the majority of nutrients are concentrated inside the mat of roots and humus that are present on or in close proximity to the soil surface. Anane-Sakyi [738] posits that there exists a positive correlation between the quantity of litter generated within a fallow system and the level of organic matter in the soil, resulting in enhanced soil fertility.

Dutta [135] posits that litter fall plays a crucial role in regulating and enhancing the microbial biomass on mining debris. The primary contributors to soil organic matter are root biomass and aboveground plant biomass, with the latter showing a strong correlation with microbial biomass [136, 423]. In a study conducted by Kimaro et al. (2007), the authors examined the nutrient use efficiency and biomass production of various tree species. The results revealed that, following a 5-year rotation, the top soils beneath *Gliricidia sepium* (Jaqua), *Acacia polyacantha* Wild, and *Acacia mangium* Wild exhibited the highest fertility levels. Specifically, the soil organic carbon content and exchangeable cation status in these soils approached levels comparable to those found in natural systems. The amounts of inorganic nitrogen and extractable phosphorus in the soil have reached a sufficient level to support the growth of maize in subsequent cultivation.

Ingram [218] has found multiple methods via which trees might augment the chemical and physical characteristics of soils. Troeh et al. [486] have also shown that the presence of vegetation on soils that have been disturbed by building operations leads to the development of soil organic matter, which subsequently enhances the hydrophysical and chemical properties of the soil. The aforementioned enhancements, as identified by Ingram [218], Asafu-Agyei [732], and Cooper et al. [774], encompass the subsequent elements.

The presence of elevated levels of organic matter, old tree root channels, and increased macro faunal activities leads to enhanced soil physical structure. This improvement is characterized by better soil aggregation and stabilization, reduced bulk density, increased available water capacity, improved infiltration, and an overall enhancement in soil texture. The formation of soil aggregates, which contribute to structural stability and favorable soil properties such as water holding capacity, permeability, aeration, rooting depth, and resistance to erosion, can be influenced by factors such as texture, clay type, organic gums, and fungal and bacterial mycelia.

Enhanced activity of soil organisms, including fungus, arthropods, termites, and worms, is facilitated by a colder and more humid microclimate. This, in turn, leads to an improved substrate for microbes, which not only fulfills their nutrient requirements but also results in the creation of growth-promoting substances. There has been an observed elevation in the level of nutritional status, enhancing the nutrient cycle by capturing nutrients that would otherwise be lost through leaching from areas beyond the immediate vicinity of the tree or crop.

Enhancement of cation exchange capacity (CEC) is sought to optimize nutrient retention and improve the efficiency of nutrient usage. The litter fall that is dominant in mostly forested sites helps in enriching the soil carbon contents which consequently improves the soil basic cations. Additionally, these basic cations are retained and precluded from being leached easily due to high rainfall conditions or wetting regimes. This is in contrast with lack of vegetation where leaching and runoff regimes increase and consequently lead to the loss of the basic cations. Subsequently, the exchangeable sites in the soil are replaced with either aluminum or proton, eventually increasing acidity and reducing the soil pH.

The mitigation of aluminum toxicity and low pH is achieved by promoting the cycling of bases and the synthesis of metabolic substances that effectively bind aluminum ions while also providing buffering capacity to the soil against abrupt fluctuations in acidity, alkalinity, and salinity. Moreover, Young [529] highlights that soils that form in the presence of natural woodland or forest, such as the renowned brown earth in temperate regions or the red earth in tropical regions, exhibit high fertility. The soil exhibits a favorable structural composition, demonstrating a notable ability to retain moisture, displaying resilience against erosion, and harboring a substantial reservoir of fertility through the presence of nutrients that are intricately bonded within organic molecules.

## 15.4 Enhancement of soil physical characteristics

The literature extensively documents the notable attributes of soil structure, porosity, moisture properties, and erosion resistance in forested areas as well as the observed degradation of these qualities following deforestation. The presence of holes within a material plays a crucial role in determining various physical features. Holes with a diameter ranging from 5 to 50 μm are responsible for determining the ability of the material to store water, while pores with a diameter exceeding 250 μm are essential for facilitating the penetration of roots [529]. Compaction was seen during the degradation of overburden and topsoil, mining operations, and reclamation activities, particularly when unfavourable moisture conditions were present and when there was insufficient time for natural soil-forming processes to reduce bulk density [523, [762]]. According to Akala and Lal [261], the sharp rise in bulk density seen at a depth of 30 cm can be attributed to the thorough grading of overburden and spoil material before the application of topsoil as well as the substantial presence of rock pieces at depths beyond 30 cm. In their study, Agodzo and Adama [713] elucidated the significance of soil water content as a critical feature governing its behavior. Moreover, bulk density serves as an indicator for issues related to root penetration, soil aeration, and water infiltration.

### 15.4.1 Enhancement of infiltration and the availability and content of soil moisture

The presence of ground vegetation, such as grasses, serves to safeguard the slope from erosion caused by the action of raindrops and subsequent runoff. Additionally, these vegetation types effectively capture and retain sediments in motion. On the other hand, the inclusion of shrubs and trees contributes to the enhancement of soil stability by means of root reinforcement. According to Morgan [341], the presence of vegetation enhances the process of water infiltration into the soil. Nevertheless, this phenomenon might give rise to complications in situations characterized by excessive levels of precipitation in terms of both quantity and intensity.

Although the decrease in runoff that occurs as a result of this phenomenon aids in the management of surface erosion, it is important to note that the augmented moisture levels in the soil may potentially intensify the occurrence of mass soil collapse [341]. According to Young [1036], the decrease in runoff can be attributed, to a limited extent, to canopy interception and direct transpiration. However, the primary factor contributing to this reduction is the enhanced soil infiltration capacity facilitated by the presence of trees. the presence of a dense surface-root system in both natural forests and plantations offer dual benefits of enhancing infiltration and preventing soil erosion.

## 15.4.2 The process of root-induced disruption of compact or hardened layers

The enhancement of surface protection results in a decrease in crusting and compaction, which subsequently leads to an increase in infiltration. Additionally, it reduces leaching and prevents erosion [218, 732; 774]. Trees are mostly utilised in the reclamation of deteriorated and contaminated mining sites because of their ability to penetrate deeper soil layers. In the reclamation of mine tailings, grasses and shrubs provide superficial benefits and are incapable of establishing roots in deeper horizons. These impacts render trees more advantageous in rehabilitating derelict mining sites.

For instance, Mensah et al. (2022) highlighted that the arsenic-cleaning process of ryegrass necessitates the extraction of roots during harvest and may be more superficial, as the roots are more lateral and cannot reach deeper horizons of the mine spoil. They further added that species employed for mine land restoration, in addition to being resilient to the local temperature and soil, should exhibit increased root length, enhanced root growth agility, and higher capacity for water and nutrient absorption.

## 15.4.3 Alteration of soil temperature extremes

Experimental evidence from research on minimum tillage suggests that the presence of ground surface litter cover can significantly mitigate the excessively high ground temperatures, which can exceed 50 °C, typically observed on bare soils in tropical regions [529]. [666] highlights the phenomenon wherein bare soil rapidly absorbs heat during the summer, resulting in elevated temperatures, and conversely experiences significant cooling during the winter. However, this effect can be mitigated by maintaining a vegetative cover on the ground surface, which acts as an insulating layer for the soil. In this scenario, the soil will not experience excessive heat or cold temperatures.

## 15.4.4 Enhancement of soil texture and soil structure

Dutta and Agrawal [136] have observed that there are notable fluctuations in silt and clay content, indicating that plantations have the potential to alter soil texture over time following their establishment and subsequent growth. the textures of mine spoils experience significant disruption as a result of the erratic piling of overburden materials. The mine spoil that underwent natural revegetation 5 years ago exhibited a composition characterized by sand, silt, and clay percentages of 61%, 25%, and 14%, respectively. The distribution of particle sizes is a crucial physical characteristic of soil that significantly impacts the success of vegetation establishment on reclaimed land. This distribution plays a pivotal role in determining several soil properties including water retention capacity, bulk density, availability of soil moisture, and nutrient content and availability.

## 15.5 Revegetation in enhancing soil chemical characteristics

### 15.5.1 Mitigating soil acidity

According to Young [529], the presence of trees has a tendency to stimulate leaching by introducing bases onto the soil surface. Nevertheless, the potential of tree litter to effectively increase the pH levels of acidic soils remains uncertain due to the substantial differences in magnitude, as indicated by Young [529]. One of the contributing factors to the aforementioned phenomenon is the insufficiency of calcium provided by trees through litter fall to effectively mitigate acidity, even by a single pH unit [529]. Young [529] posits that there exist instances in the temperate zone where trees generate acid, mor-type humus, resulting in a notable escalation in soil acidification. Nevertheless, empirical studies have also demonstrated that the alkaline substances generated via the process of trash breakdown can effectively mitigate acidification. The observed rise in pH resulting from the establishment of plantations indicates that the introduction of organic matter influences the soil's pH levels. According to Dutta and Agrawal [136], the majority of plant species utilized for revegetation purposes are dicotyledonous. As a result, these species have the potential to release a greater amount of base cations, such as $Ca^{2+}$, into the soil. Consequently, this can lead to a higher increase in soil pH compared to the initial pH of the new mining spoil. Previous studies conducted by Richart et al. [404] and Dutta and Agrawal [136] have similarly found a positive correlation between the alteration in pH levels of opencast spoil and the growth of trees. According to Ghose [177], the ideal pH range for plant nutrient availability is between 6.5 and 7.5.

### 15.5.2 Decreasing the levels of salinity or sodicity

The utilization of afforestation has proven to be an effective strategy in the reclamation of salty and alkaline soils [529]. In Karnal, India, it has been reported that the presence of *Acacia nilotica* and *Eucalyptus tereticornis* trees resulted in a decrease in top soil pH from 10.5 to 9.5 over a period of 5 years. Additionally, the electrical conductivity of the soil decreased from 4 to 2 dS/m. These changes were observed when the establishment of trees was facilitated through the application of gypsum and manure, as documented by Gill and Abrol [180].

### 15.5.3 Decrease in the rate of breakdown of organic matter

It is widely recognized that the rate of humified organic matter loss is comparatively lower in forest ecosystems as opposed to agricultural systems. One of the factors contributing to this phenomenon is the reduction in temperature caused by the shading provided by the canopy and litter cover of trees [529].

## 15.5.4 Accessibility of nitrogen

It is widely accepted that nitrogen is acquired from the atmosphere and subsequently sequestered in the soil, where it is subsequently assimilated by plants in the form of nitrates for utilization. The presence of root nodules, particularly in leguminous plants, contributes to the attainment of this objective. Leguminous cover crops, namely *Pueraria*, *Centrosema*, and *Calapogonium*, have demonstrated enhanced efficacy. Tree species such as *Leucaena leucocephala* and other leguminous tree species have demonstrated comparable levels of effectiveness. The absorption of nitrogen from the atmosphere subsequently contributes to the enhancement of soil fertility. The deposition of litter contributes to the enhancement of soil organic matter, hence creating a conducive soil environment for nitrogen fixation.

The organic matter in question contains proteins that undergo decomposition, resulting in the formation of amino acids. These amino acids are subsequently oxidized to form nitrates, a process that has been documented by [883]. Plants are capable of utilizing these nitrates for their growth and development. According to Franco and de Faria [163], legume tree species that are deemed beneficial have the potential to contribute around 12 tons of dry litter and 190 kg of nitrogen per hectare per year for the purpose of soil restoration in degraded areas. Certain leguminous plants, such as *Leucaena*, have the potential to contribute nitrogen to enrich the replenished nutrients during mineral extraction.

## 15.5.5 Accessibility of phosphorus (P)

Soils treated with organic matter exhibit increased levels of mineralized phosphorus. In a study conducted by Mbagwu et al. [907], it was shown that the application of higher rates of organic matter to soils resulted in a proportional increase in the phosphorus (P) content of the soil. Nevertheless, the cumulative phosphorus (P) accumulation in the stem and bark surpasses that found alone in the leaves. According to Ren and Yu [400], the proportion of P found in the surface litter is rather tiny compared to the overall accumulation in the forest. Previous studies have proposed that a shortage in phosphorus (P) is the causative reason for the reduced growth rate observed in *Acacia mangium* [400, 403, 516].

## 15.5.6 Exchangeable cations

They refer to positively charged ions that can be readily replaced or interchanged with other ions of the same charge in soil. Exchangeable cations abet to buffer the soil against rapid changes in the soil pH. Leaching of the basic cations such as Ca, K, Mg, and Na from the soil especially in tropical areas due to excessive rainfall and water regimes that leaches these cations from the soil exchangeable sites or from the soil colloid.

Consequently, their places on the exchange sites become available and are occupied or taken over by either protons or aluminum, which causes development of active or exchangeable acidity. Vegetation in the form of grasses, shrubs, or trees, native or introduced, in mixed cropping or monoculture systems provides literfall which decomposition and decay aid to replenish the lost or leached basic cations. Therefore, revegetation of the degraded mining sites will consequently aid to enrich or improve the soil cation exchange capacity, which is explained further in the following paragraphs.

In their study, [1014] conducted an investigation to assess the efficacy of trees in serving as a protective barrier for capturing nutrients that have leached beyond the accessible range of crop roots. Specifically, they examined alterations in exchangeable cations (calcium, magnesium, and potassium) as well as pH levels across a diverse array of alley cropping trials in the savanna region of West Africa, spanning medium- to long-term durations. It was shown that the levels of calcium content, effective cation exchange capacity, and pH in the topsoil were significantly greater in the presence of *Senna siamea* compared to L. The plant species *Leucaena leucocephala* and *Gliricidia sepium* as well as the control plots without trees were studied in locations characterized by a Bt horizon that is abundant in exchangeable calcium.

The recovery of calcium from the subsoil beneath *Senna* sp. was linked to this phenomenon. The augmentation of calcium (Ca) concentration in the uppermost layer of soil is due to the presence of *Senna* sp. The relationship between the absence of trees in the control treatment and the total quantity of dry matter applied since the initiation of the experiment was observed.

The absence of a rise in calcium (Ca) accumulation in the other species can be attributed to the probable retrieval of Ca from the topsoil itself and/or significant leaching of Ca. The deposition of calcium (Ca) in the uppermost layer of soil beneath *Senna* sp. The presence of trees had a notable impact on the pH of the topsoil, with a considerable increase observed in comparison to the *Leucaena, Gliricidia,* and control treatments without any trees.

Table 15.1 presents the potential impacts of vegetation establishment, specifically revegetation, on the process of soil rehabilitation and restoration. The study demonstrates the variations in soil qualities within the canopy of individual trees and in adjacent areas devoid of tree cover. It is evident that the soils beneath vegetation have exhibited higher levels of nitrogen (N), phosphorus (P), and potassium (K) accumulation compared to the soils in open field areas. The observed phenomenon can be attributed to the augmented availability of litter originating from vegetation.

**Tab. 15.1:** Soil properties under vegetated fields.

| Available nutrients | Soil depth | Prosopis cineraria | Prosopis juliflora | Bare field |
|---|---|---|---|---|
| kg/ha | cm | | | |
| P | 0–15 | 250 | 203 | |
| | 15–30 | 193 | 212 | 196 |
| N | 0–15 | 22 | 10 | 8 |
| | 15–30 | 10 | 5 | 4 |
| K | 0–15 | 633 | 409 | 370 |
| | 15–30 | 325 | 258 | 235 |

Source: Aggarwal (1980), cited by Young [529], page 94.

This litter adds organic matter, which subsequently undergoes decomposition, releasing nutrients that play a crucial role in enhancing soil fertility in degraded mining areas. This observation aligns with the findings of Ingram [218] and Troeh et al. [486], who documented that the augmented presence of organic material (both above- and belowground residue) in natural fallow areas is believed to play a crucial role in sustaining soil organic matter and enhancing soil productivity in degraded lands.

As an example, *Acacia albida* has been seen to exhibit significant increases in organic matter and nitrogen levels ranging from 50% to 100% beneath its canopy as well as an enhanced capacity to retain water [805]. According to Young [529], it is frequently observed that semiarid regions exhibit elevated levels of soil organic matter and nutrient concentration within the areas covered by tree canopies as compared to the surrounding open ground.

## 15.6 Significance of revegetation in enhancing soil biological characteristics

### 15.6.1 Enhancement and augmentation of soil organic matter

Organic matter is commonly defined as the component of soil that encompasses the residues of animals and plants at different degrees of decomposition. Soil organic matter consists of three primary constituents: living biota, which encompasses the roots, microbes, and other organisms inhabiting the soil; fragments of decaying plants and animal remains including fallen leaves, deceased organisms, animal excrement, and crop residues; and residues of ongoing decomposition, referred to as humus, which denotes the organic compounds that persist within the soil [958].

According to Ingram [218] and Troeh et al. [486], the presence of a greater quantity of organic matter, including both above- and belowground residue, in natural fallow areas is believed to play a crucial role in sustaining soil organic matter levels and enhancing soil productivity on degraded fields. The primary factor contributing to improvements in soil fertility is the maintenance of soil organic matter levels through the provision of litter and root residues by trees. The fundamental mover of nutrients, from which many other soil improvement activities originate, has been identified [1036]. Based on [740] findings, the chemical composition of soil organic matter is estimated to consist around 50% carbon, 5% nitrogen, 0.5% phosphorus, 39% oxygen, and 3% hydrogen. Nevertheless, it should be acknowledged that these values exhibit variability across different types of soil.

The organic content of a normally well-drained mineral soil is rather low ranging from 1% to 6% by weight in the topsoil and much lower in the subsoil. The primary impacts of soil organic matter are observed in the realm of soil physical characteristics and nutrient availability, as indicated by Young [529].

Organic matter exerts several physical effects that enhance the conditions of mineral soils. The presence of organic matter in sandy soils contributes to the enhancement of their capacity to retain water and nutrients. Furthermore, it has been shown that the application of this method results in the enhancement of clay soils through the process of loosening and enhancing their tilth [958].

The primary chemical impact pertains to the provision of nutrients. The equilibrium of the supply is maintained among primary, secondary, and micronutrients as long as they remain in organic molecule form, thereby safeguarding them from leaching, except in the case of podzols. Additionally, the release of nutrients in accessible forms occurs gradually through mineralization, as stated by Young [529].

Additional benefits of organic matter in relation to nutrient supply include the inhibition of phosphate-fixation sites, leading to enhanced phosphorus availability, as well as the formation of complexes that improve the accessibility of micronutrients. There has been a suggestion that a high level of organic matter in soil creates a favorable environment for fixation.

One further chemical phenomenon that occurs is the notable increase in cation exchange capacity (CEC) due to the presence of the clay-humus complex. This impact is especially significant in soils where the CEC of a clay mineral is naturally low, such as those predominantly composed of kaolinitic clay minerals and free iron oxides such ferralsols and acrisols [529]. Enhancing the CEC has a positive impact on the preservation of nutrients, encompassing both naturally recycled elements and those introduced through fertilizer application.

According to [958], a study conducted in 2005 found that the regular application of organic matter resulted in enhanced crop output, improved soil quality, and reduced reliance on fertilizers, particularly on sandy soils. According to [974], the primary supplier of nitrogen, a crucial nutrient for crops cultivated in nitrogen-deficient soil, is the decomposition of organic materials by soil microorganisms. According to [1036], the primary source of this organic matter is predominantly derived from tree residues. Soil organic matter is considered to be of utmost importance in the realm of soil management.

### 15.6.2 Generation of various grades of plant litter

The aforementioned phenomenon results in the gradual dispersion of nutrients that have been mineralized through the process of litter decay. In cases where the vegetation consists primarily of trees, it is seen that both woody and herbaceous residues are produced, resulting in a diverse spectrum of aboveground litter and root residues [529].

### 15.6.3 Temporal pattern of nutrient release

The varying quality of tree residues results in a differential rate of decay, leading to a gradual and staggered release of nutrients over an extended period. The management of systems allows for some control over this discharge by selecting tree species based on leaf decay rates and pruning time [529].

### 15.6.4 The impact on soil fauna

The presence of trees has a significant impact on the composition and abundance of soil fauna, typically leading to an increase in soil fertility [529]. There has been a proposed unique indirect impact indicating that the presence of shade trees in plantations can lead to a decrease in the necessity of employing chemical herbicides, which can have detrimental effects on soil fauna, due to their ability to reduce weed growth through shading. Dutta [135] posits that litter fall plays a crucial role in regulating and enhancing the microbial biomass on mining debris. The primary contributors to soil organic matter are root biomass and aboveground plant biomass, with the latter exhibiting a strong correlation with microbial biomass [136, 423]. Fig. 15.1 shows a termite mound formed in a reclaimed mine site. It is an indication of recolonization or rejuvenation of biological activities.

## 15.7 Significance of revegetation in mitigating soil erosion

Agronomic erosion control strategies leverage the protective properties of vegetation covers in order to mitigate erosion that the sole efficacious approach for erosion management and the rational means of restoring strip-mined lands to functionality is through the process of revegetation.

The primary objective of implementing erosion control measures on land that has been previously utilized for mining is to establish a stable environment conducive to the development and growth of vegetation. This is done with the intention of facilitating the reclamation of the land for agricultural or recreational purposes while si-

**Fig. 15.1:** A termite mound formed in a reclaimed mine site. It is an indication of recolonization or rejuvenation of biological activities.

multaneously reducing the extent of off-site drainage. According to Morgan [341], mine spoil banks are typically characterized by rapid erosion and limited vegetation growth due to the infertile and poisonous nature of the material.

The presence of vegetation is of significant importance in the mitigation of erosion in several environments including gullied regions, construction sites, road embankments, landslides, sandstones, mine spoils, and pipeline corridors [341]. Morgan [341] further reported that vegetation serves as a protective layer or buffer that separates the atmosphere from the soil.

According to Morgan [341], the components situated above the ground, such as leaves and stems, have a role in absorbing a portion of the energy from rainfall, running water, and wind. This absorption reduces the amount of energy directed toward the soil. Conversely, the belowground components, which consist of the root system, contribute to the mechanical strength of the soil. The soil's mechanical strength plays a crucial role in mitigating runoff and preserving soil moisture, thereby contributing to enhanced soil moisture availability.

According to [756], the implementation of crops and vegetation with substantial ground cover and an extensive root system can effectively mitigate water erosion. The production of dense sods by grass and several leguminous plants serves as a notable illustration in this regard [756]. Revegetation is implemented as a means of augmenting infiltration and mitigating runoff in the context of gully erosion control. This involves the application of grasses, legumes, shrubs, trees, or a mix thereof in the vicinity of the gullies. In certain instances, mulching is employed during the initial phases to facilitate the process [341].

The practice of tree planting, namely in afforestation, has been acknowledged as an effective approach for mitigating runoff and erosion. This is particularly true when implemented in headwater catchment areas, as it serves as a manner of flood regulation [341].

According to Morgan [341], empirical studies have demonstrated that afforestation has the potential to significantly mitigate runoff in gullied regions, with reductions ranging from 65% to 80%. Additionally, afforestation has been found to substantially decrease soil loss by 75–90%.

## 15.8 Revegetation in mitigating soil pollution with heavy metals

According to Mensah et al. [321] and Ramírez et al. [966], mine wastes are important sources of contamination and point sources for potentially toxic elements (PTEs) in soil including Al, As, Cd, Cr, Cu, Fe, Mn, Ni, Pb, Ti, V, and Zn. PTEs may be transferred into surrounding plants, food crops, soil, surface water, and plants as a result of these. Metal pollution in soils and sediments may have long-term health effects on people and ecosystems by affecting aquatic life, food safety, and water quality.

In other situations, children who frequently use metal-contaminated terrain or abandoned mine sites as play areas run the danger of ingesting metal through their diet [58, 440]. Furthermore, as noted by Mensah et al. [321], communities may pass through unfenced abandoned mine spoil sites and breathe in polluted dust or consume contaminated mine wastes. In areas with gold mining, oral consumption of soil PTEs continues to be a significant exposure pathway that affects human health [e.g., 41, 58, 321]. This could happen accidentally by ingesting dust, dirt, and soil or purposefully by engaging in geophagy, which is the deliberate consumption of soil or earth materials like clay [667]. In Latin America and Africa, pregnant women frequently engage in the practice of geophagy [96, 571]. As is known to be present in metal ores, tailing deposits, soil, and dust in gold mining sites, it is raising serious health risks for both humans and animals [96].

For example, both acute and long-term exposure to As may result in health issues for humans including skin damage, symptoms of cancer, and abnormalities in the circulatory system [35, 86]. Moreover, Wilson disease, which impairs the function of several organ systems, including the liver and the brain, is brought on by an excessive build-up of copper in the body as a result of consuming tainted food and water [1089].

Furthermore, health issues like reproductive, developmental, hepatic, hematological, and immunological health defects, abortions, infertility, birth defects, and malformations may be linked to exposure to and accumulation of Cd, Cr, Pb, Ni, and V in essential body organs [727].

However, the positive environmental effects on ecological integrity and human health resulting from improper management of mine wastes [73], runoff from stockpiles [324], and abandonment of mine spoils left untreated [321] often outweigh the benefits associated with gold mining in Ghana such as the creation of jobs and government revenues. These activities worsen the anthropogenic release and migration of PTEs into the environment, which has an impact on people's livelihoods and health.

Remediation of metal contamination issues in gold mine spoils has primarily involved soil washing and flushing [85, 509], removal and replacement of contaminated soil with clean soil [57], and chemical stabilization using oxides to reduce metal solubility [251, 376, 509]. In some cases, containment techniques have been used to lessen the migration and transfer of harmful components. These techniques include the use of built barriers, caps, and liners [515].

The two main drawbacks of these techniques are their high cost to many mining corporations and their difficulty level. Furthermore, PTEs cannot decompose like organic chemicals do; thus the clean-up usually involves either removing them [360] or employing stabilization and immobilization strategies to reduce their uptake, surface runoff, or leaching [85, 376].

In order to address the issue of heavy metal contamination in gold mine soils, phytoremediation – the use of native plants – is suggested as a practical and affordable option to environmental management [360, 964]. When selecting potential plants for contaminated site phytoremediation, some authors [e.g., 38, 667; 956; 384] have recently suggested that it may be wise to use well-adapted native species of the locality under investigation. Additionally, native and indigenous species are favored over foreign ones since they may be more suited to the local environment and fully integrated into the functioning ecosystem [335].

According to [1090], phytoremediation is still a young field of study that requires further in-depth research to fully understand its applications. For example, the potential of native plants like *Pueraria montana, Alchornea cordifolia, Lantana camara*, and *Pityrogramma calomelanos*-fern [57, [881]; Liu et al., 2019; 384] to clean mine-contaminated spoils has been investigated by Mensah et al. [333]. There is a potential of cleaning up these PTEs with native plant species. *Chromolaena odorata* (TF for As = 4.7, Cu = 5.1, Ti = 1.4, Zn = 2.0) and fern (TF for As = 1.50, Cu = 1.4, Ti = 1.8, Zn = 2.1) can be used to achieve phytoextraction. Their greater TFs for these elements suggest that *Chromolaena odorata* would be a better choice than fern. Furthermore, *Alchornea cordifolia* and *Pueraria montana* can be utilized for Cu, Ti, and Zn phytoremediation due to their TFs above 1. For Cu and Zn, *Lantana camara* could potentially be a possibility. Therefore, local plants such as *Chromolaena odorata* and *Pityrogramma calomelanos*-fern may extract As, Cu, Ti, and Zn from the multielement-contaminated mine soils and farms.

In Cd-contaminated soils in China, Liu et al. [1070] investigated *Lantana camara* exclusively for Cd remediation and discovered that the plant exhibited high Cd tolerance, with bioaccumulation and translocation values more than 1. Furthermore, *Pteris vittata* L. showed a high potential for Pb phytoextraction among the nine native plants that [1063] investigated for metal clean-up of contaminated mine tailings in Malaysia.

Thus, by reducing food chain contamination, limiting their mobility, reducing metal transfer into surface and groundwater, and guaranteeing a metal-contamination-free environment, phytoremediation lessens the negative effects of PTEs on human health. Furthermore, phytoremediation may help with carbon sequestration, regulate wind and water erosion, guarantee water quality, lessen discharge from tainted mining waste, preserve and improve soil fertility, and save soil moisture.

In Mensah et al. [323], it was shown that the incorporation of manure and iron oxide (MFE) resulted in the phytostabilization of Co, Mn, Hg, Mo, Ni, and Zn in ryegrass, as evidenced by bioconcentration factor (BCF) values exceeding 1. The results of the study indicated that the effectiveness of ryegrass phytostabilization was significantly improved with the combined application of manure and iron oxide. Therefore, it was inferred that the simultaneous utilization of manure and oxides can be employed on mine tailings to enhance the growth of ryegrass and mitigate the release of potentially toxic elements (PTEs) into the surrounding environment.

Mensah et al. [332] conducted a 60-day pot experiment to investigate the potential of ryegrass (*Lolium perenne*) aided with different soil organic and inorganic amendments in phytocleaning As in a highly contaminated gold mining spoil in southwestern Ghana. They found that ryegrass might be more useful for phytostabilization as its roots concentrate more As than its shoot. This was further supported by higher BCF above 1 as well as lower BAC and TF < 1.

## 15.8.1 Achieving mining sector sustainability with revegetation of degraded and contaminated mine sites

Restoration is a natural and intervention-assisted process aimed at promoting and facilitating the recovery of degraded, damaged, or destroyed ecosystems [939]. When sustainably carried out, this provides significant social, economic, and environmental benefits. Here, we highlight the importance of restoring degraded landscapes in Ghana caused by excessive illegal mining.

## 15.9 Social benefits

Restoration of mined sites could improve air quality, agricultural land, and water. This may help local communities near the restored sites to gain good health, enhance livelihood, and improve well-being. Studies have shown that stakeholder engagements in restoration activities enrich their knowledge and builds relationships. Formosa and Kelly [810] revealed that restoration in the Mattole River Watershed, USA, improved cooperation and collaborative partnerships, promoted skill and knowledge development, and built trust and reciprocity among community stakeholders. Also, in "Forest Landscape Restoration for Livelihoods and Well-being" by [668], the authors noted that restoration of degraded landscapes improves livelihood assets such as education, income, empowerment, roads, schools, species richness, soil stability, and water quality among others.

The restoration of degraded ecosystems plays a crucial role in the livelihood and well-being of local communities. Thus, it is necessary to integrate social dimensions into the restoration of Ghanaian mined sites, which have displaced several cocoa farmers and affected social cohesion among fringe communities. Restoration of mined sites by engaging local communities in the affected areas in Ghana could improve social capital, such as trust, cooperation, participation, knowledge, and network, which is essential for the sustainability of community development projects. It may also create awareness of the detrimental effects of mining, thereby motivating local communities to safeguard the environment, enhancing the aesthetics, promoting recreational activities, and preserving the mining landscape.

### 15.9.1 Aesthetics

Restoring mined sites raises the aesthetic values of the area, thereby providing serenity to local communities and the mining companies themselves. This brings a conducive working environment and safety to workers and residents. Enhancing the aesthetic nature of the mined areas through restoration can help the mining enterprises attain the highest social and economic benefits. The aesthetic quality of restored mined areas will impact residents' identification with the place or landscape, their emotional relationship, and their attachment to them [1091]. It is thus crucial to consider the ecological quality and aesthetic values in a restoration exercise to turn mined sites into healthy landscapes for residents.

### 15.9.2 Recreation

There is a nexus between the aesthetics of a landscape and recreation. Enhancing the aesthetic values of degraded landscapes through restoration will increase recreational activities in the area and lead to local economic growth. The restored mined sites can serve as recreational sites for those in and outside the landscape, bringing some respite to pleasure-seekers from ordinary life. Recreation activities could help build social networks and cohesion among residents through socialization. This will improve the social and economic well-being. The recreation sites will allow people to have a new experience, engage in a better life, and develop a healthy lifestyle. Recreation is important. Therefore, the restoration of mining sites is essential for this purpose. It is critical during the early childhood of children. Socialization through leisure is vital to support social-emotional maturity in healthy adults. Recreation, like any life skill, requires intentional training and support.

### 15.9.3 Land preservation and cultural values

Restoration of mined sites could restore cultural values and preserve lands to some extent in the mining communities. The cultural values include spiritual, health care, and natural resource management. The local communities have a spiritual connection with their lands, and the ecological relationship is sustainable and healthwise [873]. Land-country relations include care (Eckermann et al., [794, 1092]). The cultural values of the local community are holistic and ecological, forming a spiritual perspective of health and well-being. This encompasses the physical, mental, cultural, and spiritual dimensions and the harmonious relationship between these dimensions and the environment, ideology, politics, and social and economic situation ([911, 1007]).

## 15.10 Economic benefits

Restoration of mined sites could safeguard the economic activities of local communities fringing the degraded or damaged sites such as farming, fishing, and forestry. The restoration of degraded landscapes can generate jobs for residents in the local communities, especially the youth.

In Ghana, many young individuals, even university graduates, engage in artisanal and small-scale mining due to a lack of employment opportunities (Arthur-Homes et al., 2022), causing severe devastation to the environment. Engaging the teeming unemployed youth in the country for restoration of mined sites with allowances could provide some relief and seed money to start a business other than mining. The European Commission [801] estimated that restoration of ecosystems could create respectively up to 50,000 and 140,000 jobs and up to €4.2 and €11.1 billion of direct outputs every year.

In Australia, restoration activities are an important source of employment for the indigenous peoples [815]. Guuroh et al. [835] revealed that restoration activities at the Pamu Brekum Forest Reserve in Ghana provided indirect jobs for local communities through seedling production, tree planting, and plantation maintenance. The authors also pointed out that residents who worked to restore degraded forests improved their incomes by paying monthly salaries as temporary workers in the project and selling food crops that allowed them to meet financial obligations such as payments for health services and child education fees. Some other economic benefits of mined sites restoration have been described below.

## 15.10.1 Recovery of land value for crop production

Mining causes damage to the agricultural ecosystem, leading to a shortage of productive land for crop production. Mining and climate change have led to a reduction in agriculture productivity and increased food insecurity. Thus, restoration of mined sites could address this challenge. Mensah [335] argued that the restoration of mined lands over a long period could improve the topsoil, enhancing soil organic matter content, cation exchange capacity, available nutrients, biological activities, and physical conditions of the soil, making the land productive for agricultural activities.

## 15.10.2 Renewable energy generation

The restored mine site could be used to produce renewable energy. In countries such as the United States, Germany, Australia, and Canada, former mining sites are used for the production of renewable energies. For example, the SunMine solar power plant in Kimberley, British Columbia, is a 1 MW solar power plant operating since 2016 and is the first solar power plant connected to the grid in British Columbia [1093]. SunMine is located on land recovered after the closure of the former Suller mine; the land was freed in 2014 by Suller Mine Permit (Permit M-79) to build a solar power plant [743]. In addition, the former lignite mining plant in Espenhain, Germany, has transformed the ash deposits at the former Espenhain mine into solar power plants known as the Leipziger Land Solar Power Plant (LLSPP).

In addition, a hydroelectric power plant was built at the old Summitville Mine site in Rio Grande County, Colorado to power water treatment operations. The 560-ha Summitville mine, registered as an U.S. EPA Superfund site, is an abandoned mining operation involving gold and silver deposits, affecting surface and off-site water [1010].

# 15.11 Environmental benefits

Ghana's forest cover has declined over the years due to continuous conversion of forest lands to other land uses such as agriculture, mining, logging, and fuelwood extraction among others [724]. About 8.1 million ha of pristine forests existing in the 1990s has been reduced to about 2.1 million ha presently because of overexploitation, deforestation, and forest degradation [971; 908]. According to the Forestry Commission of Ghana (2016), about 135,000 ha of the forest is cleared annually for other land use activities. A recent documentary by Multimedia, a leading media outlet in Ghana, showed that illegal mining activities in the Apamprama Forest Reserve have led to severe deforestation and degradation in the reserve, affecting its aesthetic value. Therefore, restoration of the deforested and degraded forest lands will reverse this devastating effect of mining in the country.

## 15.11.1 Climate change mitigation

The restoration of mined sites is expected to increase the forest cover, enhance biodiversity, and improve environmental quality as well as contribute to mitigating climate change. Ontl et al. [947] claim that removing trees during mining eliminates the carbon storage capacity of forests, which has severe implications for climate change.

It is thus critical to restore mined lands to reduce and increase removals of carbon dioxide ($CO_2$) emissions into the atmosphere. Forests play a crucial role in overcoming climate crisis as carbon sinks are an important factor influencing climate change adaptation strategies ([769]:2022). Ngaba et al. [939] revealed in their meta-analysis concerning the environmental outcomes during the years 2000–2015 resulting from the "Grain for Green" Project (GFGP) implementation in the Loess Plateau (LP) that, on average, GFGP increased forest coverage by 35.7% and grassland by 1.05%. At the same time, GFGP has a positive impact on soil carbon (C) sequestration, net ecosystem production (NEP), and net primary production (NPP) from 2000 to 2015 by an average of 36%, 22.7%, and 13.5%, respectively.

## 15.11.2 Water supply

Restoration of mine sites can enhance water supply and quality in forest landscapes since forest cover increases. This is shown in a study conducted by [1034]. The authors observed that water supply increases in forests under different reforestation scenarios. Forests are a huge water source on land and thus play a vital role in water resources management [274, 878]. Healthy forests reduce direct drainage during heavy rainfall, prevent flooding, and reduce landfalls and soil losses [734]. According to Harper et al. [821], forest cover affects water content and quality by changing the water balance of the watershed and also by releasing dissolved salts into the landscape.

## 15.12 Conclusions

Restoring and enhancing soil quality is crucial in areas affected by mining activities. Mining leads to soil degradation through compaction, erosion, and leaching, which negatively impact agricultural productivity. Restoring mined lands can improve soil quality by increasing nutrient availability, organic matter content, and microbial activity. Afforestation and plantation establishment can control erosion, reduce soil loss, and improve soil texture. Trees and their debris contribute to the prevention of erosion and enhance soil fertility by providing organic matter, improving soil structure, and enhancing nutrient cycling. The presence of vegetation also enhances soil physical characteristics such as infiltration, moisture availability, and temperature regulation. Overall, restoring and enhancing soil quality through vegetation can lead to improved agricultural productivity and ecosystem health.

Albert Kobina Mensah

# Chapter 16
# Measuring and monitoring success of post-reclamation efforts

**Abstract:** A comprehensive measurement and monitoring framework is crucial for the success of reclaimed fields, focusing on restoring lost ecological functions. The restoration process involves restoring soil's physical, chemical, and biological processes, which may take several years. Assessing the success of reclamation should consider multiple factors, as relying on a single parameter alone is insufficient for monitoring progress and success in restoring degraded mine sites.

## 16.1 Introduction

The implementation of a comprehensive measurement and monitoring framework is a valuable approach to ensure the ongoing success of a reclaimed or restored field. This method facilitates the sustained progress of a restored mining site by focusing on the restoration of lost ecological functions and services. The restoration of a site is considered complete when all physical, chemical, and biological processes of the soil have been successfully revived, thereby returning it to its original or pristine condition. This may require several years to restore the original state.

According to Sheoran et al. [453], reclamation of abandoned mine land is a highly intricate procedure, and assessing its success should extend beyond merely observing the presence of vegetation at the site. The evaluation of restoration initiatives should not be restricted to a single criterion for determining success. In order to assess the condition and functionality of the soil system for ecosystem reclamation, it is necessary to take into account the multiple factors. This is because relying on a single parameter alone does not yield adequate information or provides enough evidence for monitoring progress and success of a restored degraded mine site. Here, in this chapter, some variuos parameters or variables used in appraising and monitoring the success of reclamation projects are discussed.

**Albert Kobina Mensah,** Soil Research Institute, Council for Scientific and Industrial Research, Academy Post Office, PMB, Kwadaso-Kumasi, Ghana

https://doi.org/10.1515/9783111662046-016

## 16.2 Reestablishment of indigenous species or return of the soil flora and fauna

The revitalization of vegetation occurs, leading to the reestablishment of indigenous species, including soil fauna and flora, which gradually repopulate the rehabilitated area in terms of their ecological activities and population dynamics. Hence, it is imperative to document the initial condition of the site prior to the commencement of mining activities. This tool has the capability to monitor and evaluate the advancement and effectiveness of restoration efforts.

### 16.2.1 Species richness and diversity

The enrichment, distribution, and repopulation of species, as well as their diversity, can be measured in one of the two ways: (i) either by counting the number of plants that were introduced through assisted revegetation within a specific area or (ii) by counting the number of plants that naturally rejuvenated or recolonized at the specific area at the reclaimed site.

It is possible to carry out these tasks on a regular basis, as decided by the project team or the firm. That is, to say, it is possible to determine it according to the compliance or regulatory framework that is utilized by the companies or the regulatory bodies. After that, the total count of the species that are found inside the area that has been restored or reclaimed will be compared to the species that are found in the natural forest or using an index that is recognized or approved all over the world. For example, the Shannon-Weiner (1948) index [669] can be used to evaluate or quantify both the species rich and even distribution.[1]

## 16.3 Using soil quality indicators and index

These measures of success must be evaluated by considering a combination of the physicochemical and biological attributes of the soil. Additionally, success can be determined by improvements in ecological functions and services, such as reduced runoff and erosion rates, as well as the recovery of water sources and habitat recolonization.

Numerous environmentalists have employed the term mining spoil. This term typically varies from 'mine' soil. A mine spoil refers to the overburden materials, such as tailings, and the site from where minerals were removed. They are primarily left

---

1 Diversity Indices. Accessed from: https://bio.libretexts.org/Courses/Gettysburg_College/01%3A_Ecology_for_All/22%3A_Biodiversity/22.02%3A_Diversity_Indices on 7 December 2023.

unrehabilitated, lacking organic content, poor in nutrients, and consist of rocks and loose rock fragments. The revegetated or remediated soil aims to mitigate further environmental damage.

Conversely, the mining spoil can be modified to enhance its nutrient and organic matter content for agricultural purposes. Upon the physical reclamation of mine wastes for the purpose of facilitating plant development, with or without the application of topsoil, these materials are designated as "mine soils" [986, 1072].

The sole and ultimate aim of reclamation or remediation activities is to repurpose the degraded and contaminated fields for an economic activity, protect the socio-environments in the host mining communities, prevent further spread of pollution and to ultimately restore the integrity of the land. Thus, the ecological integrity of the degraded mining sites must be reclaimed, remediated, revegetated, and eventually aim at full restoration.

A multitude of authors in earlier research endeavours have sought to establish indicators of soil quality through the assessment of diverse soil characteristics, correlating these with various management practices, productivity levels, or environmental quality considerations. The variables or indicators of soil contamination primarily consist of various soil parameters that have been assessed either through dry or wet chemistry, measured in situ at the original place of contamination or ex situ, where the contaminated soil is excavated and treated away from the main site. Arshad and Martin [1073] introduced several variables designed to establish a soil quality index for evaluating the progress of a rehabilitated mining site. The variables encompassed soil organic matter, topsoil depth, infiltration, aggregation, pH, electrical conductivity, pollutant contents, and soil respiration. These techniques thoroughly assist in evaluating the health of the soil and tracking its advancement towards the restoration of soil quality.

Dickinson et al. [1075] presented biological indicators of soil health for application in the reclamation of Brownfield sites in the U.K. They underscored that evaluations of soil health may rely on indicators of biodiversity or functional processes. The quality of the soil or the concentration of heavy metals in revegetated sites will also rely on the duration of the revegetation or remediation efforts applied to the polluted areas. There are evident correlations between the duration of restoration and the enhancement of soil processes, including microbial carbon and nitrogen levels. Therefore, the evaluation of soil quality at chronosequence sites is essential to assess and monitor the success of the reclamation efforts applied to the site. Identifying soil quality indicators is crucial for assessing the health of reclaimed mine soils to evaluate the progress of the reclamation process. An integrated 'mine soil quality index' can be built based on the measured soil quality indicator values.

According to Rodrigue and Burger [1074], the coarse fraction is a critical mine soil property that affects site quality and forest productivity. Other research (e.g., Dutta and Agrawal, [793]) utilise soil moisture as an indicative metric for the rehabilitation of degraded land usage. Littlefield et al. [1071] examined the dynamics of carbon and nutrients in reforested mining sites, considering variables such as soil CO2 respiration and organic matter. A soil quality index could be established to track the progress of soil remediation of a contaminated site by integrating physical, chemical, and biological features, while considering heavy metal pollution levels in the degraded areas. The enhancement of soil physico-chemical and biological qualities in revegetated areas signifies progress in the restoration of mining sites.

# Calculating and determining a model for soil remediation index (SRI)

Here's a hypothetical equation that incorporates various indicators of successful soil remediation:

$$SRI = a \cdot pH + b \cdot OM + c \cdot N + d \cdot P + e \cdot K - f \cdot CL - g \cdot EC + h \cdot CEC - j \cdot BD + k \cdot WHC$$
$$+ l \cdot IR + m \cdot AS + n \cdot ST - o \cdot PR + p \cdot MA + q \cdot EP + r \cdot VC + s \cdot SEA + t \cdot SR + u \cdot MFP$$

Where:
- **SRI** = Soil Remediation Index (overall measure of soil remediation success)
- **a, b, c, . . ., u** = Coefficients representing the weight of each indicator (should be determined based on empirical data)
- **pH** = Soil pH level
- **OM** = Organic Matter content
- **N** = Nitrogen level
- **P** = Phosphorus level
- **K** = Potassium level
- **CL** = Contaminant Levels (e.g., heavy metals)
- **EC** = Electrical Conductivity (soil salinity)
- **CEC** = Cation Exchange Capacity
- **BD** = Bulk Density
- **WHC** = Water Holding Capacity
- **IR** = Infiltration Rate
- **AS** = Aggregate Stability
- **ST** = Soil Texture
- **PR** = Permeability
- **MA** = Microbial Activity
- **EP** = Earthworm Population
- **VC** = Vegetation Cover
- **SEA** = Soil Enzyme Activities
- **SR** = Soil Respiration
- **MFP** = Mycorrhizal Fungi Presence

This equation is an illustrative example and can be tailored based on specific soil conditions, contaminants, and empirical data from the site being studied. The coefficients (a, b, c, . . ., u) should be determined through research to reflect the relative importance of each indicator in the context of soil remediation success.

To this end, studies have provided some insights into developing a soil quality guide for monitoring success of reclamation and remediation efforts. These studies suggest that there are primarily two steps in development of soil quality index:

The first stage which involves identifying soil qualities that elucidate soil functions. These have been provided in the equations above.

The contribution of individual soil properties to achieve the intended soil function can be quantified by assigning weights to the recognised or measured properties. These indicator weights can be determined by statistical methods such as correlation analyses, regression equations and principal component analysis.

Using the principal component analyses, the methodology outlined by others (e.g., [1076–1078] has been employed to establish the mine soil quality index (MSQI) for appraising the success of revegetation of mine contaminated sites. Here, PCA was employed to identify the relevant features and their weighting variables. Principal components (PCs) with higher percentage contributions and those with higher eigenvalues are chosen as the variables that are of significant consideration to the soil quality index.

Only the variables with significant and higher eigenvalues within the factor loadings are preserved for indexing under a certain principal component (PC), either 1, 2, 3 or above. High factor loadings are characterised as having absolute values within 10% of the maximum factor loading [1079]. The eigen values here are considered as the weights of each soil function variable, indicator or parameter (e.g., a, b, c, . . ., u = Coefficients representing the weight of each indicator; see equation above).

When multiple variables are preserved under a single principal component, correlation analysis is utilised to ascertain whether the variables may be deemed redundant and thus excluded from the soil quality index (Masto et al., 2008). Here, if the highly loaded components were uncorrelated, each was deemed significant and hence maintained in the soil quality index calculation. Among highly correlated variables, the variable with the highest absolute factor loading is selected for inclusion in the evaluation of the soil quality index.

## 16.4 Comparing the reclaimed site with natural forests

Comparisons can also be made between the restored or reclaimed site and nearby undisturbed ecosystems such as native forests. Soil quality indices, which aggregate various soil health parameters, can be utilized to assess success. The assessment of success is conducted by a comparative analysis between the restored site and a native forest, with the underlying assumption that natural forests possess inherent stability in terms of ecological activities and services.

These ecosystems are additionally presumed to be devoid of contamination and pollution due to minimal or nonexistent human interventions. Typically, natural ecological systems demonstrate superior chemical soil quality parameters including pH levels, nitrogen (N), and carbon (C) contents as well as the presence of major elements such as magnesium (Mg), calcium (Ca), potassium (K), and sodium (Na).

Native forest ecological systems also exhibit reduced exchangeable acidity, enhanced levels of basic and exchangeable cations, improved cation exchange capacity, moderate electrical conductivity, and increased availability of phosphorus (P). This is in contrast to mine-degraded sites, where these parameters are more likely to be fixed and less available. Furthermore, it is worth noting that natural forests exhibit more favorable outcomes in terms of soil physical qualities when compared to recovering reclaimed mine sites.

In comparison, it is anticipated that soil physical indicators, such as soil compaction and bulk density, infiltration rates, structure and aggregate formation, moisture content, porosity, hydraulic conductivity, and other physical characteristics, will exhibit greater improvement in natural forest areas as opposed to sites contaminated by mining activities. Similarly, it is anticipated that the soil microbial population and activities, such as soil enzymes and aeration, are predicted to exhibit superior or elevated levels in the natural forest compared to those seen in the degraded mine site.

Natural forests are also hypothesized to contain less concentrations of PTE, while the area near the gold mine tailings of a local mine in southwestern Ghana [333]. In this regard, a similar study by [735] found lower and reduced content of soil nitrogen (N) and phosphorus (P) when compared to the natural forest due to the same reason of ecosystem disruption, removal of vegetation, and loss of litter layer during mineral mining.

Mensah et al. [333] observed a notable disparity in contamination and enrichment factors of potentially toxic elements (PTEs), namely As, Cd, Pb, Ti, V, and Zn, between natural forest soils and soils obtained from the vicinity of a mine tailing in southwestern Ghana. As a result, the study revealed that forest soils exhibited a lower pollutant load index and human health hazard index compared to soils in the vicinity of the mine. Several researchers have reached the conclusion that forest ecosystems have the potential to facilitate the restoration of soils that are lacking in soil organic matter (SOM) [335, 453]. Furthermore, Petelka et al. [384] observed that the tailing soil at a gold mine in Ghana exhibits a relatively low level of SOM, suggesting the potential for improvement through the implementation of afforestation practices on the tailing area.

## 16.5 Responses and feedbacks from the host mining community

Actively involving the indigenous community or local people and considering their positive feedback and suggestions are important. Such responses and feedbacks from the community abet to build trust and lessen conflicts between the mining companies and the project affected communities. In this regard, the establishment of connections between mining operations and other community groups will not only enable the community in gaining a better understanding of the project but it can also assist the mining company to appreciate and acquaint themselves of priorities and feelings of the community regarding the restoration project. Involving community members, both women and men, in the actual planning stages of development programs will make success much more likely. Such participation should move beyond being passive to an active one. Thus, participation should move away from just informing the community after decisions have already been made at the top regarding the undertaking. Ultimately, effective and active participation in the restoration project should involve the following four steps as outlined in a Guide to Leading Practice Sustainable Development in Mining (2011):

i.    dialoguing,
ii.   working in collaborations,
iii.  building partnerships and strengthening organizations,
iv.   broadening connections with people outside the community with shared or similar interests.

## 16.6 Regulatory or compliance standards

The achievement of reclamation success can also be evaluated by comparing it to the regulatory or compliance criteria that have been set at regional, international, and national levels. Finally, the evaluation of the efficacy of plant species in the process of soil remediation by eliminating hazardous substances, commonly referred to as phytocleaning, can be regarded as a measure of achievement. The evaluation of the plant's efficiency or effectiveness can encompass various aspects, such as its ability to sequester toxic elements from the soil into the root (phytostabilization), from the soil into the shoot (phytoextraction), from the root into the shoot (translocation), or to volatilize the toxic element from the plant leaf into the atmosphere (phytovolatilization).

The achievement of success in plant or vegetation growth is contingent upon the biogeochemical properties of the soil, as discussed in the preceding chapter (Chapter 8). Success can be attained when the values of BAC, BCF, and TF surpass 1. Moreover, the effectiveness of a reclamation method will be assessed by comparing the levels of metal or element concentration to global soil average values, maximum permissible concentration, trigger action values, and other regulatory criteria.

The plant or plantation's capacity to provide sufficient biomass, its ease of coppicing following harvesting, and its potential to augment yield and improve remediation outcomes subsequent to harvesting. As an illustration, Ning et al. [357] found that the application of root cutting treatments at a rate of 10% resulted in a notable enhancement of 58.6% in the leaf biomass yield of *Celosia argentea*. This subsequently led to an improvement in its phytoremediation effectiveness and capacity to remove Cd contaminants.

It has been argued that the act of decapitation and root cutting can have an impact on various physiological characteristics of plants including their height, dry weight, and transpiration rate [357]. These factors play a significant role in determining the effectiveness of phytoremediation, thereby determining the success of post-mining revegetation efforts.

## 16.7 Socioeconomic indicators

By utilizing socioeconomic indicators, such as livelihood restoration, the rehabilitated site provides several advantages to the local community. These benefits include the provision of recreational centers, playgrounds, and parks for children as well as serving as locations for educational, research, and tourist activities.

## 16.8 Employing the use of artificial intelligence and geospatial tools

The utilization of artificial intelligence tools and geographical information system in assessing and supervising the achievement and advancement of post-reclamation cannot be disregarded. These contemporary scientific abilities and techniques allow us to systematically assess the effectiveness of reclamation activities at regular intervals. For example, drones can be employed to document the process of species regeneration and their dispersion. Illustrations can encompass the assessment of biomass productivity, species abundance and variety, plant density, plant stature, stem circumference, canopy extent, and other indicators of growth and productivity using drones or satellite imagery. In addition, satellite pictures can be utilized to monitor the distribution of vegetation, both in patches and scattered areas, as well as to track the progress of vegetation growth and changes in land cover and land use. Therefore, regions that have dispersed or patchy vegetation or those species that have been affected with pests and diseases and subsequently died can be replenished or reforested.

# 16.9 Impact of land use on reclamation success

## 16.9.1 Influence of different land uses on reclamation outcomes

The success of land reclamation efforts is closely tied to the intended post-reclamation land use, with different land uses such as forestry, agriculture, and recreation posing unique challenges and opportunities. The choice of land use has significant implications for soil management, vegetation establishment, and long-term sustainability, making it a critical factor in the reclamation process [670, 483].

## 16.9.2 Forestry as a reclamation land use

Forestry is one of the most common land uses for reclaimed mine lands, particularly in regions where mining has led to extensive deforestation. The establishment of forests on reclaimed lands offers numerous environmental benefits including soil stabilization, carbon sequestration, and habitat creation for wildlife. However, the success of forestry-based reclamation depends on several factors including the selection of appropriate tree species, soil preparation, and ongoing management [454, Berger, 2008].

### 16.9.2.1 Challenges

Species selection: One of the primary challenges in forestry reclamation is the selection of tree species that are both resilient to the altered soil conditions of reclaimed lands and capable of supporting long-term forest development. Fast-growing, hardy species such as pines or eucalyptus are often chosen for their ability to quickly establish cover and stabilize the soil. However, these species may not always be compatible with native biodiversity or long-term ecosystem sustainability [116, 671].

Soil fertility: Reclaimed soils are often deficient in nutrients and organic matter, which can hinder tree growth and forest development. The application of soil amendments and fertilizers is often necessary to support tree establishment, but these interventions can be costly and require ongoing management [762].

Long-term sustainability: Ensuring the long-term sustainability of forestry reclamation requires careful management of forest resources including thinning, pest control, and fire management. In some cases, the lack of ongoing management can lead to poor forest health, increased vulnerability to pests and diseases, and even the failure of the forest ecosystem [672].

## 16.9.3 Agriculture as a reclamation land use

Agriculture is another common land use for reclaimed mine lands, particularly in regions where land availability for farming is limited. Reclaimed lands can be converted into productive agricultural areas, providing economic benefits and supporting local food security. However, agricultural reclamation presents unique challenges related to soil quality, crop selection, and the management of potentially toxic elements (PTEs) [673].

### 16.9.3.1 Challenges

- Soil quality: The success of agricultural reclamation depends heavily on the quality of the reclaimed soil, particularly in terms of its fertility, structure, and water-holding capacity. Reclaimed soils often require significant amendments, such as the addition of organic matter, lime, and fertilizers, to restore their productivity. In some cases, the physical and chemical properties of the soil may remain suboptimal for agriculture, limiting the types of crops that can be grown (Brown & Brown, 2012).
- PTE contamination: The presence of potentially toxic elements (PTEs) in reclaimed soils poses a significant challenge for agricultural land use. Crops grown in contaminated soils may uptake PTEs, leading to food safety concerns and potential health risks for consumers. Managing PTE levels through soil remediation, crop selection, and regular monitoring is essential to ensure the safety and sustainability of agricultural practices on reclaimed lands [935].
- Crop selection: Different crops vary in their tolerance to soil conditions and their ability to uptake PTEs. Selecting crops that are well-suited to the specific conditions of reclaimed soils and that have low potential for PTE accumulation is critical for successful agricultural reclamation. However, crop options may be limited in regions with severe soil degradation or contamination.

## 16.9.4 Recreation and mixed-use land reclamation

Recreation and mixed-use developments are increasingly being considered as viable post-reclamation land uses, particularly in urban or peri-urban areas. These uses can include parks, golf courses, residential developments, and other recreational facilities. The integration of these land uses can enhance the social and economic value of reclaimed lands while providing environmental benefits such as green spaces and stormwater management.

### 16.9.4.1 Challenges

-   Soil and landscape design: The success of recreation-based reclamation often de-
    pends on the careful design of the landscape including the creation of stable land-
    forms, adequate drainage systems, and aesthetically pleasing green spaces. The
    design process must account for the limitations of reclaimed soils including their
    structure, fertility, and potential contamination.
-   Public health and safety: Ensuring public health and safety is a critical consider-
    ation in recreation-based reclamation. This includes managing the risks associ-
    ated with PTEs in soils, ensuring that recreational areas are free from hazardous
    materials, and maintaining infrastructure such as trails, playgrounds, and water
    features.
-   Long-term maintenance: The long-term success of recreation-based reclamation
    depends on ongoing maintenance including landscaping, pest control, and the
    management of recreational facilities. Without adequate maintenance, these
    areas may degrade over time, leading to reduced public use and increased man-
    agement costs.

# 16.10 Role of land use planning in reclamation success

Land use planning plays a crucial role in the success of land reclamation efforts by
ensuring that reclaimed lands are utilized in a manner that aligns with environmental, social, and economic goals. Effective land use planning involves the integration of
ecological principles, stakeholder engagement, and adaptive management strategies
to create sustainable and resilient post-reclamation landscapes.

## 16.10.1 Integrating ecological principles

Incorporating ecological principles into land use planning is essential for ensuring the
long-term sustainability of reclaimed lands. This includes considerations such as
maintaining soil health, supporting biodiversity, and enhancing ecosystem services.
For example, planning for a forestry-based reclamation should include strategies for
promoting native species diversity and maintaining soil fertility over time. Similarly,
agricultural reclamation should prioritize soil conservation practices and the use of
sustainable farming methods.

## 16.10.2 Stakeholder engagement

Successful land use planning requires the involvement of a wide range of stakeholders including local communities, government agencies, environmental organizations, and industry representatives. Engaging stakeholders in the planning process helps to ensure that reclamation efforts are aligned with local needs and priorities, increasing the likelihood of long-term success. For example, involving local farmers in the planning of agricultural reclamation projects can help identify crop preferences, market opportunities, and potential challenges.

## 16.10.3 Adaptive management and monitoring

Adaptive management is a critical component of land use planning for reclaimed lands. This approach involves continuously monitoring the outcomes of reclamation efforts and making adjustments as needed to address emerging challenges or opportunities. For example, if monitoring reveals that certain crops are not performing well in reclaimed soils, land use plans can be adjusted to introduce alternative crops or soil management practices. Adaptive management ensures that reclamation efforts remain responsive to changing conditions and can adapt to new information or technologies.

## 16.10.4 Balancing multiple land uses

In many cases, reclaimed lands may be used for multiple purposes such as a combination of forestry, agriculture, and recreation. Effective land use planning must balance these different uses to avoid conflicts and maximize the overall benefits of reclamation. For example, a mixed-use reclamation plan might include buffer zones to separate agricultural areas from recreational spaces or it might incorporate agroforestry practices that combine tree planting with crop production.

## 16.10.5 Long-term sustainability

Ultimately, the goal of land use planning in reclamation is to ensure the long-term sustainability of reclaimed lands. This involves not only restoring the land to a stable and productive state but also creating a framework for ongoing management and stewardship. Long-term sustainability is achieved when reclaimed lands continue to provide environmental, social, and economic benefits long after the initial reclamation efforts are complete.

In summary, the choice of land use plays a critical role in determining the success of land reclamation efforts. Each land use type, whether forestry, agriculture, or recreation, presents unique challenges and opportunities that must be carefully managed to achieve sustainable outcomes. Effective land use planning, which integrates ecological principles, stakeholder engagement, and adaptive management, is essential for ensuring the long-term sustainability of reclaimed mine lands.

## 16.11 Conclusions

The success of restoring a degraded mine site relies on a comprehensive measurement and monitoring framework that considers multiple factors. The reestablishment of indigenous species and the return of soil flora and fauna are important indicators of progress. Evaluating soil quality indicators and indices, comparing the reclaimed site with natural forests, and considering responses and feedback from the mining community are also crucial. Compliance with regulatory standards and the effectiveness of plant species in soil remediation are additional measures of achievement. Socioeconomic indicators, such as livelihood restoration and community benefits, should also be taken into account. Overall, a holistic approach that addresses ecological, physical, chemical, and socioeconomic aspects is necessary for successful mine site restoration.

Prince Addai and Albert Kobina Mensah

# Chapter 17
# Critical factors for driving successful restoration of degraded mine lands

**Abstract:** Restoring mine sites requires collaboration among stakeholders like government agencies, environmental experts, local communities, and industry partners. Social, environmental, and economic factors are key. Community engagement, involving indigenous groups and women, and people-driven development are essential. Choosing plant species with both economic and social benefits can also contribute to the success of these restoration initiatives. Thus, incorporating indigenous knowledge, traditional practices, and the use of indigenous knowledge in mining projects can significantly improve the success and long-term viability of these initiatives.

## 17.1 Introduction

Restoring degraded mine sites is a multidisciplinary effort that requires collaboration between various stakeholders including scientists, consultants, sociologists, government agencies, environmental experts, local communities, and industry partners to achieve successful restoration or reclamation efforts and long-term sustainability. It is important to tease out concrete factors to ensure successful restoration of the degraded and contaminated mining site. We group these factors into three major themes (outlined and summarized in Fig. 17.1 and Table 17.1):

a) Social factors
b) Environmental or ecological factors
c) Economic factors

---

**Prince Addai,** Soil Research Institute, Council for Scientific and Industrial Research, Academy Post Office, PMB, Kwadaso-Kumasi, Ghana, e-mail: addaiprince876@gmail.com, paddai@csir.org.gh, ORCID: https://orcid.org/0000-0002-7500-4271
**Albert Kobina Mensah,** Soil Research Institute, Council for Scientific and Industrial Research, Academy Post Office, PMB, Kwadaso-Kumasi, Ghana, e-mail: albert.mensah@rub.de, albertkobinamensah@gmail.com, ORCID: https://orcid.org/0000-0001-5952-3357

https://doi.org/10.1515/9783111662046-017

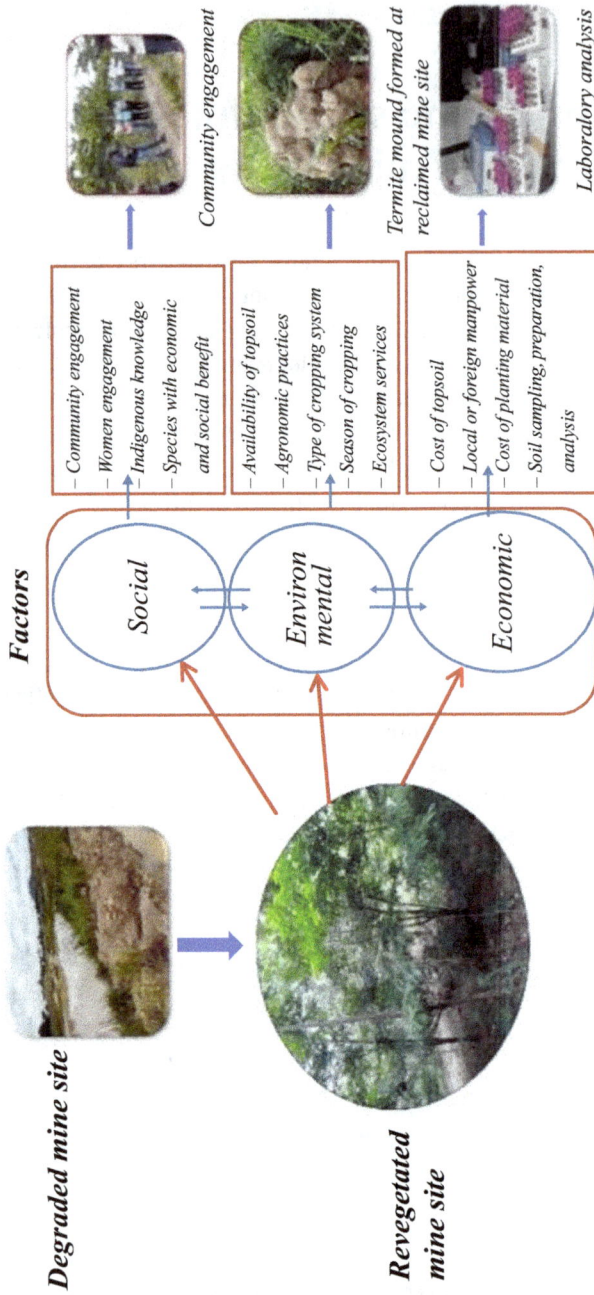

**Fig. 17.1:** A diagram conceptualizing the various factors to consider in restoring a degraded mine site.

## 17.2 Social factors

### 17.2.1 Community engagement

Broad engagement forms one of the important components of ecosystem restortaion as suggested by FAO [819]. Community engagement in mining sector is a method that aims to actively involve the community in order to attain sustainable results, foster fair decision-making procedures, and enhance the bonds and trust between mining companies and communities. By this the mining sector actively engages and collaborates well with the affected community. According to [674–675], it is recommended that stakeholder engagement be initiated at the earliest practicable stage of the consultation process.

From the aforementioned statement, this serves the purpose of identifying the social benefits provided by local communities and establishing a foundational measure for restoration planning and benchmarking. Establishing robust connections with nearby communities, native populations, and interested parties is crucial for guaranteeing the success of restoring degraded mine sites. Effective interaction with indigenous groups and local communities, traditional authority, and other key stakeholders are important factors to be incorporated into the restoration process.

The "Agenda 21" principle number 10 stated once more that environmental challenges are best managed with the cooperation of all interested individuals at the relevant level. For example, at the national level, each citizen shall have proper access to environmental information held by public authorities, including decision-making processes, and thus community engagement is a crucial issue to consider during mine site restoration [493].

In fact, international leading practice including World Banks in community development through projects calls for active community involvement in planning and implementation process. For this, indigenous people need to be encouraged and motivated to support the project with their interest and capabilities. In the opposite directions, projects that fail to capture interest of the indigenous people are likely to fail or it might attain the status of "white elephant."

In this regard, Mensah et al. [324] in their study titled, "Environmental Impacts of Mining: A Study of Mining Communities in Ghana," argued that insufficient or absent community participation in determining the social needs of host and neighboring communities can result in misaligned priorities in terms of corporate social responsibility (CSR). The researchers revealed that several projects initiated by mining firms as part of their CSR were ineffective due to insufficient participation of the impacted communities in the decision-making process. Mining firms may make sure that the restoration process supports social well-being, sustainable development, and community requirements by incorporating stakeholders and local communities.

## 17.2.2 Women engagement

The involvement of women is also an important factor to consider in every project implementation such as restoring degraded mine sites. For instance, the Rio declaration on environment and development in 1992 principle 20 stated that "women have a vital role in environmental management and development and so their full participation during restoration of degraded mine site is therefore essential to achieve sustainable restoration success" [493]. Women are more committed to the preservation and promotion of the natural environment than men. They prioritize environmental protection and future generations' well-being by actively participating in activities such as the repair of degraded mine sites.

Their involvement in restoring mine site can extend to many facets of environmental stewardship. For example, they may be tasked to take care of nursery establishment. Once the revegetation site has been established successfully and the plants have matured, it is expected that women will be able to directly gather resources for home use, such as firewood and other non-timber forest products, in the near future.

Again, Agenda 2063, aspire 6 emphasizes that "an Africa whose growth is people-driven, relying on the potential of African people, notably its women and youth, and caring for children." As a result, by 2063, Africa will be a continent in which all citizens actively participate in decision-making in all elements of development, including social, economic, political, and environmental issues. Africa will be a continent in which no child or woman is left behind [676]. This is why it is so crucial to involve women in all aspects of environmental management such restoration of degraded mine sites.

Goal 17 of Agenda 2063 calls for equal rights for men and women in every area of life. This includes making sure that women have the same chances and are represented in decision-making processes across all fields such as environmental projects. SDG 5 also talks about gender equality and giving all women and girls more power. It stresses how important it is for women to be involved in the environment especially restoration of degraded mine sites. Involving women in the restoration of degraded mine sites can result in a variety of benefits:

– Women frequently offer distinct insights on the environment and local ecosystems as well as traditional wisdom. Their active participation in the rehabilitation process has the potential to provide creative ideas and long-term practices.
– Women play an important role in many impacted communities around mining sites. Involving women in restoration efforts empowers them financially and also contributes to the overall well-being of the community.
– Women in mine site restoration programs guarantees that diverse perspectives are taken into account. Women's inclusion brings a broader spectrum of perspectives and objectives and creates effective environmental restoration methods.
– Involving women in the restoration of degraded mine sites allows them to gain socioeconomic empowerment. Employment, skill development, entrepreneurship,

and leadership roles within restoration initiatives can all contribute to their economic independence and empowerment and that authorities must eliminate barriers that may obstructs women's inclusion in land restoration initiatives [677].

## 17.2.3 Consideration of indigenous knowledge of the people

Every mining community has a wealth of local knowledge that must be considered when carrying out projects such as the restoration of deteriorated mine sites. This local experience is appreciated by the community, and incorporating it into project implementation may result in the project's success if done correctly. In this regard, principle 22 of the agenda 21 highlights the importance of acknowledging the indigenous populations, their communities, and the indigenous knowledge they employ in the stewardship and safeguarding of their watersheds.

Furthermore, the document underscores the importance of taking into account the perspectives of other local communities in the realms of environmental management and ecosystem initiatives.

It also provides pertinent positions for the utilization of local communities, and their engagements are significant in the reclamation of degraded and contaminated sites throughout the project implementation and during the closure phase of the project's life cycle.

Ultimately, the indigenous knowledge possessed by local communities ought to be utilized and enhanced by integrating it with contemporary scientific approaches in an effort to restore degraded and contaminated mining sites. For this, mining companies should recognize and duly support their identity, culture and interests and enable their effective participation in the achievement of sustainable restoration or reclamation of degraded mine lands.

An illustrative instance of a pertinent aspect to be taken into account is stated below:

– The impact of cultural and spiritual convictions: There exist mining communities that possess sacred sites and cultural values that warrant respectful consideration and should not be dismissed. Indigenous knowledge often entails the identification and demarcation of sacred sites and areas of cultural importance. The preservation of historical sites and the incorporation of cultural values during the restoration process are essential for maintaining social harmony within a community and encouraging their active involvement in restoration efforts. The people mining community may prefer to include fern as key species. This is because ferns might hold spiritual significance or be associated with certain rituals, traditions, or beliefs. Introducing ferns in the restoration process could align with local spiritual practices, symbolizing renewal, growth, or healing, thereby fostering a spiritual connection between the land and the community.

- **Ethnobotanical knowledge** is an important factor to consider. Indigenous communities often possess knowledge about the diverse uses of plant species including medicinal plants and their ecological roles. Incorporating this knowledge can help in reestablishing diverse plant communities and preserving biodiversity in the restored areas. For instance, certain species of ferns exhibit therapeutic characteristics in the context of traditional medicine.

  It is plausible that communities possess indigenous knowledge pertaining to the therapeutic use of ferns in the treatment of diverse illnesses. Thus, introduction of these species as part of restoration efforts may have the potential to establish a nearby reservoir of medicinal resources, thus making a valuable contribution to the health and overall well-being of the community.
- **Indigenous knowledge in wildlife habitat restoration:** Traditional knowledge often includes insights into maintaining habitats for various wildlife species. This knowledge can be invaluable in creating suitable environments for the return of native fauna affected by mining activities. This approach has history in fishing industry as *"open and closed season." With this approach community might at some point be allowed to visit restored degraded mines and at some points they may not be allowed depending on the conditions prevailing the time.*
- **Traditional water conservation practices:** Indigenous communities might have expertise in water conservation methods such as rainwater harvesting, building small dams or check dams, and managing water flow, which could be crucial for restoring water sources affected by mining activities.
- **Soil management techniques:** Indigenous people may have knowledge that include traditional soil conservation methods, such as using organic materials, crop rotations, or agroforestry techniques, which could be valuable in restoring soil fertility and structure in degraded mine sites.

In summary, including indigenous knowledge into the rehabilitation of degraded mine sites in Ghana is critical for developing effective and long-term plans. Collaboration with local populations and the use of their traditional practices, wisdom, and viewpoints can significantly improve the success and long-term viability of restoration initiatives.

## 17.2.4 Species with both economic and social benefits

Undoubtedly, choosing plant species to restore degraded mine sites entails taking into account plants that help local residents on an economic and social level as well as aid in the growth of the country. Plants and shrub species including ferns, *Acacia mangium, Leucaena leucocephala, Terminalia superba, Terminalia ivorensis, Tectona grandis*, and *Chromolaena odorata,* have been tested to exhibit a number of qualities that make them advantageous for the socioeconomic rehabilitation of mine sites. *Chromolaena odorata,* for instance, offers medicinal use in treating wounds in Ghana. Additionally, cash crops such as oil palm and cocoa, other food crops such as lettuce, and others have been found to tolerate toxicity to an extent and they do well in mine tailings [e.g., 6, 57, 319].

Oil palms, for instance, provide oil that can be used for either food or soap production. On the other hand, the palm bunches and fronds can be utilized by farmers to produce biochar and compost in host mining communities. These are used as organic soil amendments that improve soil quality and enrich productivity. Cocoa that is grown on land that has been reclaimed has the potential to bring in foreign currency for the nation, the pods can be utilized to generate organic soil amendments, and cocoa beans can be harvested and used to make chocolates. Figure 17.1 shows fruits harvested from oil palm successfully established from a mine tailing and Fig. 17.2 displays a cocoa farm successfully established on a mine tailing in Ghana.

Absolutely, the selection of trees for restoration efforts in degraded mine sites should indeed prioritize the involvement and agreement of indigenous communities.

The process should incorporate the indigenous knowledge and preferences regarding plant species that align with their needs and the local ecosystem. The choice of selection of plant species by the community may hover around one or more of the factors; fast growth, provision shades, adaptability to local conditions, carbon sequestration capacity, provision of firewood, timber, and ability to fix atmospheric nitrogen into the soil to increase the fertility stated of the lands and charcoal potential. These are critical considerations that may resonate with the community's needs and qualify for environmental sustainability. These qualities have been described earlier in detail in Chapter 15.

**Fig. 17.1:** Palm fruits harvested from a 5-year-old reclaimed mine site in Ghana.

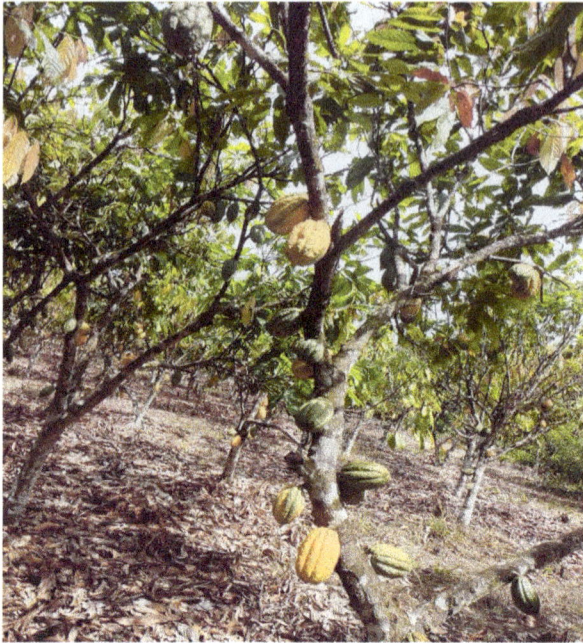

**Fig. 17.2:** Cocoa plant established from a 25-year-old reclaimed mine site in Ghana.

In this regard, the involvement of indigenous communities in the decision-making process regarding tree species selection ensures that the chosen species align not only with ecological rehabilitation goals but also with the social, economic, and cultural needs of the local population. This participatory approach promotes community ownership, increases the likelihood of project success, and fosters sustainable land management practices and account for the success of reclamation efforts.

## 17.3 Environmental factors

### 17.3.1 Environmental education and awareness initiatives

Conducting seminars and implementing awareness campaigns serve to educate the community of the significance of rehabilitation efforts and foster a collective feeling of environmental responsibility. These techniques serve to enhance democratic processes and foster a climate of acceptability toward restoration efforts. In this context, individuals see a sense of agency in expressing their viewpoints, experiencing a heightened level of respect, and fostering a strengthened sense of trust between the mining business and the local community affected by the project. In this respect, Mensah et al. [324] argued that environmental awareness campaigns and educational initiatives within diverse mining communities promote responsible and sustainable utilization of the environment amidst ongoing mining operations. This approach will establish the essential equilibrium between the pursuit of development and economic progress and the imperative environmental requirements that are indispensable for the sustenance of communal livelihoods.

## 17.3.2 Availability and quality of topsoil

The significance of topsoil in facilitating plant growth and building a sustainable ecosystem is of paramount importance. In numerous mining activities, the topmost portion of soil is either extracted or significantly disrupted, resulting in detrimental effects on its quality and composition. The presence and characteristics of appropriate topsoil, including nutrient composition, organic matter concentration, microbial activity, and texture, play a crucial role in the achievement of effective restoration results. The process of reintroducing or replacing topsoil with essential nutrients and microbes is of utmost importance for the establishment of plants and the general rehabilitation of ecosystems [678–679].

## 17.3.3 Agronomic practices

The successful restoration of mining sites relies on the implementation of effective agricultural measures. Sufficient water availability or access to irrigation infrastructure is of utmost importance, particularly in dry or semiarid locations, in order to facilitate plant development and establish a protective layer of vegetation. The combined application of organic and inorganic fertilisers is essential for addressing soil inadequacies and enhancing plant health, as emphasised by [680]. This combined strategy promotes nutrient availability and supports ecosystem stability through better soil health and effective pest management measures [681].

## 17.3.4 Choice of cropping system

The choice between monocropping, mixed cropping, or agroforestry systems on restored mine sites has significant implications for ecosystem resilience and biodiversity. Mixed cropping systems incorporating native trees, shrubs, or diverse plant species can enhance soil fertility, reduce erosion, and provide habitats for diverse wildlife. Agroforestry systems, combining trees or shrubs with crops, can improve soil structure, nutrient cycling, and microclimate regulation ([886]). In other instance the community may come and agree on mono-plantation for commercial purposes, for example, *Tectona grandis* or *Acacia mangium* plantations.

### 17.3.5 Season of restoration

In order to restore mining sites, choosing the right planting season is essential since it affects plant growth, water availability, and ecological stability. Based on the phenology of particular plant species, rainfall patterns, and local climate conditions, major and minor cropping seasons must be carefully examined. Plant growth rates, water needs, and the general effectiveness of vegetation establishment are all impacted by seasonal fluctuations. According to [684–685], planting in the dry season may necessitate more irrigation support, whereas planting during the rainy season may maximize plant life due to greater moisture availability.

## 17.4 Economic factors

Economic factors play a pivotal role in the restoration of degraded mine sites, as they determine the feasibility and sustainability of the rehabilitation process. The most important aspect of restoration is the funding sources. For better understanding and simplicity ask this question. Who pays for restoration? by [686]. Here is an expansion on the economic factors involved.

### 17.4.1 Cost of topsoil

The cost associated with acquiring topsoil for restoration purposes is a significant economic consideration. In cases where topsoil needs to be sourced or transported to the site, expenses can escalate. Often, the availability and quality of topsoil can influence its cost, with fertile and nutrient-rich topsoil being more expensive. Thus, utilizing local sources and implementing soil conservation techniques can help mitigate these costs [687–688].

### 17.4.2 Labor and manpower costs

Manpower, including both professional and unskilled labor, accounts for a significant amount of the overall restoration budget. The use of using local unskilled may cut expenses and also help community participation and empowerment. The restoration projects should also aim at training programs for people of the local community that can help them improve their abilities and find long-term work [689]. Furthermore, bringing in foreign personnel when needed might provide skills and knowledge to the restoration process.

## 17.4.3 Cost of planting materials and nursery management

It is imperative to include the costs related to the acquisition of planting materials and the management of nurseries for seedlings or plant multiplication when preparing the budget. Engaging women from the community into nursery management presents a potential avenue for employment and the enhancement of skills, consequently fostering local economic empowerment and inclusivity [690–691].

## 17.4.4 Cost of soil sampling and laboratory analyses

Conducting soil sampling and laboratory analyses to assess soil quality and nutrient levels is essential for effective restoration planning [692–693]. The costs incurred in these activities should be considered within the overall budget to ensure appropriate soil amendments and plant selection for successful rehabilitation.

## 17.4.5 Financial and human resources for implementation

Supporting small businesses, generating employment, and funding community development initiatives are crucial economic aspects of mine site restoration. Allocating resources toward these objectives not only aids in the restoration process but also contributes to broader socioeconomic development in the affected communities [491].

Overall, it is imperative to combine all the factors (social, environmental, and economic) as a holistic approach to restore the natural equilibrium in restored mine sites. Mining site restoration or rehabilitation should be a win-win situation for the environment and people and that employment of these factors not only lessen the adverse effects on the environment but also improve the quality of life in nearby communities Table 17.1.

**Tab. 17.1:** Summary of different factors and management actions that are required for successful restoration of degraded mine lands.

| Environmental factors | Possible effects and management |
|---|---|
| Irregular rainfall pattern | It may increase erosion and could destroy or deplete soil nutrients as vegetation establishment. Therefore, good drainage systems should be maintained during the restoration of degraded mine lands. |
| Loss of topsoil or organic matter | This is crucial in ecological restoration and loss of it may affect nutrient availability as well as microbial affect microbial activity |
| Soil acidity/alkalinity | It affects plant nutrient uptake and good soil amendment may be needed to balance the soil pH. [702]. |
| Prolong drought | Drought prolonging may limit effort towards restoration. Drought affects a lot of activities. E.g., It may affect vegetation establishment, microbial population, and activities. Therefore, any restorations that depend solely on natural rainfall must be carefully examined before project implementation. |
| Severe compaction | Root growth is severely restricted, affecting infiltration |
| Rock components | Can affect infiltration and hydraulic properties |
| **Socio-economic factors** | **Role effect** |
| Labor availability | This would facilitate activities such as raising seedlings, planting, fertilization, watering, weed control, bushfire control and many others |
| Funding opportunities from the government, NGOs or other international bodies such as | This is important for successful restoration. Could enhance the implementation of restoration activities (See [702]) |
| Regularization of hunters eg issuance of permits to hunt (when or when not). This part is mostly associated with some African countries. | Could disincentivize locals from hunting. It may also check the rampant setting of farmers to hunt. This may prevent the destruction of a restored ecosystem. |
| Political instability | This could affect the progress of the restoration activities if power is shifted, or low political priority is given to the restoration. |
| Incentives | Could enhance or help people to be committed to the project |
| **Landscape factors** | **Possible effects** |
| Landform (slope) | May increase run-off which will wash the topsoil, and could also affect seedling establishment |

**Tab. 17.1** (continued)

| | |
|---|---|
| Poor drainage | Could affect the hydrology of the landscape |
| Miniature voids | May cause acid drainage or, the water body may become extremely acidic due to exposure to sulphide ores |
| Presence of termite hills | May improve microbial activities. However, it must be monitored to prevent the destruction of young seedlings |
| Flat terrain | May reduce surface runoff/floods and as well improve microsite for seed establishment |
| **Ecological factors** | **Role/effects** |
| Herbivory | It may destroy young established seedlings. Therefore, proper mechanisms should be put in place to monitor activities of herbivores. |
| Presence of Invasive species | It may create competition and reduction of native species |
| Presence of native species | Good for birds and other pollinators |
| Seed source / Seeds dispersal | Species may disperse to other suitable habitats through an explosive mechanism process. This is possible if you have a reference ecosystem or an Island of reference module within the restoration area. Seed source is a very important aspect of ecological restoration. Therefore, quality seed source is prudent. |
| Allelopathy | It may affect the colonization of native seedings and reduce the establishment of native species |
| Good microsite | It may affect the germination of seeds and their establishment, especially from direct seeding through broadcasting |
| **Stakeholders (In the case of Ghana)** | **Role/effects** |
| Forest Commission/Forest Services division | This could provide expertise, and Forest guards during implementation. |
| CSIR Soil and Forest Research | They may provide expertise on methodologies and history or behaviors of different soil types for management purposes. |
| Traditional /Opinion leaders | Could provide local or indigenous knowledge |

**Tab. 17.1** (continued)

| | |
|---|---|
| NGOs | NGOs are key stakeholders in land restoration as they may promote public awareness, funding support and other training on capacity building during project initiation and presentation. |
| Environmental agencies like EPA, Wildlife and fire service | May be responsible for the Implementation of policies such as fire regulations to provide technical support |
| Farmers | May provide local knowledge and labour during restoration. Ecological local knowledge is paramount for ecological restoration (702; [802]) |
| The Municipal or District authority | May provide funding, enforcement of policies/ implementation of by-laws |
| **Risk factors** | **Possible effects** |
| Invasive species | This may surpass natural vegetation, disturb the ecological equilibrium point and impede the growth of preferred species following restoration. Therefore, this may be addressed by doing pre-restoration surveys to detect and eliminate invasive species. Employ biological control techniques, such as the introduction of natural predators. Observe and promptly eradicate emerging infestations. Implement competitive indigenous flora to inhibit invading species proliferation. |
| Climate change | This has modified precipitation patterns, elevated temperatures, and hastened soil erosion complicating ecosystem restoration efforts. To mitigate this, integrate climate-resilient indigenous plant species into restoration strategies. Employ amendments such as biochar and rock powder to enhance soil resilience. Reinstate sustainable water gathering and irrigation systems if possible. |
| Pest/disease infestation | Newly established vegetation can be destroyed by pests and diseases, which lowers ecological stability and biodiversity.Thus, choose plant species that are ecologically resistant to pests and diseases. This can be avoided using integrated pest management techniques to control it (e.g., biological controls, crop rotation). |

**Tab. 17.1** (continued)

| Rampant bushfires | This may contribute to the loss of flora and fauna diversity in restored ecosystem. Fires may severely hinder restoration efforts and destroy vegetation and soil structure. To mitigate, provide firebreaks, educate local people, introduce fire-resistant ecological plant species that are not invasive, and improve soil moisture retention. |
|---|---|
| Erratic rainfall leading to flooding | This may destroy established vegetation. Flooding destabilises plants, erodes soil, and depletes nutrients, rendering restoration unsustainable. To manage this, create water diversion systems to efficiently manage runoff. Also, to reduce erosion on slopes, use contour ploughing and terracing. Restore local wetlands to control water flow and serve as natural flood barriers (if possible) |
| **Management actions (factors)** | **Expected effects** |
| Site preparation e.g., soil testing, levelling of miniature voids, organic inputs or inorganic inputs | This may improve the microsite condition aimed at enhancing safe site |
| Artificial seeding or transplanting | This is so important in assisted restoration where natural regeneration is difficult due to the gravity of the degradation. |
| Use species that improve resource availability | Improve soil fertility |
| Eradication of invasive species | May enhance biodiversity and improve ecological integrity of the microenvironment to be restored. |
| Nursery establishment | It serves as a backup to the natural seed source and is conducive for surface for seeding |
| Improve bushfire management | This may help ensure the sustainability of the restoration project and enhance microbial populations and activities. |
| **Resources needed (Factor)** | **Purpose** |
| Machinery | Ploughing, preparation of seed bed, fertilizer application, seeding |
| Fertilizers (both organic and inorganic) | This is good for seed establishment and plant growth. |
| Labor | In restoring degraded mine lands, labour is needed during seeding, weed control, and fencing at nursery sites (where protection is needed) |

**Tab. 17.1** (continued)

| | |
|---|---|
| Incentives | Motivation for farmers/caretakers to be actively involved in the restoration project is an important component of restoration. |
| Funds (Money) | Labor costs, honorarium, fuel and maintenance are needed to support restoration initiatives. |
| Forest guards | This may help reduce or prevent encroachments during and after restoration. This would ensure the longevity of the project. (This is common in Africa) |
| **Additional information would be needed for the baseline analysis of restoration project** | |
| Soil suitability map | This is crucial in any land development project since it can assess the soil's ability to sustain certain plants and inform soil amendment techniques. |
| Landcover mapping | Land cover mapping delineates current vegetation and land-use patterns. Therefore, this can be captured in the planning to monitor changes over time. |
| History of indiscriminate bushfires | This gives prior knowledge about fire incidence in the area to be restored. It allows for the identification of fire-prone areas. This may allow you to improve resilience planning. For instance, choosing species that are resistant to fire and strategically placing firebreaks ensures that the restoration effort is sustainable. |
| Research on invasive species. | This so imperative because, comprehending invasive species, their dissemination methods, and consequences allows focused interventions to avert their establishment, alleviate harms, and restore ecological equilibrium, thus augmenting biodiversity. |
| Response of Indigenous species to climate change | This may facilitate the selection of climate-resilient species and enhance ecosystem stability. |
| Research on land tenure issue | Knowledge of land tenure is an important aspect of land restoration as it may clearly define property ownership and rights and help parties work together more effectively by avoiding conflicts that might impede restoration efforts. Since people are more likely to participate in restoration when they have certain rights to the property and its benefits, secure land tenure encourages local community involvement. (Read more on: Cliquet, 2017 and Mansourian, 2017) |

## 17.5 Conclusions

The following conclusions and recommendations are drawn from the chapter:
a)  Restoring mine sites requires collaboration among stakeholders like government agencies, environmental experts, local communities, and industry partners.
b)  Social, environmental, and economic factors are key.
c)  Community engagement, involving indigenous groups and women, and people-driven development are essential.
d)  Choosing plant species with both economic and social benefits can also contribute to the success of these restoration initiatives.
e)  Incorporating indigenous knowledge, traditional practices, and the use of indigenous knowledge in mining projects can significantly improve the success and long-term viability of these initiatives.

Emmanuel Dugan and Albert Kobina Mensah

# Chapter 18
# Management of restored mine sites

**Abstract:** Environmentally friendly soil and water management practices are essential for maintaining restored mine sites. These practices address concerns like deforestation, soil erosion, chemical contamination, and declining biological processes. Sustainable agricultural and land use practices, including engineering, agronomic, forage, and social techniques, can protect mine sites. Engineering techniques like terraces and dams prevent erosion, while agronomic techniques like crop rotation and mulching optimize productivity and soil quality. Forage and agrostological techniques use grass-legume associations and vegetation barriers to mitigate erosion and improve soil structure. Social techniques, such as implementing taboos, manage reclaimed mine sites and reduce environmental degradation. These techniques minimize harm and ensure the long-term viability of rehabilitated mine sites.

## 18.1 Introduction

Once mine site is restored, necessary steps should be taken to pursue and uphold environmentally friendly soil and water management lifestyle in order to maintain or conserve the site. This means all processes that lead to degradation – deforestation, soil erosion, compaction, chemical contamination, salinization, acidification, loss of biological activity, desertification, or even overuse of the restored site – need to be considered when making management decisions. There are several ways to conserve the restored mine sites and this can be done through sustainable agricultural or land use practices or measures aimed at preserving the restored sites. For farming, any activity that often causes loss of soil/nutrients, deforestation, and degradation should be minimized or abstained. Methods of conserving mine sites can generally be grouped under the following five major techniques:

1. Engineering or mechanical protection technique
2. Agronomic or biological technique
3. Forestry
4. Agrostological
5. Social

**Emmanuel Dugan,** Soil Research Institute, Council for Scientific and Industrial Research, Kumasi, Ghana, e-mail: emmdugan@gmail.com

**Albert Kobina Mensah,** Soil Research Institute, Council for Scientific and Industrial Research, Kumasi, Ghana, e-mail: albert.mensah@rub.de, albertkobinamensah@gmail.com, ORCID: 0000-0001-5952-3357

https://doi.org/10.1515/9783111662046-018

## 18.2 Engineering or mechanical protection technique

The practice includes various engineering techniques and structures that are adopted to supplement the biological methods when the latter alone are not sufficiently effective. It aims at the following objectives:

i.   To reduce runoff velocity and to retain it for long period so as to allow maximum water to be absorbed and held in the soil
ii.  To divide a long slope into several small parts so as to reduce the velocity of run-off water to the minimum
iii. Protection against erosion by wind and water

Examples under these are given in the following sections.

### 18.2.1 Irrigation canals and drainage

Under this, water for surface irrigation is properly channeled through canals from dams and regulated by weirs into plots or lands where the water is needed. Excess or unused water is channeled through constructed drainage system for disposal or recycling. Conveyance loses is reduced with construction of canal and drainage. Establishment of overland irrigation system such as sprinkler, where water is applied as a spray or as raindrop over the crop, and rather a targeted, precise, discrete or continuous or tiny streams and regulated through emitters, drip irrigation systems help reduce possibility of exposing the site soil to runoff and water erosion. Such systems do not only prevent overland flow of excess precipitation, and it also allows infiltration and maximum absorption of water, conserving the site soils.

### 18.2.2 Establishment of culvert or its expansion

Culvert is a tunnel constructed and encompasses a stream or a small rivulet of water in roadways or railways to provide cross drainage from one side to other. It is totally enclosed by soil or ground. It may be made from reinforced concrete, a pipe, or other material. Reinforced concrete culverts are the common types used under roadways in Ghana. Culverts are known to safely channel water under or around some obstacle, most commonly a roadway or railway. Installation of culverts reduces the velocity of stream or runoff and prevents the formation of gullies or soil erosion, which destroy roadways and settlement areas. It is typically put in areas where there is a natural flow of water and serves the purpose of the flow controller. Installing two or more culverts will double or triple down the energy of flow, thereby drastically reducing the catting and carrying capacity of water to conserve the soil. Culverts channel water away from vulnerable areas, preventing uncontrolled runoff that could erode the soil. By directing water flow in a managed way, culverts reduce the risk of soil being washed away from restored mined sites. Culverts facilitate efficient drainage, preventing water from pooling in low-lying areas. This minimizes waterlogging, which can weaken soil structure and make it prone to erosion. Uncontrolled water movement can lead to the formation of gullies, causing significant soil loss and landscape degradation. Culverts mitigate this by safely channelling water to designated areas, maintaining the land's integrity. Properly designed culverts ensure that water flow does not inundate or erode areas where vegetation is being established. Healthy vegetation is crucial for stabilizing soil and preventing further degradation on restored mined sites. Culverts regulate the speed and volume of water passing through restored areas. This helps maintain a balance between sufficient drainage and soil moisture retention, supporting plant growth and soil health. By directing water through culverts, sediment is trapped or deposited in designated collection areas rather than being carried away by runoff. This conserves the nutrient-rich topsoil essential for reclamation efforts. In restored mined sites, culverts prevent water from undermining roads, paths, or other infrastructure. Stable infrastructure ensures that reclamation activities can proceed without disruption, contributing to overall soil conservation. Culverts work in tandem with other soil conservation techniques, such as terracing, contour bunding, and vegetation planting, to manage water and soil effectively in restored landscapes. It conserves soil on restored mined sites by managing water flow, preventing erosion, supporting vegetation growth, and maintaining landscape stability. Their establishment or expansion is vital for ensuring sustainable land rehabilitation and long-term soil health.

## 18.2.3 Terracing

A terrace is a combination of ridges and channels built across slopes on a controlled grade, transforming steep land into series of level strips or platforms across the slope. The channels usually have a wide base and low height ridge. Terrace reduces slope length and excess rain water is led at a nonerosive velocity into grassed or cropped field or waterways and thus conserving the field or cropland. Terracing prevents sedimentation by slowing water flow and reduce the movement of soil particles. This minimizes sedimentation in nearby waterways, protecting aquatic ecosystems and maintaining water quality. It also creates stable, level planting surfaces that are less prone to erosion. Vegetation can establish more easily on these flat areas, and plant roots further anchor the soil, enhancing long-term stability and fertility. Terrace also enhances soil water retention he flat sections of terraces act as catchment areas, allowing water to infiltrate the soil instead of running off. This helps to retain moisture, which is essential for plant growth and soil health, particularly restored mined sites.

Depending on the soil and rainfall conditions, different types (viz. broad base, bench, zing, level, and table top, sloping outward and slopping inward) of terraces are adopted as suitable soil conservation measures. In managing restored mined sites, terracing can be designed to integrate with other reclamation strategies, such as planting native vegetation or creating drainage systems. This ensures that the restored landscape is functional, sustainable, and resilient to environmental stresses.

By incorporating terracing into soil management plans for restored mined sites, the land can recover more effectively, fostering biodiversity and reducing the risk of further degradation.

## 18.2.4 Dams/dugout

Dam is a reservoir that is constructed to retain water for activities such as irrigation, human consumption, industrial use, aquaculture, and navigability. Dams not only suppress floods but also store water for controlled distribution to prevent erosion, which, ordinarily, should have been caused by runoff. Rather, on a small scale, dugout in Ghana is a pit, depression, or trench usually meant to retain water to feed livestock and also for irrigation purposes on small-scale farms or backyard farms. Dams/dugouts collect and store surface water runoff from the surrounding landscape. This reduces the speed and volume of water flow across the restored site, minimizing soil erosion caused by heavy rainfall or sudden water movement. Dams and dugouts also control surface runoff. They collect and store surface water runoff from the surrounding landscape. This reduces the speed and volume of water flow across the restored site, minimizing soil erosion caused by heavy rainfall or sudden water movement. In terms of preventing sediment loss; by trapping water, these structures allow suspended soil particles to settle at the bottom of the reservoir. This prevents valuable topsoil from being carried away into downstream ecosystems, preserving the site's fertility and promoting vegetation growth. Stored water in dams or dugouts can be used to irrigate the restored area, ensuring that plants and newly established vegetation receive sufficient moisture. Healthy vegetation stabilizes the soil, reducing its susceptibility to erosion. Dams/dugouts slow down water flow and increase the time for water infiltration into the ground. This replenishes groundwater levels, which can be critical for long-term ecological restoration and soil hydration in the area.The presence of water bodies from dams or dugouts can create localized humid conditions that support plant growth. The surrounding vegetation anchors the soil and prevents erosion, fostering a stable and fertile environment. By incorporating dams and dugouts into soil conservation strategies, restored mined sites can achieve improved soil stability, water management, and ecological balance, ensuring sustainable land recovery.

## 18.2.5 Contour bunding

Contour bunding is the best means for arresting runoff from the watershed/farm at any particular level. The contour intercepts runoff from attaining erosive velocity and causing erosion. The velocity of rain water is slowed down, held on the field for a longer time, and then soak into the soils to benefit crop growth. Runoff in its movement from any bund to its next lower neighbor should not attain velocity to erode the soil. However, the spacing between bunds should not create inconvenience to cultivation. Contour bunds act as barriers that slow down the flow of water on slopes, reducing the velocity of surface runoff. This prevents the washing away of topsoil, which is critical for establishing vegetation and maintaining soil fertility on restored mined sites. Also by slowing water flow, contour bunding allows more time for water to seep into the ground rather than running off. This increases soil moisture, supporting plant growth and improving the soil's overall structure and stability. As water slows down behind the bunds, it deposits soil particles and nutrients that might otherwise be lost due to erosion. This accumulation of nutrient-rich soil behind the bunds creates fertile zones for vegetation establishment. Uncontrolled water runoff can lead to the formation of gullies, which exacerbate soil erosion. Contour bunding prevents this by intercepting and distributing water evenly across the slope, maintaining the landscape's integrity. The areas behind the bunds collect water and sediments, creating ideal conditions for plant growth. Vegetation roots further stabilize the soil, making it less prone to erosion and enhancing the site's ecological recovery.

Contour bunding can be combined with other reclamation strategies, such as planting native vegetation or creating drainage systems, to improve the effectiveness of soil conservation on mined sites. By stabilizing the soil and conserving moisture, contour bunding enables sustainable land use, ensuring the long-term recovery of the mined site and preventing further degradation.

In summary, contour bunding helps conserve soil in restored mined sites by reducing erosion, improving water management, supporting vegetation growth, and stabilizing the landscape, making it a vital tool for sustainable land rehabilitation.

Other mechanical methods of conservation include graded bunding, trenching, and vegetative barriers.

# 18.3 Agronomic or biological technique

Agronomic techniques are the practices and strategies used in the cultivation of crops. These techniques/practices include crop rotation, reduced tillage, mulching, cover cropping, agroforestry, and cross-slope farming; farming to increase soil organic matter content and improve soil structure and rooting depth. This is accomplished by growing secondary crops which enhance soil health. The use of these measures is entirely dependent upon the soil types, land slope, and rainfall characteristic. It plays second line of defense after mechanical or engineering measures. The goal is to maximize crop yields while conserving the soil, minimizing damage to the environment, and ensuring the sustainability of the system. Agronomic measures include contouring, strip cropping, and tillage practices to control the soil erosion. These include methods of planting, fertilization, pest control, irrigation, and harvest.

## 18.4 Forestry

Soils usually found at higher elevations, with steep and uneven slopes are less stable and highly erodible especially where we have high precipitation. Establishment of vegetation is essential in such places to avoid serious erosion as well as to maintain ecological balance. The vegetation and dried leaves on the floor intercept rain drops, reduce the impact and runoff, and improve infiltration. Forestry methods include afforestation, reafforestation, tree-crop plantations, and agroforestry. However, planting trees and shrubs take space and land for cultivation of crops or farming. Therefore, agroforestry (Figure 18.1), with forest tree and shrubs intermixed with food crops is essentially recommended to avoid serious soil erosion and maintain ecological balance and soil stability. Mixing crops (Fig. 18.2) with economic trees will not only provide food and economic benefit to the farmers, but it will also improve the productivity of the land, stabilize, and conserve the soil environment.

## 18.5 Agrostological technique

In this technique, grasses are grown to conserve the soil and its environment. The grasses possess desirable characteristics such as perennial and rhizome in nature and drought resistance, develop deep root system and good canopy, and have some economic benefits as well. They have multiplicity of uses in soil and environmental conservation. They prevent soil erosion by intercepting rainfall, binding the soil particles to improve soil structure, porosity, and infiltration, and add organic matter to the soil. They are used for stabilizing the surfaces of waterways, contour bunds, and front faces of bench terraces. Agrostological technique when applied to rehabilitate mine sites will help stabilize gully slopes and landslides. Adopting such vegetative measures to rehabilitate mountain slopes will help reducing landslides. A grass-legume association is one such technique for soil conservation in rehabilitated mine sites. The legume, such as mucuna (Fig 18.3) builds up soil fertility by fixing atmospheric nitrogen in their root nodules and also serves as a cover crop. The grass should have a massive root system with great strength and abilities to control erosion where soil erosion is a major problem to the environment. The most common grass species that is used is vetiver grass (*Chrysopogon zizanioides*).

## 18.6 Social techniques

Anthropogenic activities including mining, release of industrial waste, smelting of As ore, incineration of fossil fuel, particularly coal, utilization of As-loaded water for irrigation, and As-based pesticides, herbicides, and fertilizers [694] are mostly a responsible degradation of our land and soil resources. Thus, any social measure including taboos aimed at restricting economic or social usage of land resource will help immensely in management of reclaimed mined sites. Social taboos such as restricted or no farming in selected areas/days; no hunting in forest reserves; and no fishing in rivers are all encouraged to be introduced or in the management effort of regenerated mine sites.

**Fig. 18.1:** Agroforestry carried out in restoration project on a degraded mine site in Ghana.

**Fig. 18.2:** Cover cropping with mucuna and napia grass practiced in restoring a mine degraded land in Ghana.

**Fig. 18.3:** Mixed cropping practiced in restoring a degraded mine site in Ghana.

# 18.7 Conclusions

In conclusion, environmentally friendly soil and water management practices are crucial for the maintenance of restored mine sites. These practices effectively address various concerns such as deforestation, soil erosion, chemical contamination, and declining biological processes. By implementing sustainable agricultural and land use practices, including engineering, agronomic, forage, and social techniques, mine sites can be protected and preserved. Engineering techniques like terraces and dams prevent erosion, while agronomic techniques like crop rotation and mulching optimize productivity and soil quality. Forage and agrostological techniques utilize grass-legume associations and vegetation barriers to mitigate erosion and improve soil structure.

Joshua Aggrey and Albert Kobina Mensah

# Chapter 19
# The challenges and strategies for post-mine land restoration efforts in Ghana

**Abstract:** Gold mining in Ghana poses serious environmental and socioeconomic challenges, creating a soulless habitat for human and animal species, despite the existence of legal and regulatory frameworks. This study utilized policy analysis, participatory net-mapping, expert interviews, and focus group discussions to delve into the experiences of after-mined land reclamation strategies from four communities in the Ashanti region of Ghana. Complementarily, a survey of 100 households was also conducted. The study identified revegetation, soil management, landfilling, community participation, and integrated water management as the primary methods of reclamation. The findings reveal that there are instances of failed or unsuccessful reclamation. The lower rate of reclamation efforts in the studied communities was attributed to weak policies, a lack of education and training, limited enforcement resources, financial and logistical constraints, and key stakeholders that have been neglected in the mining and reclamation value chain. There should be an increase in capacity development for the communities. Further, all key stakeholders involved in the reclamation network must coordinate to ensure an efficient reclamation process. This coordination requires transparency, quality communication, and accommodating diverse views to achieve after-mined land reclamation success.

## 19.1 Introduction

Several studies [147, 158, 173, 176] have shown that gold mining is a vital economic contributor to the global economy, including countries like Ghana, South Africa, and Algeria. The sector, though economically viable, leads to land degradation, displacement of mining communities, pollution of surface and underground water, health-related issues, numerous death traps, and destruction of habitats [324, 462].

A recent study by the UN DESA [489] has predicted a rise in the world's population to about 10 billion, creating a spike in housing and infrastructure demand. This calls for the restoration of degraded lands in ore-rich countries, particularly Ghana, a

**Joshua Aggrey,** Nat4life Ghana Limited, 19 Kofi Annan Street, Airport Residential Area, Accra, Ghana; University of Hohenheim, Schloss Hohenheim 1, 70599 Stuttgart, Germany, e-mail: Joshua.aggrey@nat4life.org
**Albert Kobina Mensah,** Soil Research Institute, Council for Scientific and Industrial Research, Academy Post Office, Private Mail Bag, Kwadaso-Kumasi, Ghana, e-mail: albert.mensah@rub.de, albertkobinamensah@gmail.com, ORCID: https://orcid.org/0000-0001-5952-3357

https://doi.org/10.1515/9783111662046-019

middle-income nation, where a considerable portion of lands are affected by mining [147, 130, 131, 462]. Additionally, studies on how to restore mine spoils and mitigate mining-related nuisances are key. The current legal and regulatory framework shaping mining fails to capture the complexities in the sector and thus falls behind global best practices for restoring degraded mining sites. This heightens the already existing socio-environmental challenges and consequently leaves a lasting footprint on the environment and the rural well-being of the local people.

The study sets out three main objectives (i) to critically appraise regulatory instruments and policies shaping land reclamation after mining in Ghana; (ii) to analyze and document land reclamation practices implemented toward restoring degraded-mine lands in Ghana; and (iii) to examine challenges in implementing internationally successful land reclamation practices in Ghana. The study focused on employing strategies to return the disturbed soil to its initial condition or at least to a reusable state. Thus, restoration and reclamation are used interchangeably in the study.

# 19.2 Methods

## 19.2.1 Study area

The study conducted expert interviews in Accra, followed by fieldwork in the Ashanti region due to the proliferation of gold mining. Beposo, Pemenasi, Konongo Odumase, and Adwareago were the four towns selected purposively for household surveys and observations.

## 19.2.2 Research design

A comprehensive review of the literature was first conducted before interviewing key informants involved in land reclamation. Further, the study utilized snowball sampling to identify relevant actors in the land reclamation process. Included in the design were household surveys, structured and semistructured interviews, and Net-Map to extensively understand land reclamation practices.

## 19.2.3 Sampling procedure

The sample covers large-scale and small-scale mining sites, focus group discussants such as farmers, marketers (gold dealers), and opinion leaders including chiefs and Assembly men or Assembly women. The study conducted a total of 8 focus group discussions, 11 Net-Maps, 22 in-depth interviews, and 100 households, as presented in Tab. 19.1.

**Tab. 19.1:** Sources of data.

| Actors/ institutions | Net-Maps | Personal observations | FGDs | Interviews | Questionnaires |
|---|---|---|---|---|---|
| Regulating/ ministries | 2 | | | 3 | |
| NGOs | 4 | 9 | 3 | 4 | |
| Farmers | 1 | | 2 | 4 | |
| Miners | 2 | | 3 | 6 | |
| Market actors | 1 | | | 1 | |
| Household | | | | | 100 |
| Opinion leaders | 1 | | | 4 | |
| Total | 11 | 9 | 8 | 22 | 100 |

## 19.2.4 Interviews

The study employed both structured and semistructured interviews with key institutional actors involved in mining and reclamation. The interviews were conducted in both the local language and English, while clarity and reliability were ensured through member checks.

## 19.2.5 Personal observations

To obtain a holistic understanding of the reclamation practices and challenges faced by the mining communities, field visits were carried out to both completed and ongoing projects. The visits allowed for firsthand experience of both successful and unsuccessful/failed practices.

## 19.2.6 Net-mapping tool

The study utilized a participatory social network mapping tool to identify key players and understand their relationships in the reclamation process. The key actors and their linkages were then categorized focusing on knowledge flow, fund flow, and conflicts. The level of detail regarding the actors and their linkages differs from one Net-Map to another. Further, information on the challenges of implementing international best reclamation practices was obtained by discussing the map with relevant respondents.

## 19.2.7 Data analysis

Qualitatively, content analysis was utilized to identify patterns and themes that emerged from the interviews. The survey data were analyzed quantitatively using descriptive statistics. The black box approach was used to measure the success of the reclamation practices.

# 19.3 Results and discussion

## 19.3.1 Analysis of Ghana's minerals and mining policy governing mine land reclamation

The study explores policy regarding minerals and mining in Ghana with a direct focus on the practices of land reclamation. The analysis of the mining policy 2014 (Government of Ghana, 2014) not only identifies environmental issues as major challenges but also lacks details and specificity concerning land reclamation. For instance, there is a lack of details regarding capacity building and training of the miners, which consequently contributes to trial-and-error methods in reclaiming mined lands due to the lack of appropriate and requisite knowledge and information on existing best practices.

It was also noted that the EPA, the Minerals Commission, and the Forestry Commission are key regulatory bodies charged with ensuring sustainable mining practices. The review underscores that the policy falls behind successful international best practices including community involvement, topsoil storage, and water management.

Furthermore, from the policy analysis, it is expected that mining companies follow both local and international best practices on mining closure, which internationally involves not only the physical waste but also the overall ancillary impacts due to the waste and the mining activity. The current and previous impacts from mining, if adequately mitigated and controlled by a given policy and government action, then the policy, according to research as discussed in the literature, could be termed as a sound policy [695]. In this respect, Ghana's current mining policy concerning land reclamation is not a comprehensive one. This gap in the policy leads to abandoned mine sites posing death traps to communities, coupled with limited enforcement, lack of technical know-how, and disparate expectations among stakeholders that, in the long run, exacerbate environmental degradation.

## 19.3.2 Reclamation practices – findings from the field

### 19.3.2.1 Community participation

The study found local involvement to be an essential technique for achieving success-ful restoration in the community. The local community provides monitoring support by guarding against unwanted plant species that might outgrow the local plants. It is argued that controlling the reclaimed sites from foreign plant species facilitates the easy establishment of the intended habitat [50]. It was also observed that resuming farming activities including access to lands for free range mainly stimulates the com-munities for reclamation due to the scarcity of lands in the mining areas, as reiterated by Blay et al. [82]. Besides addressing the adverse effects of mining such as death traps, restoration of cultural heritage, sensitization, and job opportunities as observed by Yaro and Petursdottir [223], there is also a need for capacity building and training for the communities to achieve reclamation success.

The study revealed that the Konongo Odumase and Beposo communities had a similar average of land reclaimed (35%), whereas Adwareago and Pemenase reported 13% and 17%, respectively (see Fig. 19.1).

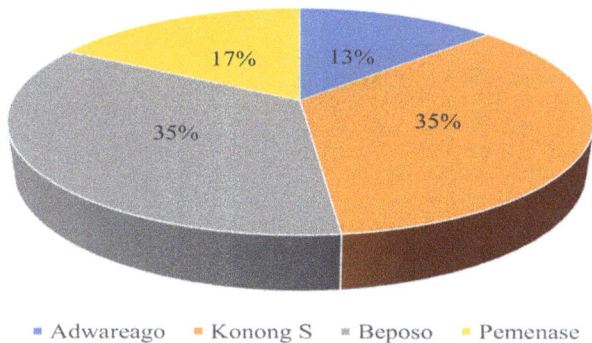

- Adwareago   - Konong S   - Beposo   - Pemenase

**Fig. 19.1:** Average land reclaimed in the community.

### 19.3.2.2 The land filling method

The study observed the use of stones, sand, and black soil or topsoil in filling the dug-out pits. According to Ghose [178], the use of topsoil is a critical factor in restoring degraded-mine sites and must be saved and managed well for its later use during the regrading of the sites. During mining, the topsoil is stripped, removed, and stockpiled for future use. On the contrary, Ghose [178] advises against the practice of stockpiling topsoil due to the detrimental effects it has on soil quality and, consequently, produc-tivity during revegetation. When there is a need for stockpiling, the topsoil ought to be judiciously positioned on a stable area that remains undisturbed and safeguarded against erosion.

The lack of topsoil, which forced people to scoop soil from nearby undisturbed lands, was observed in the study. This activity not only heightens the problem of sand winning but also results in conflicts among the community members and the land-owners. This was reported by Priyandes and Majid [396] who analyzed reclamation practices in Indonesia and concluded that access to materials for filling dug-out pits during land reclamation promoted sand mining, which often created conflict. Salifu [416] also found that sand mining for land reclamation created conflict in most mining communities in Ghana, especially in the Brong Ahafo region. The use of excavators and bulldozers dominated land reclamation in the community with varying rates of success.

### 19.3.2.3 Revegetation method

In mine land restoration, mixed species in revegetation is seen as superior to single species. This method involves mixed species planting (37%), cash crops (27%), trees only (20%), and both grass and cereals at par (5%) as shown in Fig. 19.2.

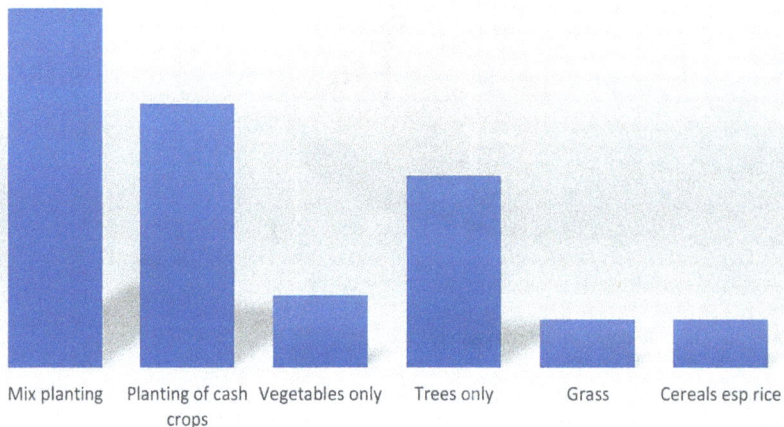

**Fig. 19.2:** Methods of different species option used in revegetation of degraded-mine sites.

This practice allows successful reclamation as the locals generate income from the sale of bamboo, for example, as already observed by FAO and INBAR [155], Lötter et al. [282], and Mensah [335]. Agroforestry, involving the integration of different tree species and food crops concurrently on the same piece of land, was common during the field observations [e.g., 82, 95]. Though this method was popular, community knowledge gaps hindered successful implementation, as noted in an interview comment by one expert:

> *The planting should involve 40 percent exotic species and 60 percent indigenous species, and in this way the indigenous forest is maintained which the rural people lack (key informant interview, 8/07/19).*

This is attributable to the lack of training personnel in the study communities. The present result confirms the findings of Blay et al. [82] that the choice of trees and crops for agroforestry should be a collective effort of the mining companies, regulatory authorities, NGOs, and the rural communities. The higher-recorded percentage level for the integrated tree and crop species (see Fig. 19.4) in the restoration activities in the study area might be the benefit derived by the locals, as put forward by others [e.g., 82, 95].

### 19.3.2.4 Soil management methods

Any technique employed in restoring a disturbed mine soil is unique and must be based on the landscape of the site [485]. The common farming methods – fertilization, mulching, liming, fallowing, and addition of soil organic manure – observed in the community were soil management practices employed for reclaiming degraded mined soil as already put forward by Tripathi et al. [485] since these practices enhance microbial activities and nutrient composition in the soil. Buckingham and Weber [95] also assessed the restoration and rehabilitation of disturbed lands and maintained that commonly adopted practices based on existing methods already known to farmers are noted to ensure reclamation success. Neina et al. [350] assessed the chemical and mineral properties of soils at Bogoso and Abosso Goldfields mines in Ghana and concluded that reclamation strategies should first target managing the soil to enhance its quality. In this way, faster establishment of fauna and flora is achieved. Another study by Mensah et al. [335] found that the addition of compost, manure, biochar, and inorganic fertilizers influences essential soil nutrients, which is critical for restoring degraded mined land. It was noted from the field observation that all the study communities lack the expertise and technical skills in assessing the level of metal residue in the soil after reclamation. Some of the comments during the experts' interviews include:

> "Research has found that mercury takes a long time to decompose, so there must be scientific check (soil scientist) to know the level of chemical residue in a reclaimed land. This will ensure healthy growth of food crops" (A representative from Solidaridad, Accra, 10/07/19). "There should be a Scientific survey (soil testing) after reclamation to determine the metal levels" (A representative of Minerals Commission, Accra, 16/07/19).

This knowledge gap significantly impacts food production in the community as found in a similar study by Yaro and Petursdottir [223]. They reported metal residue and low yield of crops on reclaimed soils in the Eastern region of Ghana. An expert reported that, "Different wastes are generated during mining and requires expertise in separating the Potential Acid Generating (PAG) from the Non-Acid Generating (NAG) and treated separately to prevent percolation and later subsidence after reclamation" (A representative from Minerals Commission, Accra, 16/07/19). The management of these wastes (acid mine drainage and heavy metals) is complicated and involves an appreciable amount of expenditure [350].

### 19.3.2.5 Integrated water management practices

The techniques employed in managing water in a disturbed ecosystem are more demanding. The interview information confirmed that most water bodies were degraded and need further treatment after reclamation. The indiscriminate use of chemicals for mining is blamed for this experience. This has been observed by others [e.g., 195, 194] that mining activities in most communities of Ghana release heavy metals, which contaminate the water bodies.

The field visit revealed a successful reclamation by Partners of Nature Africa, an NGO working with the communities, which established bamboo plants along bodies of water. The leaves of the bamboo, over time, would filter the heavy metals from the water and hasten the restoration of the degraded ecosystem. The operations manager of Partners of the organization commented:

> We planted bamboo seedlings around streams and rivers and anticipate the leaves at maturity would both filter the heavy metals and serve as canopy over the water against direct sunlight.

The effect of these plant species is tripled, particularly when other plants such as oil palm are added. This supports the findings of Lötter et al. [282] and FAO and INBAR [155] that integrating plant species on restored sites protects the soil and the water bodies.

## 19.3.3 Stakeholders and their roles in the reclamation process

The Net-Map analysis identified the Minerals Commission, Environmental Protection Agency (EPA), Forestry Commission, Water Resources Commission, District Chief Executives (DCEs), district administrators, WACAM, Tropenbos, Solidaridad, Partners of Nature Africa, Friends of the Earth, traditional authorities, smallholder farmers, traders, public research institutions (UMaT), and mining companies (large and small scale) as key actors in the reclamation network.

The household interviews supported these findings, highlighting the roles of the unit committee, chiefs, opinion leaders, individual landowners, miners, DCE, Lands Commission, EPA, NADMO, and community members (Fig. 19.3). The variation in the presence of the actors in the community, as revealed by the survey, could be attributed to how the local people perceive the roles of the actors.

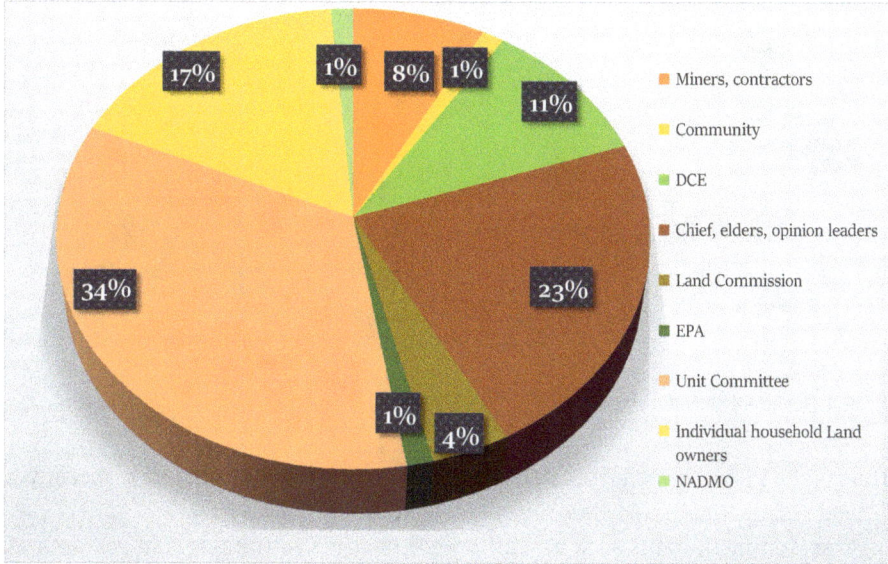

**Fig. 19.3:** Important actors identified in land reclamation.

Experts confirmed that miners are responsible for funding reclamation activities in the communities. The analysis of the Net-Map revealed that FAFA Company, legal mining companies (large-scale miners and small-scale miners), individuals, and well-wishers are major funding sources. The survey interviews showed that 55.74% (Fig. 19.4) indicated that mining companies finance the reclamation, underscoring diverse funding sources with miners as key financiers.

This finding agrees with Botchwey et al. [88], who stated that it is the prime duty of the miners to provide funds and/or undertake land reclamation after mining.

The interviews and the focus group discussion revealed the Minerals Commission, Environmental Protection Agency (EPA), Forestry Commission, Water Resources Commission, Lands Commission, DCEs, and district administrators as key regulatory bodies in the reclamation process. NGOs such as Partners of Nature Africa and Solidaridad were involved in reclaiming degraded lands after mining, while WACAM, Friends of the Earth, and Norad (Tropenbos) provide capacity building and advocacy. Traditional authorities also lease land, especially for small-scale miners.

The Precious Marketing Minerals Company (PMMC) showed no concerns for environmental impact even though they mainly purchase gold, as argued by Appiah [39] that PMMC overlooks how the activities of the miners severely disturb the landscape.

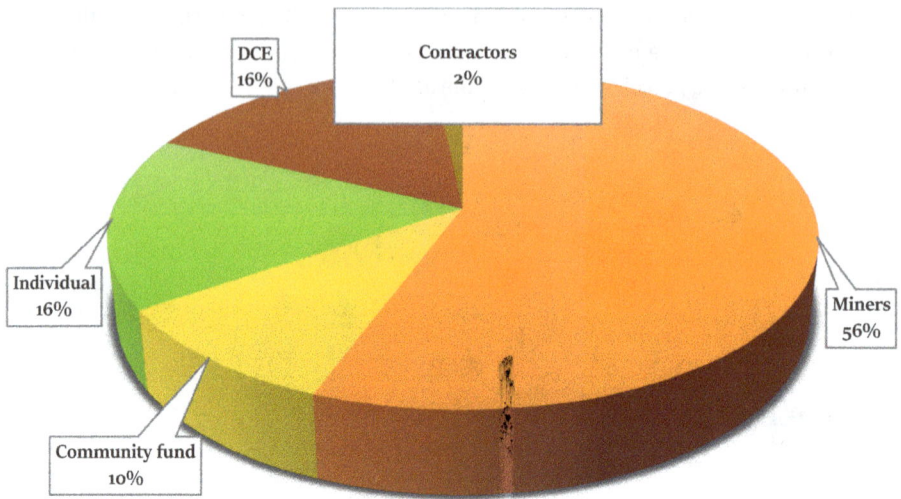

**Fig. 19.4:** Financiers of land reclamation.

"Though the PMMC is mandated to buy gold from miners they do not have incentives for land reclamation" (A key informant response, Accra, 3/07/2019).

Research institutions such as UMaT were engaged in training and research as commented by a representative of GNASSM: "For training and research programs, UMaT is an excellent actor in this area." This agrees with McQuilken and Hilson [309] that research institutions undertake research and other environmental developmental projects to enhance the reclamation of small-scale mining operations.

## 19.3.4 Institutional linkages in the reclamation process

The study explored the linkages between the key stakeholders, highlighting the fund flows, knowledge flows, and conflicts. It was also highlighted that miners must deposit funds before mining, called a reclamation bond, as default money for reclamation. The EPA and the Minerals Commission are the most connected actors in the flow of funds. Knowledge mainly flowed from regulatory bodies and NGOs to miners, with UMaT playing a significant role. The radio was the main medium of knowledge linkage in the reclamation process, as seen in the table below as seen in Table 19.2.

Conflict linkages indicate disputes among the major actors, particularly regulatory bodies, miners, and traditional authorities, due to the absence of compensation and consultation. For instance, traditional leaders are left out in the granting of licenses, which is critical for reclamation success.

**Table 19.2:** Sources of information on land reclamation.

| Source of information on land reclamation | Percent |
| --- | --- |
| Radio | 91.67 |
| TV | 4.17 |
| Acquaintance | 2.78 |
| **NGOs** | **1.39** |

Source: Own data (2019).

## 19.3.5 The underrated or neglected actors in land reclamation network

The study revealed that to achieve responsible mining and successful reclamation, some powerful but neglected actors are pivotal. These actors identified in the study include geologists, soil scientists, security agencies, media, and the church. The geologists, for example, are supposed to provide vital information on mining to enable miners to avoid the trial-and-error method of mining, which almost always leads to mined pits not being reclaimed, as seen in responses like "A big challenge is that we (miners) do not have geological information and so by trial-and-error method mined pits are abandoned without reclaiming because of money" (a key informant response, 5/08/19). As highlighted in the literature, these actors enhance rural well-being by ensuring responsible mining and reclamation by the miners [2, 127, 309, 352, 490].

## 19.3.6 Challenges in implementing land reclamation practices

A major hurdle to achieving successful reclamation is the lack of involvement of critical actors. The ability of the local community to select and combine plant species for reclaiming degraded lands is deemed effective by the role of the extension services [29, 50]. To assess the suitability of soil for food crops after reclamation, soil scientists are required to analyze the accumulation of metal residue. The study underscored some concerns indicated by the community regarding damage to their root crops because of untreated metal residues in the soil. This is consistent with the findings of Yaro and Petursdottir [223] that poor reclamation strategies utilized may fail to remove the metal residue in the soil. Other powerful actors such as the media, security agencies, and the church usually left out also accounted for the failed experiences.

Logistical problems including poor roads and trucks for transporting soil enhancers and monitoring complications were noted. Limited monitoring abilities due to constrained human resources, coupled with large and scattered mining sites, were cited by the experts as another challenge. A respondent had this to say,

*Only one person is responsible for supervising the environmental-related activities such as mining carried out in the Western region making monitoring a bit complicated and sometimes ineffective (a key informant response, 12/02/19).*

It can be inferred that monitoring capacity is below what is expected and is consistent with the findings of [342] that inadequate regulatory staff members limit their potential to check unclaimed mined lands. Furthermore, the study found insufficient education and training leading to a lack of technical expertise among the miners impeding the success of land reclamation. The experts interviewed established ineffective policies and regulations, inadequate funds, issues of transparency, and enforcement as challenges to successful reclamation. Lastly, failed reclamation efforts were observed in the studied communities due to poor coordination and/or cooperation between the regulatory bodies, local communities, and NGOs.

## 19.3.7 Measuring success of the reclamation efforts

Lands in the study community that were never restored were referred to as "wasted lands" or "lost lands." The soil structure and texture, water retention, contaminated soils, fertility, and yield per acre were criteria utilized to assess the end-use of the restored fields. Subjects were asked to compare crop yields before and after reclamation. The majority of the respondents, 69% (Fig. 19.5), indicated that the reclamation was not successful.

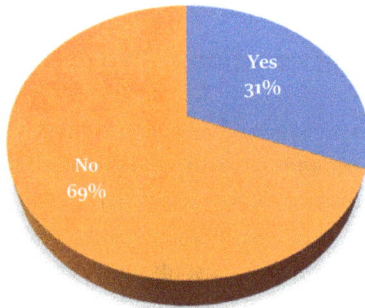

**Fig. 19.5:** Measurement of reclamation success.

The respondents further cited low yield of crops and damaged root crops as indicators of the failed reclamation efforts. The authors attributed these challenges to low levels of soil nutrients, the build-up of potentially toxic elements, and degraded topsoil. These are reinforced in the findings of others [e.g., 50, 223, 350, 6, 331, 326]. The black box approach used for the comparison had limitations including biases and recall error in relation to crop varieties, fertilizer application, climate, and the price of the produce.

## 19.4 Conclusions and recommendations

The study identified two major approaches: inaction and the implementation of best practices. Thus, less than half of the degraded lands in the study communities were reclaimed successfully. This is attributed to the inefficiency of policies, logistical challenges, insufficient human resources capacity, financial limitations, poor education, and neglect of stakeholders. The study proposes ensuring collaboration among the stakeholders including traditional leaders in decision-making. It also proposes the establishment of local offices, consolidation of regulatory bodies, implementation of capacity-building initiatives, and providing geological data to support artisanal and small-scale mining. The study further calls for research to explore plant species suitable for reclamation and to analyze soil quality and heavy metal residues at reclaimed sites.

Ferdinand Adu-Baffour*, Thomas Daum, Albert Kobina Mensah,
Konrad Martin, Akwasi Duah-Gyamfi, Frank Rasche, and Regina Birner
# Chapter 20
# A best-fit conceptual framework to enhance phytoremediation scaling

**Abstract:** Many parts of the world suffer from land contamination due to mining and other activities, which have severe impacts on human health and the environment. Traditional methods of cleaning up such sites face various challenges, so there is a growing interest in using biological solutions such as phytoremediation. Phytoremediation is a green and low-cost technology that uses plants to remove or degrade pollutants from contaminated soils. However, phytoremediation is not widely adopted, especially in the Global South, despite its potential benefits. Most of the existing research has focused on the technical aspects of phytoremediation, but there is a lack of a comprehensive framework that addresses the social, economic, political, and institutional factors that influence its success. Thus, the chapter proposes a guiding framework for the implementation of phytoremediation projects. The framework covers the technical considerations for each stage of the phytoremediation process from site selection and preparation to planting and monitoring. It also identifies the key conditions that can enable or hinder the effective and sustainable application of phytoremediation in different contexts. Finally, recommendations for regions that want to use phytoremediation as a way of harnessing the "power of plants" to restore contaminated lands are provided.

**Keywords:** Land contamination, phytoremediation, environmental consequences, global adoption, implementation framework, enabling conditions

---

*Corresponding author: Ferdinand Adu-Baffour**, Hans-Ruthenberg-Institute of Agricultural Science in the Tropics, University of Hohenheim, Stuttgart, Germany, e-mail: ferdinand.adubaffour@uni-hohenheim.de
**Thomas Daum, Konrad Martin, Frank Rasche, Regina Birner,** Hans-Ruthenberg-Institute of Agricultural Science in the Tropics, University of Hohenheim, Stuttgart, Germany
**Albert Kobina Mensah,** Soil Research Institute, Council for Scientific and Industrial Research, Academy Post Office, Private Mail Bag, Kwadaso-Kumasi, Ghana
**Akwasi Duah-Gyamfi,** Forestry Research Institute of Ghana, Council for Scientific and Industrial Research, Fumesua-Kumasi, Ghana

https://doi.org/10.1515/9783111662046-020

## 20.1 Introduction

Land suitable for agriculture, forestry, and biodiversity conservation, as well as for other social and economic purposes, is becoming scarcer worldwide due to degradation and pollution caused by mining and smelting, the use of agrochemicals, and inadequate management of sewage and waste [382]. Contaminants originating from these activities, a significant portion of which accumulate in close proximity to or within agricultural soils, also infiltrate the food chain, hence posing potential hazards to human health [154, 382]. Despite the collective endeavors of governments, scientists, commercial enterprises, affected communities, environmental NGOs, and CSOs to remediate polluted regions, these initiatives have yielded little achievement. The primary reason for this is the exorbitant expenses involved in carrying out clean-up initiatives, particularly in developing countries, where governments struggle with securing consistent funding and establishing effective institutional frameworks [365, 380, 382]. In Ghana, gold mining spoils and tailings are frequently abandoned without sufficient protective measures. These areas pose health risks to children and women, who frequently utilize them as playgrounds or as pathways to farms [334, 321].

Ex situ physicochemical remediation methods, such as soil excavation and transfer to landfills using heavy machinery, as well as the extraction or immobilization of contaminants (e.g., heavy metals) using acids or chemicals, necessitate specialized technical expertise and capabilities, intricate skills in meticulous monitoring, and ongoing measurements and evaluations, all of which incur financial implications. In addition, they have the potential to introduce secondary pollutants into soils, resulting in permanent alterations to both the soil and the ecosystem, while also compromising natural biotic activities [182]. Evidence indicates that employing such techniques may potentially diminish soil fertility [382], hence impeding the future productive utilization of restored areas. As a result, this could have an impact on economic progress, especially for impoverished agricultural rural populations who rely on this property.

Over the past 20 years, there has been significant research and strong advocacy for a novel biological method called phytoremediation, which utilizes plants to clean up polluted areas [379]. Phytoremediation, as defined by the United States Environmental Protection Agency [492], refers to the utilization of plants for the on-site treatment of polluted soils, sludges, sediments, and groundwater. This process involves the removal, degradation, or containment of contaminants. This technology harnesses the inherent capabilities of plants to absorb, transport, and store nutrients and pollutants from their roots into aboveground tissues. Alternatively, it utilizes the characteristics of microbes associated with plants in the soil to accomplish similar objectives. Certain plants possess the ability to filter, stabilize, and/or break down unwanted materials and compounds in the soil either inside or in the vicinity of their root area [389]. Research indicates that phytoremediation may serve as a more environmentally friendly option for the restoration of damaged and polluted ecosystems. Phytoremediation technology may eliminate harmful pollutants from soils without creating additional harm to ecosystems, unlike other methods such as the use of chemical reagents, encapsulation, and soil washing [465]. Research indicates that phytoremediation is a more economically efficient method with long-term environmental advantages when compared to physicochemical remediation. This is supported by studies conducted by Haslmayr et al. [200], Nascimento and Xing [349], and Wan, Lei, and Chen [506]. Despite the purported environmental and economic benefits, the technique of phytoremediation has not gained significant global commercial adoption, especially in developing regions.

Previous research on phytoremediation has primarily consisted of experimental investigations that have predominantly focused on technical elements. For instance, Guo et al. [193], Jadia and Fulekar [221], Pilon-Smits [389], Rugh [412], Tang et al. [474], Yadav, Siebel and van Bruggen [518], and Yang et al. [522] have conducted studies in this field. However, the scientific literature currently lacks information on the specific sociocultural, economic, political, and institutional strategies required for the effective, efficient, and sustainable implementation of this field. This is a difficulty, as the effective application of phytoremediation relies not only on technological answers but also on practical methods for execution and a supportive institutional and policy framework. In order to tackle these problems, a more all-encompassing strategy is required, one that takes into account the wider social, economic, and institutional frameworks in which phytoremediation is carried out.

In order to achieve this objective, we propose a new guiding framework for the execution of phytoremediation initiatives on polluted lands. This framework delineates the pertinent prerequisites necessary for the effective implementation of field operations. The analysis primarily relies on an examination of the phytoremediation literature, which encompasses successful case studies, including the practical implementation and commercialization of phytoremediation. The framework is introduced in Section 20.2, followed by an analysis of its primary technical and management aspects in Section 20.3. This includes a discussion on the preharvest factors (such as site selection, technology options, soil, crop, and agronomic management strategies) and postharvest factors (such as harvest, safe disposal of contaminated crops, and postremediation land use options) that require careful consideration and attention. Section 20.4 of the study discusses the factors that contribute to creating a favorable environment for promoting the adoption of technology in the field. The key takeaways and next steps for achieving successful implementation in a developing country are outlined in Section 20.3.11.1.

## 20.2 The conceptual framework and methodological approach

To effectively utilize phytoremediation, one must first evaluate the polluted area, select suitable plant species, employ appropriate techniques for managing the soil and plants, and follow proper procedures for harvesting, handling, and disposing of the plants (Boxes A–F) [200, 367]. In order to achieve success in these projects, it is necessary to implement these technical steps in a well-managed environment (Box G) and create favorable conditions through applied research, education, collaboration among stakeholders and institutions, exploration of economic opportunities, and the establishment of clear and enforced legislation for environmental restoration (Boxes H–K) [103, 445]. Figure 20.1 depicts the structure of the necessary processes and conditions that impact the effective implementation of phytoremediation on a large scale. This framework has been created by conducting a thorough examination of both scientific and gray literature on phytoremediation. In addition, we have utilized insights gained from practical experiences and contemplation of theoretical frameworks pertaining to the extensive implementation of phytoremediation, specifically emphasizing the Global South as the setting of our investigation.

The scientific literature search was performed utilizing respected platforms including Google Scholar, Science Web, and Scopus search engines. We employed search terms such as "field application of phytoremediation," "phytoremediation field study cases in the Global South," "phytoremediation field study cases in the Global North," "successful application cases of phytoremediation in the field," and "practical application of phytoremediation." Furthermore, we gathered pertinent information from reputable institutional websites and project implementation reports that offered valuable insights into land rehabilitation and remediation projects utilizing phytoremediation techniques. By employing this method, we were able to obtain vital information regarding implemented phytoremediation initiatives.

## 20.3 An in-depth analysis of the technical factors must be taken into account while implementing phytoremediation

Boxes A to F of the framework outline several technical factors to be taken into account at different stages of phytoremediation including preplanting, preharvest, harvesting, and postharvest. These features will be further elaborated upon in the subsequent sections. The initial stages of planting (Box A) entail the careful evaluation and selection of the site, taking into account social aspects such as the land's type, location, and value as well as biophysical elements including the contamination's extent and

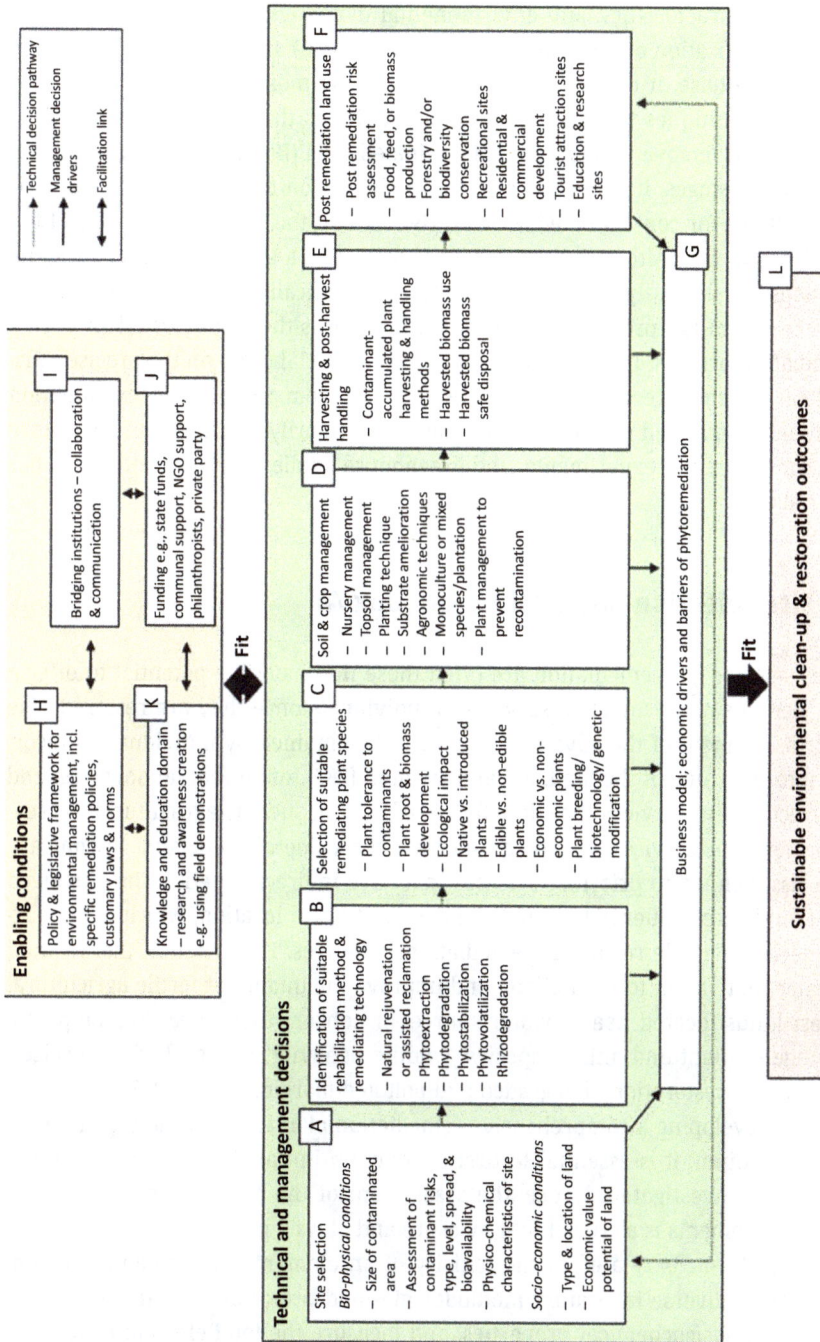

**Fig. 20.1:** Our proposed best-fit conceptual framework outlining technical factors that must be taken into account at different stages for a successful phytoremediation.

qualities, soil characteristics, and environmental dynamics. Additionally, it encompasses the identification of appropriate technology (Box B) and plants (Box C). During the preharvest phase, it is necessary to determine the most effective soil and crop management techniques to establish optimal conditions that enhance the ability of specific plants to remove toxins from contaminated soils (Box D). During the harvest and postharvest stages, it is necessary to make choices on the optimal utilization or disposal methods for contaminated biomass (Box E) and the desired future use of the restored land (Box F). Although the framework proposes a sequential procedure, the various steps can mutually influence one another. Specifically, as depicted in Fig. 20.1, the planned use of the property after cleanup influences the measures taken during the remediation process. In the subsequent sections, we elaborate on the precise intricacies of each of these technical procedures, drawing from an extensive examination of both foundational and empirical investigations, primarily conducted in laboratory settings, controlled pot experiments, and greenhouse studies, with a limited number of field trials.

## 20.3.1 Site selection and evaluation (Box A)

The lands selected for remediation are often those that have the potential to offer a range of benefits and ecosystem services to individuals once they are restored. The economic assessment of the advantages that can be obtained by individuals is incorporated into the value of the land. Ma and Swinton [287] found that landowners tend to prefer ecosystem services that directly benefit them, such as agricultural, recreational, or aesthetic services, which can increase the value of their land. In contrast, they are less inclined to prioritize services that have indirect benefits. The perceived values of lands are influenced by their kind, scarcity, and location, which in turn affect the decisions made regarding remediation initiatives. For instance, there would be a greater motivation to rehabilitate and improve the quality of fertile agricultural and forest lands located near a water source, or conservation sites that support a wider range of plant and animal species, wildlife behavior, and ecological services, compared to the restoration of degraded and polluted marginal lands [395].

Prior to developing a comprehensive remediation plan for soil or any other environmental medium, it is essential to first conduct an initial characterization of the medium being investigated. Hence, the assessment of the overall concentration of pseudo-trace elements is a crucial first measure and plays a pivotal role in identifying any potential hazards related to the toxicity of an element. Experiments are conducted utilizing diverse laboratory methodologies and protocols to characterize soil, determine its physicochemical properties, and measure the total element concentration in contaminated soils. The acquired results enable the calculation and evaluation of the danger of soil pollution and its influence on the health of those residing in close proximity to these impacted regions. Antoniadis et al. [37] present information on commonly employed indices for quantifying the toxicity levels of a polluted medium. Antoniadis et al. [37] as well as Mensah et al. [321] and [334] have identified many prevalent approaches for evaluating the potential risks associated with contaminants or elements. These methods include:

a)  Utilize the overall concentration of the metal or metalloid, such as the maximum permissible concentration or the global average concentration in soil
b)  Utilize successive extraction to isolate the bioavailable or mobile components
c)  Utilization of elemental speciation
d)  The X-ray synchrotron diffraction technique
e)  Utilizing index-based methodologies for evaluating risk
f)  Utilize spatial dispersion through the application of geographic information systems (GIS)

The usefulness of phytoremediation is restricted to particular types of contaminants and environmental circumstances, hence limiting its scope. [696] and Pathak et al. [382] found that various factors, such as soil properties (such as pH, texture, electrical conductivity, and organic matter content), contaminant characteristics (type, spread, and levels), plant species present, and climatic conditions, play a significant role in determining the appropriate site and approach (physicochemical or biological) for soil decontamination. Additional crucial soil characteristics to take into account encompass the overall concentration of the desired elements, the capacity for cation exchange, and the presence of anions such as chlorides, nitrates, phosphates, and sulfates, as noted by Addai et al. [6]. Phytoremediation, like other cleanup procedures, is limited to specific sites due to changes in space and climate and is not economically viable in all locations [280]. Specialized plant-based remediation solutions may not be feasible for sites with extremely high levels of contamination, as the survival of remediating plants may be compromised in such soil conditions. In addition, extensive regions of degraded lands with localized contamination may be more suited for physical soil reconstruction and investments aimed at restoring fertility rather than relying on phytoremediation techniques [294, 526]. Bhawan [77] and Pilon-Smits [389] state that the successful implementation of phytoremediation on a specific site relies on the characteristics of the contaminant(s) being targeted such as how easily they move, dissolve, degrade, and become available for uptake by the plant's roots in the contaminated area.

## 20.3.2 Identifying appropriate phytoremediation technology, also known as phytotechnology (Box B)

Degraded and contaminated land can undergo two approaches for recovery and regeneration. The first approach involves allowing the land to naturally recover through processes like the growth of native vegetation, with minimal or no human intervention. The second approach involves active human intervention and management techniques to rehabilitate and restore the land's quality, fertility, and ecological functionality. Phytoremediation, a component of rehabilitation, can be categorized into two forms: direct and indirect, as described by Garbisu and Alkorta [169]. Indirect bioremediation, also known as plant-assisted bioremediation, involves the collaborative efforts of plants and microbes to eliminate harmful substances in polluted soil. On the other hand, direct phytoremediation involves plants directly absorbing and decomposing pollutants within their tissues. Phytoremediation encompasses various methods, which include:

Phytoextraction is the process by which plant roots absorb pollutants and transport them to the aboveground sections of the plant [445]. Phytoextraction is a frequently used approach for remediation, involving the removal and disposal of contaminated plants. Alternatively, if the extracted toxins have commercial worth, they can be removed for commercial purposes, a process known as phytomining [308]. Pathak et al. [382] reported that over 500 plant species exhibit a remarkable ability to accumulate significant quantities of heavy metals in their aboveground components.

Phytodegradation is the process of using plants and their enzymes to break down organic contaminants, typically within the plants' tissues [308]. This method is effective for organic substances that can move throughout plants such as herbicides [389].

Phytostabilization is the process of using specific plants to decrease the movement of pollutants, particularly heavy metals, and prevent their spread (such as seeping into groundwater or entering the food chain) by reducing their ability to be absorbed by living organisms [445].

Phytovolatilization refers to the process in which plants absorb pollutants like mercury (Hg) and selenium (Se), convert them into volatile forms, and release them into the atmosphere as gases [24].

Phytostimulation, often referred to as rhizodegradation, is the process by which plants boost soil microbial activity in the root zone to break down organic pollutants such as petroleum hydrocarbons [389]. Plants secrete carbohydrates and amino acids that promote microbial activity, leading to the breakdown of organic pollutants through biodegradation.

Phytoremediation is subject to various restrictions such as its time-consuming nature and the need for meticulous agronomic care. Gerwing, and Greenberg [174] argue that phytoremediation, particularly for inorganic contaminants, can result in a substantial time delay in the reutilization of land compared to previous techniques, thus incurring a large time cost. This leads to the immobilization of capital, which must be taken into account when performing cost analyses. Furthermore, the efficacy of the technique is constrained to low-to-moderate levels of contamination, and specific plants must be chosen based on suitable root depth and contact area [511]. Ultimately, it is crucial to implement suitable techniques for crop management, harvesting, processing, and disposal in order to effectively avoid the reintroduction of pollutants into the surrounding environment.

### 20.3.3 Choice of appropriate plant species for remediation purposes (Box C)

Plant species utilized for environmental remediation must possess specific traits including the capacity to accumulate and transport pollutants, detoxify or immobilize pollutants, generate substantial biomass, withstand potentially toxic pollutant concentrations, exhibit rapid growth, and have a short life cycle [33, 306, 458]. Herbaceous plants are considered the most favored hyperaccumulating species because of their fast development, significant aboveground biomass, and broad root system, as stated by Cappa and Pilon-Smits [100]. Guerra Sierra et al. [192], Pandey and Souza-Alonso [380], and Laghlimi et al. (2015) suggest that for long-term remediation programs, a combination of grasses, shrubs, and trees, particularly leguminous ones, should be used. These plants have different functions at various stages of the remediation process and are chosen for their fast growth. According to Pandey et al. [381], native colonizer plant species that are not palatable, perennial, and economically useful are preferable for sustainable remediation compared to invasive plant species. It is important to implement strict precautions when adopting remedial species that are both invasive and beneficial, such as Pueraria phaseoloides, in order to prevent their uncontrolled proliferation [464]. To find appropriate plant species for remediation of various toxins, a thorough process of exploration, research, and development is necessary [311, 457].

In addition to choosing plant species from their original environments, plant breeding, which includes biotechnological methods, shows potential for creating more effective phytoremediators. This can be achieved by transferring genes that enable the accumulation of pollutants from low-biomass wild species to high-biomass cultivated species, or by using advanced genetic engineering techniques [280]. A significant portion of this is also accomplished by crossover, a conventional breeding method that involves gene transfer through natural means rather than more modern and complex engineering procedures. The global debate around the utilization of genetically modified products continues to be a contentious matter, as several nations maintain divergent perspectives regarding their implementation and utilization inside their respective territories [697–698, 347]. Despite the apparent efficacy of genetically altered products for phytoremediation in the lab, their acceptance in the field is directly affected. Research has shown that by manipulating the genetic makeup of plants, it is possible to enhance metal accumulation per plant by a factor of 2–3 [156, 388]. Assuming similar outcomes can be observed in real-world scenarios, this has the potential to significantly decrease the expense of phytoremediation. Moreover, the implementation of novel genetic modification techniques has resulted in the development of plants that possess the ability to eliminate mercury (Hg) in a manner that is not possible for other plants. This particular attribute of genetic modification enhances its overall significance [388].

## 20.3.4 Conditions pertaining to the management of soil and crops (Box D)

Phytoremediation is the deliberate manipulation of interactions between plants and soil to address environmental issues [109]. The efficacy of the technique relies on the correct management of topsoil [335] and the suitable implementation of agronomic and crop management practices [244]. Rathore et al. [398] found that the ability of plants to clean up heavy metal-contaminated soils, known as phytoremediation capacity, can be improved by using agricultural techniques like adding organic matter and applying soil amendments such as synthetic organic chelates, bio-inoculants, and sulfur (which helps move the metals). Additionally, the prompt emergence and establishment of crops heavily rely on the use of fertilizer, effective weed control, and the provision of supplementary irrigation.

Regarding degraded mined lands, conventional methods like topsoil replacement, fertilization, and the introduction of remediating plant species are applicable for vegetation restoration. However, these approaches encounter difficulties due to the frequent lack of salvaged topsoil and its limited availability, often necessitating costly transportation. Consequently, there is a growing preference for contemporary approaches that immediately reintroduce plant life to mined regions without topsoil [335].

An ameliorative approach entails conducting a thorough analysis of the soil and making chemical modifications to specifically address issues such as adjusting soil pH, reducing the solubility of heavy metals, and providing essential nutrients for the establishment and growth of plants. Mensah [335] proposes the addition of mulches to a depth of 0.15 m, with recommended application rates of at least 20 t/ha straw or 80 t/ha sawdust. This practice can effectively reduce salt accumulation during drying cycles and promote strong vegetation cover. In contrast, an adaptive strategy centers around the identification, specification, and establishment of plants that are well-suited and resilient to the specific conditions of the site, known as ecotypes.

Improving soil conditions alone may not be enough; it may be necessary to design agricultural systems that facilitate biological rehabilitation. In order to promote plant growth, it is frequently required to employ a multiphase planting strategy, particularly when dealing with deteriorated soils that may not be able to support immediate growth [495]. Plants may require initial care and growth in a nursery environment before being moved to the cleanup site [379, 501]. Various cropping systems have demonstrated the ability to improve the effectiveness of phytoremediation in diverse situations. Brereton et al. [93], Kumar et al. [257], and Wang et al. [541] demonstrate that utilizing the unique capabilities of different crop species through mixed cropping, especially when the crops have complementary remedial functions, is a superior strategy for enhancing tolerance to various soil challenges compared to mono-cropping. Dhillon and Dhillon [123] and Bian et al. [78] suggest using intercropping and crop rotation cropping systems as a means of implementing phytoremediation. These systems ensure that the soil remains covered and contaminants are continuously absorbed, making it a long-term strategy. Tang et al. [475] emphasize that when it comes to agronomic and crop management, it is crucial to take into account soil types and carefully choose the most suitable plant species or cultivar for optimal results.

## 20.3.5 Collection, processing, and proper disposal of plants that have been contaminated (Box E)

For phytoremediation to be effective, it is essential to manage and dispose of the plant debris contaminated with pollutants properly, to prevent further environmental damage [179]. Despite the abundant literature on the technical progress and understanding of the technology (see Sections 3.1–3.4), there is still a knowledge gap on how to handle and dispose of the phytoremediating plants safely. Sas-Nowosielska et al. [420] proposed several methods for disposing of the contaminated crops such as composting, compacting, burning, ashing, pyrolysis, direct disposal, and liquid extraction. Among these methods, incineration is recommended as the most practical and eco-friendly option for the final disposal.

Other ways of getting rid of waste, such as burning, breaking down with heat, and using solvents, require a lot of skill and money. Liu and Tran [279] showed that using nanomaterials made by living things, burning, turning into gas, or heating without oxygen could reduce the amount of plant waste faster. The methods of making and using nanomaterials are less likely to cause more pollution, but they are still costly and complex. Phytomining or agromining is a newer method that involves taking out heavy metals from plants after burning them. This method could produce valuable materials such as cobalt, selenium, manganese, gold, silver, thallium, and nickel that can be used again (270, 435, 462). However, phytomining is still being tested and needs more research to make plants absorb more metals. To sum up, it is important to deal with and get rid of polluted plants properly to make phytoremediation work well. More research is needed to explore and improve these methods.

## 20.3.6 Postremediation land use (Box F)

The choice of an appropriate remediation technique is influenced by the desired future use of the land following the remediation process. This decision necessitates the allocation of time, effort, and resources [200]. The selection of phytoremediation techniques, appropriate choice of crops for remediation, and design of the cropping system may differ depending on whether the land is utilized for food or fodder production, biomass generation, or as forests and areas for biodiversity protection. To prevent the recurrence of contaminants and mitigate potential health risks to humans and wildlife, it is essential to regularly monitor and maintain the land, conducting periodic testing and assessments, regardless of its intended use [379].

## 20.3.7 Executive choices-economic drivers and barriers of phytoremediation (Box G)

Phytoremediation is an environmentally friendly alternative to traditional soil clean-up methods, but its implementation in the field has been limited due to uncertain effectiveness, lengthy remediation periods, and lack of economic incentives [64]. Conventional hyperaccumulating plants, which contain pollutants like heavy metals, need to be disposed of properly and do not generate immediate economic benefits [366]. However, using economically valuable crops that yield valuable biomass can incentivize the adoption of phytoremediation [278, 366]. "Sustainable phytoremediation" or phytomanagement refers to identifying and promoting commercial opportunities arising from the ecological benefits of phytoremediation [380]. High-value crops such as bioenergy crops, aromatic oil-producing crops, and timber crops can be effective for remediation purposes and generate economic profits [380]. Before utilizing biomass from contaminated sites, a comprehensive ecotoxicological risk assessment study should be conducted to evaluate energy potential and minimize the release of contaminants [379].

The success of implementing sustainable land management approaches like phytoremediation depends on land ownership rights and the distribution of costs and benefits [10]. In Ghana, informal artisanal and small-scale mining activities have led to the destruction and pollution of rural lands, particularly in areas with shared family lands and ambiguous ownership structures [11, 61]. The economic value and intended use of restored lands influence the willingness to invest in land restoration projects, but this becomes challenging in areas with unclear ownership and use rights [11].

### 20.3.7.1 Facilitating factors for achieving optimal phytoremediation results and wider implementation

When considering the most suitable approach, the successful implementation and practical use of phytoremediation technology rely on specific favorable circumstances. Lessons are derived from case studies conducted in nations in the Global North, specifically the United States and Canada, which are significant markets for phytoremediation [380, 425], while recognizing certain contextual variations. In the specified study situations (see Table 2 for specific examples), significant resources and funds have been dedicated not only to the investigation and advancement of optimal application methods but also to the practical implementation of the technology in real-world settings. This section elaborates on each of these facilitating circumstances (Boxes H–K), culminating in a concise overview of insights gained for emerging regions.

## 20.3.8  Securing financial resources for the expenses associated with remediation (Box H)

Phytoremediation, an emerging technique for environmental restoration, requires financial support for research and implementation [715; 506]. While it is more cost-effective than traditional alternatives, it does take time to achieve desired outcomes [715; 506]. In developed countries, state support for university research and funding of remediation initiatives, along with subsidies for practical field research, has facilitated progress in phytoremediation [494]. Private sector participation, with support from organizations like Ecolotree, Phytokenetics, and Applied Natural Sciences, is also crucial [494]. The Environmental Protection Agency (USEPA) in the United States has been involved in early applications-oriented research [699]. Private enterprises can receive funding from the USEPA and other government agencies through competitive grants [495]. Federal institutions in the United States provide financial and legal incentives to encourage private sector engagement in environmental cleanup and redevelopment [62]. In Europe, there is a preference for cautious and fundamental research, while the United States focuses on entrepreneurial and application-driven research [297, 445].

In developing areas, the lack of reliable financial frameworks for research and implementation hinders the expansion of inexpensive environmental restoration methods [494]. Limited government assistance places the burden on environmental nongovernmental organizations (NGOs) to bridge the gap [494]. In Ghana, organizations like A Rocha and Tropenbos have led initiatives to restore and detoxify degraded lands used for small-scale mining [494]. These projects aim to prepare the lands for future economic utilization and receive funding from entities like the Norwegian Agency for Development Cooperation [494]. Economic tree species such as *Terminalia superba*, *Terminalia ivorensis*, *Senna siamea*, *Acacia mangium*, *Khaya species*, and Bambusa long internode are focused on in these restoration efforts [494].

*Lessons for upscaling:* Financial and legal incentives, such as remediation grants, tax breaks, loans, and legal indemnities, can be utilized as direct or indirect means to facilitate the widespread implementation and advancement of phytoremediation in developing regions. In countries belonging to the Global North, such as the United States, conventional government assistance serves as a significant financial resource for university research and restoration initiatives, while the involvement of the private sector is also crucial. Insufficient finance has hindered the progress of phytoremediation in developing areas. However, environmental nongovernmental organizations (NGOs) have the potential to advocate for affordable remediation methods in some situations.

## 20.3.9 Policy and legislative framework for the remediation of contaminated land (Box I)

Mench et al. [311] stress the importance of thoroughly evaluating all factors, including current policies, social norms, and financial resources, to effectively implement sustainable environmental management strategies in response to different pollution levels. The effectiveness of encouraging and implementing phytoremediation is frequently observed in developed countries with well-established environmental policy frameworks, as exemplified by Luo et al. [286]. Policy makers, regulators, and legislators can establish comprehensive and context-specific frameworks by enacting targeted remediation rules that address polluted sites with specific pollutants. These frameworks can analyze effective approaches to tackle crucial technical challenges, such as cutting-edge biotechnologies and techniques for managing and safely disposing of polluted biomass. They also take into account facilitating factors, such as offering financial or legal incentives to promote remediation efforts, which have implications for the implementation of phytoremediation. These regulations also enable the targeted resolution of certain pollution issues while considering the trade-offs and advantages of possible remedies. Regrettably, numerous nations in the Global South suffer from a lack of well-defined remediation strategies and instead depend on generic environmental standards and legislation [11, 208].

The efficacy of preventing detrimental activities and fostering eco-conscious behaviors is contingent upon the implementation of pragmatic and easily understandable regulatory benchmarks that can be universally applied [512]. Nevertheless, in numerous developing nations, the structured establishments formed by the government do not consistently correspond with the unspoken regulations and traditions that influence the conduct of local operational entities located in rural communities. Even if there is alignment, there can still be inadequate enforcement [11, 118]. The lack of alignment within institutions is further intensified as it pertains to land management, due to a frequent lack of comprehension about traditional, community-based customs and regulations. Hence, it is imperative to undertake comprehensive research on indigenous institutions in order to gain a deeper comprehension of their impact on the formation and execution of environmental policies.

Regulatory agencies typically exercise prudence [129]. Hence, for phytoremediation to gain widespread acceptance as a remediation technique, regulators must be persuaded of its effectiveness [715]. Ensuring compliance is crucial in the environmental protection industry, so the technology must demonstrate its effectiveness in satisfying both state and federal criteria for environmental quality [445]. Achieving a harmonious equilibrium between minimizing substantial financial expenses and guaranteeing the safeguarding of human health and the environment necessitates co-operative efforts among environmental engineers, scientists, policy makers, regulatory agencies, environmental nongovernmental organizations (NGOs), civil society organizations (CSOs), and active participation from the general public [106, 496]. Empirical study is crucial for gaining a deeper comprehension of the ecological and human health implications of phytoremediation. Therefore, policies and regulations supporting phytoremediation must find a reasonable equilibrium between cost and efficacy.

*Lessons derived for developing regions*: These indicate the necessity of implementing tailored remediation policies that are suitable for the local circumstances. These policies should take into account crucial factors such as the utilization of biotechnology and the proper management of polluted biomass. Additionally, they should address barriers and offer incentives to encourage the broader adoption of phytoremediation. In order to guarantee the efficacy of phytoremediation in satisfying both state and federal environmental quality standards, it is imperative for environmental laws and regulations to strike a balance between the economic costs and the effectiveness of phytoremediation. A comprehensive understanding of the impact of indigenous institutions on the creation and execution of environmental policies is essential.

## 20.3.10 Domain of knowledge and education (Box J)

Through the utilization of research, education, practical demonstrations, and easily accessible information, phytoremediation has the potential to become an essential instrument in the process of environmental restoration. Box J in Fig. 20.1 integrates the essential elements of research and education to advance and advocate for the widespread implementation of effective phytoremediation techniques. The next sections provide an in-depth examination of these crucial factors, presenting significant observations and viewpoints.

### 20.3.10.1 The foundation of phytoremediation lies in both fundamental and applied research

The increasing fascination with phytoremediation has been fueled by fundamental and, more recently, practical research conducted over the years. [700] state that applied studies have commonly encompassed hydroponic research, small pot experiments with soils from polluted sites, and greenhouse experiments including varying doses of induced pollutants for toxicity and remediation investigations. In recent years, developed countries in Europe, North America, and Asia have started doing field research on phytoremediation, specifically focusing on heavy metals, organic pollutants, salts, and radionuclides [716]. These studies have examined the practical applications of genetically engineered plants as effective remediation agents. They have investigated various aspects, including the ability of these plants to remove contaminants, the ways in which different plants absorb pollutants, and the impact of soil amendments on the effectiveness of phytoremediation. Lee et al. [266] and Singh et al. [457] suggest that these studies have shown improved efficiency in using phytoremediation. Although these advancements are now inadequate to solve the numerous specific research problems of phytoremediation, they are important for achieving widespread implementation of the technology in polluted locations. Although there have been ongoing research contributions, practical application attempts in most parts of the developing South still face severe difficulties. Lee et al. [266] attribute this to a discrepancy between researchers and the individuals who are actively engaging in phytoremediation but generally lack the necessary technical expertise for its implementation.

A frequently overlooked yet crucial piece of information, found lacking in both scholarly publications and business reports, is comprehensive cost-benefit statistics [174]. These are frequently withheld or concealed due to concerns over confidentiality. According to Lee et al. [266], cost-benefit studies play a crucial role in assessing the practicality of implementing phytoremediation on a large scale. These studies are especially significant in developing regions such as sub-Saharan Africa, where communities often face resource limitations when it comes to physicochemical remediation methods. According to 715, while comparing remediation projects published by the USEPA, including those utilizing phytoremediation, it was shown that phytoremediation is 50% more cost-effective than soil excavation and ex situ treatment. Nevertheless, the criteria for phytoremediation can differ greatly across regions due to factors such as site locations, degradation levels, types of contaminants, soil composition, climate conditions, and the duration required for successful phytoremediation. These variables have an impact on the costs and benefits of remediation over time. Hence, the advantages of this technology are limited not only by the understanding and accessibility of resources but also by the fundamental comprehension and accessibility of information required for decision-making.

### 20.3.10.2 Lessons learned for developing regions

The practical use of phytoremediation still faces shortcomings, mostly due to a deficiency of technical comprehension. The effectiveness of this technique relies on the availability of necessary resources, fundamental expertise, and data obtained from studies, all of which are vital for making informed decisions and maximizing its benefits. The absence of cost-benefit data impedes the evaluation of feasibility, as factors such as site and climatic circumstances, as well as contaminant kinds, can significantly impact the cost-effectiveness of phytoremediation.

### 20.3.10.3 Enhancing overall awareness, knowledge, and abilities through education and sensitization

According to Marmiroli et al. [297], training and education are crucial for enhancing understanding and acceptability of phytoremediation, as there is currently a dearth of public awareness regarding the usage of this method. The authors of this assessment on the current and future potential of phytoremediation recommend implementing initiatives that focus on spreading awareness, providing education, and offering training. These initiatives aim to enhance the knowledge and confidence of the general public and other stakeholders in innovative and sustainable technologies such as phytoremediation. In order to get broader public recognition, it is imperative to not only evaluate the advantages of the technology but also comprehensively outline the associated hazards. According to Sharma and Pandey [445], when discussing risks, it is important to consider the alternative scenario of leaving the contaminated site as it is or using different methods for rehabilitation.

The USEPA has created an open-access website (https://www.clu-in.org/databases) to collect and share information on various greenhouse applications of phytotechnology. The website aims to identify and distribute project solutions and lessons learned that can be useful for new sites. The information compiled includes data on full-scale, field-scale, and large-scale applications of phytotechnology. The data is collected from industry and peer-reviewed journals, conference papers, as well as from technology providers and site owners. The project profiles provide comprehensive details on the site's background, the contaminants treated, the plant species utilized, the processes of phytotechnology employed, the date of planting, the size and location of the project, the associated costs, the monitoring and performance outcomes, as well as the contacts and references for further information. By implementing frequent updates, this platform actively promotes public awareness and education, functions as a networking tool, and supports the analysis of trends in the utilization of phytotechnologies.

As emphasized by [1013], the acceptance of phytoremediation, like any other novel technology, hinges on its ability to demonstrate success. Field demonstrations and field days have been widely recognized in the world of agricultural extension as effective means of providing empirical evidence and educating the public about the significant advantages of green innovation. Field demonstrations can serve as a tool for scientists, technical experts, and extension agents to bring about positive changes in the behavior of rural populations, create ideal learning environments, and facilitate meaningful communication and interaction. This is particularly relevant in rural local communities [241]. An illustration of this is the on-field demonstration projects conducted by U.S. researchers, which have played a significant role in establishing the favorable reputation that phytoremediation enjoys today [1013].

### 20.3.10.4 Developing regions can promote phytoremediation

By focusing on education, training, and information dissemination, we can enhance public acceptance and comprehension. In order to achieve broader recognition and approval, it is necessary to methodically delineate the potential dangers and advantages. Field demonstrations and training can serve as effective means of educating and raising awareness about new breakthroughs such as phytoremediation, particularly in rural regions. The excellent reputation that phytoremediation enjoys now is largely due to the success of on-field demonstration programs.

## 20.3.11 Institutional intermediaries (Box K)

Effective communication and networking among experts from different fields are crucial for the successful implementation of phytoremediation [297]. Collaboration among scientists, research institutes, and private firms is necessary for the development and innovation of customized plants, planting systems, and soil supplements for restoring polluted regions [297, 389]. In the United States, partnerships between research organizations, industry, and government institutions have played a significant role in the progress and successful introduction of phytoremediation technology [62, 305].

In Europe, the issue of patent ownership of intellectual property rights poses a significant obstacle to the commercialization of phytoremediation technology. In contrast, the United States has effectively addressed this issue through active cooperation between specialized private phytoremediation companies, universities, and government institutions [305].

In the Global South, such as sub-Saharan Africa, social networking and collaboration are important for land rehabilitation including phytoremediation. The institutional, regulatory, and economic systems governing natural resource and environmental management are not as developed in these regions [365]. Collaboration between various stakeholders, such as local chiefs, elders, district assembly representatives, state regulators, private companies, and environmental NGOs, can encourage active participation in land reclamation and remediation efforts at the local level, leading to sustainable community land management [12].

Overall, effective communication, networking, and collaboration among experts and stakeholders are essential for the progress, innovation, and successful implementation of phytoremediation in different contexts.

### 20.3.11.1 Key insights for developing regions

Establishing phytoremediation plant systems in sub-Saharan Africa requires efficient communication and collaboration among scientists, research institutions, and private enterprises. The successful execution of these systems requires participation from national and local governments as well as NGOs. Sustainable land restoration and remediation initiatives in rural communities require active involvement from local stakeholders including community chiefs and opinion leaders. This is crucial in informal customary land tenure structures, where governance is often guided by traditional rules.

## 20.4 Conclusion

Phytoremediation is a promising and affordable solution for soil remediation, particularly in regions with soil contamination and soil degradation. However, its global application and commercialization are limited, with most successes in the Global North. This chapter provides guidance on technical, social, economic, and institutional aspects for successful phytoremediation projects. It emphasizes the importance of collaboration among scientists, state institutions, NGOs and private enterprises, financial incentives, and community engagement. The strategic use of economically valuable plant species for biomass also enhances the efficacy of phytoremediation. By promoting this sustainable solution, regions can work toward a greener future for remediation and environmental restoration.

Albert Kobina Mensah

# Chapter 21
# Phytostabilization of Co, Hg, Mo, and Ni by ryegrass with manure and iron oxides reduced environmental concerns

**Abstract:** The study investigates the soil contamination risks of metals and metalloids in abandoned gold mine soil in southwestern Ghana, focusing on the potential of ryegrass-assisted phytoremediation to reclaim the site. A 60-day pot experiment was conducted to determine the effectiveness of soil amendments, including compost, iron oxide, and poultry manure, in reducing environmental risks. The study found that Hg contributed the most to the total pollution load index. When assisted with iron oxide, ryegrass demonstrated phytostabilization potential for metals and can be used to reduce their associated environmental and human health impacts. The addition of iron oxide increased uptake into the root due to sorbed metals on manure surfaces, dissolution by erosion, rainfall, runoff, mineralization of organic manure, or soluble metals washed down the soil profile or into groundwater. The study suggests that ryegrass-assisted phytoremediation could be a viable solution to address these environmental and human health concerns.

**Keywords:** Gold mining, human health impacts, phytoremediation, potentially toxic elements, ryegrass, soil pollution

## 21.1 Introduction

Metal and metalloid contamination in mine tailings, and soils and farms near abandoned gold mine land, is a major source of concern to residents because such pollution ultimately affects the food crops they eat with consequences on their health. The contamination of these sites with potentially toxic elements (PTEs) is thus attributed to either mobilization by runoff and/or by the action of water erosion, rainfall, or via deposition by wind into the surrounding environments. Additionally, contamination of the sites with PTEs could be attributed to either illegal gold mining activities taking place in such surroundings, indiscriminate disposal of mine wastes, poor handling of mine wastes, or dry deposition from the mining spoil. These artisanal miners usually

**Albert Kobina Mensah,** Soil Research Institute, Council for Scientific and Industrial Research, Kumasi, Ghana, e-mail: albert.mensah@rub.de, albertkobinamensah@gmail.com,
ORCID: https://orcid.org/0000-0001-5952-3357

https://doi.org/10.1515/9783111662046-021

operate with no license or approval from regulatory institutions. This action further leaves room for production without considering effects on environmental safety and their health. The lack of vegetation cover of mine sites and the abandonment of mine lands without protection may further exacerbate the widespread pollution of sites and farms in gold mining areas with toxic elements. These have deleterious effects on food crops grown, available water sources, and agricultural soils found in the area, with consequent effects on the health of the people [313, 333]. In gold mining areas, these were further confirmed with findings from the health impact calculations. In the soil-to-human health appraisal, the hazard quotient (HQ) and hazard index (HI) in all sites, especially for children and women, were above the critical threshold of 1; thus it indicates the health risk concerns at these sites. There is the possibility of using plant species to reclaim these heavily laden metal and metalloid-polluted sites; a strategy referred to as phytoremediation [38, 289, 330, 332].

A sustainable mine is one that respects the community where mineral extraction occurs, protects the environment, and maximizes profits while gaining economic returns. [704] defines sustainability in the mining sector as the design, construction, operation, and closure of mines that respect and respond to the social, environmental, and economic needs of present generations and anticipate future generations. [705] suggests that for a mine to be sustainable, immediate negative effects must be corrected through remediation, and socioeconomic benefits should be designed to bridge to a more sustainable future for the local community. Remediation measures are needed to repair land damage during and at the closure phase of the mine project life cycle, reduce environmental impacts, and protect the health of humans and animals in mining areas. Remediation of degraded sites can reduce toxic elements' migration into watercourses, offer pollution protection, improve soil quality, and mitigate human health concerns.

[703] capture phytoremediation as a green technology that involves using plants to remediate toxic compounds. They added that the approach is a cost-effective, socially acceptable, and environmentally friendly technology for soil and groundwater cleanup. [701] proposed that when dealing with more heavily polluted sites (e.g., mine tailings), plants that do not transport the metals to the shoots, but instead bind them in the root or the rhizosphere, are preferred; a strategy termed phytostabilization [702]. Candidate plants for phytostabilization (e.g., using grasses) should have an extensive root system, and other factors to be considered in choosing species are outlined by Alvarenga et al. [26]. The success of phytostabilization efforts may be enhanced through combination with immobilization techniques to detoxify the element in soil; a strategy called assisted or aided phytostabilization [717]. These can be achieved via lone or combined addition of soil amendments, for example, compost, biochar, iron oxides, manure, and phosphates [332, 318]. Addition of the soil amendments generally modifies the soil physicochemical and biological properties, quickens the revegetation process, and facilitates the remediation efforts. Consequently, the soil amendments can enhance soil quality, boost plant growth, and increase crop

yield, and they may aid in cleaning the soil and ameliorating the presence of toxic elements in contaminated soil and water [323].

There is a need to explore the potential of many plant species in order to widen the selection options that could be available for stabilizing the mobility of PTEs in mine-contaminated sites and to ultimately lessen their associated environmental and human health impacts. In Ghana, environmental degradation associated with gold mining has gained center stage in media discussions and has attracted many political tensions [52, 204, 374]. Therefore, any study that seeks to provide environmental solutions to the land degradation problems would be a step in the right direction. Such studies would aid in restoring the many degraded lands, protecting the integrity of the ecosystems, reinstating livelihoods, and ultimately helping to safeguard the health of the people in gold mining hotspots in the country.

However, many of the previous studies have either overly focused on trees and their role in restoring soil quality of the degraded mine land [335]; shrubs such as *Chromolaena odorata* and *Pityrogramma calomelanos* [34, 333]; or have basically involved collecting the plant species growing around the derelict lands for laboratory analyses of their heavy metal contents [e.g., 34, 333, 384] without any concrete pot or field experimental trials. Besides, other studies that experimentally tried the potential of ryegrass for remediation [e.g., 26, 166, 235, 236, 332] restricted the plant to its ability to stabilize As without investigating its phytoremediation potential for other toxic elements of concern such as Co, Hg, Mo, Ni, Pb, Sb, and Se.

In light of this, I conducted a 60-day pot experimental trial to assess the effect of various organic and inorganic amendments on the phytoavailability of Co, Hg, Mo, Ni, Pb, Sb, and Se in ryegrass grown in a multimetal-contaminated gold mine tailing in southwestern Ghana. My aim was to (i) quantify the effects of compost, manure, and iron oxides applied alone or in combination on the contents of Co, Hg, Mo, Ni, Pb, Sb, and Se in mine tailing; (ii) evaluate the Co, Hg, Mo, Ni, Pb, Sb, and Se phytoremediation potential of ryegrass grown in an abandoned gold mining site, as aided by compost, manure, and iron oxides; and (iii) ascertain whether the lone and/or combined application of soil amendments done to augment the phytoremediation success increases or decreases the environmental risks.

## 21.2 Materials and methods

### 21.2.1 Study area and soil amendments' characterization

Soil samples were obtained from a deserted gold mining location in Prestea, a prominent gold mining city in the southwestern region of Ghana. Comprehensive information regarding the physical and chemical characteristics of the site, the methodology used for soil sampling, and the prevailing climatic conditions may be found in the publications by Mensah et al. [318, 322, 332] and Mensah et al. [320, 319]. The soil utilized in the pot experiment exhibited the following characteristics (mean values): sand content of 63%, silt content of 34%, clay content of 3%, pH level of 7.1, total carbon content of 1.2%, arsenic concentration of 5,104.0 mg/kg, iron concentration of 23.48 g/kg, and phosphorus concentration of 0.3 g/kg [321, 329]. The study included compost, iron oxides, and poultry manure as soil ameliorants. A mixture of garden soil, residential biowaste, plant litter, and partially decomposed materials was utilized to make the compost.

In addition, I obtained the manure from a poultry farm in Witzenhausen, Germany, and the iron oxide ($Fe_2O_3$ – oxide red) was manufactured by ABC Beton Netherlands. The amendments, with the exception of iron oxide, were pulverized, made uniform, and filtered through a 0.63-mm sieve after being dried in an oven at 80 °C. Their pH, EC, total carbon and nitrogen, total elements, and heavy metals were evaluated. The pH meter was utilized to quantify the pH of both the soil and the amendment, using a soil to 0.01 M $CaCl_2$ solution ratio of 1:5. The total carbon and nitrogen levels were determined using the procedure outlined by [706]. The total element concentrations were obtained using the nitric acid digestion approach, which involved heating at 120 °C for 15 min and combining 10 mL of concentrated $HNO_3$ with 10 mL of deionized water [1094]. Table 21.1 presents data regarding the physicochemical properties of the soil ameliorants.

**Tab. 21.1:** Physico-chemical properties of the mine soil and total elements in the amendments used in the study.

| Parameter | Unit | Soil | Compost | Manure |
|---|---|---|---|---|
| pH | – | 7 | 6 | 7 |
| EC | μS/cm | 632 | 881 | 2,290 |
| Al | | 3,155 | 5137 | 2257 |
| As | | 5104 | 4 | 2 |
| Ca | | 11597 | 15880 | 88646 |
| Fe | | 23,480 | 8734 | 1449 |
| K | | 388 | 3161 | 17859 |
| Mg | mg/kg | 6,184 | 4664 | 5432 |
| Na | | 2891 | 1204 | 2847 |
| P | | 300 | 1126 | 9521 |
| Mn | | 321 | 471 | 392 |
| Zn | | 5 | 166 | 234 |

Source: Mensah et al. (2022b).

## 21.2.2 Experimental layout and design

A controlled greenhouse experiment was conducted at the Botanical Garden of Ruhr University Bochum, Germany. The studies were conducted from January 26 to April 9,

2021. The greenhouse's average temperature ranged from 20 to 25 °C. The amendments were meticulously mixed with 3 kg of soil that had been air-dried and had a particle size of less than 2 mm. The mixture was then placed in plastic pots with a volume of 3,000 cm³. Amendments were applied either individually or in combination at a rate of 5% by weight of the amendment relative to the weight of the soil. The selection of soil amendments at a 5% application rate was based on the successful rates of application observed in prior studies [e.g., 326, 332]. The experiment contained various treatments: control soil (unamended) denoted as C, compost only denoted as Com, manure only denoted as M, iron oxides only denoted as FE, compost and manure denoted as CM, compost and iron oxides denoted as CFE, and manure and iron oxides denoted as MFE. When amendments were combined, each amendment was added in equal rates and quantities. Table 21.2 contains the specific information regarding the treatments. To promote plant germination and growth, all treatments were treated with inorganic NPK fertilizer at a rate of 5 g/kg (0.5%). The composition of the inorganic NPK fertilizer is as follows: 15% of ammonium nitrogen, 15% of $P_2O_5$ (neutral-ammonium citrate and water-soluble phosphate), 15% of water-soluble $K_2O$, and 11% of S. The application rates for the inorganic fertilizer were determined based on the typical range of application used by farmers in Ghana, which is between 200 and 560 kg/ha (i.e., 200–560 kg/ha).

**Tab. 21.2:** Details of treatment ratios used in the pot experiments.

| Treatment | Acronym | Application rate (dw/dw) | Replication |
|---|---|---|---|
| Mine soil (no amendment) | C | 100% | 3 |
| Mine soil + compost only | Com | 95% + 5% | 3 |
| Mine soil + manure only | M | 95% + 5% | 3 |
| Mine soil + iron oxide only | FE | 95% + 5% | 3 |
| Mine soil + compost + manure | CM | 95% + 2.5% + 2.5% | 3 |
| Mine soil + compost + iron oxides | CFE | 95% + 2.5% + 2.5% | 3 |
| Mine soil + manure + iron oxides | MFE | 95% + 2.5% + 2.5% | 3 |

A total of 21 pots were organized into seven treatments, with three replications for each treatment, and one plant variety (7 treatments × 3 replications × 1 plant; Table 21.2). Subsequently, the pots were carefully maintained in a greenhouse with strict regulations, ensuring that they were kept at a water-holding capacity of 20%. The treatments were assigned to a randomized complete block design and subjected to a 14-day incubation period. Following a 2-week incubation period, 1.5 g of ryegrass seeds was planted in each pot. The plants were then subjected to postseeding agronomic treatments, which were detailed in a previous study conducted by Gadepalle et al. [173]. The seeds were sown by evenly dispersing them on the soil surface, lightly burying them with soil, and promptly watering them after planting. The soil moisture in the pots was maintained at 20% of its maximum water-holding capacity, and the pots were watered with 200 mL of water every day.

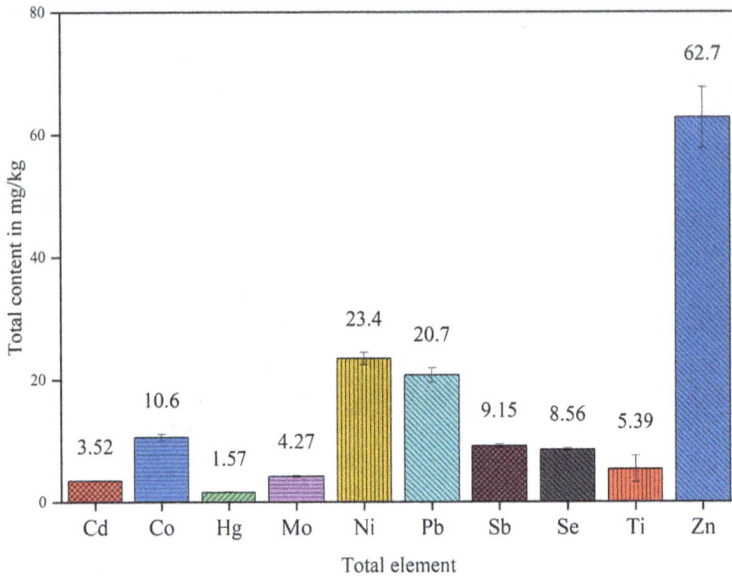

**Fig. 21.1:** Initial total concentrations of potentially toxic elements in the collected mine soil.

## 21.3 Harvesting and determination of PTE contents in soil and plants after harvesting

### 21.3.1 Soil analyses

After harvest, the soil in the pots was collected, air-dried for 48 h, passed through a 2 mm sieve, and prepared for analyses. Soil analyses included measurements for soil moisture, pH, EC, DOC, anions ($F^-$, $Cl^-$, $NO_3^-$, $PO_4^{3-}$, and $SO_4^{2-}$), and total elements (Al, Ca, Fe, K, Mg, Mn, Na, and P). For the soil pH, EC, DOC, and anions, 10 g of the wet soil was taken into a 100-mL PE bottle, and 50 mL of deionized water was added. The contents were shaken for 2 h at room temperature and then filtered through 0.45-μm filter paper. From the filtrates, pH, EC, DOC, and anions were determined. The soil moisture content was determined using the gravimetric water content method after oven-drying at 105 °C for 48 h. Soil pH was measured using a pH meter (Sentix 41, WTW GmbH, Weilheim, Germany), EC with an EC meter (TetraCon 325, WTW GmbH, Weilheim, Germany), DOC, total carbon, and nitrogen were measured by an elemental analyzer (Vario max cube, Elementar Analysesysteme GmbH, Hanau, Germany), and anions were determined using ion chromatography (Metrohm, 881 Compact IC Pro).

## 21.3.2 Soil contamination assessment

I employed three indices to evaluate the extent of PTE pollution in the uppermost layer of soil (0–20 cm): contamination factor (CF), enrichment factor (EF), and geo-accumulation index ($I_{geo}$). These indicators have been effectively utilized in previous research to monitor the level of PTE contamination and evaluate the risk to human health in soils along rivers in Germany [405], garden vegetable soils [37], children's playgrounds in Beijing [225], temperate and arid regions in Germany and Egypt [440], around an abandoned iron-ore mine in north-central Nigeria [862], around a lead/zinc smelter in southwest China [271], around an industrial area in Greece [37, 35], and in mine tailings in Ghana [321].

### 21.3.2.1 Soil contamination factor (CF)

$$CF = C_s / C_{ref}$$

where CF denotes the soil contamination factor, $C_s$ denotes the total element content in soil (mg/kg), and $C_{ref}$ denotes the element's reference/background value in uncontaminated soil (world soil average values were obtained from Kabata-Pendias [230]). The reference background values in mg/kg for the various PTEs, as shown in Fig. 21.1, were used as follows: Co = 8.0, Hg = 0.5, Mo = 4.0, Ni = 29.0, Pb = 27.0, Sb = 5.0, and Se = 5. 275, provided the reference value for Al. This index further categorizes soil contamination into four levels: low contamination (when CF = 1), moderate contamination (CF = 1–3), significant contamination (CF = 3–6), and very high contamination (CF ≥ 6) [321, 405, 440, Shaheen et al., 2019, 439].

### 21.3.2.2 Pollution load index (PLI)

According to Rinklebe et al. [405] and Shaheen et al. [453], PLI is the product of the individual CF values of all studied PTEs:

$$PLI = (CF_{s,1} \times CF_{s,2} \times \cdots \times CD_{s,n})^{1/n}$$

where $CF_{s,1}$, $CF_{s,2}$, and $CF_{s,n}$ are the contamination factors of elements 1, 2, . . ., $n$ under consideration. A pollution load index >1 indicates that the area under study has significant PTE soil contamination.

### 21.3.2.3 Enrichment factor (EF)

The enrichment factor assists in determining whether PTE pollution in soil is geogenic or anthropogenic. Because EF assumes that the contents of elements such as Al and Fe

occur in natural media and are mostly geogenic, Al or Fe is used as normalizer element in calculating enrichment [271, 440]. As a result, soil enrichment with higher
PTE contents (higher EF values) indicates pollution from anthropogenic sources. According to Antoniadis et al. ([1095–1096]), EF classifies contamination as minor (if EF
= 1.5–3), moderate (EF = 3–5), severe (EF = 5–10), or very severe (EF > 10):

$$EF = (C_s/Al_s)/(C_{ref}/Al_{ref})$$

where $Al_s$ is the total content of Al in the contaminated soil and $Al_{ref}$ is the total content of Al in the uncontaminated/background reference soil [e.g., 271, 414, 440].

### 21.3.2.4 Geo-accumulation index ($I_{geo}$)

The geo-accumulation index, as per Klubi et al. [245] is given as

$$I_{geo} = \log_2 \left( \frac{C_s}{1.5C_{refs}} \right)$$

## 21.3.3 Plant analyses

The plants were carefully pulled out of the pot by hand after 60 days of the experiment. Scissors were used to divide the harvested plants into above- and belowground
biomass. They were then rinsed and washed thoroughly in distilled water. Following
the measurement of the roots' and shoots' fresh weights, the roots' and shoots' dry
matter weights were determined by drying them in a 60 °C oven until a stable dry
weight was reached. By subtracting the oven-dried weight from the fresh weight, the
dry matter weight (biomass) of the roots and shoots was calculated. Afterward, the
plant samples were reduced to a fine powder in a stainless grinder, saved in Ziploc
bags, and kept for analysis. The content of the PTEs in plant biomass was determined
using the same methods employed for the soil analyses.

### 21.3.3.1 Determination of the phytoremediation efficiency of the ryegrass

The phytoavailability and absorption capability of plants used for phytoremediation
are estimated using soil-to-plant transfer factors/plant contamination indices. Thus,
we assessed the phytoavailability and uptake efficiency of ryegrass using plant absorption capabilities, soil-to-plant transfer abilities, and root-to-shoot translocation indicators [333].

#### 21.3.3.1.1 Soil-to-plant transfer coefficients

Using the bioconcentration factor (BCF) and bioaccumulation coefficient, it was possible to compute the accumulation of PTEs (Pb, Hg, Ni, Mo, Co, Sb, and Se) in the roots and shoots of ryegrass. The ratio of the soil's Pb, Hg, Ni, Mo, Co, Sb, and Se content to that of the ryegrass is determined here via soil-to-plant transfer coefficients. Thus, we assess the ryegrass's capacity to take up Pb, Hg, Ni, Mo, Co, Sb, and Se from the mine soil and decontaminate the site. The relative concentration of Pb, Hg, Ni, Mo, Co, Sb, and Se in the roots divided by the total concentration of Pb, Hg, Ni, Mo, Co, Sb, and Se in the soil is known as the BCF [333]. The BAC is the ratio of Pb, Hg, Ni, Mo, Co, Sb, and Se in the ryegrass shoots to the total concentration of Pb, Hg, Ni, Mo, Co, Sb, and Se in the soil. Here, the ryegrass may qualify as a phytoremediation plant for any of the PTEs if their BCF and BAC values are above 1. Plants that depict their BCFs and BACs below 1 tend to limit soil-to-root and root-to-shoot transfers and are thus inappropriate for phytoextraction or stabilization of particular PTEs. The equations used are as follows:

$$BAC = \frac{\text{PTE content in shoot}}{\text{Total PTE content in soil}}$$

$$BCF = \frac{\text{PTE content in root}}{\text{Total PTE content in soil}}$$

#### 21.3.3.1.2 Translocation factor (TF)

The studied PTEs (Pb, Hg, Ni, Co, Mo, Sb, and Se) were tested for their movement from the root to the shoot of ryegrass by TF. We performed this to quantify the ability of the ryegrass to remove PTE pollutants from the mine soil through either their shoot or root. In this instance, a TF > 1 qualifies the ryegrass for removing the pollutants. The following equation, as reported in Marchiol et al. [37] and used by others [e.g., 36, 37], was employed:

$$TF = \frac{\text{PTE content in shoot}}{\text{PTE content in root}} \tag{2.3}$$

### 21.3.4 Quality control, data treatment, and statistical analyses

I used standard solutions, blanks, and replicate measurements for quality control. Also, I used standard reference materials – SRM 2709a for baseline trace element concentrations (National Institute of Standards and Technology, USA), and European Commission Community Bureau of References (reference material No. 679, cabbage powder batch number 345, and No. 414, trace elements in plankton). The calibration range of our ICP-OES ranged from 0 to 2 mg/L.

I performed a one-way ANOVA among the means of different treatment rates (control, compost, manure, and iron oxide). Additionally, multiple range tests were performed using Tukey's honestly significant difference (HSD) test among the means of treatments at $P < 0.05$. I used OriginPro 2022b (Origin Lab Corporation, Northampton, USA) software to analyze our data as well as to create figures.

## 21.4 Results and discussion

### 21.4.1 Pseudo-total element contents in the mine soil

The elemental characteristics of the mine soil used for the experiment are as follows: Pb = 31.4 mg/kg, Se = 25.5 mg/kg, Co = 17.8 mg/kg, Ni = 14.4 mg/kg, Sb = 9.9 mg/kg, Mo = 2.7 mg/kg, and Hg = 1.2 mg/kg (Fig. 21.1). The elemental composition of the mine soil ranged from 1.2 to 31.4 mg/kg, with Pb having the greatest concentration and Hg having the lowest. Thus, it may be concluded that the concentrations of the selected PTEs in the mine tailings were in the descending order: Pb > Se > Co > Ni > Sb > Mo > Hg. All the elements contained in the mine soil exceeded their respective world soil averages recommended by Kabata-Pendias [230]. For instance, Co exceeded its recommended threshold in many folds by 122.5%, Hg by 140%, Pb by 116%, Sb by 98%, and Se by 410%. This implies that in increasing order of contamination, the elements in the mine tailings may be categorized as Sb < Pb < Co < Hg < Se.

Thus, Se may be the most highly contaminated element in the mine tailing, while Sb may provide the least contamination among the studied PTEs. Nickel and Mo contents in the mine tailing fell below the recommended threshold of 29 and 4 mg/kg, respectively. In mine tailings, high contents of Se have previously been reported by other scientists [e.g., 110, 476]; however, those works reported relatively lower contents than those found in our present study. Additionally, Se contents in my sample were higher than the trigger action range values (TAV = 3–10 mg/kg) according to data compiled from different sources in the European Union by Kabata-Pendias [230]. Moreover, Se contents found in our study were within the range of 0.1–435 mg/kg reported in other studies (e.g., Favorito et al., [804]; Xing et al., [1027]). It should be emphasized that Se pollution from mining operations is a major concern as an emerging contaminant. It is capable of entering surface water and groundwater, soil, and plants, where it can bioaccumulate and pose serious health and environmental risks ([875, 952]). Excessive levels of Se, for example, can cause selenosis in humans, alkali disease in grazing animals, and larval and developmental deformities and mortality in aquatic organisms [151]. Furthermore, Etteieb et al [151] reported that soil selenium levels ranging from 3 to 9 mg/kg caused alkali disease in grazing animals. Alkali disease causes dullness, lack of vitality, emaciation, rough coat, hoof sloughing, joint and

bone erosion, anemia, lameness, liver cirrhosis, and decreased reproductive performance in animals.

Mercury levels in the mine soil exceeded those reported for loam/silt soils
(0.5 mg/kg), were closer to those recommended for clay soils (1 mg/kg), but fell short
of the maximum allowable concentration suggested by Kabata-Pendias [230]. Furthermore, the contents are typical of gold mine soil from artisanal and small-scale gold
mining. There have been reports of anthropogenic effluent release from an ongoing
artisanal gold mining operation in the mine soil and its surroundings [333]. Mercury
is a key element in the artisanal small-scale gold mining sector for amalgamating gold
in order to recover it from impurities. Mercury is classified as a pollutant due to poor
handling and indiscriminate disposal from mine fields [57]. In gold mining communities, Hg is recognized as a major contributor to point and nonpoint source contamination of rivers, soils, and the food chain [207, 353]. According to [744], inorganic Hg(II)
and inorganic forms of Hg in soil are linked to complex health problems due to their
ability to infiltrate the food chain and bioaccumulate even at low exposure levels. The
Hg contents observed in this study are higher than those found in pristine soil around
the world (i.e., 0.01–0.2 mg/kg) by Adriano [9]. Furthermore, the contents are higher
than those found in river sediments collected at artisanal gold mining sites in Senegal
[353], but they are lower by many orders of magnitude than those found in 2015 in the
same river sediments by [940] and in soil samples collected from the Tongguan gold
mining area in China by Feng et al. (maximum value = 76 mg/kg) in 2006.

Cobalt is the third most abundant element among the studied PTEs. The contents
measured in our study are higher than the minimum value (3.2 mg/kg) reported by
Shaheen et al. [440] but less than the maximum and the average values reported in
the same study that compared PTE contents in multicontaminated soils from Egypt
and Germany. Additionally, the contents are lower than the maximum allowable concentration range (20–50 mg/kg) and that of the trigger action value (30–100 mg/kg) reported by Kabata-Pendias [230].

For Pb and Sb, their contents were within their respective maximum allowable
concentrations and their trigger action values. Particularly, for Pb, its concentrations
in the mine tailings were higher than the minimum allowable limit (10 mg/kg) but
lower than the maximum allowable limit (300 mg/kg). Antimony is classed as a priority
contaminant by both the US Environmental Protection Agency (USEPA) and the Council of the European Union (EU) ([707]; Hua et al., 2021). Antimony fell below the World
Health Organization (WHO)'s maximum acceptable value (36 mg/kg) in soil [708, 1097].
Excessive ingestion of Sb by humans can lead to nausea, diarrhea, skin rashes, and respiratory disorders (Hua et al., 2021). According to the International Agency for Research
on Cancer (IARC), Sb trioxide ($Sb_2O_3$) is a potentially carcinogenic substance to humans (IARC, 1989).

In conclusion, it is assumed that all the studied PTEs may be of geogenic origin.
This assumption is supported by the significant positive relationship of all the studied
PTEs with Fe, as given in the Appendix: Fig. S1: Se versus Fe ($r$ = +0.50, $p < 0.01$); Co

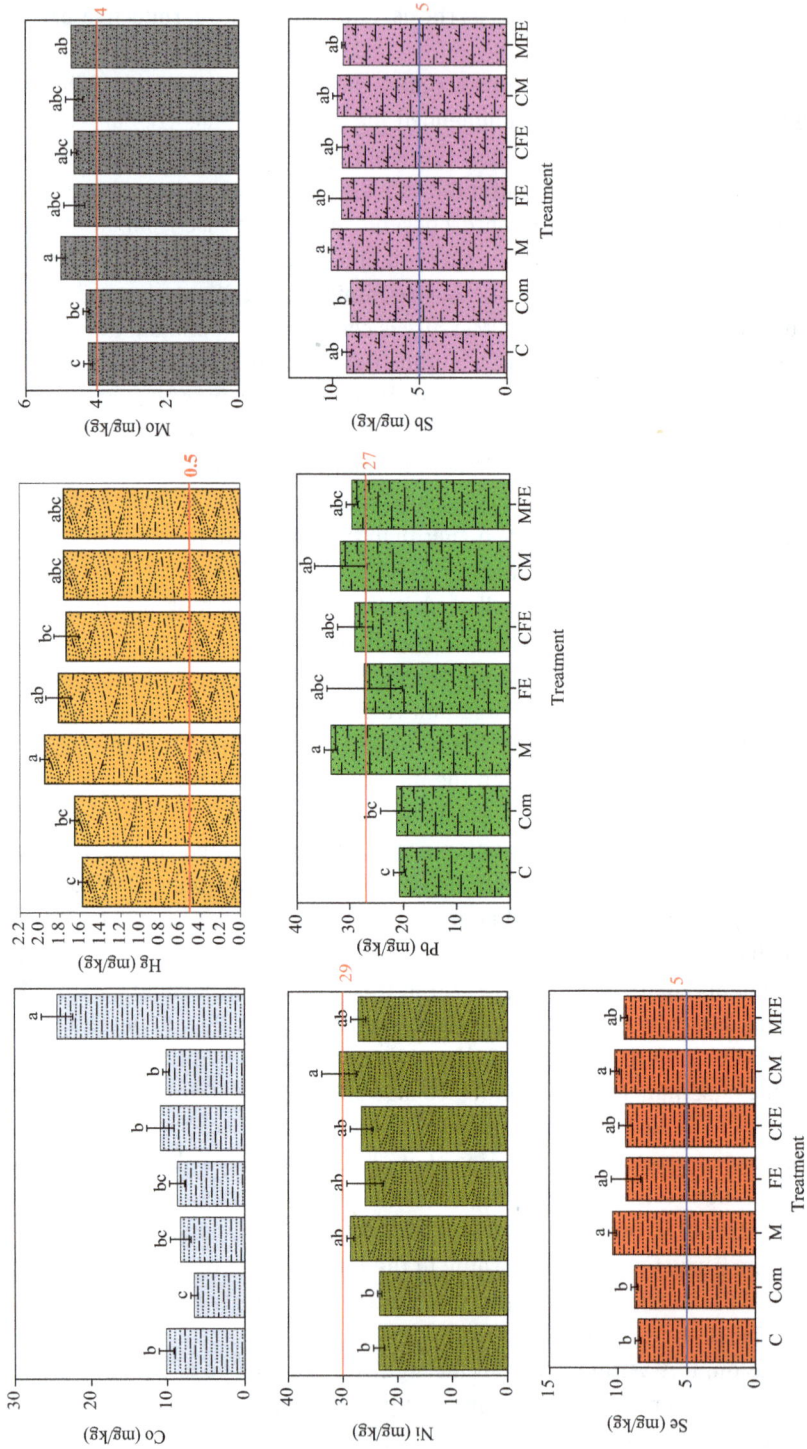

**Fig. 21.2:** Concentrations of PTEs in the differently treated or amended mine-contaminated soil at the end of the 60-day pot experiments.

versus Fe ($r$ = +0.80, $p$ < 0.01); Hg versus Fe ($r$ = +0.40, $p$ < 0.05); Mo versus Fe ($r$ = +0.40, $p$ < 0.05); Ni versus Fe ($r$ = +0.60, $p$ < 0.01); and Sb versus Fe ($r$ = +0.40, $p$ < 0.05). Thus, Sb, Pb, Co, Hg, and Se may be held with Fe oxides in the mine spoil soil either through coprecipitation, adsorption, surface complexation, ion exchange, or occlusion in the soil crystal lattice, as indicated by Liu et al. [276]. In this regard, the elements may be released to contaminate the nearby food chain and surface and groundwater during either dissolution of the soil mineral [189] or due to redox-induced changes in the mine tailings [e.g., 38, 408, 437, 433]. Furthermore, these toxic elements may be dispersed downstream due to the weathering processes that take place in the mine soil as explained by Mensah et al. [321].

## 21.4.2 Soil contamination assessment using contamination factor, enrichment factor, geo-accumulation index, and pollution load index

Contamination factor (CF) values of Hg, Se, and Sb obtained from the site (Table 21.3) revealed that there is enrichment beyond their reference background values. The CF values of Hg, with a mean of 17.41, suggest an extremely worrisome level of contamination at the site. In this respect, Wang et al. [507] predicted that mercury contamination of soil will result in an elevated exposure of mercury to local residents and will consequently pose a direct threat to the health of these people. The CF values found for Hg in this study surpassed by many folds than those (mean CF = 0.25) found by Klubi et al. [246] for sediments in wetlands, but fell below those (median CF was 37.8) reported by 414, for multicontaminated soils along the Central Elbe River in Germany. Several studies have reported higher CF values of Pb in mine tailings compared to our findings [548, 1080, 1081].

Loska et al. [898] and Saha et al. [414] found that mine soils showed extremely high enrichment (EF > 40) to very high enrichment (EF = 20–40). Rinklebe et al. [405] found similar levels in abandoned mine soils, which were indications of recent deposition. Intense rainfall, flooding, downstream transport of metal-contaminated fine sediments, dust from unpaved roads, local runoff from polluted surface soils, and lack of vegetation may be the causes of the recent deposition. Donkor et al. [789] found similar Hg EF results in soil sediments from Ghana's Pra basin affected by artisanal gold mining, with Hg EF values ranging from mild to severe enrichment during the rainy season.

The site had moderate to extremely high enrichment of Co, Hg, Mo, Ni, Pb, Se, and Sb over their EF thresholds. These indicate various metal/metalloid contamination at the site similar to other mine-contaminated sites described by others. According to [1041], EF values above 1.5 may indicate that a large fraction of trace metals comes from anthropogenic sources. This anthropogenic enrichment may be sourced from one or a combination of the following:

**Tab. 21.3:** Soil contamination assessment as determined by contamination factor (CF), geo-accumulation index (Igeo), enrichment factor (ER), and pollution load index (PLI).

| Element | Contamination indices | Minimum | Maximum | Mean | SD | Median |
|---------|----------------------|---------|---------|------|-----|--------|
| Co | CF | 0.9 | 37.9 | 2.4 | 4.6 | 1.6 |
| | Igeo | −0.8 | 4.7 | 0.2 | 0.8 | 0.1 |
| | ER | | | | | |
| Hg | CF | 0.5 | 18.0 | 17.4 | 2.6 | 17.9 |
| | Igeo | −1.6 | 3.6 | 3.5 | 0.7 | 3.6 |
| | ER | 2.0 | 18.5 | 7.9 | 3.5 | 7.1 |
| Ni | CF | 0.5 | 1.7 | 1.1 | 0.3 | 1.1 |
| | Igeo | −1.5 | 0.2 | −0.5 | 0.4 | −0.4 |
| | ER | 0.0 | 70.6 | 33.3 | 21.7 | 39.2 |
| Se | CF | 0.0 | 8.5 | 5.0 | 3.1 | 6.4 |
| | Igeo | −7.5 | 2.5 | 0.4 | 3.0 | 2.1 |
| | ER | 0.0 | 4.6 | 2.1 | 1.4 | 2.5 |
| Mo | CF | 0.0 | 0.7 | 0.3 | 0.2 | 0.4 |
| | Igeo | −9.6 | −1.1 | −3.4 | 2.8 | −1.9 |
| | ER | 0.3 | 38.3 | 20.5 | 6.4 | 18.9 |
| Sb | CF | 0.0 | 4.1 | 3.0 | 0.5 | 3.0 |
| | Igeo | −5.0 | 1.5 | 0.9 | 0.8 | 1.0 |
| | ER | 0.1 | 26.1 | 14.1 | 4.5 | 13.9 |
| Pb | CF | 0.0 | 3.5 | 2.1 | 0.5 | 2.0 |
| | Igeo | −6.9 | 1.2 | 0.3 | 1.0 | 0.4 |
| | ER | 0.1 | 26.1 | 11.0 | 4.6 | 9.8 |
| | PLI | 2.0 | 74.4 | 31.3 | 11.8 | 32.4 |

i) many years of mine site abandonment without reclamation or vegetation cover;
ii) continuous stockpiling and deposition of mine soils in mounds;
iii) ongoing artisanal gold mining activities in the surrounding area;
iv) earthmoving and turning of the spoil material with earthmoving equipment when a local mining company operates on the abandoned mine site;
v) indiscriminate release and disposal of mine effluents from illegal mining in the area;
vi) poor storage and handling of mine-contaminated wastes.

Mean $I_{geo}$ values were negative for Ni and Mo. The mean $I_{geo}$ of 0.33 for Pb indicated unpolluted Pb in the mine soil. As for Hg, its enrichment from illegal mining may be the cause of the abandoned site's geo-accumulation index of 3.5. The Se $I_{geo}$ averages a value of 0.4. A maximum Se $I_{geo}$ of 2.5 ($5 < I_{geo}$) indicates low contamination concerns as per Yang et al. (2019). The mean $I_{geo}$ of Sb was 0.9, indicating low natural enrichment.
The site's mean PLI (31.3) was manyfold higher, indicating the possibility of multiple element contamination. Multiple element contamination is indicated by the pollutant load index (PLI) [37, 319, 443]. Due to high pollution and PLI values, geogenic and accelerated anthropogenic activities may contribute to trace element accumulation

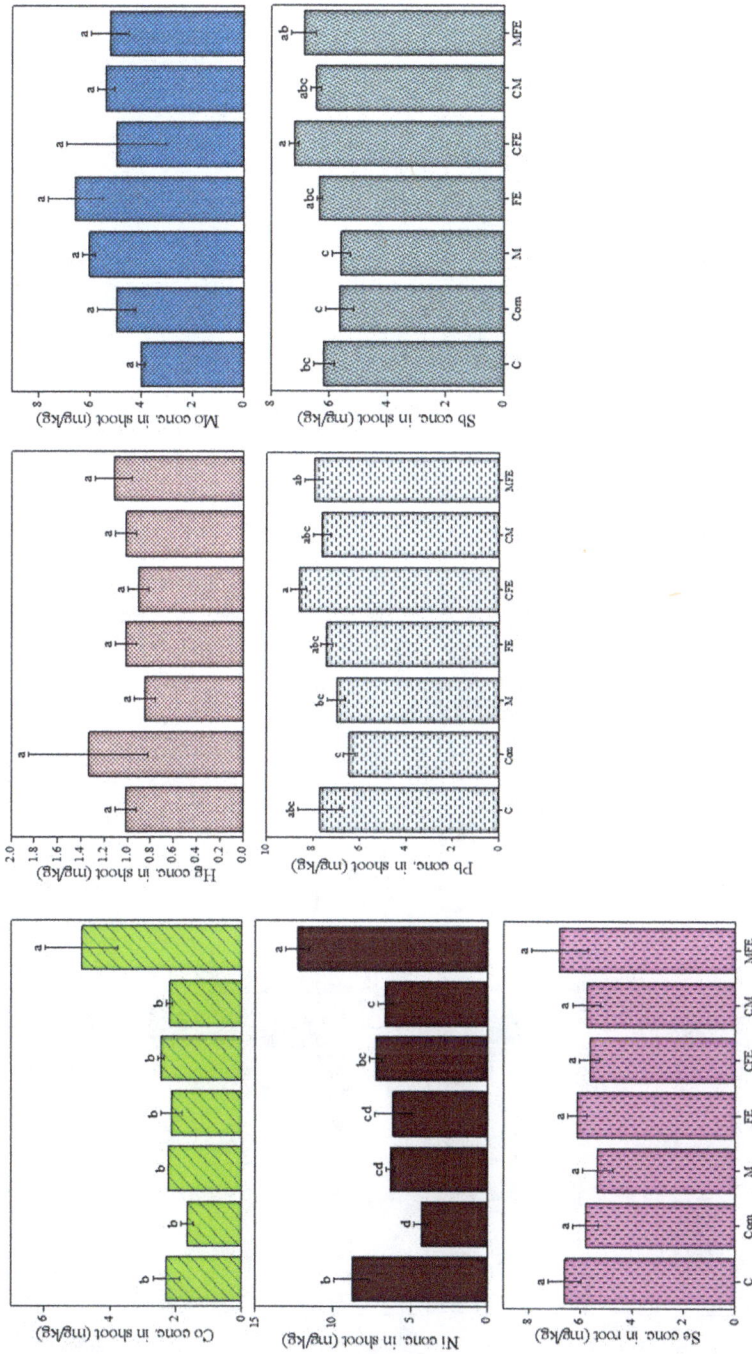

**Fig. 21.3:** Concentration of the studied PTEs in the (a) plant root and (b) plant shoot of ryegrass at the end of the 60-day pot experiments.

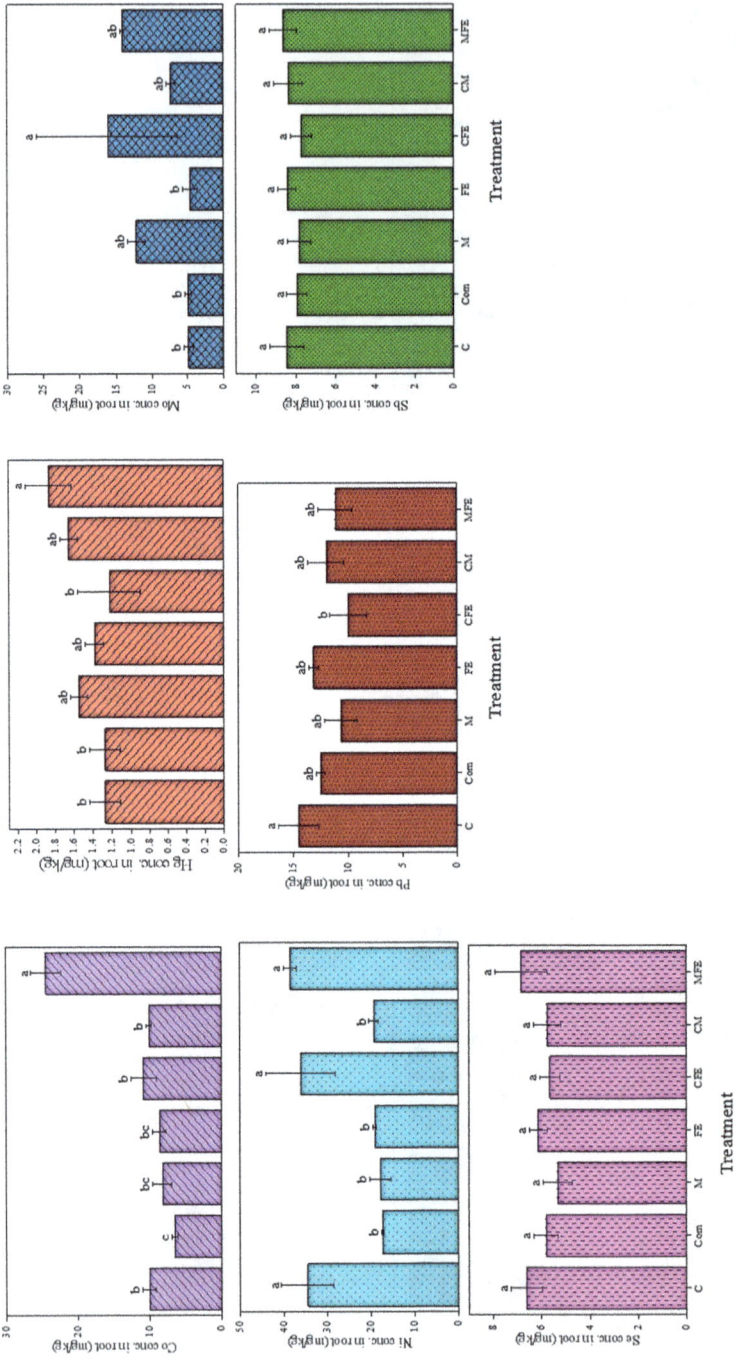

**Fig. 21.3** (continued)

[256]. The pollution load index values in our study are greater than Krishnakumar et al. [256] for multielement contaminated agricultural locations in India (PLI = 0.97–1.99) and exceeded Klubi et al. [246]'s PLI < 1 for wetland soils in Ghana.

## 21.4.3 Impacts of soil amendments on the potentially toxic elements contents in the mine soil

The changes in the soil PTEs upon the addition of the soil amendments are presented in Fig. 21.2. The concentration of Pb in mine tailings was significantly increased by the sole addition of manure from 20.72 mg/kg in the control to 33.60 mg/kg with a percentage increase of 62%. Mine soil amended with CM, MFE, and CFE increased Pb concentration from 20.72 mg/kg in the control to 31.73, 29.49, and 28.99 mg/kg, respectively. There were no statistical differences among the other treatments, but significant differences existed between the control and the combined addition of compost and manure. Similarly, Se concentration in the amended tailings ranged from 8.56 to 10.40 mg/kg, with the highest Se content recorded in the mine amended with manure, showing a percentage increase of 21%. Mine tailings amended with CM increased Se concentration from 8.56 to 10.24 mg/kg, which is statistically the same as MFE-amended tailings (9.55 mg/kg). The lowest Se content was recorded in mine tailings with no addition of soil amendments. Thus, the addition of manure and/or compost significantly increased the Pb and Se contents in the mine tailing relative to the control. The Pb contents in the lone addition of manure and/or compost-amended mine tailing were above the recommended threshold values (i.e., Pb = 27 mg/kg; Se = 5 mg/kg) provided by Kabata-Pendias [230] but below the prescribed recommendation (i.e., Pb = 300 mg/kg) set by the European Union Directive (CEC, 1986) cited in Antoniadis et al. (2021). The increment in the Pb and Se contents in the manure and/or compost treatment may be related to the high Al, Fe, Mn, P, and EC contents in both the manure and the compost. These assumptions are further verified by the significant positive relationships between these parameters and Pb and Se contents in the amended soils (shown in Fig. S1), meaning the Pb and Se contents may increase as these parameters also increase.

Lead is associated with plant photosynthesis, transpiration, and stomatal conductance at soil concentrations above 300 mg/kg [813], but our work indicated Pb concentrations lower than these thresholds. Nevertheless, human health impacts of Pb above are reported by others [e.g., 521, 527] in China, where the metal together with Cd occupies the eighth position among the top inorganic contaminants in the country. There, the metal accumulates in their staple rice food crop and potentially impacts more than 60% of the population who depend on rice. The positive correlation between Pb/Se and Al/Fe/Mn further demonstrates that manure and/or compost provides electrostatic sites where Pb may be held either via coprecipitation, surface complexation, or occlusion into the inner sphere complexes. In this respect, whilst some fractions are fixed in unavailable forms, certain aspects of the total Pb held onto the manure/compost-amended mine tailings may become mobilized during dynamic alterations in the soil properties such as the pH, organic carbon, and P content or the redox potential as also found by others [e.g., 70, 69]. In a leaching column experiment using cow manure and bone meal, Houben et al. [851] observed increased Pb leaching. The explanation was that high DOC contents in the soil enhanced the formation of organometallic complexes, which consequently facilitated Pb mobility and release. Furthermore, the association of Pb and Se with Al/Fe/Mn may give a clue about the geogenic nature of the contamination in the mine tailing, as mentioned by others in similar studies [190, 225, 245]. The increased Se in the manure-treated mine tailings also agrees with observations made by Kabata-Pendias [228] that the application of farmyard manure is a good means of increasing the contents of Se in Se-deficient soil. Thus, it may be concluded that a single addition of the manure or in a mixture with compost may be used to hasten the phytoextraction success of plant species revegetated on the multicontaminated gold mine soil. This success of the phytoremediation project can be augmented or accelerated if the sulfates and phosphates contents are increased. It is thus argued that phosphates and sulfates lessen the adsorption of Se onto the surfaces of Fe oxides, and they are effective in releasing and mobilizing up to 90% of the soil-adsorbed selenite and selenates [228].

Cobalt content in our amended mine tailings was significantly affected by the addition of compost, combined compost and iron oxide, combined compost and manure, and combined manure and iron oxide (Fig. 21.2). Cobalt content in amended tailings ranged between 9.87 and 13.95 mg/kg. Particularly, combined manure and iron oxide significantly increased the Co in the control from 10.61 to above 25 mg/kg, an increased enrichment content of more than 150%. Additionally, the mixture of compost and iron oxide raised the Co content in the mine tailing above the world soil average value (8 mg/kg) by some 213%. Mine tailings amended with CM after the experiment, for instance, significantly increased cobalt content in mine tailings from 10.61 to 13.95 mg/kg (+32% increase), whilst Com decreased it from 10.61 to 9.87 mg/kg (−7% decrease). These percentage increases and decreases may be related to their respective total Fe and P contents in the amendments used, as demonstrated by the significant positive relationships with Co (Fig. 21.4). Kabata-Pendias [228] reported that Fe is strongly associated with Co, as they are mostly hidden in various Fe minerals, and they closely resemble Fe and Mn. In addition, these two elements (Fe/Mn) largely control and govern the fate of weathering, distribution, and dissolution of Co in soils and sediments. Thus, we assume that the high Fe contents in the iron oxide and manure, for instance, might have provided available sorption sites for Co in their amorphous components and the crystallized fractions. In other instances, the Co might have been complexed or coprecipitated with the phosphate components in these treatments. Attachment of Co with the amorphous and crystallized fractions of the Fe and coprecipitation with P in these amended mine tailings eventually immobilizes Co and consequently reduces their availability and release. Further, the decrease in the Co content in the compost-amended mine tailing implies that they may be lost either during decomposition or mineralization of the compost or liberated from Fe/P during dissolution. In effect, we hypothesize that the lone addition of compost may enhance the liberation of Co in the mine tailing and increase ecotoxicological effects, whilst a mixture of manure and iron oxide may sequester Co and reduce its associated environmental and human health risks. Additionally, the addition of compost may be employed to enhance the absorption of Co in phytoremediation projects.

Ni contents in our amended tailings ranged between 23.31 and 30.67 mg/kg, with the highest value observed with the addition of CM and the lowest in the Com (Fig. 21.2). Combined addition of compost and manure significantly raised the Ni contents by 31%, relative to the control. Furthermore, the concentration of Sb in the amended mine tailings ranged from 8.93 to 10.05 mg/kg. Sb content in the amended tailings was increased upon the addition of manure. For instance, at the end of the experiment, the addition of M increased the Sb content from 9.15 to 10.05 mg/kg. The positive significant relation between Ni/Sb and Al/Fe/Mn/P/EC/chloride/nitrate and sulfate (Fig. 21.5) implies that Ni/Sb may be attached to the manure and compost added to the mine tailing, further increasing their contents. For instance, increased EC content of the mine tailing by the addition of compost and manure may increase the salinity, consume protons, raise the pH, and ultimately immobilize Ni/Sb and reduce its availability. In a similar study, Rizwan [410] observed that the addition of compost and manure to a Ni-contaminated field significantly increased the soil pH and EC and consequently decreased the available Ni. The association of Ni with oxide and/or oxy-hydroxide minerals when being deposited in the soil was further confirmed by Grybos et al. [187]. They concluded that the dissolution of these minerals under reducing conditions leads to the release of Ni in soil solution. Our findings are in contrast with those made by Rizwan [410] when they found that compost and manure addition to a Ni-contaminated soil significantly decreased available Ni in the postharvest soil of a maize crop. Addition of combined compost and manure contributed to raising the Ni in the mine tailing above the recommended threshold reported by Kabata-Pendias [230], but the concentration in our study is far below those reported for agricultural flood plain soils by Shaheen et al. [436]. The highest Ni contamination, up to 26,000 mg/kg, was reported for top soils near the Ni–Cu smelter at Sudbury, Canada [778].

Mo concentration in the amended tailings was significantly increased upon the addition of soil amendments in the manure and/or iron oxide amended samples. For example, manure-amended tailings increased the Mo content to 5.01 mg/kg relative to the control (4.27 mg/kg). For Hg, concentrations ranged between 1.57 and 1.95 mg/kg. Mercury concentrations found in our treatments and the control were above the average background concentrations (0.06 mg/kg) reported by Wang et al. [507]. The addition of manure significantly increased the Hg content in the control by nearly 24%. The increments in the Hg and Mo content in the manure-amended mine soil may be related to the contents of Al/Fe/Mn/EC/chloride/nitrate and sulfate as demonstrated by their significant correlations (Fig. 21.5). The implications of these relationships have

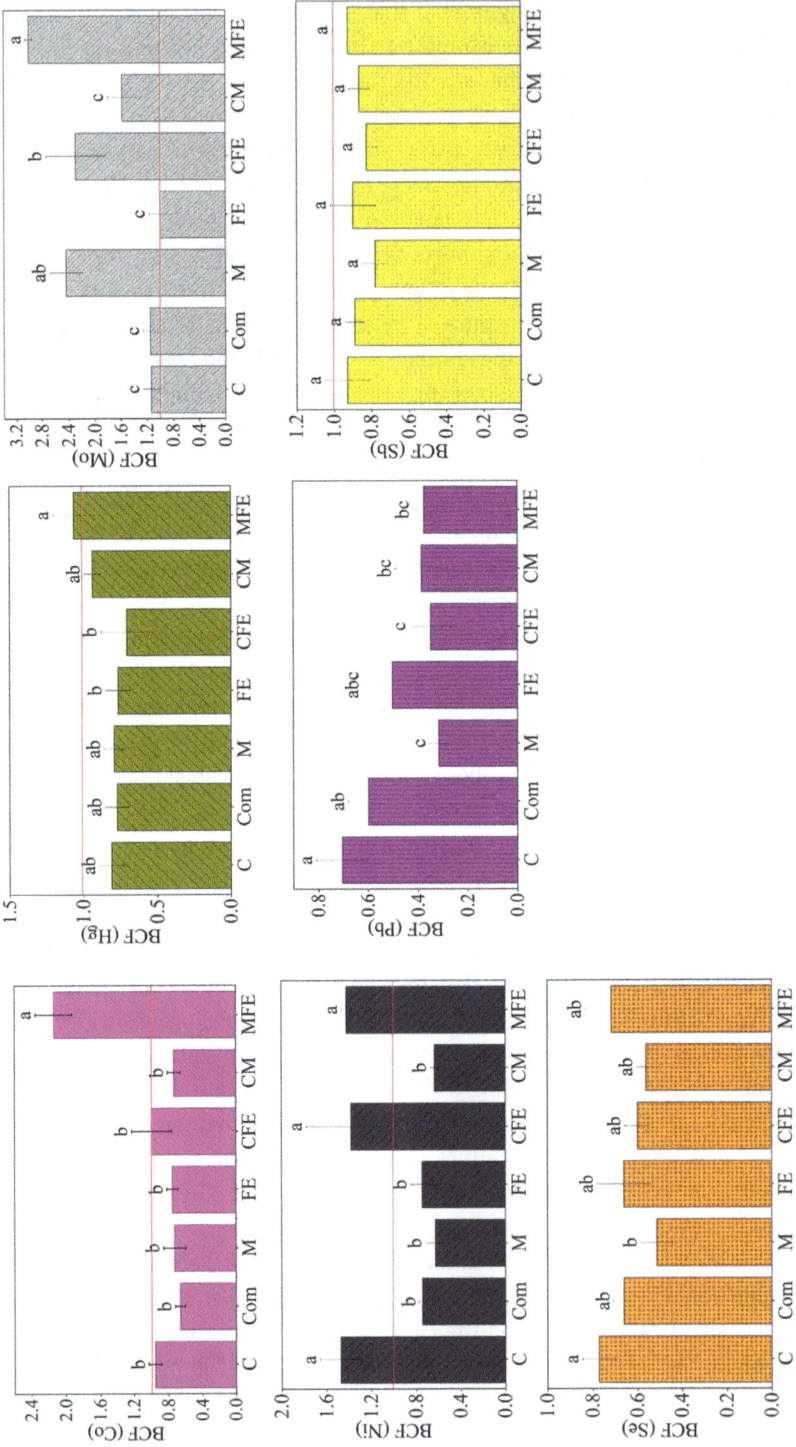

**Fig. 21.4:** Bioconcentration factor of the ryegrass as augmented with soil ameliorants at the end of the 60-day pot experiment.

been explained earlier, and the influence of Al/Fe/Mn in controlling the mobility of many metals, nonmetals, and rare earth elements is reiterated by others [e.g., 187, 188]. Concentrations of Mo in the amended-mine soils were slightly above the recommended threshold reported by Kabata-Pendias [230] but fell below those reported for biochar-amended mine soils by El-Naggar et al. [150].

## 21.4.4 Accumulation of potentially toxic element contents in plant root and shoot of ryegrass

PTEs accumulated in the roots and shoots are demonstrated in Fig. 21.3a and b. Lead contents in the roots of the ryegrass ranged from 9.97 to 14.46 mg/kg, with the highest concentrations reported in the control and the lowest concentrations found in CFE-amended mine tailing. Thus, the addition of CFE to the mine tailing reduced Pb root uptake in ryegrass by 46%. We assume this reduction to be associated with the contents of C, C/N, and Fe in the root, as demonstrated by the significant relationships with Pb contents in the roots (Fig. 21.6b). For example, the application of CFE may lead to the formation of ferric soluble salts (e.g., $FeSO_4$), as also opined by Mensah et al. [332]. This root chemistry may provide available sorption spaces to sequester more Pb and consequently reduce their availability for uptake.

Moreover, a high C/N ratio implies high recalcitrance of the root to decompose, and thus the rate of mineralization of the organic compost is reduced [326]. Consequently, there is a lower rate of release of any attached Pb to the organic matter and hence the resultant reduction in Pb for root uptake. This explanation is further confirmed by the negative correlation between C content in the root and Pb uptake in the root ($r = 0.6$, $P < 0.01$; Fig. 21.6b), meaning high C will lead to low Pb and vice versa. Thus, root carbon may become a carrier for many PTEs and either limit or encourage their uptake [332]. Here, the root C might have carried the root Pb and limited its uptake in the root and subsequently its upward uptake into the shoot. Also, Pb concentrations in the shoot varied between 6.45 and 8.59 mg/kg (Fig. 21.6b), with CFE treatments having the highest value and the lowest observed in the Com. None of the treatments significantly differed from that of the control in the shoot. Additionally, the concentration of Pb in the root is higher than that in the shoot, implying that Pb may be more sequestered in the root of ryegrass. The concentrations of Pb in both the root and shoot of ryegrass found in our study are higher than those (Pb in root = 0.51 mg/kg; shoot = 0.65 mg/kg) reported for *Alopecurus pratensis* used in a phytoremediation project by Antoniadis et al. [38].

Cobalt content in the roots varied from 6.51 to 24.37 mg/kg (Fig. 21.3a), with the lower concentrations observed in compost-amended mine tailings and the maximum concentrations being found in MFE, where there was a 141% increase in cobalt content compared to the control. The reduction in the Co content in the Com-amended soil may be attributed to its organic matter content, which may bind the Co in the root rhizosphere and ultimately restrict its uptake into the root. Additionally, as demonstrated by the correlation results, we observed a significant positive correlation between the root Ca ($r = 0.7$, $P < 0.01$) and Mg ($r = 0.6$, $P < 0.05$) with root Co (Fig. 21.6b). In this regard, Ca and Mg in the compost may provide a liming effect to the mine tailing to increase the pH and reduce its acidity [26]. The increased pH consequently induces more negative charges (e.g., $OH^-$ and $CO_3^{2-}$) [472], which may fix Co and thus reduce their availability for uptake by the root.

The use of liming materials in assisted-rehabilitation and phytostabilization of multimetal contaminated mine sites using ryegrass (*Lolium perenne* L.) is well reiterated by Alvarenga et al. [26]. Cobalt concentrations in the shoots varied from 1.65 to 4.85 mg/kg (Fig. 21.3b). An average concentration of Co in plant tissues ranges from 0.03 to 0.55 mg/kg of dry matter [989]. Plants absorb cobalt from soil most often in the form of divalent cobalt ions. Cobalt concentrations in the roots found in our study are higher than those observed in the roots of oats (mean Co = 2.3 mg/kg) grown on contaminated mine sites and amended with organic manure (e.g., 256). MFE addition significantly raised the Co content in the shoot by 112% when compared to the control, whereas Com addition lowered the Co content in the shoot by 28% (not significant). The Co concentrations in the shoot were lower than those in the root. These observations are also confirmed by, for example, Kosiorek and Wyszkowski [252], when they reported that roots accumulate higher content of Co than the aerial parts.

Mercury concentrations in the roots ranged from 1.28 to 1.87 mg/kg, with the lowest concentrations observed in the control and Com-amended mine tailings, and the highest concentrations observed in MFE, which had a rise in Hg concentration of 46% above the control. We may assume that the increase in the Hg uptake in the root is associated with the MFE amendments due to its relationship with the root P ($r = +0.40$, $P < 0.05$), C ($r = -0.50$, $P < 0.05$), C/N ($r = -0.90$, $P < 0.01$), and that with Mn in root ($r = +0.65$, $P < 0.01$) (Fig. 21.6a). For instance, root Hg might have either formed complexes or coprecipitated with the phosphates, organic matter, or Mn in the root rhizosphere. Consequently, the Hg became available for root uptake during dissolution upon irrigation or due to redox-induced changes that might have occurred in the mine tailings. These mechanisms may liberate the Hg attached to the P, C, or the Mn for possible root absorption. In this regard, Wang et al. [507] reported that the Mn oxide-bound fraction of soil mercury may represent a pool of potentially bioavailable metal which could be transformed into a more available form and subsequently taken up by plants. Moreover, Beckers et al. [66] found that various redox-induced changes due to flooding in an Hg-polluted floodplain soil resulted in substantial Hg releases from the lower course of the Wupper River in Germany. Also mentioned was the stabilization of mercury in mercury mining wastes by the application of chemically bonded phosphate ceramics (CBPC) technology [e.g., 507]. Mercury concentrations in the shoots varied from 0.85 mg/kg in the manure-amended mine tailing to 1.33 mg/kg in the compost-amended treatments. None of the treatments differed significantly from the control. In general, the amount of Hg accumulated in the roots was higher than that in the shoot; this is consistent with similar observations made by others [e.g., 298, 765, 931]. 303, further explained that Hg is more accumulated in the root because metal uptake by the roots is rapid compared with the transport to other plant tissues. It is also thus argued that Hg accumulation in the shoot and other parts of the plant is limited due to the retention of the metal/metalloid in the roots. Mercury concentrations in the roots found in our study are below those observed for *Jatropha curcas* (1.07–7.25 mg/kg) cultivated on gold mine soil in the USA by Marrugo-Negrete et al. [298].

The mean Ni content in the root varied from 17.40 to 38.51 mg/kg, with the highest concentration in the mine soil treated with MFE (Fig. 21.3a). There were no significant differences between mine soil treated with CFE, MFE, and the control. It is also worth noting that mine soil altered with CM, FE, M, and Com significantly reduced Ni in the root from 34.61 mg/kg to 19.36 mg/kg, 19.15 mg/kg, 17.92 mg/kg, and 17.39 mg/kg, respectively (Fig. 21.3a). These imply that these amendments may be used to assist phytoremediation success in sequestering Ni contents in the mine soil and reduce the associated environmental and human health risks. Nickel concentrations in the shoot varied from 4.32 to 12.27 mg/kg, with the lowest value reported in the Com-modified mine tailing and the highest concentration recorded in the MFE. With the exception of MFE (12.27 mg/kg, Fig. 21.6b) which significantly increased Ni concentration, all other treatments significantly lowered Ni concentrations in the shoot compared to the control. Similar to earlier observations, the increased Ni shoot uptake in the MFE may be associated with Ni availability during dissolution from the fraction that is complexed/coprecipitated with the root C/Mn. These observations are further confirmed by the significant relationships in Fig. 21.6b between the shoot Ni and C ($r = -0.80$, $P < 0.01$) and with shoot Mn ($r = +0.50$, $P < 0.05$).

Concentrations of Sb in the roots ranged from 7.73 to 8.68 mg/kg (Fig. 21.3a), with the lowest concentration observed in CFE-amended mine tailings and the highest in MFE-treated mine tailings. None of the treatments differed significantly from the control. Concentrations in the shoots ranged from 5.60 to 6.88 mg/kg (Fig. 21.3b), with higher accumulation in MFE-treated mine soil. The addition of CFE and MFE differed significantly from the control at $P < 0.05$. The concentration of Sb in the root was higher than in the shoot. Selenium levels in the roots ranged from 5.33 to 6.83 mg/kg (Fig. 21.3a), with the lowest quantity found in mine tailings amended with manure and the highest in MFE. None of the treatments differed significantly from the control. The concentrations in shoots varied between 3.15 and 4.11 mg/kg (Fig. 21.3b), with the highest accumulation in MFE-treated mine soil.

Molybdenum content in the root ranged from 4.32 to 5.01 mg/kg (Fig. 21.3a), with the lowest concentration observed in mine tailings treated with Com and the highest concentration in CFE-treated mine soil. Mo contents in shoots ranged from 3.84 to 6.03 mg/kg (Fig. 21.3b), with the highest content found in FeO-treated mine soils. There were no significant differences between the various treatments and the control.

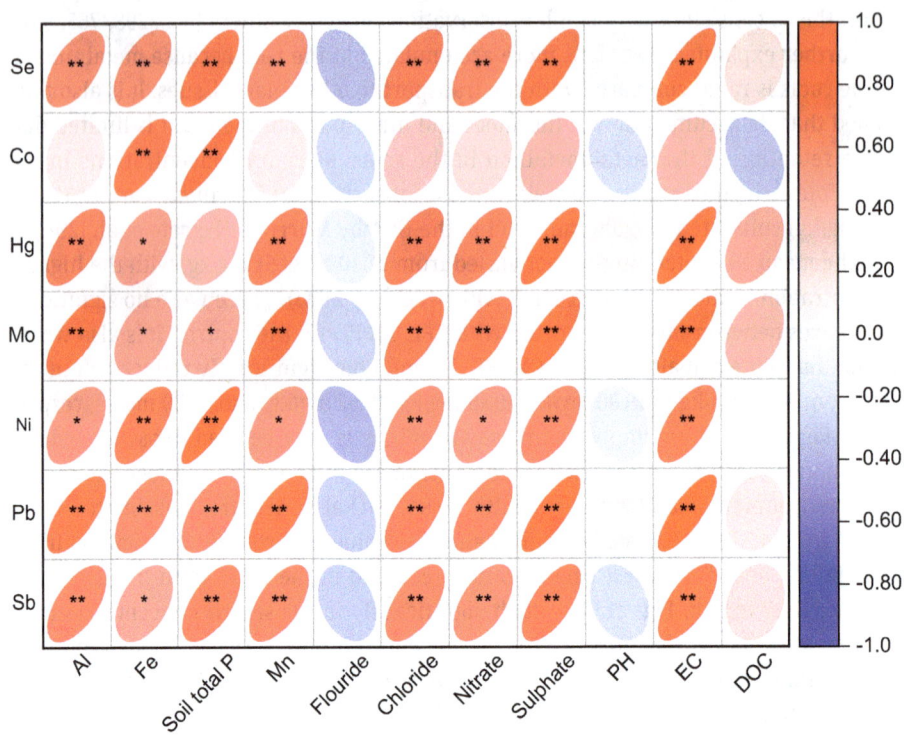

* p<=0.05 ** p<=0.01

**Fig. 21.5:** Correlation matrices between the studied elements concentrations in the soil against selected soil governing parameters.

## 21.4.5 Phytoremediation potential of ryegrass using bioconcentration factor (BCF) and translocation factor (TF)

After the end of the trial, the calculated BCF from Co ranged from 0.66 to 2.14; the highest BCF value was found in mine soil amended with MFE (BCF = 2.14, +125% relative to the control), and the least value was observed in the Com (see Fig. 21.4). All the other amended mine soils show BCF values less than one (BCF < 1) except MFE. For TF, none of the treatments differed significantly from the control and all values fell below one (Table 21.4). Also, for Hg, only MFE amended mine gave BCF greater than one (BCF > 1), and the rest of the amended mine tailings gave BCF values less than one. Moreover,

**Tab. 21.4:** Translocation factor of Co, Hg, Mo, Ni, Pb, Sb and Se from the plant root into the shoot.

| Treatment | Statistics | Co | Hg | Mo | Ni | Pb | Sb | Se |
|---|---|---|---|---|---|---|---|---|
| C | Mean | 0.19a | 0.76a | 0.82bc | 0.22cd | 0.48a | 0.69b | 0.45b |
| | SD | 0.03 | 0.02 | 0.12 | 0.04 | 0.02 | 0.07 | 0.03 |
| Com | Mean | 0.25a | 1.09a | 1.00b | 0.25bcd | 0.52a | 0.71b | 0.59ab |
| | SD | 0.03 | 0.55 | 0.11 | 0.02 | 0.02 | 0.02 | 0.09 |
| M | Mean | 0.28a | 0.55a | 0.50d | 0.35a | 0.66a | 0.71a | 0.59ab |
| | SD | 0.05 | 0.05 | 0.04 | 0.04 | 0.06 | 0.02 | 0.08 |
| FE | Mean | 0.25a | 0.73 | 1.43a | 0.32abc | 0.56a | 0.76b | 0.57ab |
| | SD | 0.03 | 0.06 | 0.10 | 0.06 | 0.02 | 0.05 | 0.05 |
| CFE | Mean | 0.23a | 0.76a | 0.37d | 0.21d | 0.88a | 0.93a | 0.71a |
| | SD | 0.04 | 0.14 | 0.06 | 0.05 | 0.17 | 0.08 | 0.08 |
| CM | Mean | 0.22a | 0.62a | 0.74c | 0.34ab | 0.64 | 0.77b | 0.65ab |
| | SD | 0.01 | 0.08 | 0.07 | 0.01 | 0.09 | 0.05 | 0.05 |
| MFE | Mean | 0.20a | 0.61a | 0.37d | 0.32abc | 0.72a | 0.80ab | 0.61ab |
| | SD | 0.06 | 0.13 | 0.04 | 0.02 | 0.08 | 0.02 | 0.11 |

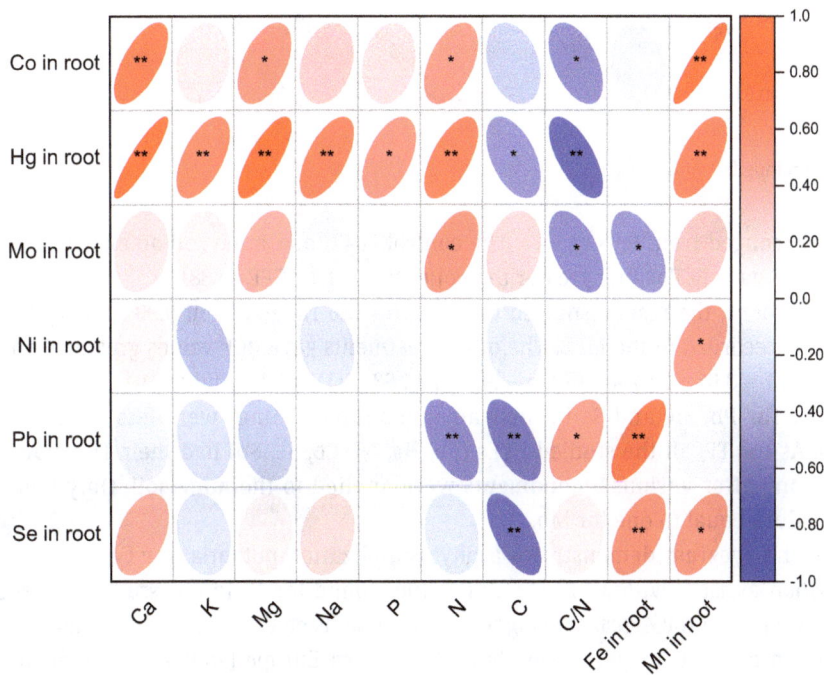

\* $p <= 0.05$ \*\* $p <= 0.01$

**Fig. 21.6:** Correlation matrices between the studied element concentrations in the (a) plant root against selected controlling plant root parameters and (b) plant shoot against selected plant shoot governing parameters.

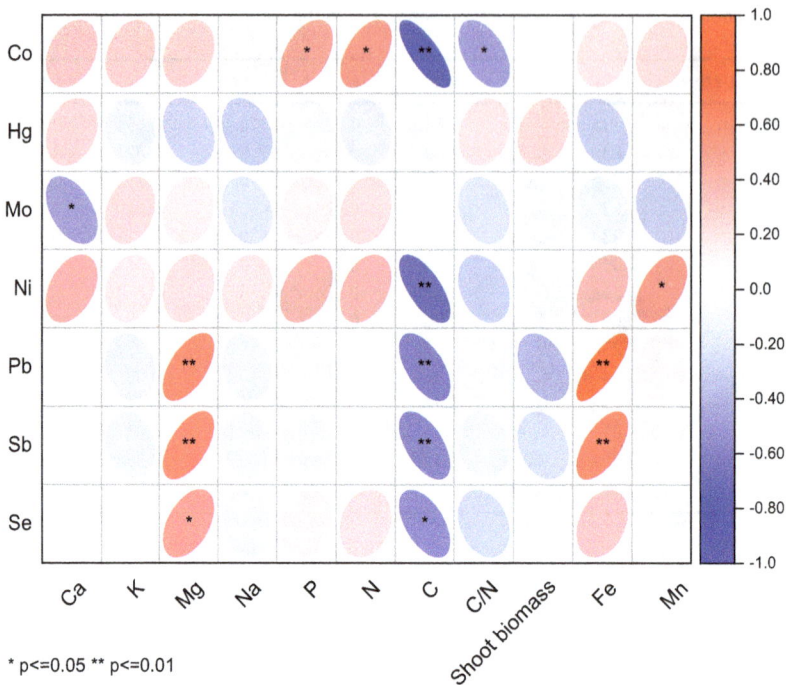

* p<=0.05 ** p<=0.01

**Fig. 21.6** (continued)

for Ni, BCF ranged from 0.63 to 1.47. The control (1.47) and mine soil amended with MFE and CFE gave BCF values greater than one (MFE = 1.42, CFE = 1.38).

Furthermore, the results showed that BCF for Mo ranged from 0.99 to 3.01; the highest was recorded in the MFE. The other treatments gave BCF values greater than one (i.e., MFE = 3.01, M = 2.44, CFE = 2.29, CM = 1.58, COM = 1.15, and C = 1.14). Finally, BCF values for Pb, Sb, and Se in their amended mine tailings were less than one (BCF < 1). As for TF, all the studied PTEs (Pb, Hg, Ni, Co, Se, Sb) had their TF values less than one after various amendments were applied to the mine soil. Only Com gave TF values equal to one for Mo.

Thus, the ryegrass demonstrates a phytostabilization potential for Co, Hg, Mo, and Ni when assisted with MFE. The BCF values found for Ni in our study are less than those reported for *Artemisia vulgaris* by Antoniadis et al. [38]. *Artemisia vulgaris* is a common herbaceous perennial plant of northern Europe [780], a heavy metal-tolerant species [946], and used as a medicinal herb for pain relief. Similar to our findings, Marrugo-Negrete et al. [298] reported Hg TF values for a 4-month remediation trial with *Jatropha curcas* to be below 1 and BCF values at the end of the 4-month trial to be above 1 (i.e., BCF = 1.43). The same study reported very low BCFs, mostly below 1; these values are way below those found in our study. Many previous studies have suggested that plant species with high BCF above 1 and TFs less than 1 could be considered suitable candidates for phytostabilization (e.g., Meng et al., [717, 915, 1083]). Additionally, plants regarded as suitable candidates for phytostabilization must limit the metal/metalloid accumulation to the shoot [914]. This process is achieved through siderophore secretion (phytochelatins), where the element Hg is easily incorporated due to its sulfur amino acid content (cysteine). This Hg is unavailable for translocation to the stems, resulting in high Hg concentrations in the roots [1082].

## 21.5 Conclusions

I investigated the soil contamination risks of Co, Hg, Mo, Ni, Pb, Sb, and Se in an abandoned gold mine soil in southwestern Ghana as well as the possibility of reclaiming the site using ryegrass-assisted phytoremediation. I conducted a 60-day pot experiment to see if applying soil amendments alone or in combination to boost phytoremediation success increases or decreases environmental risks. When assisted with MFE, ryegrass demonstrated phytostabilization potential for Co, Hg, Mo, and Ni and thus can be used to reduce their associated environmental and human health impacts. The addition of MFE increased uptake into the root or shoot via sorption on the manure/ Fe surfaces, dissolution by the action of rainfall or runoff or mineralization of the organic manure. The depth of the roots and whether they are lateral or deep/tap affect phytostabilization. Tap root systems are deeper and can provide a deeper search for nutrients and water while also assisting in the metal/metalloid amelioration. I propose a longer-term field study to investigate the effects of the treatments on PTE uptake. Furthermore, future research should consider the effects of flooding on mobilization of the studied PTEs as affected by the various treatments. Chemical speciation of the various studied PTEs in the amended mine soil should also be investigated.

Orlando Boafo and Albert Kobina Mensah

# Chapter 22
# Identifying research gaps for future reclamation studies

**Abstract:** This chapter highlights significant research gaps in the existing literature regarding land reclamation, soil quality, and potentially toxic elements (PTEs), underscoring the necessity for focused research to improve reclamation methodologies. The primary areas of focus are the long-term sustainability of reclaimed soils, the interactions between PTEs and soil microbiota, and the effectiveness of phytoremediation in various environments. The socioeconomic impacts of reclamation and the effects of land use changes following reclamation require further investigation. This research integrates ecological theories and frameworks to enhance understanding of reclamation processes, thereby informing more effective and sustainable land management strategies.

## 22.1 Significant gaps in current literature

The literature on land reclamation, soil quality, and potentially toxic elements (PTEs) is extensive and has significantly enhanced our understanding of managing and restoring degraded lands, especially those impacted by mining activities. Nonetheless, despite these advancements, significant gaps persist that hinder the effectiveness and sustainability of existing reclamation practices. The identified gaps indicate potential avenues for further research aimed at resolving outstanding challenges and enhancing the efficacy of land reclamation initiatives.

### 22.1.1 Long-term sustainability of reclaimed soils

Although numerous studies have investigated the short- to medium-term outcomes of land reclamation, there is a significant gap in research regarding the long-term sustainability of reclaimed soils. The persistence of soil quality improvements and the long-term behavior of PTEs in reclaimed soils are inadequately studied. Inquiries regarding the evolution of soil structure, fertility, and biological activity over decades,

**Orlando Boafo,** CSIR-College of Science and Technology, Soil Health and Environmental Resource Management Programme, Accra, Ghana
**Albert Kobina Mensah,** Council for Scientific and Industrial Research-Soil Research Institute, Kumasi, Ghana, e-mail: albert.mensah@rub.de, albertkobinamensah@gmail.com, ORCID: https://doi.org/0000-0001-5952-3357

https://doi.org/10.1515/9783111662046-022

alongside the migration or stabilization of PTEs within the soil profile, are essential for evaluating the genuine sustainability of reclamation initiatives. Longitudinal research on reclaimed sites is essential to address this gap and elucidate long-term outcomes [889].

### 22.1.2 Interactions between PTEs and soil microbiota

The interactions between PTEs and soil microbiota are intricate and remain inadequately elucidated. Soil microorganisms are essential for nutrient cycling, organic matter decomposition, and the immobilization or mobilization of PTEs. The mechanisms through which PTEs affect microbial communities, and the reciprocal influences, remain inadequately understood. Further investigation is required to examine the influence of various PTEs on soil microbial diversity, activity, and functions, as well as the subsequent effects of these changes on soil health and plant growth in reclaimed lands [831].

### 22.1.3 Effectiveness of phytoremediation across various environmental conditions

Phytoremediation has been extensively researched as a strategy for addressing PTE contamination in soils; however, the effectiveness of this method across various environmental conditions and soil types remains inadequately understood. Climate, soil pH, organic matter content, and the presence of competing plant species are factors that can influence the effectiveness of phytoremediation. Further research is required to evaluate phytoremediation strategies across diverse environmental contexts, especially in areas characterized by extreme weather conditions or distinct soil properties. Furthermore, research investigating the long-term efficacy of phytoremediation and the behavior of PTEs within plant biomass is essential to confirm this method as a viable sustainable remediation strategy [23].

### 22.1.4 Socioeconomic impacts of land reclamation

The literature has focused less on the socioeconomic aspects of land reclamation, such as the effects on local communities, economic development, and land use planning, in comparison to the environmental and technical dimensions. Analyzing the impact of land reclamation projects on livelihoods, property values, and community well-being is crucial for ensuring that these efforts achieve environmental success while also promoting social equity and economic viability. Further investigation is required to assess the social and economic implications of reclamation projects, especially in areas where mining operations have resulted in considerable social effects [917].

## 22.1.5 Impact of land use changes on reclaimed lands

The impact of post-reclamation land use changes on the stability and sustainability of reclaimed lands necessitates additional research. Land reclamation typically aims to restore land for specific purposes, including forestry or agriculture. However, subsequent alterations in land use, such as urban development or intensive farming, can significantly affect soil quality and the mobility of PTEs. Investigating the impact of various land use trajectories on the long-term results of reclamation projects is essential for informing land use planning and management in reclaimed regions [552].

# 22.2 Addressing research gaps via targeted studies

To address the identified gaps in the literature, targeted studies must focus on specific areas where knowledge is deficient. The studies should focus on delivering practical insights to enhance the effectiveness and sustainability of existing land reclamation practices.

## 22.2.1 Longitudinal studies on soil quality and the behavior of potentially toxic elements (PTEs)

Longitudinal studies that monitor soil quality and the behavior of PTEs over extended periods are essential to address the gap in understanding the long-term sustainability of reclaimed soils. Studies should monitor alterations in soil structure, fertility, and biological activity, along with the disposition of PTEs in both soil and plant biomass. These studies may include field experiments, laboratory analyses, and modeling to forecast long-term outcomes and determine optimal practices for maintaining soil health and stability [997].

### 22.2.1.1 Research on interactions between soil, potentially toxic elements, and microbiota

Additional research on the interactions between soil microbiota and PTEs is essential for the advancement of more effective remediation strategies. Research must concentrate on elucidating the responses of various microbial communities to PTE contamination, strategies for managing these communities to improve PTE immobilization or degradation, and the impact of microbial activity on the overall health of soil in reclaimed lands. Advancements in metagenomics and soil microbiome analysis offer novel insights into these intricate interactions [831].

### 22.2.1.2 Context-specific phytoremediation investigations

Research should prioritize context-specific studies to enhance the effectiveness of phytoremediation by testing various strategies across diverse environmental conditions. These studies should investigate the utilization of native and non-native plant species, the effects of soil amendments on phytoremediation efficacy, and the long-term stability of PTEs within plant biomass. Tailoring phytoremediation practices to specific environmental contexts enables the development of more reliable and scalable remediation techniques [23].

### 22.2.1.3 Socioeconomic research on the impacts of reclamation

Comprehensive studies are necessary to evaluate the socioeconomic impacts of land reclamation projects. These studies must assess the impact of reclamation on local economies, land values, employment opportunities, and community well-being. Furthermore, research should investigate the design of various reclamation approaches to optimize social benefits and address the needs and priorities of local communities. The integration of social science methodologies, including surveys and interviews, into reclamation research offers significant insights into the human aspects of land reclamation [917].

### 22.2.1.4 Studies on land use transitions following reclamation

The investigation of land use changes following reclamation is essential for comprehending their effects on the long-term efficacy of reclamation initiatives. Research should examine the impact of various land use practices, including urban development, intensive agriculture, and conservation, on soil quality, the mobility of PTEs, and ecosystem health in reclaimed areas. This research can guide land use planning and ensure that reclaimed lands are managed to sustain their environmental and economic value over time [552].

# 22.3 Areas necessitating additional enquiry

In addition to the identified research gaps, there are overarching domains within land reclamation and soil science that necessitate further exploration to enhance the discipline and refine existing methodologies.

## 22.3.1 Development of new soil amendments

There is a need for research focused on the development and testing of novel soil amendments that enhance soil quality and immobilize PTEs. Proposed amendments may encompass innovative materials like biochar, nanomaterials, or engineered clays, which have the potential to improve soil properties and decrease the bioavailability of PTEs. Research should concentrate on elucidating the mechanisms underlying these amendments, assessing their long-term efficacy, and evaluating their environmental consequences. Future scientific research should prioritize studies on enhanced weathering that utilize basalt, biochar co-compost, iron oxides, sewage sludge, bio-deposit, and activated carbon [909].

## 22.3.2 Climate change and reclamation

The influence of climate change on land reclamation initiatives is a developing area of concern that necessitates further investigation. Alterations in temperature, precipitation patterns, and extreme weather events may influence the efficacy of reclamation projects by modifying soil processes, vegetation dynamics, and the behavior of PTEs. Research should investigate the impact of climate change on reclamation outcomes and identify adaptive strategies to enhance the resilience of reclaimed lands in a changing climate [878].

## 22.3.3 Integration of remote sensing and GIS in reclamation monitoring

The use of remote sensing and geographic information systems (GIS) in the management and oversight of reclaimed lands presents significant opportunities for improvement. Research should focus on developing and applying remote sensing technologies to assess soil quality, vegetation cover, and the distribution of PTEs across landscapes. These systems provide real-time data that improve the efficiency and effectiveness of reclamation initiatives. Remote sensing and GIS can be employed to evaluate and track the advancement of revegetation initiatives. This may involve the use of drones and satellite imagery to monitor vegetation growth, evaluate the recovery of flora and fauna, analyze samples and data collected from diverse sampling sites, and assess erosion rates in restored watersheds [996].

## 22.3.4 Interdisciplinary approaches to reclamation

There is a necessity for increased interdisciplinary research that combines soil science, ecology, social sciences, and engineering to effectively tackle the intricate challenges associated with land reclamation. Collaborative research involving experts from diverse fields can enhance reclamation practices by integrating environmental, social, and economic considerations comprehensively [552].

## 22.3.5 Integrating theoretical frameworks into contemporary reclamation research

### 22.3.5.1 Ecological theories and soil science models serve as foundational elements

Integrating ecological theories and soil science models into contemporary reclamation studies is crucial for establishing a solid theoretical framework that elucidates the intricate relationships among land reclamation, soil quality, and PTEs. These frameworks will guide the interpretation of research findings and structure the study's approach to addressing identified research gaps [909].

### 22.3.5.2 Ecological succession theory

Ecological succession theory serves as a foundational framework for comprehending ecological recovery processes after disturbances, including those resulting from mining activities. This theory posits that ecosystems undergo a sequence of stages after a disturbance, transitioning from early successional communities characterized by pioneer species to more intricate and stable ecosystems over time. This theory is pertinent to land reclamation, as it establishes a foundation for forecasting the ecological recovery trajectory on reclaimed lands [878].

Ecological succession theory serves as a framework for selecting plant species in reclamation efforts, allowing for predictions regarding the establishment and interactions of various species over time. This also enhances the understanding of the evolution of soil quality, encompassing factors such as organic matter content and nutrient availability, throughout the progression of succession. Utilizing this theory enables researchers to more accurately predict the challenges and opportunities linked to various phases of ecosystem recovery, thereby informing more effective reclamation strategies [964].

## 22.3.6 Models of soil formation and pedogenesis

Soil formation models, including Jenny's State Factor Model, offer a framework for comprehending the development and transformation of soils over time, influenced by five primary factors: climate, organisms, relief (topography), parent material, and time. This model is effective for analyzing reclaimed soils that have experienced considerable disturbance and are currently undergoing processes of soil formation and modification. Integrating soil formation models into reclamation studies enhances the contextual understanding of changes in soil quality on reclaimed lands. The model elucidates the influence of climate and topography on the physical and chemical properties of reclaimed soils, as well as the role of biological processes, including plant root growth and microbial activity, in the evolution of soil structure and fertility over time. This theoretical framework facilitates a refined analysis of soil quality data and guides the formulation of reclamation strategies that correspond with natural soil formation processes [882].

### 22.3.6.1 Contaminant transport and fate models

Analyzing the behavior of PTEs in reclaimed soils necessitates the use of contaminant transport and fate models, which delineate the movement of contaminants within the soil environment and the various transformation processes they undergo. These models generally account for factors including advection, dispersion, sorption, and degradation, offering insights into the mobility, bioavailability, and persistence of PTEs in soils. The integration of contaminant transport and fate models within the literature review facilitates a comprehensive analysis of the behavior of PTEs under varying soil conditions and the impact of reclamation practices on their fate. These models can predict the potential for PTE leaching into groundwater, assess the effectiveness of soil amendments in immobilizing contaminants, and evaluate the long-term stability of PTEs in reclaimed soils. The application of these models enables the identification of the most effective strategies for managing PTEs and mitigating their associated environmental and health risks [997].

### 22.3.6.2 Ecosystem services framework

The ecosystem services framework offers a comprehensive view of the advantages ecosystems confer to humans, encompassing provisioning services (e.g., food and water), regulating services (e.g., climate regulation and water purification), and supporting services (e.g., nutrient cycling and soil formation). This framework is pertinent to land reclamation as it emphasizes the extensive environmental and societal advantages of rehabilitating degraded lands. Incorporating the ecosystem services framework into the literature review enables the study to assess the effectiveness of reclamation efforts, considering not only soil quality and PTE management but also the restoration of ecosystem functions and services. This approach highlights the significance of evaluating the long-term sustainability and multifaceted effects of reclamation projects, particularly their roles in biodiversity, climate resilience, and human well-being [794].

## 22.4 Theoretical frameworks pertinent to the understanding of land reclamation, soil quality, and potentially toxic elements (PTEs)

Multiple theoretical frameworks are essential for comprehending the connections among land reclamation, soil quality, and PTEs. The frameworks offer conceptual tools essential for analyzing the intricate interactions present in reclaimed ecosystems and for devising effective reclamation strategies.

### 22.4.1 The framework for disturbance and recovery

The disturbance and recovery framework is essential for analyzing ecosystem responses and recovery processes following significant disruptions, including mining activities. This framework examines the magnitude and frequency of disturbances, the resilience of the ecosystem, and the mechanisms of recovery. It evaluates soil degradation from mining, the success of reclamation efforts in facilitating recovery, and the prospects for long-term ecosystem stability [1037].

### 22.4.2 The framework for soil health

The soil health framework emphasizes the ability of soil to operate as a crucial living system that supports plants, animals, and humans. The assessment includes physical, chemical, and biological indicators of soil quality, such as soil structure, nutrient cycling, and microbial activity. This framework is significant for evaluating the effectiveness of land reclamation initiatives, as it offers a thorough assessment of soil recovery and the restoration of ecological functions. The incorporation of the soil health framework into reclamation research facilitates a comprehensive assessment of reclamation outcomes and underscores the significance of preserving soil health for sustainable land use [129].

### 22.4.2.1  Risk assessment frameworks for PTEs

Risk assessment frameworks for PTEs offer a systematic method for evaluating potential hazards linked to PTE contamination in soils. These frameworks generally include the identification of contamination sources, assessment of exposure pathways, evaluation of the toxicity of PTEs, and determination of the likelihood and severity of adverse effects on human health and the environment. Incorporating risk assessment frameworks into the research enables a systematic analysis of the risks associated with PTEs in reclaimed soils and identifies the most critical factors influencing these risks. This method guides the creation of specific remediation strategies and risk management practices. The USEPA [1030] proposed this framework for monitoring the effects of soil contamination through ingestion, dermal contact, and inhalation. The proposal includes models for assessing soil health in indoor environments, industrial locations, dust exposure, and soil contamination in children's school parks and playgrounds [1030].

### 22.4.2.2  Adaptive management framework

The adaptive management framework represents an iterative method for managing complex environmental systems, defined by continuous monitoring, evaluation, and modification of management practices in response to new information and evolving conditions. This framework is pertinent to land reclamation, as the effectiveness of reclamation initiatives can fluctuate based on site-specific conditions and the dynamic characteristics of ecosystems. The incorporation of the adaptive management framework into the literature review underscores the significance of flexibility and responsiveness in reclamation planning, facilitating the ongoing enhancement of management practices to attain long-term sustainability [868].

## 22.5 Conclusion

Despite notable advancements in land reclamation, several essential research gaps persist that must be addressed to enhance the effectiveness and sustainability of existing practices. The identified gaps encompass the necessity for long-term investigations into soil quality and the behavior of PTEs, an enhanced comprehension of soil–microbiota–PTE interactions, context-specific research on phytoremediation, and the assessment of socioeconomic impacts. Targeted and interdisciplinary research is essential for addressing these gaps, advancing the field, and ensuring the long-term success of land reclamation efforts ([997]; 917; [909]).

Albert Kobina Mensah

# Chapter 23
# General conclusions and summaries

The presence of contaminants in soils is often associated with a decrease in organic matter content, which can be attributed to the inhibitory effects of elevated metal concentrations, which negatively impact soil biological processes, hamper vegetation growth, and impede the synthesis of organic matter. Remediation is a remedial action by reducing or eliminating the transfer of metals and metalloids into the food chain to safeguard food quality and protect public health of humans.

Land or soil rehabilitation refers to the process of returning the ecosystem to its previous state, accompanied by the reintroduction of vegetation or the establishment of alternative land utilization on the rehabilitated landform. The potential impacts of mine wastes are diverse, including soil erosion, contamination of air and water, toxicity, geo-environmental disasters, reduction in biodiversity, and economic decline. In conclusion, the process of mining has a significant impact on the physical, chemical, and biological properties of soil, and the use of revegetation methods can help mitigate these effects.

The availability and quality of topsoil is essential for plant growth and building a sustainable ecosystem. Mining activities often extract or disrupt the topmost portion of soil, resulting in detrimental effects on its quality and composition. Reintroducing or replacing topsoil with essential nutrients and microbes is crucial for plant establishment and overall ecosystem rehabilitation. Selecting plant species for the restoration of degraded mine sites requires considering the economic and social benefits of these plants. Indigenous communities should be involved in the decision-making process, considering factors such as fast growth, shade provision, adaptability to local conditions, carbon sequestration capacity, firewood provision, and charcoal potential. This participatory approach promotes community ownership, increases project success, and fosters sustainable land management practices.

In conclusion, a holistic approach combining social, environmental, and economic factors is necessary for successful mining site restoration or rehabilitation. This approach should be a win-win situation for the environment and people, reducing adverse effects on the environment and improving the quality of life in nearby communities. Ultimately, the purpose of this book was to evaluate the impact of revegetation on the enhancement and restoration of soil quality in degraded mining areas. The underlisted summaries and conclusions have been derived:

**Albert Kobina Mensah,** Soil Research Institute, Council for Scientific and Industrial Research, Kumasi, Ghana, e-mail: albert.mensah@rub.de, albertkobinamensah@gmail.com, ORCID: https://orcid.org/0000-0001-5952-3357

https://doi.org/10.1515/9783111662046-023

- This book in the opening chapters provides a thorough examination and categorization of the various scientific approaches in the assessment of risks associated with PTEs, consolidating them into a single document.
- The aim was to provide a thorough examination of various approaches that can be employed to accurately detect and evaluate the degree of pollution in mining sites before implementing remediation and rehabilitation strategies.
- These methodologies included determination of element concentration, risk assessment indices, sequential extraction analyses, geospatial analyses, conducting redox chemistry experiments, speciation analysis using advanced techniques via synchrotron radiation science, conducting incubation experiments, utilizing pot experiments, and performing size fractionation experiments.
- Synchrotron radiation research is an advanced technique in dry chemistry that has been employed in investigations related to PTE, with particular emphasis on As speciation.
- There is a need to expand the utilization of synchrotron-based approaches in research and studies pertaining to the contamination of persistent toxic elements as well as in the broader context of reclaiming mining site.
- The process of mining leads to soil degradation by causing the destruction of soil structure, accelerating soil erosion, promoting excessive leaching, inducing compaction, reducing soil pH, accumulating heavy metals in soils, depleting organic matter, decreasing the availability of plant nutrients, reducing cation exchange capacity, and diminishing microbial activity.
- The findings in this book suggest that the implementation of revegetation using forest vegetation is an effective approach for the restoration of depleted soil fertility. This restoration is achieved through enhancements in soil organic matter content, availability of nutrients, cation exchange capacity, promotion of biological activities, and improvement in the physical properties of the soil.
- Based on the analysis provided in the book, it can be deduced that the implementation of revegetation techniques has the potential to enhance the soil quality of damaged mined lands.
- However, it is important to note that the process of restoring soil quality to levels comparable to the original state may necessitate a prolonged duration.
- Several tree species show promise for revegetation purposes including *Acacia*, *Leucaena*, and other legume trees. These species are known for their ability to tolerate acidic conditions and contribute a significant quantity of organic matter to the soil.
- The implementation of a mixed cropping system, which involves the combination of multiple species such as grasses and shrubs as cover crops, can offer a viable approach to addressing the challenges of erosion and soil degradation on mining sites. By utilizing grasses and shrubs as cover crops, a stable ground can be established, effectively mitigating erosion.

- Additionally, the inclusion of trees with tap roots that possess the ability to pene-trate deeper into the soil horizon further enhances the sustainability of this recla-mation method. This approach holds promise for the restoration of degraded and contaminated mining sites.
- Restoring degraded mine sites is a multidisciplinary effort that requires collabo-ration between various stakeholders including government agencies, environ-mental experts, local communities, and industry partners to achieve successful restoration as well as reclamation and long-term sustainability.

Thus, an abandoned mine field post-mining may present the following socio-environ-mental impacts (these lists are not exhaustive):

*Soil and water erosion*

*Generation of dusts*

*Air pollution*

*Soil contamination e.g. arsenic, mercury, copper, zinc, etc.*

*Loss of arable lands*

*Food chain contamination*

*Surface and groundwater pollution*

*Siltation of water bodies*

*Mine spillage, e.g. cyanide and mercury*

*Loss of aquatic biodiversity*

*Removal of vegetation*

*Health implications on women and children.*

*Spread of diseases, e.g. skin diseases, airborne diseases.*

*Economic impacts on the people as they spend huge sums of money in water treatment*

*Loss of livelihoods*

Mining companies can do the following to achieve sustainability for the community:

*Design and implementation of environmental management tools and practices (e.g. audits, EMS, reviews, etc.)*

*Installation of pollution abatement technology*

*Implementation of environmental management programs (e.g. training and edu-cation)*

*Reclamation of the mined field*

*Impact assessment and appraisals*

*Financing and construction of community infrastructure (e.g. water, electricity, roads, schools)*

*Development of important community programs*

*Creation and development of alternative livelihood programs (e.g. investments in youth and the local people)*

*Contribution to local universities and research units*

*Creation of relationships built on respect and trust. E.g. ensuring transparency over revenue flows are an important antidote to mistrust*

*Independent social audits can play a key role in mediation between the mine and the local community*

*Implementation of resettlement plans for affected communities.*

*Production of sustainability reports*

*Financing of road development*

*Construction of housing complexes*

*Financial donations to medical facilities and fire protection*

*Construction of educational facilities*

*Construction of recreational facilities, e.g. children playground*

*Providing students with scholarships*

*Building and establishing a mine training institute to accelerate training for natives*

*Establishing a small business development organisation to train local entrepreneurs for the business world*

*Working with government agencies to develop pre-school, adult literacy and other educational services*

Overall, running a soil pollution and remediation project may comprise the following stages depending on the project goals or target.

1. **Establishing and identifying the problem**

   This is the first and most important phase of any restoration/remediation project. This stage is very essential for outlining the project's scope and significance. This stage entails comprehensively analysing the pollution or degradation context, stating the impacts, and assessing the nature and magnitude of contamination/ degradation. At this stage, the participating project stakeholders have firmly established the problems driving the project's initiation. This may be done in the following ways.

   a) *Background to the project*: This point helps to elucidate the project's background and thus necessitates an examination of the pollutions or degradation history and its context. This phase focuses on characterising the category of contaminants; this includes groups such as heavy metals, hydrocarbons, or pesticides and their possible sources, either by natural occurrence or human-induced variables resulting from industrial effluent, agricultural runoff, or mining operations, as listed by [732]. Collecting this information helps determine the baseline parameters of the project areas for detailed planning and implementation.

   b) *Problem statement and justification:* The problem statement prior to the restoration project initiation is important because it clearly states the problems to be addressed. In other words, it presents the actual problem at the contaminated sites that may need or require redress. This may outline challenges associated with the environment, economy or society as a result of negative ramifications of the pollution. It explains the importance of soil health and the detrimental effects of contamination on ecosystems, food safety, and human health.

In this context, it is essential to clarify the necessity of a remediation project by outlining the reasons for intervention and detailing potential benefits, such as improved soil quality, reduced environmental hazards, and sustainable land use practices. Additionally, articulating the issue and providing a rationale for the necessity of the project or the need for addressing the problem helps to establish a clear context. This approach helps frame the issue and its suggested remedies within the broader landscape of scientific enquiry, highlighting the potential application of scientific methods to address the challenges presented.

c) *Project objectives and hypotheses:* This entails specifying precise, quantifiable goals for the project, such as reducing contaminant levels below standard limits or restoring soil fertility. According to [865], setting up a realistic goal is imperative for re-establishing ecological species and functional ecosystems. It is also important to state the restoration/remediation project hypotheses. This may offer a scientific framework for evaluating possible remedies, such as determining whether additions to soil amendments might mitigate toxicity in the contaminated sites or boost microbial activity for soil restoration.

d) *Do a desktop review and establish how the project will proceed.* This approach helps to choose the appropriate and suitable remediation option (s). A comprehensive review of existing literature in the form of articles, books, case studies, and remediation technologies is essential. This step helps determine the most appropriate gaps and cost-effective remediation strategies, such as bioremediation, chemical stabilisation, or soil washing. This may lead to developing an itinerary or roadmap for implementation to ensure that the project has structured strategies, including identifying key stages, stakeholders, and resources required.

This step facilitates the identification of completed works and the gaps that require further attention. It facilitates advance planning for remediation projects and fosters awareness of the strengths, weaknesses, opportunities, and threats related to the chosen remediation strategy. Furthermore, it is essential to assess and evaluate the advantages and disadvantages; identify successful cases that can be scaled; and recognise potential failures linked to a specific method that can be mitigated or avoided. Additionally, implement strategies proactively to mitigate potential environmental and human health impacts, along with their associated risks and uncertainties.

2. **Assessment of materials and methods to employ**

This stage has to involve identifying the most effective methodologies and instruments/equipment used for the remediation of soil pollutant contamination. It includes the processes of sample collection, site characteriszation, laboratory analysis, and risk assessment to guarantee that the selected methodologies are both effective and tailored to the specific site.

a) **Soil collection/sampling**

For the restoration project to be successful, methods of soil sampling must be well stated and taken into consideration in order to have a representative sample. It is therefore important to note that the implementation of a systematic sampling methodology is essential for the precise evaluation of contamination levels. Decisions must be made regarding the depth of sampling, specifically whether to concentrate on the surface (topsoil) or subsoil, as well as the selection of the sampling methodology, which may include stratified, random, or composite sampling techniques. For instance, composite sampling amalgamates several subsamples to yield a representative depiction of contamination levels. Thematic maps may be generated using geospatial tools such as ArcGIS (version 10.71), which effectively depict the spatial distribution of potentially toxic elements (PTEs) at the polluted site (see work on polluted soil by Mensah et al., [938]). These maps offer significant insights for the strategic targeting of remediation efforts.

b) **Soil/study area characteriszation**

Prior to initiation of the restoration/remediation project, it is advisable to characterise the soil type/social demographics in the project area to help in understanding the broader environmental and social context of the project. The very key parameters should include soil edaphic properties (e.g., pH, texture, organic matter content), the demographic profile of the community, and major economic activities like farming or mining. Data on this is fundamental for assessing how soil contamination/degradation impacts livelihoods and identifying community-specific remediation priorities (see [702] for more details).

c) **Laboratory analyses of collected samples and any soil amendments to use.**

Laboratory analysis of collected soil and amendment samples assists in restoration projects. Determining the types and levels of pollutants in collected samples is crucial. Many studies used analytical methods such as gas chromatography, atomic absorption spectroscopy (AAS), or inductively coupled plasma mass spectrometry (ICP-MS) to measure pollutants. Also, any soil amendments (like iron oxides, compost, lime, or biochar) must be evaluated to see if they are right for the job, and the right amounts must be found in order to get the best restoration.

d) **Risk assessment involves the identification and quantification of pollutants and types in the contaminated site.**

The process of risk assessment involves the identification and quantification of various pollutants and types present in the contaminated site. A comprehensive risk assessment determines the extent of pollution by identifying and measuring contaminants present at the location. This phase evaluates parameters such as the toxicity, mobility, and bioavailability of contaminants to assess possible consequences for human health, ecosystems, and agricultural productivity. Instruments like soil quality indices or risk quotients can serve this function. By comprehending the risks, the project can prioritise repair measures and establish definitive success criteria.

e) *Soil Amendment Trials and Feasibility Testing*

Soil amendment trials are an important initiative in remediation projects to correct treatment lapses before scaling up. Testing different soil amendments in small-scale trials ensures their efficacy for the specific site conditions. This step involves evaluating their ability to immobilise or degrade contaminants, improve soil health, and support vegetation growth. The trials provide critical data for scaling up the chosen solutions.

f) *Assessment of Socioeconomic Variables*

Remediation/restoration solutions and options should mostly correspond with the community's socioeconomic circumstances [1055]. Evaluating the financial impact on local stakeholders, prospective employment prospects, and enduring advantages guarantees the project's feasibility and sustainability. Community engagement at this stage promotes acceptability and guarantees that the solutions meet their particular needs ([1055], [819]).

3. **Impact assessment on environment and humans**

This phase assesses the wider ramifications of soil pollution by evaluating the environmental and human health concerns linked to potentially toxic elements (PTEs). It includes geochemical fractionation, bioavailability assessments, and pollutant speciation to deliver a thorough understanding of the effects.

a) **Quantification of geochemical fractions of the identified PTEs in the mine soils**

By means of quantifying the geochemical fractions of PTEs, one can ascertain their distribution among several chemical forms, including residual, carbonate-bound, or exchangeable fractions. Understanding the transportable and bioavailable nature of the pollutants under different environmental conditions depends on this investigation. For example, a 2018 work by Liu et al.

presented a sequential extraction technique extensively applied in the evaluation of metal partitioning in soils. These fractions might draw attention to certain pollutants that demand first attention and cause immediate hazards. Furthermore, geochemical fractionation may help predict long-term soil contamination behaviour.

b) **Bioavailability of the pollutants in environmental media**

Evaluating PTE bioavailability in soil, water, or air helps one understand how easily these pollutants might find their way into biological systems, including humans, animals, and plants. The soil pH, organic matter content, and microbial activity all affect bioavailability. For instance, studies by [732] and Mensah et al. (2022) stress the need to assess the bioavailable percentages of metals instead of total concentrations because they directly interact with ecological, biological, and health matters. Thus, drawing up remediation plans that reduce exposure routes depends on this characteriszation and assessment.

c) **Speciation of the dominant pollutants where necessary**
The main goal of speciation analysis is to find the chemical forms of major pollutants, like arsenic species (arsenate or arsenite) or mercury species (elemental mercury or methylmercury). Various species have various toxicity levels, motility, and environmental persistence; hence this stage is crucial. Often used for this aim are methods including high-performance liquid chromatography (HPLC) or X-ray absorption spectroscopy (XAS). A study on Molybdenum (Mo) by Yang & Wang (2025) emphasises the need for speciation in comprehending metal behaviour and developing actionable remediation strategies.

4. **Choosing the soil remediation method and implementation**
Based on site-specific conditions and the type of pollutants, the most suitable remediation procedures to address the contamination are selected and applied at this stage. Three main types of remediation approaches are physical, chemical, and biological. Every one of these options has its own set of benefits, drawbacks, and optimal uses.

a) **Physical method, e.g., excavation, and clean soil, either in situ or ex situ:**
Physical remediation approaches either remove or contain polluted soil. To physically remove contaminated soil and replace it with clean soil, either on-site (in situ) or off-site (ex situ), a common method is excavation. Sites with extremely concentrated pollution benefit from this method; however, it can be disruptive and expensive. Physical approaches are your best bet if you need to quickly remove heavy metal contamination. Khalil and Hassan [892] on their reviews on removing metals from the soil by physical means. On the other hand, hazardous waste facilities face significant limitations due to their high disposal and transportation costs.

b) **The chemical method involves the use of either organic or inorganic amendments:** Chemical remediation employs the use of soil amendments like lime, phosphate, or organic compounds (e.g., biochar and compost). These amendments help stabilise or neutralise pollutants. These strategies can halt the mobility and bioavailability of contaminants, thereby mitigating environmental and human health risks. Lime can immobilise heavy metals by elevating soil pH, but biochar improves adsorption and microbial activity. Research by Beesley et al. [68] revealed the effectiveness of biochar for metal stabilisation and soil health enhancement. This approach is economically and ecologically sustainable, particularly when combined with other remediation methods.

c) **Biological e.g., phytoremediation:** Phytoremediation is a form of biological remediation that uses plants to remove, stabilise, or break down pollutants ([948; 1000]). Hyperaccumulators are plants that can take in large amounts of heavy metals from the ground. Though it may take more time than physical or chemical procedures to accomplish the same goals, phytoremediation is better for the environment and looks good doing it. For example, studies by Mensah (2022) and Montreemuk et al. [948] show that phytoremediation is a good option for large areas with mild to moderate contamination.

d) **Use of soil amendments:** For this approach, decisions could be made about the choice of amendments to employ in remediating the contaminated sites based on their availability, cost, soil condition, and many other factors. This is particularly relevant in situations where assisted phytoremediation or revegetation are under consideration. Examples of soil amendments include biochar, compost, iron oxides, manure, zeolites, clays, biochar-co-compost, red mud, fly ash, biosolids, basalt, etc.

Using standard laboratory methods, initial chemical characteriszation of the soil and amendments can be carried out for soil parameters such as the pH; total nitrogen, carbon and sulphur; ash content (for biochar); total elements; PTEs; specific surface area; and characteriszation/functional groups on the amendments.

5. **Project appraisal:** This is essentially the crux of the project, determining whether the cleanup initiative will succeed or fail. It empowers planners to enhance strategies that will ensure project success and to implement corrective actions to mitigate the risk of project failure. This component will address the cost-benefit evaluations of the cleanup alternatives and execution. This may include the application and execution of the selected project and methodology, along with its potential for profit or loss.

Project appraisal consists of two primary components: project feasibility and viability. The feasibility analysis entails assessing whether the projects and their proposed plans are possible in terms of planned resources and available budgets. This important part of project evaluation makes planning easier by letting planners find possible risks early on, and which plans to be put in place to deal with risks and uncertainties. Feasibility appraisal assists in delineating project expenses, encompassing labour, materials, technology, and regulatory compliance, ensuring sufficient funding and preventing unforeseen contingencies. It evaluates environmental legislation and legal obligations, assuring adherence and averting potential penalties or delays. Project viability component involves assessing the likelihood of making profits by analysing anticipated costs, revenues, and market conditions. This aids in the prediction of the project's financial viability, enabling sponsors to evaluate the possible return on investment and prevent the misallocation of resources on potentially unprofitable ventures. The viability of a remediation aid evaluates the project's financial worthiness, assisting stakeholders in making educated judgements regarding future actions.

Project appraisal abets identify project risks and uncertainties. In this context, risks are anticipated obstacles that may emerge from the execution of the project. Potential risks may encompass environmental degradation resulting from construction activities, soil excavation, and the stockpiling or transportation of clean soil from locations other than the original sites. The introduction of alien species into remediated sites can be anticipated, and mitigating techniques can be preemptively devised. Uncertainties are unpredictable events that may occur during the execution of the undertaking. Initially—be it technological, regulatory, environmental, or financial. Soil remediation operations encounter considerable uncertainties, such as contamination, technology effectiveness, regulatory adherence, site conditions, data precision, environmental consequences, funding limitations, weather variability, stakeholder perceptions, long-term monitoring, legal liabilities, and technological progress. These uncertainties may escalate expenses, prolong timelines, and generate legal complications.

By doing feasibility and profitability analyses ahead of time and before full operation and implementation of the soil remediation project, project managers are able to assure companies that the exercise is done or executed with a solid understanding of the feasibility, the technicalities and the profit and loss component of the remediation projects. This holistic approach minimizes the chance of failure and increases the likelihood of delivering a successful and sustainable project. Planning, flexibility, and adaptive management practices that allow for project needs and styles to evolve in response to problems can help mitigate these difficulties, minimises incurring losses and maximising positive gains and profits. Further, regular stakeholder and community engagements can aid strengthen these efforts, help manage any risks and uncertainties and ultimately contributes to achieving sustainability of remediation efforts.

6. **Establishing field experimental trials: this can begin by first doing some microcosm trials in the lab and in the greenhouse.** Possibly doing microcosm studies in a lab to simulate or copy field trials and test the effectiveness of remediation methods in a controlled setting is the first step in setting up experiments. When it comes to cadmium and zinc mobility, for example, Cojocaru & Macoveanu used greenhouse and microcosm experiments to decontaminate polluted soil. Hartley and Lepp [201] and Mensah et al. (2024) studies showed that adding iron to plants could help stop them from taking in arsenic and other metals. For this, the results will enable researchers to monitor plant-soil interactions in semi-controlled conditions. Also, a field experiment demonstrated that compost and biochar were effective in stabilising potentially toxic metals in agricultural soils [853]. Researchers make sure remediation solutions are scientifically sound, eco-friendly, and applicable to large-scale projects by combining findings from various phases.

7. **Training and capacity building**: Training and capacity building are critical elements of soil remediation/restoration projects because they may contribute to long-term success and sustainability. This stage lists key stakeholders, such as local communities, researchers, and technical personnel, with the necessary knowledge and skills to effectively manage ecosystem restoration and sustainable land management projects and implement remediation strategies. Read more about the role of stakeholders under principle number one of the international standards for ecological restoration [832]. The study by Gann et al. [832] unequiv-

ocally asserts that stakeholders are empowered to positively or negatively impact a project, underscoring the importance of their engagement through capacity building.

Capacity-building initiatives include workshops, seminars, and practical training sessions aimed at educating participants on restoration strategies, including phytoremediation, soil amendment applications, and environmental monitoring (see [702]). Gann et al. 832 added that participatory approaches to training programmes enhance local ownership and facilitate the adoption of sustainable land management practices. These activities enable communities to assess soil health, control pollution sources, and sustain restored ecosystems over time.

Alongside technical training, public awareness campaigns can inform communities about the risks associated with soil contamination and the significance of remediation efforts. According to Young and FAO [819], these campaigns make sure that cleanup projects are in line with local priorities and get more people involved in the community. This shows how important stakeholder participation is for good environmental management.

Furthermore, purposeful training for scientists and policymakers may improve institutional capacity to plan and implement global restoration initiatives. For instance, programmes like the GRO Land Restoration Training Programmes in the UNESCO category II by the Icelandic government in the last decades have demonstrated effectiveness in developing expertise in ecosystem restoration and sustainable land management. So, spending money on building people's skills helps restoration and cleanup projects have long-lasting effects, giving people the tools they need to run restoration projects well.

8. **Project Beneficiaries, e.g., community engagements and periodic interaction with the community:** If we want restoration and reclamation efforts to have an effect and be inclusive, we need to find out who will benefit from the initiative and get them involved [832]. Project ownership and sustainability are both improved by community engagement since it helps to better understand local needs and encourages affected communities to participate in decision-making [702]. Regularly involving communities in activities like stakeholder meetings and participatory workshops can help build trust and make sure that remediation solutions are in line with local goals, like improving the health of the soil for farming or lowering health risks. FAO [819] highlights the importance of participatory approaches in involving beneficiaries in the standard of practice of the ecosystem restoration guild. It encourages mutual learning and gives communities agency over their own resource management. Remediation programmes can improve livelihoods, ecological health, and food security in the long run by addressing beneficiaries' individual concerns.

9. **Indicate and draw up time schedules, including duration and activities to be carried out in each time phase:** In order to effectively manage time and allocate resources, time plans are necessary for describing the order and length of project

activities. A well-structured timetable organizes tasks such as site assessment, remediation, monitoring, and reporting into distinct stages of the project and assigns specific due dates. Gantt charts and other project management software, such as Microsoft Project, can help visualize the schedule and make tracking easier. Precise schedules are necessary to monitor phytoremediation trials and ensure goal fulfillment within specific project schedules. Project completion on time, improved food security, and better accountability are all results of a well-defined timeline.

10. **Draw a budget for the project:** A project budget offers a comprehensive analysis of expected expenses, promoting financial clarity and efficient resource distribution. Essential budget categories may include personnel salaries, equipment and materials, transportation, laboratory analyses, and community engagement activities (Mirsal, [944]; [1021]). Therefore, it is important to allocate an adequate budget for soil amendments, such as iron or biochar, in restoration or remediation trials. Allocate contingency funds for the sustainability of the project.

11. **Identify the project team, their expertise and responsibilities:** For every restoration project to be successful, an assemblage of a multidisciplinary team with expertise in fields such as soil science, environmental engineering, toxicity, and community involvement is crucial for the project. It's important to clarify each expert's role in the project plan to avoid confusion and conflict. For instance, a soil scientist might evaluate levels of degradation, disturbances, and contamination and recommend changes while a project or environmental engineer draws up a plan or project drawings to be followed.

As indicated by [941] and [954], community project liaison personnel would guarantee efficient connection with local people to handle their issues and goals. Clearly identifying roles guarantees effective teamwork and reduces project execution's overlaps and duplications [954]. Consequently, the meticulous selection of project team members with explicitly defined roles and duties enhances the project's long-term viability.

In summary, executing a soil remediation project necessitates expertise from diverse disciplines, including soil scientists, geographers, soil chemists, public health scientists, agricultural economists, agricultural extensionists, crop scientists, agricultural engineers, biological scientists, environmental scientists, project managers, and others. A varied team focused on objectives, complemented by a public relations officer to share findings and engage the media, will ultimately enhance the success of remediation and restoration activities.

# References

[1]     Abbaslou, H. & Bakhtiari, S. (2017). Phytoremediation potential of heavy metals by two native pasture plants (Eucalyptus grandis and Ailanthus altissima) assisted with AMF and fibrous minerals in contaminated mining regions.

[2]     Abissath, M. (2017). Galamsey, land reclamation project – don't ignore the Ghanaian media. All Africa global media. Retrieved from https://allafrica.com/stories/201711231031.html

[3]     Abreu, M. M., Tavares, M. T. & Batista, M. J. (2008). Potential use of Erica andevalensis and Erica australis in phytoremediation of sulphide mine environments: São Domingos, Portugal. Journal of Geochemical Exploration, 96(2–3), 210–222.

[4]     Acosta, J. A., Abbaspour, A., Martínez, G. R., Martínez-Martínez, S., Zornoza, R., Gabarrón, M. & Faz, A. (2018). Phytoremediation of mine tailings with Atriplex halimus and organic/inorganic amendments: A five-year field case study. Chemosphere, 204, 71–78.

[5]     Acosta, J. A., Arocena, J. M. & Faz, A. (2015). Speciation of arsenic in bulk and rhizosphere soils from artisanal cooperative mines in Bolivia. Chemosphere, 138, 1014–1020. https://doi.org/10.1016/j.chemosphere.2014.12.050.

[6]     Addai, P., Mensah, A. K., Sekyi-Annan, E. & Adjei, E. O. (2023). Biochar, compost and/or NPK fertilizer affect the uptake of potentially toxic elements and promote the yield of lettuce grown in an abandoned gold mine tailing. Journal of Trace Elements and Minerals, 4, 100066.

[7]     Addo-Fordjour, P., Obeng, S., Anning, A. K. & Addo, M. G. (2013). Floristic composition and structure of a disturbed dry semi-deciduous forest in Ghana. International Journal of Biodiversity and Conservation, 5(2), 78–88. https://doi.org/10.5897/IJBC12.097.

[8]     Adelesi, O. O. (2019). Economic Analysis of Small Holder Aquaculture Farmers: The Case of Nigeria.

[9]     Adriano, D. C. (2001). Trace Elements in Terrestrial Environments: Biogeochemistry, Bioavailability, and Risks of Metals (2nd ed.). Springer, https://doi.org/10.1007/978-1-4612-2162-4.

[10]    Adu-Baffour, F., Daum, T., Obeng, E. A., Birner, R. & Bosch, C. (2023). Making land rehabilitation projects work in small-scale mining areas: Insights from a case study in Ghana Hohenheim working. Papers on Social and Institutional Change in Agricultural Development, https://490c.uni-hohenheim.de/en/75736.

[11]    Adu-Baffour, F., Daum, T. & Birner, R. (2021). Governance challenges of small-scale gold mining in Ghana: Insights from a process net-map study. Land Use Policy, 102, 105271.

[12]    Adu-Baffour, F., Daum, T., Asantewaa Obeng, E., Birner, R. & Bosch, C. (2023). Making Land Rehabilitation Projects Work in Small-Scale Mining Areas: Insights from a Case Study in Ghana (pp. 015–2023). Stuttgart.

[13]    Afonso, T. F., Demarco, C. F., Pieniz, S., Camargo, F. A., Quadro, M. S. & Andreazza, R. (2019). Potential of Solanum viarum Dunal in use for phytoremediation of heavy metals to mining areas, southern Brazil. Environmental Science and Pollution Research, 26, 24132–24142.

[14]    Agbotui, P., Anornu, G., Agbotui, T., Gyabaah, F., Amankwah-Minkah, A., Brookman-Amissah, M., Blankson-Darku, D. & Sallah, J. (2021). Risk-based contaminated land management policy mindset: A way out for Ghana's environmental challenges. African Geographical Review, 00(00), 1–14. https://doi.org/10.1080/19376812.2021.1875853.

[15]    Akabzaa, T. & Darimani, A. (2001). Impact of mining sector investment in Ghana: A study of the Tarkwa mining region (A draft report) prepared by for Sapri. 47–61.

[16]    Akoto, O. & Adiyiah, J. (2015). Heavy metal pollution in surface soils in the vicinity of abundant railway servicing workshop in Kumasi, Ghana. International Journal of Environmental Research and Public Health, 12(10), 12692–12702. https://doi.org/10.3390/ijerph121012692.

[17]    Akoto, O., Bruce, T. N. & Darko, G. (2014). Heavy metals pollution profiles in streams serving the Owabi reservoir. Environmental Monitoring and Assessment, 186(5), 3573–3580. https://doi.org/10.1007/s10661-014-3635-5.

https://doi.org/10.1515/9783111662046-024

[18]    Al-Abed, S. R., Jegadeesan, G., Purandare, J. & Allen, D. (2007). Arsenic release from iron rich mineral processing waste: Influence of pH and redox potential. Chemosphere, 66(4), 775–782.

[19]    Alaboudi, K. A., Ahmed, B. & Brodie, G. (2018). Phytoremediation of Pb and Cd contaminated soils by using sunflower (Helianthus annuus) plant. Annals of agricultural sciences, 63(1), 123–127.

[20]    Mensah, A. K, Shaheen, S. M., Rinklebe, J. & Stefanie Heinze, B. M. (2022). Phytoavailability and uptake of arsenic in ryegrass affected by various amendments in soil of an abandoned gold mining site. Environmental Research, 214(July), 113729. https://doi.org/10.1016/j.envres.2022.113729.

[21]    Albert, K. M. (2015). Role of revegetation in restoring fertility of degraded mined soils in Ghana: A review. International Journal of Biodiversity and Conservation, 7(2), 57–80. https://doi.org/10.5897/ijbc2014.0775.

[22]    Alexander, M. J. (1990). Reclamation after tin mining on the Jos plateau, Nigeria. Geographical J, 156, 44–50.

[23]    Ali, H., Khan, E. & Sajad, M. A. (2013). Phytoremediation of heavy metals – concepts and applications. Chemosphere, 91(7), 869–881. https://doi.org/10.1016/j.chemosphere.2013.01.075.

[24]    Ali, H., Khan, E. & Anwar Sajad, M. (2013). Phytoremediation of heavy metals-concepts and applications. Chemosphere, 91(7), 869–881.

[25]    Alloway, B. J. (2013). Heavy Metals in Soils: Trace Metals and Metalloids in Soils and their Bioavailability (3rd ed.). Springer, https://doi.org/10.1007/978-94-007-4470-7.

[26]    Alvarenga, P., Gonçalves, A. P., Fernandes, R. M., De varennes, A., Vallini, G., Duarte, E. & Cunha-Queda, A. C. (2008). Evaluation of composts and liming materials in the phytostabilization of a mine soil using perennial ryegrass. Science of the Total Environment, 406(1–2), 43–56. https://doi.org/10.1016/j.scitotenv.2008.07.061.

[27]    Álvarez-Mateos, P., Alés-Álvarez, F. J. & García-Martín, J. F. (2019). Phytoremediation of highly contaminated mining soils by Jatropha curcas L. and production of catalytic carbons from the generated biomass. Journal of environmental management, 231, 886–895.

[28]    An, J., Jeong, B. & Nam, K. (2019). Evaluation of the effectiveness of in situ stabilization in the field aged arsenic-contaminated soil: Chemical extractability and biological response. Journal of Hazardous Materials, 367, 137–143.

[29]    Anaeto, E., Ohajianya, O., Korie, D. & P, U. (2016). Determinants of farmers' use of land reclamation practices in oil producing areas of Imo state, Nigeria: Application of multivariate logistic analysis. Review of Plant Studies, 3(2), 7–12. https://doi.org/10.18488/journal.69.2016.32.7.12.

[30]    Andreazza, R., Bortolon, L., Pieniz, S., Barcelos, A. A., Quadro, M. S. & Camargo, F. A. (2015). Phytoremediation of vineyard copper-contaminated soil and copper mining waste by a high potential bioenergy crop (Helianthus annus L.). Journal of plant nutrition, 38(10), 1580–1594.

[31]    Andreazza, R., Bortolon, L., Pieniz, S., Bento, F. M. & Camargo, F. A. D. O. (2015). Evaluation of two Brazilian indigenous plants for phytostabilization and phytoremediation of copper-contaminated soils. Brazilian Journal of Biology, 75, 868–877.

[32]    Anning, A. K. & Akoto, R. (2018). Assisted phytoremediation of heavy metal contaminated soil from a mined site with Typha latifolia and Chrysopogon zizanioides. Ecotoxicology and Environmental Safety, 148, 97–104.

[33]    Anning, A. K., Korsah, P. E. & Addo-Fordjour, P. (2013). Phytoremediation of wastewater with Limnocharis flava, Thalia geniculata and Typha latifolia in constructed wetlands. International Journal of Phytoremediation, 15(5), 452–464.

[34]    Ansah, S. (2016). Evaluation of soil quality and heavy metals in Manihot esculenta and Musa paradisiaca grown on reclaimed mined lands in the Bogoso/Prestea area, Ghana. A thesis submitted to the department of environmental science college of science. Partial Fulfilment of the Requirements for the Award of Degree of Master of Science in Environmental Science.

[35]  Antoniadis, V., Golia, E. E., Liu, Y. T., Wang, S. L., Shaheen, S. M. & Rinklebe, J. (2019). Soil and maize contamination by trace elements and associated health risk assessment in the industrial area of Volos, Greece. Environment International, 124, 79–88. https://doi.org/10.1016/j.envint.2018.12.053.

[36]  Antoniadis, V., Levizou, E., Shaheen, S. M., Ok, Y. S., Sebastian, A., Baum, C., Prasad, M. N. V., Wenzel, W. W. & Rinklebe, J. (2017a). Trace elements in the soil-plant interface: Phytoavailability, translocation, and phytoremediation – a review. Earth-Science Reviews, 171(June), 621–645. https://doi.org/10.1016/j.earscirev.2017.06.005.

[37]  Antoniadis, V., Shaheen, S. M., Boersch, J., Frohne, T., Du Laing, G. & Rinklebe, J. (2017b). Bioavailability and risk assessment of potentially toxic elements in garden edible vegetables and soils around a highly contaminated former mining area in Germany. Journal of Environmental Management, 186, 192–200. https://doi.org/10.1016/j.jenvman.2016.04.036.

[38]  Antoniadis, V., Shaheen, S. M., Stärk, H. J., Wennrich, R., Levizou, E., Merbach, I. & Rinklebe, J. (2021). Phytoremediation potential of twelve wild plant species for toxic elements in a contaminated soil. Environmental International, 146, https://doi.org/10.1016/j.envint.2020.106233.

[39]  Appiah, M. K. (2018). Promoting small-scale mining sector businesses and the role of institutions-a conflict prevention and resolution study in Ghana. Zepplin University.

[40]  Aragon, F. M. & Rud, J. P. (2013). Modern industries, pollution and agricultural productivity: Evidence from Ghana. Draft: International Growth Centre, 44(March), 1–51. Retrieved from http://www.theigc.org/sites/default/files/Aragon and Rud final.pdf.

[41]  Armah, F. A. & Gyeabour, E. K. (2013). Health risks to children and adults residing in riverine environments where surficial sediments contain metals generated by active gold mining in Ghana. Toxicological Research, 29(1), 69–79. https://doi.org/10.5487/TR.2013.29.1.069.

[42]  Armah, F. A., Obiri, S., Yawson, D. O., Afrifa, K. A., Yengoh, G. T. & Olsson, J. A. (2011). Journal of environmental planning and assessment of legal framework for corporate environmental behaviour and perceptions of residents in mining communities in Ghana. Journal of Environmental Planning and Management, 54(2), 193–209. https://doi.org/10.1080/09640568.2010.505818.

[43]  Armah, F. A., Quansah, R. & Luginaah, I. (2014). A systematic review of heavy metals of anthropogenic origin in environmental media and biota in the context of gold mining in Ghana. International Scholarly Research Notices, 2014, 1–37. https://doi.org/10.1155/2014/252148.

[44]  Aryee, B. N. A., Ntibery, B. K. & Atorkui, E. (2003). Trends in the small-scale mining of precious minerals in Ghana: A perspective on its environmental impact. Journal of Cleaner Production, 11(2), 131–140. https://doi.org/10.1016/S0959-6526(02)00043-4.

[45]  Asante, F. A. & Sasu, F. M. (2018). Ecological restoration of degraded landscapes: A case study from Ghana. Environmental Science & Policy, 88, 1–12. https://doi.org/10.1016/j.envsci.2018.06.005.

[46]  Asante, K. A., Agusa, T., Subramanian, A., Ansa-Asare, O. D., Biney, C. A. & Tanabe, S. (2013). Contamination status of arsenic and other trace elements in drinking water and residents from Tarkwa, a historic mining township in Ghana. Chemosphere, 88(4), 377–382. https://doi.org/10.1016/j.chemosphere.2012.03.077.

[47]  Asare, R. & Darko, J. A. (2016). Small-scale mining and land rehabilitation in Ghana: A review. Journal of Mining and Environment, 7(4), 707–721.

[48]  Asare, R., Mantey, S. & Darko, J. A. (2018). Reclamation practices at small-scale mining sites in Ghana: A case study. Journal of Sustainable Mining, 17(4), 206–213. https://doi.org/10.1016/j.jsm.2018.12.001.

[49]  Asase, A. & Tetteh, D. A. (2010). The role of sacred groves in biodiversity conservation in Ghana. Biodiversity and Conservation, 19(1), 299–313. https://doi.org/10.1007/s10531-009-9711-5.

[50]  Asiedu, J. B. (2013). Technical report on reclamation of small-scale surface mined lands in Ghana: A landscape perspective. American Journal of Environmental Protection, 1(2), 28–33. https://doi.org/10.12691/env-1-2-3.

[51] Awotwi, A., Anornu, G. K., Quaye-Ballard, J. A. & Annor, T. (2018). Monitoring land use and land cover changes due to extensive gold mining, urban expansion, and agriculture in the Pra River Basin of Ghana, 1986–2025. Land degradation & development, 29(10), 3331–3343.

[52] Ayelazuno, J. A. & Mawuko-yevugah, L. (2019). Large-scale mining and ecological imperialism in Africa: The politics of mining and conservation of the ecology in Ghana. Journal of Political Ecology, 26(June), 244–262. https://doi.org/10.2458/v26i1.22962.

[53] Ayhan, H. Ö. & Işiksal, S. (2004). Memory recall errors in retrospective surveys: A reverse record check study. Quality and Quantity, 38(5), 475–493. https://doi.org/10.1007/s11135-005-2643-7.

[54] Baath, E. & Anderson, T. H. (2003). Comparison of soil fungal/bacterial ratios in a pH gradient using physiological and PLFA-based techniques. Soil Biology and Biochemistry, 35(7), 955–963. https://doi.org/10.1016/S0038-0717(03)00154-8.

[55] Banerjee, R., Goswami, P., Lavania, S., Mukherjee, A. & Lavania, U. C. (2019). Vetiver grass is a potential candidate for phytoremediation of iron ore mine spoil dumps. Ecological Engineering, 132, 120–136.

[56] Bank, M. S., Rinklebe, J., Feng, X., Xu, X. & Lin, C. (2019). Science of the total environment mercury cycling and bioaccumulation in a changing environment. Science of the Total Environment, 670, 345. https://doi.org/10.1016/j.scitotenv.2019.03.271.

[57] Bansah, K. J. & Addo, W. K. (2016). Phytoremediation potential of plants grown on reclaimed spoil lands. Ghana Mining Journal, 16(1), 68–75.

[58] Bansah, K. J., Dumakor-dupey, N. K., Stemn, E. & Galecki, G. (2018). Mutualism, commensalism, or parasitism? Perspectives on tailings trade between large-scale and artisanal and small-scale gold mining in Ghana. Resources Policy, 57(March), 246–254. https://doi.org/10.1016/j.resourpol.2018.03.010.

[59] Bansah, K. J., Sakyi-Addo, G. B. & Dumakor-Dupey, N. (2016). Post-mining reclamation in artisanal and small-scale mining. ICANM2016 Proceedings (August 1–3, 2016, Montreal, Canada), (2010), 172–179. https://doi.org/10.13140/RG.2.1.2559.7685

[60] Barbafieri, M., Dadea, C., Tassi, E., Bretzel, F. & Fanfani, L. (2011). Uptake of heavy metals by native species growing in a mining area in Sardinia, Italy: Discovering native flora for phytoremediation. International Journal of Phytoremediation, 13(10), 985–997.

[61] Barenblitt, A., Payton, A., Lagomasino, D., Fatoyinbo, L., Asare, K., Aidoo, K., Pigott, H., Kofi Som, C., Smeets, L., Seidu, O. & Wood, D. (2021). The large footprint of small-scale artisanal gold mining in Ghana. Science of the Total Environment, 781, 146644.

[62] Bartsch, C. & Wells, B. (2005). State Brownfield Financing Tools and Strategies.

[63] Baysal, A., Ozbek, N. & Akm, S. (2013). Determination of trace metals in wastewater and their removal processes. Wastewater – Treatment Technologies and Recent Analytical Developments, https://doi.org/10.5772/52025.

[64] Beans, C. (2017). Phytoremediation advances in the lab but lags in the field. Proceedings of the National Academy of Sciences of the United States of America, 114(29), 7475–7477.

[65] Bech, J., Duran, P., Roca, N., Poma, W., Sánchez, I., Roca-Pérez, L. . . . Poschenrieder, C. (2012). Accumulation of Pb and Zn in Bidens triplinervia and Senecio sp. spontaneous species from mine spoils in Peru and their potential use in phytoremediation. Journal of Geochemical Exploration, 123, 109–113.

[66] Beckers, F., Mothes, S., Abrigata, J., Zhao, J., Gao, Y. & Rinklebe, J. (2019). Mobilization of mercury species under dynamic laboratory redox conditions in a contaminated floodplain soil as affected by biochar and sugar beet factory lime. Science of the Total Environment, 672, 604–617. https://doi.org/10.1016/j.scitotenv.2019.03.401.

[67] Beesley, L., Marmiroli, M., Pagano, L., Pigoni, V., Fellet, G., Fresno, T., Vamerali, T., Bandiera, M. & Marmiroli, N. (2013). Biochar addition to an arsenic contaminated soil increases arsenic concentrations in the pore water but reduces uptake to tomato plants (Solanum lycopersicum L.).

Science of the Total Environment, 454–455, 598–603. https://doi.org/10.1016/j.scitotenv.2013. 02.047.

[68]  Beesley, L., Moreno-Jiménez, E., Gomez-Eyles, J. L., Harris, E., Robinson, B. & Sizmur, T. (2011). A review of biochars' potential role in the remediation, revegetation and restoration of contaminated soils. In Environmental Pollution (Vol. 159, Issue 12, pp. 3269–3282). Elsevier Ltd, https://doi.org/10.1016/j.envpol.2011.07.023.

[69]  Beiyuan, J., Awad, Y. M., Beckers, F., Tsang, D. C. W., Ok, Y. S. & Rinklebe, J. (2017). Mobility and phytoavailability of As and Pb in a contaminated soil using pine sawdust biochar under systematic change of redox conditions. Chemosphere, 178, 110–118. https://doi.org/10.1016/j.chemosphere. 2017.03.022.

[70]  Beiyuan, J., Awad, Y. M., Beckers, F., Wang, J., Tsang, D. C. W., Ok, Y. S., Wang, S. L., Wang, H. & Rinklebe, J. (2020). (Im)mobilization and speciation of lead under dynamic redox conditions in a contaminated soil amended with pine sawdust biochar. Environment International, 135 (September 2019), 105376. https://doi.org/10.1016/j.envint.2019.105376.

[71]  Bell, A., Ward, P., Tamal, M. E. H. & Killilea, M. (2019). Assessing recall bias and measurement error in high-frequency social data collection for human-environment research. Population and Environment, 40(3), 325–345. https://doi.org/10.1007/s11111-019-0314-1.

[72]  Bempah, C. K., Ewusi, A., Obiri-yeboah, S., Asabere, B., Mensah, F., Boateng, J. & Voigt, H. (2013). Distribution of arsenic and heavy metals from mine tailings dams at Obuasi municipality of Ghana. American Journal of Engineering Research, 2(05), 61–70.

[73]  Bempah, C. K., Ewusi, A., Obiri-Yeboah, S., Asabere, S. B., Mensah, F., Boateng, J. & Hans-Jürgen, V. (2013). Distribution of arsenic and heavy metals from mine tailings dams at Obuasi municipality of Ghana. American Journal of Engineering Research, 2(5), 61–70.

[74]  BenDor, T., Lester, T. W., Livengood, A., Davis, A. & Yonavjak, L. (2015). Estimating the size and impact of the ecological restoration economy. PLOS ONE, 10(6), e0128339. https://doi.org/10.1371/ journal.pone.0128339.

[75]  Bharti, S. & Banerjee, T. K. (2012). Phytoremediation of the coalmine effluent. Ecotoxicology and Environmental Safety, 81, 36–42.

[76]  Bhattacharyya, R., Chatterjee, D., Nath, B., Jana, J., Jacks, G. & Vahter, M. (2003). High arsenic groundwater: Mobilization, metabolism and mitigation – an overview in the Bengal delta plain. Molecular and Cellular Biochemistry, 253, 347–355.

[77]  Bhawan, P. 2011. A State-of-the-Art Report on Bioremediation, Its Application to Contaminated Sites in India. (Vol. 33).

[78]  Bian, F., Zhong, Z., Chengzhe, L., Zhang, X., Lijian, G., Huang, Z., Gai, X. & Huang, Z. (2021). Intercropping improves heavy metal phytoremediation efficiency through changing properties of rhizosphere soil in bamboo plantation. Journal of Hazardous Materials, 416(January), 125898.

[79]  Birner, R. & Poku, A.-G. (2017). Governance of emerging biomass-based value webs in Africa: Case studies from Ghana.

[80]  Bisaglia, M. & Bubacco, L. (2020). Copper ions and Parkinson's disease: Why is homeostasis so relevant?. Biomolecules, 10(2), https://doi.org/10.3390/biom10020195.

[81]  Bissen, M., Frimmel, F. H. & Ag, C. (2003). Arsenic – a review. Part I: Occurrence, toxicity, speciation, mobility. Acta Hydrochimica et Hydrobiologica, 31(1), 9–18.

[82]  Blay, D., Appiah, M., Damnyag, L., Dwomoh, F. K., Luukkanen, O. & Pappinen, A. (2008). Involving local farmers in rehabilitation of degraded tropical forests: Some lessons from Ghana. Environment, Development and Sustainability, 10(4), 503–518. https://doi.org/10.1007/s10668-006- 9077-9.

[83]  Boateng, E., Ntow, W. J. & Acquah, S. O. (2014). Reclamation strategies for abandoned small-scale gold mining sites in the southwestern region of Ghana. International Journal of Environmental Science and Development, 5(5), 458–464. https://doi.org/10.7763/IJESD.2014.V5.523.

[84]    Boateng, T. K., Opoku, F., Acquaah, S. O. & Akoto, O. (2017). Heavy metal contamination assessment of groundwater quality: A case study of Oti landfill site, Kumasi. Environmental Monitoring and Assessment, 189(9), 413. https://doi.org/10.1007/s10661-017-6140-4.

[85]    Bolan, N., Kunhikrishnan, A., Thangarajan, R., Kumpiene, J., Park, J., Makino, T., Kirkham, M. B. & Scheckel, K. (2014). Remediation of heavy metal(loid)s contaminated soils – To mobilize or to immobilize?. Journal of Hazardous Materials, 266, 141–166. https://doi.org/10.1016/j.jhazmat.2013. 12.018.

[86]    Bortey-Sam, N., Nakayama, S. M. M., Ikenaka, Y., Akoto, O., Baidoo, E., Yohannes, Y. B., Mizukawa, H. & Ishizuka, M. (2015). Human health risks from metals and metalloid via consumption of food animals near gold mines in Tarkwa, Ghana: Estimation of the daily intakes and target hazard quotients (THQs). Ecotoxicology and Environmental Safety, 111, 160–167. https://doi.org/10.1016/j.ecoenv.2014.09.008.

[87]    Bortey-Sam, N., Nakayama, S. M. M., Ikenaka, Y., Akoto, O., Baidoo, E., Yohannes, Y. B., Mizukawa, H. & Ishizuka, M. (2015). Human health risks from metals and metalloid via consumption of food animals near gold mines in Tarkwa, Ghana: Estimation of the daily intakes and target hazard quotients (THQs). Ecotoxicology and Environmental Safety, 111, 160–167. https://doi.org/10.1016/j. ecoenv.2014.09.008.

[88]    Bose-O'Reilly, S., Schierl, R., Nowak, D., Siebert, U., William, J. F., Owi, F. T. & Ir, Y. I. (2016). A preliminary study on health effects in villagers exposed to mercury in a small-scale artisanal gold mining area in Indonesia. Environmental Research, 149, 274–281. https://doi.org/10.1016/j.envres. 2016.04.007.

[89]    Botchwey, G., Crawford, G., Loubere, N. & Lu, J. (2018). South-south irregular migration: The impacts of China's informal gold rush in Ghana. International Migration, 57(4), https://doi.org/10. 1111/imi.12518.

[90]    Bradshaw, A. D. (1996). Underlying principles of restoration. Canadian Journal of Fisheries and Aquatic Sciences, 53(suppl. 1), 3–9. https://doi.org/10.1139/cjfas-53-s1-3.

[91]    Bradshaw, A. D. & Chadwick, M. J. (1980). The Restoration of Land: The Ecology and Reclamation of Derelict and Degraded Land. University of California Press.

[92]    Brady, N. C. & Weil, R. R. (2008). The nature and Properties of Soils (14th ed.). Pearson Prentice Hall.

[93]    Brady, N. C. & Weil, R. R. (2010). Elements of the Nature and Properties of Soils (3rd ed.). Pearson.

[94]    Brereton, N. J. B., Gonzalez, E., Desjardins, D., Labrecque, M. & Pitre, F. E. (2020). Co-cropping with three phytoremediation crops influences rhizosphere microbiome community in contaminated soil. Science of the Total Environment, 711, 135067.

[95]    Bronick, C. J. & Lal, R. (2005). Soil structure and management: A review. Geoderma, 124(1), 3–22. https://doi.org/10.1016/j.geoderma.2004.03.005.

[96]    Buckingham, K. & Weber, S. (2015). Restoration, Management and Rehabilitation of Degraded Secondary Tropical Forests Case Studies of Ghana. Assessing the ITTO Guidelines for the restoration, Management And Rehabilitation of Degraded and Secondary Tropical forests: Case studies of Ghana (p. 30). Indonesia and Mexico. Retrieved from http://www.itto.int/partner/ id=4690.

[97]    Bundschuh, J., Schneider, J., Alam, M. A., Niazi, N. K., Herath, I., Parvez, F., Tomaszewska, B., Guilherme, L. R. G., Maity, J. P., López, D. L., Cirelli, A. F., Pérez-Carrera, A., Morales-Simfors, N., Alarcón-Herrera, M. T., Baisch, P., Mohan, D. & Mukherjee, A. (2021). Seven potential sources of arsenic pollution in Latin America and their environmental and health impacts. Science of the Total Environment, 780(May), 146274. https://doi.org/10.1016/j.scitotenv.2021.146274.

[98]    Buscaroli, A., Zannoni, D., Menichetti, M. & Dinelli, E. (2017). Assessment of metal accumulation capacity of Dittrichia viscosa (L.) Greuter in two different Italian mine areas for contaminated soils remediation. Journal of Geochemical Exploration, 182, 123–131.

[99] Cagnin, R. C., Quaresma, V. S., Chaillou, G., Franco, T. & Bastos, A. C. (2017). Arsenic enrichment in sediment on the eastern continental shelf of Brazil. Science of the Total Environment, 607–608, 304–316. https://doi.org/10.1016/j.scitotenv.2017.06.162.

[100] Cantamessa, S., Massa, N., Gamalero, E. & Berta, G. (2020). Phytoremediation of a highly arsenic polluted site, using Pteris vittata L. and arbuscular mycorrhizal fungi. Plants, 9(9), 1211.

[101] Cappa, J. J. & Pilon-Smits, E. A. H. (2014). Evolutionary aspects of elemental hyperaccumulation. Planta, 239(2), 267–275.

[102] Caravaca, F., Hernandez, M. T., Garia, C. & Roldan, A. (2002). Improvement of rhizosphere aggregates stability of afforested semi- arid, plant species subjected to mycorrhizal inoculation and compost addition. Geoderma, 108, 133–144.

[103] Cardoso, E. J. B. N., Vasconcellos, R. L. F., Bini, D., Miyauchi, M. Y. H., dos Santos, C. A., Alves, P. R. L. . . . Nogueira, M. A. (2013). Soil health: Looking for suitable indicators. What should be considered to assess the effects of use and management on soil health?. Scientia Agricola, 70(4), 274–289. https://doi.org/10.1590/S0103-90162013000400009.

[104] Carlon, C. (ed.). (2007). Derivation Methods of Soil Screening Values in Europe: A Review and Evaluation of National Procedures towards Harmonisation. Ispra.

[105] Carson, R. (1962). Silent Spring. Houghton Mifflin.

[106] Case, D. D. (1990). The community's toolbox: the idea, methods and tools for participatory assessment, monitoring and evaluation in community forestry. viii + 146.

[107] Catney, P., Eiser, D., Henneberry, J. & Stafford, T. (2006). Democracy, trust and risk related to contaminated sites in the UK. In Dixon, T., Raco, M., Catney, P. & Lerner, D. N. (eds.). Sustainable Brownfield Regeneration: Liveable Places from Problem Spaces. Oxford: Blackwell Publishing Ltd.

[108] Chakraborti, D., Rahman, M. M., Murrill, M., Das, R., Siddayya, Patil, S. G., Sarkar, A., Dadapeer, H. J., Yendigeri, S., Ahmed, R. & Das, K. K. (2013). Environmental arsenic contamination and its health effects in a historic gold mining area of the Mangalur greenstone belt of Northeastern Karnataka, India. Journal of Hazardous Materials, 262, 1048–1055. https://doi.org/10.1016/j.jhazmat.2012.10.002.

[109] Chamba-Eras, I., Griffith, D. M., Kalinhoff, C., Ramírez, J. & Gázquez, M. J. (2022). Native hyperaccumulator plants with differential phytoremediation potential in an artisanal gold mine of the Ecuadorian Amazon. Plants, 11(9), 1186.

[110] Chaney, R. L. & Baklanov, I. A. (2017). Phytoremediation and Phytomining: Status and Promise (Vol. 83). Elsevier Ltd.

[111] Chapman, E. E. V., Moore, C. & Campbell, L. M. (2019). Native plants for revegetation of mercury- and arsenic-contaminated historical mining waste – Can a low-dose selenium additive improve seedling growth and decrease contaminant bioaccumulation?. Water, Air, and Soil Pollution, 230(9), https://doi.org/10.1007/s11270-019-4267-x.

[112] Cheng, H., Hu, Y., Luo, J., Xu, B. & Zhao, J. (2009). Geochemical processes controlling fate and transport of arsenic in acid mine drainage (AMD) and natural systems. Journal of hazardous materials, 165(1–3), 13–26.

[113] Chugh, Y. P. (2017). Concurrent mining and reclamation for underground coal mining subsidence impacts in China. In Land Reclamation in Ecological Fragile Areas (pp. 315–332). CRC Press.

[114] Chugh, Y. P. (2018). Concurrent reclamation: Planning for sustainability in the mining sector. International Journal of Mining, Reclamation and Environment, 32(1), 1–18.

[115] Clewell, A. F. & Aronson, J. (2013). Ecological Restoration: Principles, Values, and Structure of an Emerging Profession (2nd ed.). Island Press.

[116] Cooke, J. A. & Johnson, M. S. (2002). Ecological restoration of land with particular reference to the mining of metals. Environmental Reviews, 10(1), 41–71. https://doi.org/10.1139/a02-002.

[117]   Cooke, J. A. & Johnson, M. S. (2002). Ecological restoration of land with particular reference to the mining of metals and industrial minerals: A review of theory and practice. Environmental Reviews, 10(1), 41–71. https://doi.org/10.1139/a01-014.

[118]   Coulter, M. A. (2016). Minamata Convention on Mercury. International Legal Materials, 55(3), 582–616. https://doi.org/10.5305/intelegamate.55.3.0582.

[119]   Cramb, R. A. & Wills, I. R. (1990). The role of traditional institutions in rural development: Community-based land tenure and government land policy in Sarawak, Malaysia. World Development, 18(3), 347–360.

[120]   Daly, H. E. (1990). Sustainable development: From concept and theory to operational principles. Population and Development Review, 16, 25–43.

[121]   Danila, V., Kumpiene, J., Kasiuliene, A. & Vasarevičius, S. (2020). Immobilisation of metal(loid)s in two contaminated soils using micro and nano zerovalent iron particles: Evaluating the long-term stability. Chemosphere, 248. https://doi.org/10.1016/j.chemosphere.2020.126054.

[122]   DeSisto, S. L., Jamieson, H. E. & Parsons, M. B. (2016). Subsurface variations in arsenic mineralogy and geochemistry following long-term weathering of gold mine tailings. Applied Geochemistry, 73, 81–97. https://doi.org/10.1016/j.apgeochem.2016.07.013.

[123]   DeSisto, S. L., Jamieson, H. E. & Parsons, M. B. (2017). Arsenic mobility in weathered gold mine tailings under a low-organic soil cover. Environmental Earth Sciences, 76(22), 1–16. https://doi.org/10.1007/s12665-017-7041-7.

[124]   Dhillon, S. K. & Dhillon, K. S. (2009). Phytoremediation of selenium-contaminated soils: The efficiency of different cropping systems. Soil Use and Management, 25(4), 441–453.

[125]   dla piper. (2012). Mining in Africa – a legal overview. Retrieved from http://www.dlapiper.com/~/media/files/insights/publications/2012/11/

[126]   Dobson, M. C. & Moffat, A. J. (1995). A re-evaluation of objections to tree planting on containment landfills. Waste Management & Research, 13(6), 579–600. https://doi.org/10.1016/S0734-242X(96)90008-4.

[127]   Document, F. (n.d.). Agenda 2063.

[128]   Doddamani, A., Angadi, J. G., Govinda, G. V., Biradar, B. N. & Binkadakatti, J. (2011). Factors influencing the knowledge and adoption of land reclamation practices among the farmers of Malaprabha command area. International Journal of Science and Nature, 2(2), 376–378.

[129]   Doran, J. W. & Zeiss, M. R. (2000). Soil health and sustainability: Managing the biotic component of soil quality. Applied Soil Ecology, 15(1), 3–11. https://doi.org/10.1016/S0929-1393(00)00067-6.

[130]   Doremus, H. & Dan Tarlock, A. (2005). Science, Judgment and Controversy in Natural Resource Regulation. Public Land and Resources Law Reveiw, 26(June), 1.

[131]   Dorleku, M. K., Nukpezah, D. & Carboo, D. (2018). Effects of small-scale gold mining on heavy metal levels in groundwater in the Lower Pra Basin of Ghana. Applied Water Science, 8(5), 1–11. https://doi.org/10.1007/s13201-018-0773-z.

[132]   Doso, S. J., Cieem, G., Ayensu-ntim, A., Twumasi-ankrah, B. & Barimah, P. T. (2015). Effects of loss of agricultural land due to large-scale gold mining on agriculture in Ghana: The case of the western region. British Journal of Research, 2(6), 196–221.

[133]   Drewniak, L. & Sklodowska, A. (2013). Arsenic-transforming microbes and their role in biomining processes. Environmental Science and Pollution Research, 20(11), 7728–7739. https://doi.org/10.1007/s11356-012-1449-0.

[134]   Du Laing, G., Rinklebe, J., Vandecasteele, B., Meers, E. & Tack, F. M. G. (2009). Trace metal behaviour in estuarine and riverine floodplain soils and sediments: A review. Science of the Total Environment, 407(13), 3972–3985. https://doi.org/10.1016/j.scitotenv.2008.07.025.

[135]   Durante-Yánez, E. V., Martínez-Macea, M. A., Enamorado-Montes, G., Combatt Caballero, E. & Marrugo-Negrete, J. (2022). Phytoremediation of soils contaminated with heavy metals from gold mining activities using Clidemia sericea D. Don. Plants, 11(5), 597.

[136]  Dutta, R. K. (1999). Performance and impact of selected exotic plant species on coal mine spoil. Ph.D. Thesis. Banaras Hindu University, Varanasi, India.

[137]  Dutta, R. K. & Agrawal, M. (2002). Effect of tree plantations on the soil characteristics and microbial activity of coal mine spoil land. Tropical Ecology, 43(2), 315–324.

[138]  Dybowska, A., Farago, M., Valsami-Jones, E. & Thornton, I. (2005). Operationally defined associations of arsenic and copper from soil and mine waste in south-west England. Chemical Speciation and Bioavailability, 17(4), 147–160. https://doi.org/10.3184/095422906783438811.

[139]  Susan, E., Singh, S. & D'Souza, S. F. (2007). Phytoremediation of metals and radionuclides. In Singh, S. N. & Tripathi, R. D. (eds.). Environmental Bioremediation Technologies (pp.189–209). Berlin Heidelberg: Springer.

[140]  Ebbs, S. D. & Kochian, L. V. (1998). Phytoextraction of zinc by oat (Avena sativa), barley (Hordeum vulgare), and Indian mustard (Brassica juncea). Environmental science & technology, 32(6), 802–806.

[141]  Edraki, M., Baumgartl, T., Manlapig, E., Bradshaw, D., Franks, D. M. & Moran, C. J. (2014). Designing mine tailings for better environmental, social and economic outcomes: A review of alternative approaches. Journal of Cleaner Production, 84(1), 411–420. https://doi.org/10.1016/j.jclepro.2014.04.079.

[142]  Eisler, R. (1988). Arsenic hazards to fish, wildlife, and invertebrates: A synoptic review. Biology Reports, 85, 1–12.

[143]  El Rasafi, T., Pereira, R., Pinto, G., Gonçalves, F. J., Haddioui, A., Ksibi, M. . . . Marques, C. R. (2021). Potential of Eucalyptus globulus for the phytoremediation of metals in a Moroccan iron mine soil – A case study. Environmental Science and Pollution Research, 28, 15782–15793.

[144]  El-naggar, A., Chang, S. X., Cai, Y., Han, Y. & Wang, J. (2021). Mechanistic insights into the ( im) mobilization of arsenic, cadmium, lead, and zinc in a multi-contaminated soil treated with different biochars. Environment International, 156(November 2020), https://doi.org/10.1016/j.envint.2021.106638.

[145]  El-naggar, A., Hou, D., Sarmah, A. K. & Moreno-jim, E. (2021). (Im) mobilization of arsenic, chromium and nickel in soils via biochar : A. 286(May), https://doi.org/10.1016/j.envpol.2021.117199.

[146]  El-Naggar, A., Shaheen, S. M., Hseu, Z. Y., Wang, S. L., Ok, Y. S. & Rinklebe, J. (2019). Release dynamics of As, Co, and Mo in a biochar treated soil under pre-definite redox conditions. Science of the Total Environment, 657, 686–695. https://doi.org/10.1016/j.scitotenv.2018.12.026.

[147]  El-Naggar, A., Shaheen, S. M., Hseu, Z. Y., Wang, S. L., Ok, Y. S. & Rinklebe, J. (2019). Release dynamics of As, Co, and Mo in a biochar treated soil under pre-definite redox conditions. Science of the Total Environment, 657, 686–695. https://doi.org/10.1016/j.scitotenv.2018.12.026.

[148]  El-Naggar, A., Shaheen, S. M., Ok, Y. S. & Rinklebe, J. (2018). Biochar affects the dissolved and colloidal concentrations of Cd, Cu, Ni, and Zn and their phytoavailability and potential mobility in a mining soil under dynamic redox-conditions. Science of the Total Environment, 624, 1059–1071. https://doi.org/10.1016/j.scitotenv.2017.12.190.

[149]  Emmanuel, A. Y., Jerry, C. S. & Dzigbodi, D. A. (2018). Review of environmental and health impacts of mining in Ghana. Journal of Health and Pollution, 8(17), 43–52. https://doi.org/10.5696/2156-9614-8.17.43.

[150]  Escobar, M., Hue, N. & Cutler, W. G. (2006). Recent developments on Arsenic: Contamination and remediation. Recent Research Developments in Bioenergetics, 4, 1–32.

[151]  Esdaile, L. J. & Chalker, J. M. (2018). The mercury problem in artisanal and small-scale gold mining. Chemistry – A European Journal, 24(27), 6905–6916. https://doi.org/10.1002/chem.201704840.

[152]  Eshun, P. A. (2005). Sustainable small-scale gold mining in Ghana: Setting and strategies for sustainability. Geological Society Special Publication, 250, 61–72. 76. https://doi.org/10.1144/GSL.SP.2005.250.01.07.

[153]   Etteieb, S., Magdouli, S., Zolfaghari, M. & Brar, S. K. (2020). Monitoring and analysis of selenium as an emerging contaminant in mining industry: A critical review. Science of the Total Environment, 698, 134339. https://doi.org/10.1016/j.scitotenv.2019.134339.

[154]   Evans, R. (2000). The Environmental Impact of Mining: Past, Present, and Future. Palgrave Macmillan.

[155]   Fageria, N. K. & Baligar, V. C. (2008). Ameliorating soil acidity of tropical Oxisols by liming for sustainable crop production. Advances in Agronomy, 99, 345–399. https://doi.org/10.1016/S0065-2113(08)00408-9.

[156]   FAO & ITPS. (2015). Intergovernmental Technical Panel on Soils. Status of the World's Soil Resources.

[157]   FAO and INBAR. (2018). Bamboo for land restoration. Retrieved from http://www.inbar.int/sites/default/files/resources/

[158]   Fasani, E., Manara, A., Martini, F., Furini, A. & DalCorso, G. (2018). The potential of genetic engineering of plants for the remediation of soils contaminated with heavy metals. Plant Cell and Environment, 41(5), 1201–1232.

[159]   Fergusson, J. E. (1990). The Heavy Elements: Chemistry, Environmental Impact and Health Effects. Pergamon Press, Oxford, UKFitz, W. J., & Wenzel, W. W. (2002). Arsenic transformations in the soil-rhizosphere-plant system: Fundamentals and potential application to phytoremediation. Journal of Biotechnology, 99(3), 259–278. https://doi.org/10.1016/S0168-1656(02)00218-3

[160]   Festin, E. S., Tigabu, M., Chileshe, M. N., Syampungani, S. & Odén, P. C. (2019). Progresses in restoration of post-mining landscape in Africa. Journal of Forestry Research, 30(2), 381–396. https://doi.org/10.1007/s11676-018-0621-x.

[161]   Fischer, D. & Glaser, B. (2012). Synergisms between Compost and Biochar for Sustainable Soil Amelioration. Management of Organic Waste, https://doi.org/10.5772/31200.

[162]   Fitz, W. J. & Wenzel, W. W. (2002). Arsenic transformations in the soil–rhizosphere–plant system: Fun- damentals and potential application to phytoremediation. Journal of biotechnology, 99, 259–278.

[163]   Fliepbach, A., Martens, R. & Reber, H. (1994). Soil microbial biomass and activity in soils treated with heavy metal contaminated sewage sludge. Soil Biology and Biochemistry, 26, 1201–1205.

[164]   Fosu, S., Owusu, C., Ntsiful, F., Ackah, K., Fosu, S. & Owusu, C. (2020). Determining acid and metalliferous drainage potential of waste rock on a mine. Ghana Mining Journal, 20(2), 49–59.

[165]   Franco, A. A. & Faria de, S. M. (1996). The contribution of N2-fixing tree legumes to land reclamation and sustainability in the tropics. Soil Biology and Biochemistry, 29(5/6), 897–903.

[166]   Frimpong, K. A., Amoakwah, E., Osei, B. A. & Arthur, E. (2016). Changes in soil chemical properties and lettuce yield response following incorporation of biochar and cow dung to highly weathered acidic soils. Journal of Organic Agriculture and Environment, 4(1), 28–39.

[167]   Frohne, T., Rinklebe, J., Diaz-Bone, R. A. & Du Laing, G. (2011). Controlled variation of redox conditions in a floodplain soil: Impact on metal mobilization and biomethylation of arsenic and antimony. Geoderma, 160(3–4), 414–424. https://doi.org/10.1016/j.geoderma.2010.10.012.

[168]   Gadepalle, V. P., Ouki, S. K., Van Herwijnen, R. & Hutchings, T. (2008). Effects of amended compost on mobility and uptake of arsenic by rye grass in contaminated soil. Chemosphere, 72(7), 1056–1061. https://doi.org/10.1016/j.chemosphere.2008.03.048.

[169]   Gadepalle, V. P., Ouki, S. K., Van Herwijnen, R. & Hutchings, T. (2008). Effects of amended compost on mobility and uptake of arsenic by rye grass in contaminated soil. Chemosphere, 72(7), 1056–1061. https://doi.org/10.1016/j.chemosphere.2008.03.048.

[170]   Gankhuyag, U. & Gregoire, F. (2018). Managing mining for sustainable development. Retrieved from http://transparency.org.au/our-work/mining-for-sustainable-development/

[171] Gao, P. Z., Mao, L., Zhi, Y.-E. & Shi, W.-J. (2010). Assessment of effects of heavy metals combined pollution on soil enzyme activities and microbial community structure: modified ecological dose–response model and PCR-RAPD. Environmental Earth Sciences, 60, 603–612.

[172] Garbisu, C. & Alkorta, I. (2001). Phytoextraction: A cost-effective plant-based technology for the removal of metals from the environment. Bioresource Technology, 77(3), 229–236.

[173] García-Sánchez, A., Alonso-Rojo, P. & Santos-Francés, F. (2010). Distribution and mobility of arsenic in soils of a mining area (Western Spain). Science of the Total Environment, 408(19), 4194–4201. https://doi.org/10.1016/j.scitotenv.2010.05.032.

[174] Gautam, M. & Agrawal, M. (2017). Phytoremediation of metals using vetiver (Chrysopogon zizanioides (L.) Roberty) grown under different levels of red mud in sludge amended soil. Journal of Geochemical Exploration, 182, 218–227.

[175] Gebel, T. (1997). Arsenic and antimony: Comparative approach on mechanistic toxicology. Chemico-Biological Interactions, 107(3), 131–144.

[176] George, M. W. (2016). GOLD (Data in metric tons 1 of gold content unless otherwise noted).

[177] Gerhardt, K. E., Gerwing, P. D. & Greenberg, B. M. (2017). Opinion: Taking Phytoremediation from Proven Technology to Accepted Practice. Plant Science, 256, 170–185.

[178] Gersztyn, L., Karczewska, A. & Gałka, B. (2013). Influence of pH on the solubility of arsenic in heavily contaminated soils. Environmental Protection and Natural Resources, 24(3), 7–11. https://doi.org/10.2478/oszn-2013-0031.

[179] Ghana Chamber of Mines. (2015). Mining in Ghana – What future can we expect? (July), 13–39. Retrieved from http://www.icmm.com/document/9151

[180] Ghose, M. K. (2004). Effect of opencast mining on soil fertility. J. Sci. Industrial Res, 63, 1006–1009.

[181] Ghose, M. K. (2001). Management of topsoil for geoenvironmental reclamation of coal mining areas. Environmental Geology, 40(11–12), 1405–1410. https://doi.org/10.1007/s002540100321.

[182] Ghosh, M. & Singh, S. (2005). A review on phytoremediation of heavy metals and utilization of its by-products. Asian Journal on Energy and Environment, 6(4), 214–231.

[183] Gill, H. S. & Abrol, I. P. (1986). Salt-affected soils and their amelioration through afforestation. In Prinsley, R. T. & Swift, M. J. (eds.). Amelioration of Soil by Trees: A Review of Current Concepts and Practices (pp. 43–56). London: Commonwealth Science Council.

[184] Goldberg, S. (2011). Chemical equilibrium and reaction modeling of arsenic and selenium in soils. In Dynamics and Bioavailability of Heavy Metals in the Rootzone (pp. 65–92).

[185] Gomes, M. A. D. C., Hauser-Davis, R. A., De souza, A. N. & Vitória, A. P. (2016). Metal phytoremediation: General strategies, genetically modified plants and applications in metal nanoparticle contamination. Ecotoxicology and Environmental Safety, 134, 133–147.

[186] Goswami, S. & Das, S. (2015). A study on cadmium phytoremediation potential of Indian mustard, Brassica juncea. International Journal of Phytoremediation, 17(6), 583–588.

[187] Grant, C. D., Loneragan, W. A. & Koch, J. M. (2007). Functional attributes of restored plant communities: A biodiversity and ecosystem function approach. Restoration Ecology, 15(4), 560–569. https://doi.org/10.1111/j.1526-100X.2007.00258.x.

[188] Gregorich, E. G., Carter, M. R., Angers, D. A., Monreal, C. M. & Ellert, B. H. (1994). Towards a minimum data set to assess soil organic matter quality in agricultural soils. Canadian Journal of Soil Science, 74(4), 367–385. https://doi.org/10.4141/cjss94-051.

[189] Daniel, G., Lindemann-Matthies, P., Markus Huppenbauer, D. G., Lindemann-Matthies, P. & Huppenbauer, Á. M. (2011). Ethical discourse on the use of genetically modified crops: A review of academic publications in the fields of ecology and environmental ethics. Journal of Agricultural and Environmental Ethics, 25(3), 265–293.

[190] Grybos, M., Davranche, M., Gruau, G. & Petitjean, P. (2007). Is trace metal release in wetland soils controlled by organic matter mobility or Fe-oxyhydroxides reduction?. Journal of Colloid and Interface Science, 314(2), 490–501. https://doi.org/10.1016/j.jcis.2007.04.062.

[191]   Grybos, M., Davranche, M., Gruau, G., Petitjean, P. & Pédrot, M. (2009). Increasing pH drives organic matter solubilization from wetland soils under reducing conditions. Geoderma, 154(1–2), 13–19. https://doi.org/10.1016/j.geoderma.2009.09.001.

[192]   Gu, S., Gruau, G., Dupas, R., Petitjean, P., Li, Q. & Pinay, G. (2019). Respective roles of Fe-oxyhydroxide dissolution, pH changes and sediment inputs in dissolved phosphorus release from wetland soils under anoxic conditions. Geoderma, 338(June 2018), 365–374. https://doi.org/10.1016/j.geoderma.2018.12.034.

[193]   Gu, X., Lin, C., Wang, B., Wang, J. & Ouyang, W. (2022). A comprehensive assessment of anthropogenic impacts, contamination, and ecological risks of toxic elements in sediments of urban rivers: A case study in Qingdao, East China. Environmental Advances, 7, 100143. https://doi.org/10.1016/j.envadv.2021.100143.

[194]   Guba, E. G. & Lincoln, Y. S. (1989). Fourth Generation Evaluation. Newbury Park, CA: Sage Publications.

[195]   Guerra, S., Beatriz, E., Muñoz Guerrero, J. & Sokolski, S. (2021). Phytoremediation of heavy metals in tropical soils an overview. Sustainability (Switzerland), 13(5), 1–25.

[196]   Guo, Y., Sommer, N., Martin, K. & Rasche, F. (2023). Rhizophagus irregularis improves Hg tolerance of Medicago truncatula by Upregulating the Zn Transporter Genes ZIP2 and ZIP6. Mycorrhiza, 33(1–2), 23–32.

[197]   Gyamfi, E., Appiah-Adjei, E. K. & Adjei, K. A. (2019). Potential heavy metal pollution of soil and water resources from artisanal mining in Kokoteasua, Ghana. Groundwater for Sustainable Development, 8, 450–456. https://doi.org/10.1016/j.gsd.2019.01.007.

[198]   Hadzi, G. Y., Essumang, D. K. & Adjei, J. K. (2015). Distribution and risk assessment of heavy metals in surface water from pristine environments and major mining areas in Ghana. Journal of Health and Pollution, 5(9), 86–99. https://doi.org/10.5696/2156-9614-5-9.86.

[199]   Han, Q., Schaefer, W. & Barry, N. (2013). Land reclamation using waste as fill material: A case study in Jakarta. International Journal of Environmental and Ecological Engineering, 7(6), 318–327.

[200]   Harris, J. A., Birch, P. & Palmer, J. (1996). Land restoration and reclamation: Principles and practice. Addison Wesley Longman.

[201]   Hartley, W. & Lepp, N. W. (2008). Remediation of arsenic contaminated soils by iron-oxide application, evaluated in terms of plant productivity, arsenic and phytotoxic metal uptake. Science of the Total Environment, 390(1), 35–44. https://doi.org/10.1016/j.scitotenv.2007.09.021.

[202]   Hartley, W., Dickinson, N. M., Riby, P., Leese, E., Morton, J. & Lepp, N. W. (2010). Arsenic mobility and speciation in a contaminated urban soil are affected by different methods of green waste compost application. Environmental Pollution, 158(12), 3560–3570. https://doi.org/10.1016/j.envpol.2010.08.015.

[203]   Haslmayr, H.-P., Meißner, S., Langella, F., Baumgarten, A. & Geletneky, J. (2014). Using microbes for the regulation of heavy metal mobility at ecosystem and landscape scale establishing best practice for microbially aided phytoremediation. Environmental Science and Pollution Research, 21, 6765–6774.

[204]   Havlin, J. L., Tisdale, S. L., Nelson, W. L. & Beaton, J. D. (2016). Soil fertility and fertilizers: An Introduction to Nutrient Management (8th ed). Pearson.

[205]   Hays, S. P. (1987). Beauty, health, and Permanence: Environmental Politics in the United States, 1955-1985. Cambridge University Press.

[206]   Hilson, G. (2002). An overview of land use conflicts in mining communities. Land Use Policy, 19(1), 65–73. https://doi.org/10.1016/S0264-8377(01)00043-6.

[207]   Hilson, G. (2017). Shootings and burning excavators : Some rapid re fl ections on the Government of Ghana ' s handling of the informal Galamsey mining 'menace'. Resources Policy, 54(August), 109–116. https://doi.org/10.1016/j.resourpol.2017.09.009.

[208] Hilson, G. & Murck, B. (2000). Sustainable development in the mining industry: Clarifying the corporate perspective. Resources Policy, 26(4), 227–238.

[209] Hilson, G. & Yakovleva, N. (2007). Strained relations: A critical analysis of the mining conflict in Prestea, Ghana. Political Geography, 26(1), 98–119. https://doi.org/10.1016/j.polgeo.2006.09.001.

[210] Hilson, G., Zolnikov, T. R., Ortiz, D. R. & Kumah, C. (2018). Formalizing artisanal gold mining under the Minamata convention: Previewing the challenge in Sub-Saharan Africa. Environmental Science and Policy, 85(March), 123–131. https://doi.org/10.1016/j.envsci.2018.03.026.

[211] Hilson, G. (2002). Small-scale mining in Africa: Tackling pressing environmental problems with improved strategy. Journal of Environment & Development, 11(2), 149–174.

[212] Hoang Ha, N. T., Sakakibara, M., Sano, S., Hori, R. S. & Sera, K. (2009). The potential of Eleocharis acicularis for phytoremediation: case study at an abandoned mine site. Clean–Soil, Air, Water, 37(3), 203–208.

[213] Hobbs, R. J. & Harris, J. A. (2001). Restoration ecology: Repairing the Earth's ecosystems in the new millennium. Restoration Ecology, 9(2), 239–246. https://doi.org/10.1046/j.1526-100X.2001.009002239.x.

[214] Honorene, J. (2017). Understanding the of Triangulation in Research (pp.91–95).

[215] Horneck, D. A., Sullivan, D. M., Owen, J. S. & Hart, J. M. (2011). Soil Test Interpretation Guide. Oregon State University Extension Service.

[216] Hou, D., O'Connor, D., Igalavithana, A. D., Alessi, D. S., Luo, J., Tsang, D. C. W., Sparks, D. L., Yamauchi, Y., Rinklebe, J. & Ok, Y. S. (2020). Metal contamination and bioremediation of agricultural soils for food safety and sustainability. Nature Reviews Earth & Environment, 1(7), 366–381. https://doi.org/10.1038/s43017-020-0061-y.

[217] Hou, D., Ok, Y. S., Tsang, D. C. W. & Bolan, N. S. (2020). Biochar for environmental management: An introduction. Journal of Hazardous Materials, 393, 122380.

[218] Hou, Y., Zhao, Y., Lu, J., Wei, Q., Zang, L. & Zhao, X. (2023). Environmental contamination and health risk assessment of potentially toxic trace metal elements in soils near gold mines – A global meta-analysis. Environmental Pollution, 330, 121803. https://doi.org/10.1016/j.envpol.2023.121803.

[219] Hussain, M. M., Bibi, I., Shahid, M., Shaheen, S. M., Shakoor, M. B., Bashir, S., Younas, F., Rinklebe, J. & Niazi, N. K. (2019). Biogeochemical cycling, speciation and transformation pathways of arsenic in aquatic environments with the emphasis on algae. Comprehensive Analytical Chemistry, 85, 15–51. https://doi.org/10.1016/bs.coac.2019.03.007.

[220] IARC (International Agency for Research on Cancer). (2012). A Review of Human Carcinogens: Arsenic, Metals, Fibres, and Dusts. Lyon: World Health Organization Press.

[221] Ingram, J. (1990). The Role of Trees in Maintaining and Improving Soil Productivity.A Review of the Literature.Commonwealth Science Council (pp. 1–19). London.

[222] Insam, H. & Domsch, K. H. (1988). Relationship between soil organic carbon and microbial biomass on chronosequences of reclaimed sites. Microbial Ecology, 15(2), 177–188. https://doi.org/10.1007/BF02011711.

[223] Irunde, R., Ijumulana, J., Ligate, F., Maity, J. P., Ahmad, A., Mtamba, J., Mtalo, F. & Bhattacharya, P. (2022). Arsenic in Africa: Potential sources, spatial variability, and the state of the art for arsenic removal using locally available materials. In Groundwater for Sustainable Development. Elsevier B.V, https://doi.org/10.1016/j.gsd.2022.100746.

[224] Jadia, C. D. & Fulekar, M. H. (2008). Phytoremediation: The application of vermicompost to remove zinc, cadmium, copper, nickel and lead by sunflower plant. Environmental Engineering and Management Journal, 7(5), 547–558.

[225] James, B. R. & Riha, S. J. (1986). pH buffering in forest soil organic horizons: relevance to acid precipitation. Journal of Environmental Quality, 15, 229–234.

[226] Jemilatu Yaro, I. & Thorunn Petursdottir, S. (2014). The perception of farmers in Akyem Adukrom, eastern region of Ghana, on Using Reclaimed Mined-Out Areas for Crop Production. Retrieved from www.unulrt.is/static/fellows/document/Yaro2014.pdf

[227] Jiménez, M. N., Bacchetta, G., Casti, M., Navarro, F. B., Lallena, A. M. & Fernández-Ondoño, E. (2011). Potential use in phytoremediation of three plant species growing on contaminated mine-tailing soils in Sardinia. Ecological engineering, 37(2), 392–398.

[228] Jin, Y., O'Connor, D., Ok, Y. S., Tsang, D. C. W., Liu, A. & Hou, D. (2019). Assessment of sources of heavy metals in soil and dust at children's playgrounds in Beijing using GIS and multivariate statistical analysis. Environment International, 124, 320–328. https://doi.org/10.1016/j.envint.2019.01.024.

[229] Johnson, D. B. & Hallberg, K. B. (2005). Acid mine drainage remediation options: A review. Science of the Total Environment, 338(1–2), 3–14. https://doi.org/10.1016/j.scitotenv.2004.09.002.

[230] Johnson, M. S. & Bradshaw, A. D. (1979). Restoration of mined lands – revegetation and the value of bracken as a nurse plant. Journal of Applied Ecology, 16(2), 371–382. https://doi.org/10.2307/2402531.

[231] Kabata-Pendias, A. (2010). Trace elements in soils and plants: Fourth edition. In Trace Elements in Soils and Plants (Fourth Edition). https://doi.org/10.1201/b10158.

[232] Kabata-Pendias, A. (2010). Trace Elements in Soils and Plants (4th ed, Vol. 53, Issue 9). CRC Press Taylor & Francis Group 6000 Broken Sound Parkway NW.

[233] Kabata-Pendias, A. (2011). Trace Elements in Soils and Plants (4th ed, Vol. 53, Issue 9). CRC Press Taylor & Francis Group 6000 Broken Sound Parkway NW.

[234] Kabata-Pendias, A. & Mukherjee, A. B. (2007). Trace Elements from Soil to Human. Springer.

[235] Kakaire, J., Mensah, A. K. & Menya, E. (2016). Factors affecting adoption of mulching in Kibaale sub-catchment, South Central Uganda. International Journal of Sustainable Agricultural Management and Informatics, 2(1), https://doi.org/10.1504/IJSAMI.2016.077268.

[236] Kandeler, E., Lurienegger, G. & Schwarz, S. (1997). Influence of heavy metals on the functional diversity of soil microbial communities. Biology and Fertility of Soils, 23, 299–306.

[237] Karak, T., Abollino, O., Bhattacharyya, P., Kishore, K. D. & Paul, R. K. (2011). Fractionation and speciation of arsenic in three tea gardens soil profiles and distribution of As in different parts of tea plant (Camellia sinensis L.). Chemosphere, 85, 948–960.

[238] Karami, N., Clemente, R., Moreno-Jiménez, E., Lepp, N. W. & Beesley, L. (2011). Efficiency of green waste compost and biochar soil amendments for reducing lead and copper mobility and uptake to ryegrass. Journal of Hazardous Materials, 191(1–3), 41–48. https://doi.org/10.1016/j.jhazmat.2011.04.025.

[239] Karczewska, A., Gałka, B., Dradrach, A., Lewińska, K., Mołczan, M., Cuske, M., Gersztyn, L. & Litak, K. (2017). Solubility of arsenic and its uptake by ryegrass from polluted soils amended with organic matter. Journal of Geochemical Exploration, 182, 193–200. https://doi.org/10.1016/j.gexplo.2016.11.020.

[240] Karczewska, A., Gałka, B., Dradrach, A., Lewińska, K., Mołczan, M., Cuske, M., Gersztyn, L. & Litak, K. (2017). Solubility of arsenic and its uptake by ryegrass from polluted soils amended with organic matter. Journal of Geochemical Exploration, 182, 193–200. https://doi.org/10.1016/j.gexplo.2016.11.020.

[241] Karczewska, A., Krysiak, A., Mokrzycka, D., Jezierski, P. & Szopka, K. (2013). Arsenic distribution in soils of a former As mining area and processing. Polish Journal of Environmental Studies, 22(1).

[242] Karlen, D. L., Andrews, S. S. & Doran, J. W. (2001). Soil quality: Current concepts and applications. Advances in Agronomy, 74, 1–40. https://doi.org/10.1016/S0065-2113(01)74029-1.

[243] Kazapoe, R. W., Arhin, E. & Amuah, E. E. Y. (2021). Known and anticipated medical geology issues in Ghana. Ecofeminism and Climate Change, ahead-of-p(ahead-of-print), https://doi.org/10.1108/efcc-06-2020-0020.

[244]  Kemper, W. D. & Rosenau, R. C. (1986). Aggregate stability and size distribution. In Methods of Soil Analysis: Part 1—Physical and Mineralogical Methods (Vol. 5, pp. 425–442). American Society of Agronomy, Soil Science Society of America.

[245]  Khan, A., Pervaiz, U., Maula Khan, N., Ahmad, S. & Nigar, S. (2009). Effectiveness of demonstration plots as extension method adopted by AKRSP for agricultural technology dissemination in district chitral. Sarhad Journal of Agriculture, 25(2), 313–320.

[246]  Khan, K. M., Chakraborty, R., Bundschuh, J., Bhattacharya, P. & Parvez, F. (2020). Health effects of arsenic exposure in Latin America: An overview of the past eight years of research. Science of the Total Environment, 710, 136071. https://doi.org/10.1016/j.scitotenv.2019.136071.

[247]  Khan, S., Naushad, M., Lima, E. C., Zhang, S., Shaheen, S. M. & Rinklebe, J. (2021). Global soil pollution by toxic elements: Current status and future perspectives on the risk assessment and remediation strategies – A review. Journal of Hazardous Materials, 417(February), 0–2. https://doi.org/10.1016/j.jhazmat.2021.126039.

[248]  Kidd, P., Mench, M., Álvarez-López, V., Bert, V., Dimitriou, I., Friesl-Hanl, W., Herzig, R., Olga Janssen, J., Kolbas, A., Müller, I., Neu, S., Renella, G., Ruttens, A., Vangronsveld, J. & Puschenreiter, M. (2015). Agronomic practices for improving gentle remediation of trace element-contaminated soils. International Journal of Phytoremediation, 17(11), 1005–1037.

[249]  Klubi, E., Abril, J. M., Nyarko, E. & Delgado, A. (2018). Impact of gold-mining activity on trace elements enrichment in the West African estuaries: The case of Pra and Ankobra rivers with the Volta estuary (Ghana) as the reference. Journal of Geochemical Exploration, 190(March), 229–244. https://doi.org/10.1016/j.gexplo.2018.03.014.

[250]  Klubi, E., Adotey, D. K., Addo, S. & Abril, J. M. (2021). Assessment of metal levels and pollution indices of the Songor Wetland, Ghana. Regional Studies in Marine Science, 46, 101875. https://doi.org/10.1016/j.rsma.2021.101875.

[251]  Ko, B. G., Anderson, C. W., Bolan, N. S., Huh, K. Y. & Vogeler, I. (2008). Potential for the phytoremediation of arsenic-contaminated mine tailings in Fiji. Soil Research, 46(7), 493–501.

[252]  Kobina, A., Marschner, B., Wang, J. & Bundschuh, J. (2022). Reducing conditions increased the mobilisation and hazardous effects of arsenic in a highly contaminated gold mine spoil. 436 (March), 0–2. https://doi.org/10.1016/j.jhazmat.2022.129238.

[253]  Kolay, A. K. (2000). Basic Concepts of Soil Science (2nd edition) (pp. 78–80, 90–91, 102–0105, 138). New Delhi, India: New Age International Publishers.

[254]  Komárek, M., Vaněk, A. & Ettler, V. (2013). Chemical stabilization of metals and arsenic in contaminated soils using oxides - A review. Environmental Pollution, 172, 9–22. https://doi.org/10.1016/j.envpol.2012.07.045.

[255]  Komárek, M., Vaněk, A. & Ettler, V. (2013). Chemical stabilization of metals and arsenic in contaminated soils using oxides – A review. Environmental Pollution, 172, 9–22. https://doi.org/10.1016/j.envpol.2012.07.045.

[256]  Kosiorek, M. & Wyszkowski, M. (2019). Remediation of cobalt-polluted soil after application of selected substances and using oat (Avena sativa L.). Environmental Science and Pollution Research, 26(16), 16762–16780. https://doi.org/10.1007/s11356-019-05052-x.

[257]  Kossoff, D. & Hudson-Edwards, K. A. (2012). Arsenic in the environment Chapter 1. In Santini, J. M. & Ward, S. M. (eds.). The Metabolism of Arsenite, Arsenic in the Environment (Vol. 5, pp. 1–23).

[258]  Kpan, J. D. (2013). Environmental impacts of mining activities and their remediation strategies. Environmental Science and Technology, 47(9), 4180–4186.

[259]  Krebs, C. J. (2006). Ecological Methodology (2nd ed). Benjamin Cummings.

[260]  Krishnakumar, A., Das, R., Aditya, S. K. & Anoop Krishnan, K. (2022). Enrichment of potential toxic elements and environmental health implications: A study of the tropical agricultural soils in

southern Western Ghats, India. Environmental Quality Management, 31(3), 393–402. https://doi. org/10.1002/tqem.21792.

[261] Kumar, A., Kumar Maiti, S., Tripti, M. N. V. P. & Shekhar Singh, R. (2017). Grasses and legumes facilitate phytoremediation of metalliferous soils in the vicinity of an abandoned chromite–asbestos mine. Journal of Soils and Sediments, 17(5), 1358–1368.

[262] Lago-Vila, M., Arenas-Lago, D., Rodríguez-Seijo, A., Andrade, M. L. & Vega, F. A. (2019). Ability of Cytisus scoparius for phytoremediation of soils from a Pb/Zn mine: Assessment of metal bioavailability and bioaccumulation. Journal of environmental management, 235, 152–160.

[263] Laine, J. E., Bailey, K. A., Rubio-Andrade, M., Olshan, A. F., Smeester, L., Drobná, Z., Herring, A. H., Stýblo, M., García-Vargas, G. G. & Fry, R. C. (2015). Maternal arsenic exposure, arsenic methylation efficiency, and birth outcomes in the Biomarkers of Exposure to AR-senic (BEAR) pregnancy cohort in Mexico. Environmental Health Perspectives, 123(2), 186–192. https://doi.org/10.1289/ehp. 1307476.

[264] Lal, R. (1997). Degradation and resilience of soils. Philosophical Transactions of the Royal Society of London, Series B: Biological Sciences, 352(1356), 997–1010. https://doi.org/10.1098/rstb.1997.0072.

[265] Lal, R. (2001). Soil degradation by erosion. Land Degradation & Development, 12(6), 519–539. https://doi.org/10.1002/ldr.472.

[266] Lal, R. (2004). Soil carbon sequestration impacts on global climate change and food security. Science, 304(5677), 1623–1627. https://doi.org/10.1126/science.1097396.

[267] Lal, R. (2015). Restoring soil quality to mitigate soil degradation. Sustainability, 7(5), 5875–5895. https://doi.org/10.3390/su7055875.

[268] Lam, E. J., Cánovas, M., Gálvez, M. E., Montofré, Í. L., Keith, B. F. & Faz, Á. (2017). Evaluation of the phytoremediation potential of native plants growing on a copper mine tailing in northern Chile. Journal of Geochemical Exploration, 182(210–217).

[269] Lebrun, M., Van Poucke, R., Miard, F., Scippa, G. S., Bourgerie, S., Morabito, D. & Tack, F. M. G. (2020). Effects of carbon-based materials and redmuds on metal(loid) immobilization and growth of Salix dasyclados Wimm. On a former mine Technosol contaminated by arsenic and lead. Land Degradation and Development, March, 1–15. https://doi.org/10.1002/ldr.3726.

[270] Lee, J., Kaunda, R. B., Sinkala, T., Workman, C. F., Bazilian, M. D. & Clough, G. (2021). Phytoremediation and phytoextraction in sub-Saharan Africa: Addressing economic and social challenges. Ecotoxicology and Environmental Safety, 226(July), 112864.

[271] Lehmann, J., Gaunt, J. & Rondon, M. (2006). Bio-char sequestration in terrestrial ecosystems–a review. Mitigation and adaptation strategies for global change, 11(2), 403–427.

[272] Lemonte, J. J., Stuckey, J. W., Sanchez, J. Z., Tappero, R., Rinklebe, J. & Sparks, D. L. (2017). Sea level rise induced arsenic release from historically contaminated coastal soils. Environmental Science and Technology, 51(11), 5913–5922. https://doi.org/10.1021/acs.est.6b06152.

[273] Li, L., Ren, J. L., Yan, Z., Liu, S. M., Wu, Y., Zhou, F., Liu, C. G. & Zhang, L. (2014). Behaviour of arsenic in the coastal area of the Changjiang (Yangtze River) Estuary: Influences of water mass mixing, the spring bloom and hypoxia. Continental Shelf Research, 80, 67–78.

[274] Li, N., Kang, Y., Pan, W., Zeng, L., Zhang, Q. & Luo, J. (2015). Concentration and transportation of heavy metals in vegetables and risk assessment of human exposure to bioaccessible heavy metals in soil near a waste-incinerator site, South China. Science of the Total Environment, 521–522, 144–151. https://doi.org/10.1016/j.scitotenv.2015.03.081.

[275] Li, P., Lin, C., Cheng, H., Duan, X. & Lei, K. (2015). Contamination and health risks of soil heavy metals around a lead/zinc smelter in southwestern China. Ecotoxicology and Environmental Safety, 113, 391–399. https://doi.org/10.1016/j.ecoenv.2014.12.025.

[276] Li, X., Poon, C. S. & Liu, P. S. (2001). Heavy metal contamination of urban soils and street dusts in Hong Kong. Applied Geochemistry, 16, 1361–1368.

[277] Lillesand, T. M., Kiefer, R. W. & Chipman, J. (2015). Remote Sensing and Image Interpretation (7th ed). John Wiley & Sons.

[278] Lin, J., Gupta, S., Loos, T. K. & Birner, R. (2019). Opportunities and challenges in the Ethiopian bamboo sector: A market analysis of the bamboo-based value web. Sustainability (Switzerland), 11(6), https://doi.org/10.3390/su11061644.

[279] Lindsay, M. B. J., Moncur, M. C., Bain, J. G., Jambor, J. L., Ptacek, C. J. & Blowes, D. W. (2015). Geochemical and mineralogical aspects of sulfide mine tailings. Applied Geochemistry, 57, 157–177. https://doi.org/10.1016/j.apgeochem.2015.01.009.

[280] Liu, G., Wang, J., Liu, X., Liu, X., Li, X., Ren, Y., Wang, J. & Dong, L. (2018). Partitioning and geochemical fractions of heavy metals from geogenic and anthropogenic sources in various soil particle size fractions. Geoderma, 312(February), 104–113. https://doi.org/10.1016/j.geoderma.2017.10.013.

[281] Liu, J., Schulz, H., Brandl, S., Miehtke, H., Huwe, B. & Glaser, B. (2012). Short-term effect of biochar and compost on soil fertility and water status of a Dystric Cambisol in NE Germany under field conditions. Journal of Plant Nutrition and Soil Science, 175(5), 698–707.

[282] Liu, Z. Q., Li Li, H., Jie Zeng, X., Cheng, L., Jing Ying, F., Jun Guo, L., Mwangi Kimani, W., Li Yan, H., Zhen Yan, H., Qing Hao, H. & Chun Jing, H. (2020). Coupling phytoremediation of cadmium-contaminated soil with safe crop production based on a sorghum farming system. Journal of Cleaner Production, 275, 123002.

[283] Liu, Z. & Quang Tran, K. (2021). A review on disposal and utilization of phytoremediation plants containing heavy metals. Ecotoxicology and Environmental Safety, 226, 112821.

[284] Lone, M. I., Zhen-li, H., Stoffella, P. J. & Yang, X.-E. (2008). Phytoremediation of heavy metal polluted soils and water: Progresses and perspectives. Journal of Zhejiang University, 9(3), 210–220.

[285] Loos, T. K., Hoppe, M., Dzomeku, B. M. & Scheiterle, L. (2018). The potential of plantain residues for the ghanaian bioeconomy-assessing the current fiber value web. Sustainability 79 (Switzerland), 10(12), https://doi.org/10.3390/su10124825.

[286] Lötter, L., Stronkhorst, L. D. & Smith, H. J. (2009). Report: Sustainable Land Management Practices of South Africa. Water.

[287] Lou, C., Liu, X., Nie, Y. & Emslie, S. D. (2015). Fractionation distribution and preliminary ecological risk assessment of As, Hg and Cd in ornithogenic sediments from the Ross Sea region, East Antarctica. Science of the Total Environment, 538, 644–653. https://doi.org/10.1016/j.scitotenv.2015.08.102.

[288] Lu, N., Li, G., Sun, Y., Wei, Y., He, L. & Li, Y. (2021). Phytoremediation potential of four native plants in soils contaminated with Lead in a mining area. Land, 10(11), 1129.

[289] Lu, X. & Wang, H. (2012). Microbial oxidation of sulphide tailings and the environmental consequences. Elements, 8(2), 119–124. https://doi.org/10.2113/gselements.8.2.119.

[290] Luo, Q., Catney, P. & Lerner, D. (2009). Risk-based management of contaminated land in the UK: Lessons for China?. Journal of Environmental Management, 90(2), 1123–1134.

[291] Ma, S. & Swinton, S. M. (2011). Valuation of ecosystem services from rural landscapes using agricultural land prices. Ecological Economics, 70(9), 1649–1659.

[292] Madejon, E., De mora, A. P., Felipe, E., Burgos, P. & Cabrera, F. (2006). Soil amendments reduce trace element solubility in a contaminated soil and allow re-growth of natural vegetation. Environmental Pollution, 139, 40–52.

[293] Mahar, A., Wang, P., Ali, A., Awasthi, M. K., Lahori, A. H., Wang, Q., Li, R. & Zhang, Z. (2016). Challenges and opportunities in the phytoremediation of heavy metals contaminated soils: A review. Ecotoxicology and Environmental Safety, 126, 111–121. https://doi.org/10.1016/j.ecoenv.2015.12.023.

[294]   Mahdieh, M., Yazdani, M. & Mahdieh, S. (2013). The high potential of Pelargonium roseum plant for phytoremediation of heavy metals. Environmental monitoring and assessment, 185, 7877–7881.

[295]   Malayeri, B. E., Chehregani, A., Mohsenzadeh, F., Kazemeini, F. & Asgari, M. (2013). Plants growing in a mining area: Screening for metal accumulator plants possibly useful for bioremediation. Toxicological & Environmental Chemistry, 95(3), 434–444.

[296]   Mamindy-Pajany, Y., Hurel, C., Marmier, N. & Roméo, M. (2011). Arsenic (V) adsorption from aqueous solution onto goethite, hematite, magnetite and zero-valent iron: Effects of pH, concentration and reversibility. Desalination, 281(1), 93–99. https://doi.org/10.1016/j.desal.2011.07.046.

[297]   Mamindy-Pajany, Y., Hurel, C., Marmier, N. & Roméo, M. (2011). Arsenic (V) adsorption from aqueous solution onto goethite, hematite, magnetite and zero-valent iron: Effects of pH, concentration and reversibility. Desalination, 281(1), 93–99. https://doi.org/10.1016/j.desal.2011.07.046.

[298]   Mandal, S. K., Ray, R., González, A. G., Pokrovsky, O. S., Mavromatis, V. & Jana, T. K. (2019). Accumulation, transport and toxicity of arsenic in the Sundarbans mangrove, India. Geoderma, 354, 113891. https://doi.org/10.1016/j.geoderma.2019.113891.

[299]   Mantey, J., Nyarko, K. & Owusu-Nimo, F. (2016). Costed Reclamation and Decommissioning Strategy for Galamsey Operations in 11 Selected MDAs of the Western Region. Ghana.

[300]   Marcelo-Silva, J., Ramabu, M. & Siebert, S. J. (2023). Phytoremediation and nurse potential of aloe plants on mine tailings. International Journal of Environmental Research and Public Health, 20(2), 1521.

[301]   Marchiol, L., Assolari, S., Sacco, P. & Zerbi, G. (2004). Phytoextraction of heavy metals by canola (Brassica napus) and radish (Raphanus sativus) grown on multicontaminated soil. Environmental Pollution, 132(1), 21–27. https://doi.org/10.1016/j.envpol.2004.04.001.

[302]   Marmiroli, N., Marmiroli, M. & Maestri, E. (2006). Phytoremediation and phytotechnologies: A review for the present and the future. Soil and Water Pollution Monitoring, Protection and Remediation, 403–416.

[303]   Marrugo-Negrete, J., Durango-Hernández, J., Pinedo-Hernández, J., Olivero-Verbel, J. & Díez, S. (2015). Phytoremediation of mercury-contaminated soils by Jatropha curcas. Chemosphere, 127, 58–63.

[304]   Marschner, P. (2012). Marschner's Mineral Nutrition of Higher Plants (3rd ed). Academic Press. https://doi.org/10.1016/C2009-0-63043-9.

[305]   Martiñá-Prieto, D., Cancelo-González, J. & Barral, M. T. (2018). Arsenic mobility in as-containing soils from geogenic origin: Fractionation and leachability. Journal of Chemistry, https://doi.org/10.1155/2018/7328203.

[306]   Martínez, C. E., Sauvé, S., Jacobson, A. & McBride, M. B. (1999). Thermally induced release of adsorbed Pb upon aging ferrihydrite and soil oxides. Environmental science & technology, 33(12), 2016–2020.

[307]   Matschullat, J. (2000). Arsenic in the geosphere – a review. Science of the Total Environment, 249, 297–312.

[308]   Mborah, C., Bansah, K. J. & Boateng, M. K. (2015). Evaluating alternate post-mining land-uses: A review. Environment and Pollution, 5(1), 14. https://doi.org/10.5539/ep.v5n1p14.

[309]   McBride, M. B. (1994). Environmental Chemistry of Soils. Oxford University Press.

[310]   Marmiroli, N., & McCutcheon, S. C. (2003). Making phytoremediation a successful technology. Phytoremediation: Transformation and control of contaminants, 85–119.

[311]   McGrath, S. P., Zhao, J. & Lombi, E. (2002). Phytoremediation of metals, metalloids, and radionuclides. Advances in Agronomy, 75, 1–56.

[312]   McIntyre, T. (2003). Phytoremediation of heavy metals from soils. Advances in Biochemical Engineering/Biotechnology, 78, 97–123.

[313] McIntyre, T. C. (2003). Databases and protocol for plant and microorganism selection: Hydrocarbons and metals. In McCutcheon, S. C. & Schnoor, J. L. (eds.). Phytoremediation: Transformation and Control of Contaminants (Vol. 1). New York: Wiley.

[314] McQuilken, J. & Hilson, G. (2016b). Artisanal and small-scale gold mining in Ghana. Evidence to inform an 'action dialogue. Pubs.Iied.Org, https://doi.org/10.13140/RG.2.2.36435.99368.

[315] Meharg, A. A. & Macnair, M. R. (1992). Suppression of the high affinity phosphate uptake system: A mechanism of arsenate tolerance in Holcus lanatus L. Journal of Experimental Botany, 43(4), 519–524. https://doi.org/10.1093/jxb/43.4.519.

[316] Mench, M., Lepp, N., Bert, V., Paul Schwitzguébel, J., Gawronski, S. W., Schröder, P. & Vangronsveld, J. (2010). Successes and limitations of phytotechnologies at field scale: Outcomes, assessment and outlook from COST action 859. Journal of Soils and Sediments, 10(6), 1039–1070.

[317] Mendoza-Hernández, J. C., Vázquez-Delgado, O. R., Castillo-Morales, M., Varela-Caselis, J. L., Santamaría-Juárez, J. D., Olivares-Xometl, O. . . . Pérez-Osorio, G. (2019). Phytoremediation of mine tailings by Brassica juncea inoculated with plant growth-promoting bacteria. Microbiological Research, 228, 126308.

[318] Mensah, A. K. (2015). Effects of illegal small-scale mining on land in the western region of Ghana. Research Journal of Environmental and Earth Sciences, 7(3), 150–156.

[319] Mensah, A. K. (2015). Role of revegetation in restoring fertility of degraded mined soils in Ghana: A review. International Journal of Biodiversity and Conservation, 7(2), 57–80. https://doi.org/10.5897/ ijbc2014.0775.

[320] Mensah, A. K. & Frimpong, K. A. (2018). Biochar and/or compost applications improve soil properties, growth, and yield of maize grown in acidic rainforest and coastal savannah soils in Ghana. International Journal of Agronomy, 2018, 1–8. https://doi.org/10.1155/2018/6837404.

[321] Mensah, A. K. & Frimpong, K. A. (2018). Biochar and/or compost applications improve soil properties, growth, and yield of maize grown in acidic rainforest and coastal savannah soils in Ghana. International Journal of Agronomy, 2018, 1–8. https://doi.org/10.1155/2018/6837404.

[322] Mensah, A. K., Bernd Marschner, K. B., Eric Stemn, S. M. S. & Rinklebe, J. (2023). Arsenic in gold mining wastes: An environmental and human health threat in Ghana. In Global Arsenic Hazards. Springer International Publishing, https://doi.org/10.1007/978-3-031-16360-9.

[323] Mensah, A. K., Mahiri, I. O., Owusu, O., Mireku, O. D., Wireko, I. & Kissi, E. A. (2015). Environmental impacts of mining: A study of mining communities in Ghana. Applied Ecology and Environmental Sciences, 3(3), 81–94. https://doi.org/10.12691/aees-3-3-3.

[324] Mensah, A. K., Marschner, B. & Shaheen, S. M. (2022). Biochar, compost, iron oxide, manure, and inorganic fertilizer affect bioavailability of arsenic and improve soil quality of an abandoned arsenic-contaminated gold mine spoil. Ecotoxicology and Environmental Safety, 234(234), https://doi.org/10.1016/j.ecoenv.2022.113358.

[325] Mensah, A. K., Marschner, B., Antoniadis, V., Stemn, E., Shaheen, S. M. & Rinklebe, J. (2021). Science of the Total Environment Human health risk via soil ingestion of potentially toxic elements and remediation potential of native plants near an abandoned mine spoil in Ghana. Science of the Total Environment, 798, 149272. https://doi.org/10.1016/j.scitotenv.2021.149272.

[326] Mensah, A. K., Marschner, B., Antoniadis, V., Stemn, E., Shaheen, S. M. & Rinklebe, J. (2021). Human health risk via soil ingestion of potentially toxic elements and remediation potential of native plants near an abandoned mine spoil in Ghana. Science of the Total Environment, 798, 149272. https://doi.org/10.1016/j.scitotenv.2021.149272.

[327] Mensah, A. K., Marschner, B., Shaheen, S. M., Wang, J., Wang, S. L. & Rinklebe, J. (2020). Arsenic contamination in abandoned and active gold mine spoils in Ghana: Geochemical fractionation, speciation, and assessment of the potential human health risk. Environmental Pollution, 261, 114116. https://doi.org/10.1016/j.envpol.2020.

[328] Mensah, A. K., Marschner, B., Shaheen, S. M., Wang, J., Wang, S. L. & Rinklebe, J. (2020). Arsenic contamination in abandoned and active gold mine spoils in Ghana: Geochemical fractionation, speciation, and assessment of the potential human health risk. Environmental Pollution, 261, https://doi.org/10.1016/j.envpol.2020.114116.

[329] Mensah, A. K., Marschner, B., Wang, J., Shaheen, S. M. & Rinklebe, J. (2020). Mobilization, Release and Speciation of Arsenic in an As-contaminated Gold Mine Spoil Under Varied Soil Redox Conditions. https://doi.org/10.5194/egusphere-egu2020-1360.

[330] Mensah, A. K., Shaheen, S. M., Rinklebe, J., Heinze, S. & Marschner, B. (2022). Phytoavailability and uptake of arsenic in ryegrass affected by various amendments in soil of an abandoned gold mining site. Environmental Research, 214, 113729.

[331] Mensah, A. K., Xavier, F. & Tuokuu, D. (2023). Polluting our rivers in search of gold: how sustainable are reforms to stop informal miners from returning to mining sites in Ghana? May 1–16. https://doi.org/10.3389/fenvs.2023.1154091.

[332] Mensah, A. K., Mahiri, I. O., Owusu, O., Mireku, O. D., Wireko, I. & Kissi, E. A. (2015). Environmental impacts of mining : A study of mining communities in Ghana. Applied Ecology and Environmental Sciences, 3(3), 81–94. https://doi.org/10.12691/aees-3-3-3.

[333] Mensah, A. K. (2015). Role of revegetation in restoring fertility of degraded mined soils in Ghana: A review. International Journal of Biodiversity and Conservation, 7(2), 57–80. https://doi.org/10.5897/IJBC2014.0775.

[334] Mensah, A. K. & Frimpong, K. A. (2018). Biochar and/or compost applications improve soil properties, growth, and yield of maize grown in acidic rainforest and coastal savannah soils in Ghana. International Journal of Agronomy, 2018, 1–8. https://doi.org/10.1155/2018/6837404.

[335] Mensah, A. K., Bernd Marschner, K. B., Eric Stemn, S. M. S. & Rinklebe, J. (2023). Arsenic in gold mining wastes: An environmental and human health threat in Ghana. In Global Arsenic Hazards. Springer International Publishing, https://doi.org/10.1007/978-3-031-16360-9.

[336] Mensah, A. K., Bernd Marschner, K. B., Eric Stemn, S. M. S. & Jörg, R. (2023). Arsenic in gold mining wastes: An environmental and human health threat in Ghana. In Global Arsenic Hazards (First pp. 0–13). Springer Nature. https://doi.org/10.1007/978-3-642-03503-6.

[337] Mensah, A. K., Marschner, B., Shaheen, S. M., Wang, J., Wang, S. L. & Rinklebe, J. (2020). Arsenic contamination in abandoned and active gold mine spoils in Ghana: Geochemical fractionation, speciation, and assessment of the potential human health risk. Environmental Pollution, 261.

[338] Mensah, A. K., Marschner, B., Antoniadis, V., Stemn, E., Shaheen, S. M. & Rinklebe, J. (2021). Human health risk via soil ingestion of potentially toxic elements and remediation potential of native plants near an abandoned mine spoil in Ghana. Science of the Total Environment, 798, 149272.

[339] Mensah, A. K., Marschner, B. & Shaheen, S. M. (2022). Biochar, compost, iron oxide, manure, and inorganic fertilizer affect bioavailability of arsenic and improve soil quality of an abandoned arsenic-contaminated gold mine spoil. Ecotoxicology and Environmental Safety, 234(234), https://doi.org/10.1016/j.ecoenv.2022.113358.

[340] Mensah, A. K., Marschner, B. & Shaheen, S. M. (2022). Ecotoxicology and Environmental Safety Biochar, compost, iron oxide, manure, and inorganic fertilizer affect bioavailability of arsenic and improve soil quality of an abandoned arsenic-contaminated gold mine spoil. Ecotoxicology and Environmental Safety, 234(234), https://doi.org/10.1016/j.ecoenv.2022.113358.

[341] Mensah, A. K., Marschner, B., Antoniadis, V., Stemn, E., Shaheen, S. M. & Rinklebe, J. (2021). Science of the Total Environment Human health risk via soil ingestion of potentially toxic elements and remediation potential of native plants near an abandoned mine spoil in Ghana. Science of the Total Environment, 798, 149272. https://doi.org/10.1016/j.scitotenv.2021.149272.

[342] Mensah, A. K., Marschner, B., Shaheen, S. M., Wang, J., Wang, S. L. & Rinklebe, J. (2020). Arsenic contamination in abandoned and active gold mine spoils in Ghana: Geochemical fractionation,

speciation, and assessment of the potential human health risk. Environmental Pollution, 261, https://doi.org/10.1016/j.envpol.2020.114116.

[343] Mensah, A. K. (2015). Role of revegetation in restoring fertility of degraded mined soils in Ghana: A review. International Journal of Biodiversity and Conservation, 7(2), 57–80.

[344] Mikesell, J. L. (1994). State lottery sales and economic activity. National Tax Journal, 47(1), 165–171.

[345] Minerals Commission. (2014). Minerals and Mining Policy of Ghana: Ensuring Mining Contributes to Sustainable Development. Accra: Minerals Commission.

[346] Mkandawire, M., Taubert, B. & Dudel, E. G. (2004). Capacity of Lemna gibba L. (Duckweed) for uranium and arsenic phytoremediation in mine tailing waters. International Journal of Phytoremediation, 6(4), 347–362.

[347] MLNR. (2017). Project Appraisal and Implementation Document (PAID), Multi-Sectoral Mining Integrated Project (MMIP) (g, Vol. 233). Ministry of Lands and Natural Resources, Government of Ghana.

[348] Mohammadi, Z., Claes, H., Cappuyns, V., Nematollahi, M. J., Helser, J., Amjadian, K. & Swennen, R. (2020). High geogenic arsenic concentrations in travertines and their spring waters: Assessment of the leachability and estimation of ecological and health risks. Journal of Hazardous Materials, 124429. https://doi.org/10.1016/j.jhazmat.2020.124429.

[349] Morgan RPC. (2005). Soil erosion and Conservation (3rd edition, Vol. 59, pp. 169–170). 190–91, U.K: Blackwell Publishing Company Ltd.

[350] Morrison-Saunders, A., McHenry, M. P., Sequeira, A. R., Gorey, P., Mtegha, H. & Doepel, D. (2016). Integrating mine closure planning with environmental impact assessment: Challenges and opportunities drawn from African and Australian practice. Impact Assessment and Project Appraisal, 34(2), 117–128. https://doi.org/10.1080/14615517.2016.1176407.

[351] Mugica-Alvarez, V., Cortés-Jiménez, V., Vaca-Mier, M. & Domínguez-Soria, V. (2015). Phytoremediation of mine tailings using Lolium multiflorum. The Journal of Environment & Development, 6(4), 246.

[352] Mummey, D. L., Stahl, P. D. & Buyer, J. S. (2002). Microbial biomarkers as an indicator of ecosystem recovery following surface mine reclamation. Applied Soil Ecology, 21(3), 251–259. https://doi.org/10.1016/S0929-1393(02)00090-2.

[353] Katherine, M., Hartt, C. M. & Pohlkamp, G. (2015). Social media discourse and genetically modified organisms. The Journal of Social Media in Society, 4(1), 38–65.

[354] Murphy, R. & Strongin, D. R. (2009). Surface reactivity of pyrite and related sulfides. Surface Science Reports, 64(1), 1–45. https://doi.org/10.1016/j.surfrep.2008.09.002.

[355] Muzhinji, N. & Ntuli, V. (2020). Genetically modified organisms and food security in Southern Africa: conundrum and discourse (pp. 25–35).

[356] Nakamura, K. (2011). Biomimetic and Bio-Inspired Catalytic System for Arsenic Detoxification: Bio-Inspired Catalysts with Vitamin-B12 Cofactor, on Biomimetics. Pramatarova, D. L. (ed.). INTECH, 10.5772/19616. ISBN: 978-953-307-271-5.

[357] Nascimento, C. & Xing, B. (2006). Phytoextraction: A review on enhanced metal availability and plant accumulation. Scientia Agricola, 63(3), 299–311.

[358] Neina, D., Van Ranst, E. & Verdoodt, A. (2019). Chemical and mineralogical properties of post-mining sites in two gold mining concessions in Ghana. Ghana Journal of Science Technology and Development, 6(1), 1–14, Retrieved from. www.gjstd.org.

[359] Nelson, D. W. & Sommers, L. E. (1996). Total carbon, organic carbon, and organic matter. In Methods of Soil Analysis: Part 3 – Chemical Methods (pp. 961–1010). Soil Science Society of America.

[360] Ngigi, M. E. (2006). Sustainable development and environment: An Investigation on the Role of the Roman Catholic Church and Inter Aid Kenya in Kiambogo Area, Nakuru District. A Research Project Submitted in Partial Fulfilment of the Requirements for The Award of a Degree of Master of Arts.

[361] Niane, B., Guédron, S., Feder, F., Legros, S., Ngom, P. M. & Moritz, R. (2019). Impact of recent artisanal small-scale gold mining in Senegal: Mercury and methylmercury contamination of terrestrial and aquatic ecosystems. Science of the Total Environment, 669, 185–193. https://doi.org/10.1016/j.scitotenv.2019.03.108.

[362] Niazi, N. K., Bibi, I., Shahid, M., Ok, Y. S., Burton, E. D., Wang, H., Shaheen, S. M., Rinklebe, J. & Lüttge, A. (2018). Arsenic removal by perilla leaf biochar in aqueous solutions and groundwater: An integrated spectroscopic and microscopic examination. Environmental Pollution, 232, 31–41. https://doi.org/10.1016/j.envpol.2017.09.051.

[363] Ning, R. Y. (2002). Arsenic removal by reverse osmosis. Desalination, 143(3), 237–241.

[364] Ning, W., Li, W., Pi, W., Xu, Y., Cao, M. & Luo, J. (2021). Effects of decapitation and root cutting on phytoremediation efficiency of Celosia argentea. Ecotoxicology and Environmental Safety, 215, 112162.

[365] Ning, Z., Shaheen, S. M., Rinklebe, J., Ok, Y. S. & Tsang, D. C. W. (2021). Root cutting enhances phytoextraction of cadmium by Celosia argentea in contaminated soil. Environmental Science and Pollution Research, 28(1), 112–121.

[366] Novo, L. A., Covelo, E. F. & González, L. (2013). Phytoremediation of amended copper mine tailings with Brassica juncea. International Journal of Mining, Reclamation and Environment, 27(3), 215–226.

[367] Nsiah, P. K. & Schaaf, W. (2019). Effect of topsoil stockpiling on soil properties and organic amendments on tree growth during gold mine reclamation in Ghana. Journal American Society of Mining and Reclamation, 8(1), 45–68. https://doi.org/10.21000/jasmr19010045.

[368] Nwaichi, E. O., Wegwu, M. O. & Onyeike, E. N. (2009). Phytoextracting cadmium and copper using Mucuna pruriens. African Journal of Plant Science, 3(12), 277–282.

[369] O'Day, P. A. (2006). Chemistry and mineralogy of arsenic. Elements, 2(2), 77–83. https://doi.org/10.2113/gselements.2.2.77.

[370] Obeng, E. A., Oduro, K. A., Obiri, B. D., Abukari, H., R. T., Djagbletey, G. D. . . . . Appiah, M. (2019). Impact of illegal mining activities on forest ecosystem services: local communities' attitudes and willingness to participate in restoration activities in Ghana. Heliyon, 5(10), https://doi.org/10.1016/j.heliyon.2019.e02617.

[371] Obiri, S., Dodoo, D. K., Essumang, D. K. & Armah, F. A. (2010). Cancer and no-cancer risk assessment from exposure to arsenic, copper, and cadmium in borehole, tap, and surface water in the Obuasi Municipality, Ghana. Human Ecological Risk Assessment, 16, 651–665.

[372] Ocansey, I. T. (2013). Davies, O., Maier, D. & John Fage, E. B. Mining Impacts on Agricultural Lands and Food Security: Case Study of Towns in and aroud Kyebi in the Eastern Region of Ghana (pp. 1–49). 2019 Ghana. Retrieved November 9, 2019, from Encyclopedia Britannica website https://www.britannica.com/place/Ghana.

[373] Odoh, C. K., Zabbey, N., Sam, K. & Nwadibe Eze, C. (2019). Status, progress and challenges of phytoremediation – An African scenario. Journal of Environmental Management, 237, 365–378.

[374] Oh, K., Tao, L., Cheng, H., Xuefeng, H., Chiquan, H., Yan, L. & Shinichi, Y. (2013). Development of profitable phytoremediation of contaminated soils with biofuel crops. Journal of Environmental Protection, 04(04), 58–64.

[375] Oh, K., Cao, T., Tao, L. & Cheng, H. (2014). Study on application of phytoremediation technology in management and remediation of contaminated soils. Journal of Clean Energy Technologies, 2(3), 216–220.

[376] Olivares, A. R., Carrillo-González, R., González-Chávez, M. D. C. A. & Hernández, R. M. S. (2013). Potential of castor bean (Ricinus communis L.) for phytoremediation of mine tailings and oil production. Journal of environmental management, 114, 316–323.

[377] Opoku-Agyeman, M., Boateng, C. O. & Quarshie, B. (2015). Agroforestry practices and land restoration in Ghana. Journal of Agricultural Research, 7(4), 100–110.

[378]  Osei, J., Fianko, J. R. & Osae, S. (2015). Assessment of heavy metals pollution in soils from selected waste dumpsites in the Accra metropolis, Ghana. Environmental Earth Sciences, 74(6), 5177–5189. https://doi.org/10.1007/s12665-015-4566-5.

[379]  Osei, P. D. (2016). Assessment of Rehabilitated Surface Mine Lands Using Geospatial technology.

[380]  Ostrom, E. (2014). Collective action and the evolution of social norms. Journal of Natural Resources Policy Research, 6(4), 235–252, 81. https://doi.org/10.1080/19390459.2014.935173.

[381]  Ouedraogo, J. V. (2006). Intoxication a' l'arsenic dans la région du Nord: deux morts et onze forages Bsous embargo. Sidwaya, 5786, 20–21.

[382]  Owusu, O., Bansah, K. J. & Mensah, A. K. (2019). "Small in size, but big in impact": Socio-environmental reforms for sustainable artisanal and small-scale mining. Journal of Sustainable Mining, 18(1), 38–44. https://doi.org/10.1016/j.jsm.2019.02.001.

[383]  Palansooriya, K. N., Shaheen, S. M., Chen, S. S., Tsang, D. C. W., Hashimoto, Y., Hou, D., Bolan, N. S., Rinklebe, J. & Ok, Y. S. (2020). Soil amendments for immobilization of potentially toxic elements in contaminated soils: A critical review. Environment International, 134(November 2019), 105046. https://doi.org/10.1016/j.envint.2019.105046.

[384]  Palansooriya, K. N., Shaheen, S. M., Chen, S. S., Tsang, D. C. W., Hashimoto, Y., Hou, D., Bolan, N. S., Rinklebe, J. & Ok, Y. S. (2020). Soil amendments for immobilization of potentially toxic elements in contaminated soils: A critical review. Environment International, 134(November 2019), 105046. https://doi.org/10.1016/j.envint.2019.105046.

[385]  Palansooriya, K. N., Shaheen, S. M., Chen, S. S., Tsang, D. C. W., Hashimoto, Y., Hou, D., Bolan, N. S., Rinklebe, J. & Ok, Y. S. (2020). Soil amendments for immobilization of potentially toxic elements in contaminated soils: A critical review. Environment International, 134(November 2019), 105046. https://doi.org/10.1016/j.envint.2019.105046.

[386]  Palansooriya, K. N., Shaheen, S. M., Chen, S. S., Tsang, D. C. W., Hashimoto, Y., Hou, D., Bolan, N. S., Rinklebe, J., Violante, O., Cozzolino, A., Perelomov, V., Caporale, L., Caporale, A. G. & Pigna, M. (2010). Mobility and bioavailability of heavy metals and metalloids in soil environments. Journal of Soil Science and Plant Nutrition, 10(3), 268–292. https://doi.org/10.4067/S0718-95162010000100005.

[387]  Pan, P., Lei, M., Qiao, P., Zhou, G., Wan, X. & Chen, T. (2019). Potential of indigenous plant species for phytoremediation of metal (loid)-contaminated soil in the Baoshan mining area, China. Environmental Science and Pollution Research, 26, 23583–23592.

[388]  Pandey, V. C. & Bajpai, O. (2018). Phytoremediation: From Theory Toward Practice. Elsevier Inc.

[389]  Pandey, V. C. & Souza-Alonso, P. (2018). Market Opportunities: In Sustainable Phytoremediation. Elsevier Inc.

[390]  Pandey, V. C., Narayan Pandey, D. & Singh, N. (2015). Sustainable phytoremediation based on naturally colonizing and economically valuable plants. Journal of Cleaner Production, 86, 37–39.

[391]  Pathak, S., Vikram Agarwal, A. & Chandra Pandey, V. (2020). Phytoremediation—a Holistic Approach for Remediation of Heavy Metals and Metalloids. INC.

[392]  Paul, E. A. (2014). Soil Microbiology, Ecology, and Biochemistry (4th ed). Elsevier.

[393]  Petelka, J., Abraham, J., Bockreis, A., Deikumah, J. P. & Zerbe, S. (2019). Soil Heavy Metal(loid) Pollution and Phytoremediation Potential of Native Plants on a Former Gold Mine in Ghana. Water, Air, and Soil Pollution, 230(11), https://doi.org/10.1007/s11270-019-4317-4.

[394]  Pierzynski, G. M., Sims, J. T. & Vance, G. F. (2000). Soils and Environmental Quality (2nd ed). CRC Press.

[395]  Pigna, M., Cozzolino, V., Violante, A. & Meharg, A. A. (2009). Influence of phosphate on the arsenic uptake by wheat (Triticum durum L.) irrigated with arsenic solutions at three different concentrations. Water, Air, and Soil Pollution, 197(1–4), 371–380. https://doi.org/10.1007/s11270-008-9818-5.

[396]  Pigna, M., Cozzolino, V., Violante, A. & Meharg, A. A. (2009). Influence of phosphate on the arsenic uptake by wheat (Triticum durum L.) irrigated with arsenic solutions at three different

concentrations. Water, Air, and Soil Pollution, 197(1–4), 371–380. https://doi.org/10.1007/s11270-008-9818-5.

[397]   Pilon-Smits, E. & Pilon, M. (2002). Phytoremediation of metals using transgenic plants. Critical Reviews in Plant Sciences, 21(5), 439–456.

[398]   Pilon-Smits, E. (2005). Phytoremediation. Annual Review of Plant Biology, 56, 15–39.

[399]   Plater, Z. J. B. (1987). Environmental Law and Policy: Nature, Law, and Society. Aspen Publishers.

[400]   Poku, A., Gupta, S. & Birner, R. (2016). "Solidarity in a competing world — fair use of resources" Governance Challenges in the Liberalisation of the Commercial Maize Seed System in Africa: A Case Study of Ghana.

[401]   Polishchuk, E. V., Merolla, A., Lichtmannegger, J., Romano, A., Indrieri, A., Ilyechova, E. Y., Concilli, M., De Cegli, R., Crispino, R., Mariniello, M., Petruzzelli, R., Ranucci, G., Iorio, R., Pietrocola, F., Einer, C., Borchard, S., Zibert, A., Schmidt, H. H., Di Schiavi, E. & Polishchuk, R. S. (2019). Activation of autophagy, observed in liver tissues from patients with wilson disease and From ATP7B-deficient animals, protects hepatocytes from copper-induced apoptosis. Gastroenterology, 156(4), 1173–1189, e5. https://doi.org/10.1053/j.gastro.2018.11.032.

[402]   Posada-Ayala, I. H., Murillo-Jiménez, J. M., Shumilin, E., Marmolejo-Rodríguez, A. J. & Nava-Sánchez, E. H. (2016). Arsenic from gold mining in marine and stream sediments in Baja California Sur, Mexico. Environmental Earth Sciences, 75(11), https://doi.org/10.1007/s12665-016-5550-4.

[403]   Pourret, O. & Hursthouse, A. (2019). It's time to replace the term "heavy metals" with "potentially toxic elements" when reporting environmental research. International journal of environmental research and public health, 16(22), 4446.

[404]   Power, A. G. (2010). Ecosystem services and agriculture: Tradeoffs and synergies. Philosophical Transactions of the Royal Society B: Biological Sciences, 365(1554), 2959–2971.

[405]   Priyandes, A. & Majid, M. (2009). Impact of reclamation activities on the environment: A case study of reclamation on the northern coast of Batam. Journal Alam Bina, 15(1), 21–34.

[406]   Raskin, I., Smith, R. D. & Salt, D. E. (1997). Phytoremediation of metals: Using plants to remove pollutants from the environment. Current Opinion in Biotechnology, 8(2), 221–226. https://doi.org/10.1016/S0958-1669(97)80005-2.

[407]   Rathore, S. S., Shekhawat, K., Dass, A., Kandpal, B. K. & Singh, V. K. (2019). Phytoremediation mechanism in Indian mustard (Brassica Juncea) and Its enhancement through agronomic interventions. Proceedings of the National Academy of Sciences India Section B – Biological Sciences, 89(2), 419–427.

[408]   Rehman, M. U., Khan, R., Khan, A., Qamar, W., Arafah, A., Ahmad, A., Ahmad, A., Akhter, R., Rinklebe, J. & Ahmad, P. (2021). Fate of arsenic in living systems: Implications for sustainable and safe food chains. Journal of Hazardous Materials, 417(September 2020). 126050. https://doi.org/10.1016/j.jhazmat.2021.126050.

[409]   Ren, H. & Yu, Z. (2008). Biomass changes of an Acacia mangium plantation in Southern China. Tropical Forest Sci, 20(2), 105–110.

[410]   Rhoades, J. D. (1996). Salinity: Electrical conductivity and total dissolved solids. In Methods of soil analysis: Part 3. In Chemical Methods (pp. 417–435). Soil Science Society of America.

[411]   Rhoades, J. D., Kandiah, A. & Mashali, A. M. (1999). The use of saline waters for crop production. FAO Irrigation and Drainage Paper 48, Food and Agriculture Organization of the United Nations. https://www.fao.org/3/X661E/X661E00.htm

[412]   Ribet, J. & Drevon, J. J. (1996). The phosphorus requirement of N2-fixing and urea-fed Acacia mangium. New Phytologist, 132, 383–390. http://dx.doi.org/10.1111/j.1469-8137.1996.tb01858.x.

[413]   Richart, S. I., Nancy, J. H., David, T., John, R. T., Mark, S. & Kothleanc, Z. (1987). Old field succession on Minnesota sand plain. Ecol, 68, 12–26. http://dx.doi.org/10.2307/1938801.

[414] Rinklebe, J., Antoniadis, V., Shaheen, S. M., Rosche, O. & Altermann, M. (2019). Health risk assessment of potentially toxic elements in soils along the Central Elbe River, Germany. Environment International, 126(January), 76–88. https://doi.org/10.1016/j.envint.2019.02.011.

[415] Rinklebe, J., Shaheen, S. M. & Frohne, T. (2016). Amendment of biochar reduces the release of toxic elements under dynamic redox conditions in a contaminated floodplain soil. Chemosphere, 142, 41–47. https://doi.org/10.1016/j.chemosphere.2015.03.067.

[416] Rinklebe, J., Shaheen, S. M. & Yu, K. (2016). Release of As, Ba, Cd, Cu, Pb, and Sr under pre-definite redox conditions in different rice paddy soils originating from the U.S.A. and Asia. Geoderma, 270, 21–32. https://doi.org/10.1016/j.geoderma.2015.10.011.

[417] Rinklebe, J., Shaheen, S. M., El-Naggar, A., Wang, H., Du Laing, G., Alessi, D. S. & Sik Ok, Y. (2020). Redox-induced mobilization of Ag, Sb, Sn, and Tl in the dissolved, colloidal and solid phase of a biochar-treated and un-treated mining soil. Environment International, 140(April), 105754. https://doi.org/10.1016/j.envint.2020.105754.

[418] Rinklebe, J., Shaheen, S. M., Schröter, F. & Rennert, T. (2016). Exploiting biogeochemical and spectroscopic techniques to assess the geochemical distribution and release dynamics of chromium and lead in a contaminated floodplain soil. Chemosphere, 150, 390–397. https://doi.org/10.1016/j.chemosphere.2016.02.021.

[419] Rizwan, M. (2016). Contrasting effects of biochar, compost and farm manure on alleviation of nickel toxicity in maize (Zea mays L.) in relation to plant growth, photosynthesis and metal uptake. https://doi.org/10.1016/j.ecoenv.2016.07.023

[420] Rosli, R. A., Harumain, Z. A., Zulkalam, M. F., Hamid, A. A., Sharif, M. F., Mohamad, M. A. & Shahari, R. (2021). Phytoremediation of arsenic in mine wastes by Acacia mangium. Remediation Journal, 31(3), 49–59.

[421] Rugh, C. L. (2001). Mercury detoxification with transgenic plants and other biotechnological breakthroughs for phytoremediation. Vitro Cellular and Developmental Biology – Plant, 37(3), 321–325.

[422] Sadler, R., Olszowy, H., Shaw, G., Biltoft, R. & Connell, D. (1994). Soil and water contamination by arsenic from a tannery waste. Water, Air, and Soil Pollution, 78, 189–198.

[423] Saha, S., Hazra, G. C., Saha, B. & Mandal, B. (2015). Assessment of heavy metals contamination in different crops grown in long-term sewage-irrigated areas of Kolkata, West Bengal, India. Environmental Monitoring and Assessment, 187(1), https://doi.org/10.1007/s10661-014-4087-9.

[424] Sai Kachout, S., Ennajah, A., Guenni, K., Ghorbel, N. & Zoghlami, A. (2024). Potential of halophytic plant Atriplex hortensis for phytoremediation of metal-contaminated soils in the mine of Tamra. Soil and Sediment Contamination: An International Journal, 33(2), 139–154.

[425] Salifu, B. (2016). Implications of Sand Mining on the Environment and Livelihoods in Brong Ahafo Region. Ghana: Kwame Nkrumah University of Science and Technology.

[426] Salt, D. E., Smith, R. D. & Raskin, I. (1998). Phytoremediation. Annual Review of Plant Physiology and Plant Molecular Biology, 49(1), 643–668. https://doi.org/10.1146/annurev.arplant.49.1.643.

[427] Salt, D. E., Smith, R. D. & Raskin, I. (1998). Phytoremediation. Annual Review of Plant Biology, 49, 643–668. https://doi.org/10.1146/annurev.arplant.49.1.643.

[428] Sasmaz, A. & Sasmaz, M. (2009). The phytoremediation potential for strontium of indigenous plants growing in a mining area. Environmental and Experimental Botany, 67(1), 139–144.

[429] Sasmaz, A., Dogan, I. M. & Sasmaz, M. (2016). Removal of Cr, Ni and Co in the water of chromium mining areas by using Lemna gibba L. and Lemna minor L. Water and Environment Journal, 30(3–4), 235–242.

[430] Sas-Nowosielska, A., Kucharski, R., Małkowski, E., Pogrzeba, M., Kuperberg, J. M. & Kryński, K. (2004). Phytoextraction crop disposal – an unsolved problem. Environmental Pollution, 128(3), 373–379.

[431] Schiffer, E. (2007). Net-maptoolbox: Influence mapping of social networks. In International Food Policy Research Institute. Washington, DC, USA.

[432]   Schmidt, E. L., Quinlivan, W. E. & Alexander, M. (1996). Soil organic matter formation as influenced by clay type. Soil Science Society of America Journal, 30(1), 185–189.

[433]   Schnurer, J., Clarholm, M. & Rosswal, T. (1985). Microbial biomass and activity in agricultural soil with different organic matter contents. Soil Biology and Biochemistry, 17, 611–618.

[434]   Schueler, V., Kuemmerle, T. & Schröder, H. (2011). Impacts of surface gold mining on land use systems in Western Ghana. Ambio, 40(5), 528–539. https://doi.org/10.1007/s13280-011-0141-9.

[435]   Schwitzguébel, J. P., Van der lelie, D., Baker, A., Glass, D. J. & Vangronsveld, J. (2002). Phytoremediation: European and American trends: Successes, obstacles and needs. Journal of Soils and Sediments, 2(2), 91–99.

[436]   Selvam, A. & Wong, J. W. C. (2008). Phytochelatin systhesis and cadmium uptake of Brassica napus. Environmental technology, 29(7), 765–773.

[437]   Selvam, A. & Wong, J. W. C. (2009). Cadmium uptake potential of Brassica napus cocropped with Brassica parachinensis and Zea mays. Journal of Hazardous Materials, 167(1–3), 170–178.

[438]   Shaheen, S. M. & Tsadilas, C. D. (2015). Influence of phosphates on fractionation, mobility, and bioavailability of soil metal(loid)s. In Phosphate in Soils: Interaction with Micronutrients, Radionuclides and Heavy Metals (Issue Adriano 2001, pp. 169–202).

[439]   Shaheen, S. M., Ali, R. A., Abowaly, M. E., Rabie, A. E. M. A., El Abbasy, N. E. & Rinklebe, J. (2018). Assessing the mobilization of As, Cr, Mo, and Se in Egyptian lacustrine and calcareous soils using sequential extraction and biogeochemical microcosm techniques. Journal of Geochemical Exploration, 191(May), 28–42. https://doi.org/10.1016/j.gexplo.2018.05.003.

[440]   Shaheen, S. M., Antoniadis, V., Kwon, E., Song, H., Wang, S. L., Hseu, Z. Y. & Rinklebe, J. (2020). Soil contamination by potentially toxic elements and the associated human health risk in geo- and anthropogenic contaminated soils: A case study from the temperate region (Germany) and the arid region (Egypt). Environmental Pollution, 262, https://doi.org/10.1016/j.envpol.2020.114312.

[441]   Shaheen, S. M., Rinklebe, J. & Selim, M. H. (2015). Impact of various amendments on immobilization and phytoavailability of nickel and zinc in a contaminated floodplain soil. International Journal of Environmental Science and Technology, 12(9), 2765–2776. https://doi.org/10.1007/s13762-014-0713-x.

[442]   Shaheen, S. M., Rinklebe, J. & Selim, M. H. (2015). Impact of various amendments on immobilization and phytoavailability of nickel and zinc in a contaminated floodplain soil. International Journal of Environmental Science and Technology, 12(9), 2765–2776. https://doi.org/10.1007/s13762-014-0713-x.

[443]   Shaheen, S. M., Rinklebe, J., Frohne, T., White, J. R. & DeLaune, R. D. (2014a). Biogeochemical factors governing Cobalt, Nickel, Selenium, and Vanadium dynamics in periodically flooded Egyptian North Nile delta rice soils. Soil Science Society of America Journal, 78(3), 1065–1078. https://doi.org/10.2136/sssaj2013.10.0441.

[444]   Shaheen, S. M., Rinklebe, J., Rupp, H. & Meissner, R. (2014b). Lysimeter trials to assess the impact of different flood-dry-cycles on the dynamics of pore water concentrations of As, Cr, Mo and V in a contaminated floodplain soil. Geoderma, 228–229, 5–13. https://doi.org/10.1016/j.geoderma.2013.12.030.

[445]   Shaheen, S. M., Shams, M. S., Khalifa, M. R., El-Dali, M. A. & Rinklebe, J. (2017). Various soil amendments and environmental wastes affect the (im)mobilization and phytoavailability of potentially toxic elements in a sewage effluent irrigated sandy soil. Ecotoxicology and Environmental Safety, 142(April), 375–387. https://doi.org/10.1016/j.ecoenv.2017.04.026.

[446]   Shaheen, S. M. & Rinklebe, J. (2015). Impact of emerging and low-cost alternative amendments on the (im)mobilization and phytoavailability of Cd and Pb in a contaminated floodplain soil. Journal of Environmental Management, 151, 160–167.

[447]  Shaheen, S. M. & Rinklebe, J. (2017). Redox-induced mobilization of copper, selenium, and zinc in deltaic soils originating from Mississippi (U.S.A.) and Nile (Egypt) River Deltas. Journal of Soils and Sediments, 17, 2099–2110.

[448]  Shaheen, S. M., Rinklebe, J. & Tsang, D. C. W. (2015). Immobilization of nickel and zinc in a contaminated floodplain soil using biochar supported zero-valent iron nanoparticles. Environmental Science and Pollution Research, 22(24), 19728–19737.

[449]  Shaheen, S. M., Rinklebe, J. & Tsang, D. C. W. (2017). Redox-induced mobilization of copper, selenium, and zinc in deltaic soils originating from Mississippi (U.S.A.) and Nile (Egypt) River Deltas. Environmental Pollution, 230, 1018–1029.

[450]  Shaheen, S. M., Ali, R. A., Abowaly, M. E., Rabie, A. E. M. A., El Abbasy, N. E. & Rinklebe, J. (2018). Assessing the mobilization of As, Cr, Mo, and Se in Egyptian lacustrine and calcareous soils using sequential extraction and biogeochemical microcosm techniques. Journal of Geochemical Exploration, 191(May), 28–42. https://doi.org/10.1016/j.gexplo.2018.05.003.

[451]  Shaheen, S. M., Antoniadis, V., Kwon, E., Song, H., Wang, S. L., Hseu, Z. Y. & Rinklebe, J. (2020). Soil contamination by potentially toxic elements and the associated human health risk in geo- and anthropogenic contaminated soils: A case study from the temperate region (Germany) and the arid region (Egypt). Environmental Pollution, 262, https://doi.org/10.1016/j.envpol.2020.114312.

[452]  Shaheen, S. M., Antoniadis, V., Kwon, E., Song, H., Wang, S. L., Hseu, Z. Y. & Rinklebe, J. (2020). Soil contamination by potentially toxic elements and the associated human health risk in geo- and anthropogenic contaminated soils: A case study from the temperate region (Germany) and the arid region (Egypt). Environmental Pollution, 262, https://doi.org/10.1016/j.envpol.2020.114312.

[453]  Shaheen, S. M., Balbaa, A. A., Khatab, A. M. & Rinklebe, J. (2017). Compost and sulfur affect the mobilization and phyto-availability of Cd and Ni to sorghum and barnyard grass in a spiked fluvial soil. Environmental Geochemistry and Health, 39(6), 1305–1324. https://doi.org/10.1007/s10653-017-9962-1.

[454]  Shaheen, S. M., Frohne, T., White, J. R., DeLaune, R. D. & Rinklebe, J. (2017). Redox-induced mobilization of copper, selenium, and zinc in deltaic soils originating from Mississippi (U.S.A.) and Nile (Egypt) River Deltas: A better understanding of biogeochemical processes for safe environmental management. Journal of Environmental Management, 186, 131–140. https://doi.org/10.1016/j.jenvman.2016.05.032.

[455]  Shaheen, S. M., Kwon, E. E., Biswas, J. K., Tack, F. M. G., Ok, Y. S. & Rinklebe, J. (2017). Arsenic, chromium, molybdenum, and selenium: Geochemical fractions and potential mobilization in riverine soil profiles originating from Germany and Egypt. Chemosphere, 180, 553–563. https://doi.org/10.1016/j.chemosphere.2017.04.054.

[456]  Shaheen, S. M., Rinklebe, J., Frohne, T., White, J. R. & DeLaune, R. D. (2014). Biogeochemical factors governing cobalt, nickel, selenium, and vanadium dynamics in periodically flooded Egyptian North Nile Delta Rice Soils. Soil Science Society of America Journal, 78(3), 1065–1078. https://doi.org/10.2136/sssaj2013.10.0441.

[457]  Sharma, P. & Pandey, S. (2014). Status of phytoremediation in world scenario. International Journal of Environmental Bioremediation & Biodegradation, 2(4), 178–191.

[458]  Sharma, V. K. & Sohn, M. (2009). Aquatic arsenic: Toxicity, speciation, transformations, and remediation. Environment International, 35, 743–759.

[459]  Shen, Z., Xu, D., Li, L., Wang, J. & Shi, X. (2019). Ecological and health risks of heavy metal on farmland soils of mining areas around Tongling City, Anhui, China. Environmental Science and Pollution Research, 26(15), 15698–15709. https://doi.org/10.1007/s11356-019-04463-0.

[460]  Sheoran, V., Sheoran, A. S. & Poonia, P. (2010). Soil reclamation of abandoned mine land by Revegetation. A Review, Intl J. Soil, Sediment and Water, 3, 1–21. Available at http://scholarworks.umass.edu/intljssw/vol3/iss2/13.

[461]  Sheoran, V. (2010). Soil reclamation of abandoned mine land by revegetation: A review. 3(2), 13.

[462] Sheoran, V., Sheoran, A. S. & Poonia, P. (2009). Phytomining : A Review. Minerals Engineering, 22(12), 1007–1019.

[463] Sheoran, V., Sheoran, A. S. & Poonia, P. (2010). Soil reclamation of abandoned mine land by revegetation: A review. International journal of soil, sediment and water, 3(2), 13.

[464] Sheoran, V., Sheoran, A. S. & Poonia, P. (2011). Role of hyperaccumulators in phytoextraction of metals from contaminated mining sites: A review. Critical Reviews in Environmental Science and Technology, 41(2), 168–214. https://doi.org/10.1080/10643380902718418.

[465] Sheoran, V., Sheoran, A. S. & Poonia, P. (2010). Soil reclamation of abandoned mine land by revegetation: A review. International Journal of Soil, Sediment and Water, 3(2), 1–21.

[466] Shrestha, R. K. & Lal, R. (2011). Changes in physical and chemical properties of soil after surface mining and reclamation. Geoderma, 161(3–4), 168–176. https://doi.org/10.1016/j.geoderma.2010.12.001.

[467] Shrivastava, A., Ghosh, D., Dash, A. & Bose, S. (2015). Arsenic contamination in soil and sediment in India: Sources, effects, and remediation. Current Pollution Reports, 1(1), 35–46. https://doi.org/10.1007/s40726-015-0004-2.

[468] Singh, P. K. & Singh, R. S. (2016). Environmental and social impacts of mining and their mitigation. National Seminar (September).

[469] Singh, R. P., Dhania, G., Sharma, A. & Jaiwal, P. K. (2007). Biotechnological approaches to improve phytoremediation efficiency for environment contaminants. In Singh, S. N. & Tripathi, R. D. (eds.). Environmental Bioremediation Technologies. Springer.

[470] Singh, S. N. & Tripathi, R. D. (2007). Environmental Bioremediation Technologies. Singh, S. & Tripathi, R. edited by. Berlin Heidelberg: Springer.

[471] Six, J., Conant, R. T., Paul, E. A. & Paustian, K. (2002). Stabilization mechanisms of soil organic matter: Implications for C-saturation of soils. Plant and Soil, 241(2), 155–176. https://doi.org/10.1023/A:1016125726789.

[472] Smedley, P. L. & Kinniburgh, D. G. (2002). A review of the source, behaviour and distribution of arsenic in natural waters. Applied Geochemistry, 17(5), 517–568.

[473] Smedley, P. L., Knudsen, J. & Maiga, D. (2007). Arsenic in groundwater from mineralised Proterozoic basement rocks of Burkina Faso. Applied Geochemistry, 22(5), 1074–1092.

[474] Snapir, B., Simms, D. M. & Waine, T. W. (2017). Mapping the expansion of galamsey gold mines in the cocoa growing area of Ghana using optical remote sensing. International Journal of Applied Earth Observation and Geoinformation, 58(June), 225–233. https://doi.org/10.1016/j.jag.2017.02.009.

[475] Sparling, G. P. (1997). Soil microbial biomass, activity and nutrient cycling as indicators of soil health. In Biological Indicators of Soil Health (pp. 97–119). CAB International.

[476] Ssenku, J. E., Ntale, M., Backeus, I. & Oryem-Origa, H. (2017). Phytoremediation potential of Leucaena Leucocephala (Lam.) de Wit. for heavy metal-polluted and heavy metal-degraded environments. Phytoremediation Potential of Bioenergy Plants, 189–209.

[477] Stanković, D. & Devetaković, J. (2016). Application of plants in remediation of contaminated sites. Reforesta, 1(1), 300–320.

[478] Stojanović, M. D., Mihajlović, M. L., Milojković, J. V., Lopičić, Z. R., Adamović, M. & Stanković, S. (2012). Efficient phytoremediation of uranium mine tailings by tobacco. Environmental chemistry letters, 10, 377–381.

[479] Study, C., Guide, F. T., Alpers, C. N., Myers, P. A. & Millsap, D. (2014). Arsenic associated with historical gold mining in the Sierra Nevada foothills. Reviews in Mineralogy and Geochemistry, 79, 553–587.

[480] Su, C., Jiang, L. & Zhang, W. (2014). A review on heavy metal contamination in the soil worldwide: Situation, impact and remediation techniques. Environmental Skeptics and Critics, 3(2), 24–38.

[481]  Subhashini, V. & Swamy, A. V. V. S. (2014). Phytoremediation of metal (Pb, Ni, Zn, Cd And Cr) contaminated soils using Canna indica. Current World Environment, 9(3), http://dx.doi.org/10.12944/CWE.9.3.26.

[482]  Tack, F. M. G. (2010). Trace elements: General soil chemistry, principles and processes. Trace Elements in Soils, https://doi.org/10.1002/9781444319477.ch2.

[483]  Tack, F. M. G. & Egene, C. E. (2018). Potential of biochar for managing metal contaminated areas, in synergy with phytomanagement or other management options. Biochar from Biomass and Waste: Fundamentals and Applications, 91–111. https://doi.org/10.1016/B978-0-12-811729-3.00006-6.

[484]  Tack, F. M. G. (2010). Trace elements: general soil chemistry, principles and processes. In Hooda, P. (ed.). Trace Elements in Soils. Wiley-Blackwell (pp. 9–37).

[485]  Tang, J., Liao, Y., Yang, Z., Chai, L. & Yang, W. (2016). Characterization of arsenic serious-contaminated soils from Shimen realgar mine area, the Asian largest realgar deposit in China. Journal of Soils and Sediments, 16(5), 1519–1528. https://doi.org/10.1007/s11368-015-1345-6.

[486]  Tang, L., Luo, W., Chen, W., Zhenli, H., Kumar Gurajala, H., Hamid, Y., Deng, M. & Yang, X. (2017). Field crops (Ipomoea Aquatica Forsk. and Brassica Chinensis L.) for phytoremediation of cadmium and nitrate co-contaminated soils via rotation with sedum alfredii hance. Environmental Science and Pollution Research, 24(23), 19293–19305.

[487]  Tang, Y. T., Hao Bo Deng, T., Qi Hang, W., Zhong Wang, S., Liang Qiu, R., Bin Wei, Z., Fang Guo, X., Qi Tang, W., Lei, M., Tong Bin Chen, G. E., Sterckeman, T., Simonnot, M. O. & Morel, J. L. (2012). Designing cropping systems for metal-contaminated sites: A review. Pedosphere, 22(4), 470–488.

[488]  Teixeira, R. A., Fernandes, A. R., Ferreira, J. R., Vasconcelos, S. S. & Braz, A. M. D. S. (2018). Contamination and soil biological properties in the Serra Pelada mine – Amazonia, Brazil. Revista Brasileira de Ciencia Do Solo, 42, 1–15. https://doi.org/10.1590/18069657rbcs20160354.

[489]  Tetteh, E. N., Vincent, L. & Partey, S. T. (2015). Effect of duration of reclamation on soil quality indicators of a surface-mined acid forest oxisol in South-Western Ghana article in West African. Journal of Applied Ecology, Retrieved from https://www.researchgate.net/publication/290443578.

[490]  Tetteh, E., Ampofo, K. & Logah, V. (2015). Adopted practices for mined land reclamation in Ghana: A case study of AngloGold Ashanti Iduapriem Mine Ltd. Journal of Science and Technology (Ghana), 35(2), 77. https://doi.org/10.4314/just.v35i2.8.

[491]  Tetteh, I. K., Asiedu, A. K. & Safo, E. K. (2018). A review of land reclamation techniques in mining areas of Ghana. Environmental Monitoring and Assessment, 189(5), 1–12. https://doi.org/10.1007/s10661-018-6659-3.

[492]  Tisdall, J. M. & Oades, J. M. (1982). Organic matter and water-stable aggregates in soils. Journal of Soil Science, 33(2), 141–163. https://doi.org/10.1111/j.1365-2389.1982.tb01755.x.

[493]  Tonelli, F. M. P., Bhat, R. A., Dar, G. H. & Hakeem, K. R. (2022). The history of phytoremediation. Phytoremediation, 1–18. https://doi.org/10.1016/b978-0-323-89874-4.00018-2.

[494]  Tordoff, G. M., Baker, A. J. M. & Willis, A. J. (2000). Current approaches to the revegetation and reclamation of metalliferous mine wastes. Chemosphere, 41, 219–228.

[495]  Tordoff, G. M., Baker, A. J. M. & Willis, A. J. (2000). Current approaches to the revegetation and reclamation of metalliferous mine wastes. Chemosphere, 41(1–2), 219–228. https://doi.org/10.1016/S0045-6535(99)00414-2.

[496]  Tordoff, G. M., Baker, A. J. M. & Willis, A. J. (2000). Current approaches to the revegetation and reclamation of metalliferous mine wastes. Chemosphere, 41(1–2), 219–228.

[497]  Toth, G., Hermann, T., Da Silva, M. R. & Montanarella, L. (2007). Contaminated sites in Europe: A review of the current situation based on data collected through a European network. Journal of Environmental and Public Health, 2007, 1–11. https://doi.org/10.1155/2007/24042.

[498]  Tripathi, N., Singh, R. S. & Hills, C. D. (2016). Reclamation of mine-impacted land for ecosystem recovery. Reclamation of Mine-Impacted Land for Ecosystem Recovery, https://doi.org/10.1002/9781119057925.

[499]  Troeh, F. R., Hobbs, J. A. & Donahue, R. L. (1980). Soil and Water Conservation for Production and Environmental Protection. Eaglewood Cliffs NJ: Prentice- Hall, Inc.

[500]  Tschakert, P. & Singha, K. (2007). Participatory GIS for development and planning: Spatial insights from local knowledge in dryland West Africa. Geografiska Annaler: Series B, Human Geography, 89(1), 41–59. https://doi.org/10.1111/j.1468-0467.2007.00236.x.

[501]  Tuokuu, F. X. D., Gruber, J. S., Idemudia, U. & Kayira, J. (2018). Challenges and opportunities of environmental policy implementation: Empirical evidence from Ghana's gold mining sector. Resources Policy, 59(May), 435–445. https://doi.org/10.1016/j.resourpol.2018.08.014.

[502]  UN DESA. (2019). World population prospects 2019. Department of economic and social affairs. World Population Prospects 2019, 141, 49–78. Retrieved from http://www.ncbi.nlm.nih.gov/pubmed/12283219.

[503]  UNDP. (2017). Nkitahodie (E N G A G E M E N T) with Political Parties 2017 Paris Climate Change Agreement-Implications for Ghana: An Interactive Policy Dialogue Accra-Ghana 15 December Artisanal and Small-Scale Mining Legal Regime. Ghana: Policy Options for Address, Retrieved from www.undp.org.

[504]  United Nations. (2019). World population prospects 2019: Data booklet. Department of Economic and Social Affairs Population Division, 1–25. Retrieved from https://population.un.org/wpp/Publications/Files/WPP2019_DataBooklet.pdf.

[505]  United States Environmental Protection Agency Office of Solid Waste and Emergency Response Technology Innovation Office. (1999). Phytoremediation Resource Guide.

[506]  United, T. & Conference, N. (1992). The Rio Declaration on Environment and Development (1992).

[507]  USEPA. 2014. "Lead | US EPA – Environmental protection agency." United States Environmental Protection Agency. Retrieved February 25, 2023 (https://www.epa.gov/).

[508]  USEPA. 2020a. "CLU-IN | Databases > Phytotechnology Project Profiles > Search Results." United States Environmental Protection Agency. Retrieved February 25, 2023 (https://clu-in.org/products/phyto/search/phyto_list.cfm).

[509]  USEPA. (2020b). Superfund Community Involvement Handbook. United Nations Environmental Programme.

[510]  van oosten, C., Gunarso, P., Koesoetjahjo, I. & Wiersum, F. (2014). Governing forest landscape restoration: Cases from Indonesia. Forests, 5(6), 1143–1162. https://doi.org/10.3390/f5061143.

[511]  Jaco, V., Herzig, R., Weyens, N., Boulet, J., Adriaensen, K., Ruttens, A., Thewys, T., Vassilev, A., Meers, E., Nehnevajova, E., Vangronsveld, J., Weyens, N., Boulet, J., Adriaensen, K., Ruttens, A., Thewys, T., Herzig, R., Nehnevajova, E., Vassilev, A., Meers, E., Van Der Lelie, D. & Mench, M. (2009). Phytoremediation of contaminated soils and groundwater: Lessons from the field. Environmental Science and Pollution Research 2009, 16(7), 765–794.

[512]  Vargas, C., Pérez-Esteban, J., Escolástico, C., Masaguer, A. & Moliner, A. (2016). Phytoremediation of Cu and Zn by vetiver grass in mine soils amended with humic acids. Environmental Science and Pollution Research, 23, 13521–13530.

[513]  Violante, A., Cozzolino, V., Perelomov, L., Caporale, A. G. & Pigna, M. (2010). Mobility and bioavailability of heavy metals and metalloids in soil environments. Journal of Soil Science and Plant Nutrition, 10(3), 268–292. https://doi.org/10.4067/S0718-95162010000100005.

[514]  Violante, A., Cozzolino, V., Perelomov, L., Caporale, A. G. & Pigna, M. (2010). Mobility and bioavailability of heavy metals and metalloids in soil environments. Journal of Soil Science and Plant Nutrition, 10(3), 268–292. https://doi.org/10.4067/S0718-95162010000100005.

[515] Volk, T. A., Abrahamson, L. P., Nowak, C. A., Smart, L. B., Tharakan, P. J. & White, E. H. (2006). The development of short-rotation willow in the Northeastern United States for bioenergy and bioproducts, agroforestry and phytoremediation. Biomass and Bioenergy, 30(8–9), 715–727.

[516] Von der heyden, B. P., Benoit, J., Fernandez, V. & Roychoudhury, A. N. (2020). Synchrotron X-ray radiation and the African earth sciences: A critical review. Journal of African Earth Sciences, 104012, https://doi.org/10.1016/j.jafrearsci.2020.104012.

[517] Walker, G. P., Chappel, M. J., Chapman, R. B. & McNeil, R. D. (2009). Integrating pests, people, and policy into future reclamation efforts. Ecological Applications, 19(2), 431–437. https://doi.org/10. 1890/07-1816.1.

[518] Walker, L. R. (2005). Restoration of ecosystems: An introduction. Cambridge University Press.

[519] Walker, L. R. & Shiels, A. B. (2013). Tropical Ecology. Cambridge University Press.

[520] Wan, X., Lei, M. & Chen, T. (2016). Cost–benefit calculation of phytoremediation technology for heavy-metal-contaminated soil. Science of the Total Environment, 563–564, 796–802.

[521] Wang, J., Feng, X., Anderson, C. W. N., Xing, Y. & Shang, L. (2012). Remediation of mercury contaminated sites – A review. Journal of Hazardous Materials, 221–222, 1–18. https://doi.org/10. 1016/j.jhazmat.2012.04.035.

[522] Wang, J., Shaheen, S. M., Rinklebe, J., Ok, Y. S. & Tsang, D. C. W. (2021). Biochar for metal(loid) immobilization in contaminated soils: Current understanding and future prospects. Journal of Hazardous Materials, 401, 123399.

[523] Wang, L., Rinklebe, J., Tack, F. M. G. & Hou, D. (2021). A review of green remediation strategies for heavy metal contaminated soil. Soil Use and Management, 1–28. https://doi.org/10.1111/sum.12717.

[524] Wang, L., Rinklebe, J., Tack, F. M. G. & Hou, D. (2021). A review of green remediation strategies for heavy metal contaminated soil. Soil Use and Management, March, 1–28. https://doi.org/10.1111/ sum.12717.

[525] Wang, S., Liu, Y., Kariman, K., Jialin, L., Zhang, H., Fangbai, L., Chen, Y., Chongjian, M., Liu, C., Yuan, Y., Zhu, Z. & Rengel, Z. (2021). Co-cropping Indian mustard and silage maize for phytoremediation of a cadmium-contaminated acid paddy soil amended with peat. Toxics, 9(5), 91.

[526] Wang, X., Chen, C. & Wang, J. (2017). Cadmium phytoextraction from loam soil in tropical Southern China by Sorghum bicolor. International Journal of Phytoremediation, 19(6), 572–578.

[527] Watts, J. D., Tacconi, L., Irawan, S. & Wijaya, A. H. (2019). Village transfers for the environment: Lessons from community-based development programs and the village fund. Forest Policy and Economics, 108(January), 101863.

[528] Wenzel, W. W., Kirchbaumer, N., Prohaska, T., Stingeder, G., Lombi, E. & Adriano, D. C. (2001). Arsenic fractionation in soils using an improved sequential extraction procedure. Analytica Chimica Acta, 436(2), 309–323. https://doi.org/10.1016/S0003-2670(01)00924-2.

[529] Wenzel, W. W., Kirchbaumer, N., Prohaska, T., Stingeder, G., Lombi, E. & Adriano, D. C. (2001). Arsenic fractionation in soils using an improved sequential extraction procedure. Analytica Chimica Acta, 436(2), 309–323. https://doi.org/10.1016/S0003-2670(01)00924-2.

[530] Williamson, J. C. & Johnson, D. B. (1991). Microbiology of soils at opencast sites: II. Population transformations occurring following land restoration and the influence of rye grass/ fertilizer amendments. J. Soil Sci, 42, 9–16.

[531] Wuana, R. A. & Okieimen, F. E. (2011). Heavy metals in contaminated soils: A review of sources, chemistry, risks and best available strategies for remediation. ISRN Ecology, 2011, 1–20. https://doi.org/10.5402/2011/402647.

[532] Xu, D., Yang, Z. & He, Q. (1998). Above the ground biomass production and nutrient cycling of middle-age (pp. 592–598).

[533] S, Y. (2020). Soil amendments for immobilization of potentially toxic elements in contaminated soils: A critical review. Environment International, 134(November 2019). 105046. https://doi.org/10. 1016/j.envint.2019.105046.

[534] Yadav, B. K., Siebel, M. A. & Van bruggen, J. J. A. (2011). Rhizofiltration of a heavy metal (Lead) containing wastewater using the Wetland plant Carex pendula. Clean – Soil, Air, Water, 39(5), 467–474.

[535] Yan-Chu, H. (1994). Arsenic distribution in soils. In Nriagu, J. O. (ed.). Arsenic in the Environment. Part I: Cycling and Characterization (pp.17–50). New York: Wiley.

[536] Yang, X., Hinzmann, M., Pan, H., Wang, J., Bolan, N., Daniel, C., Tsang, W., Sik, Y., Wang, S., Shaheen, S. M. & Wang, H. (2022). Pig carcass-derived biochar caused contradictory effects on arsenic mobilization in a contaminated paddy soil under fluctuating controlled redox conditions. Journal of Hazardous Materials, 421(July 2021), 0–2. https://doi.org/10.1016/j.jhazmat.2021.126647.

[537] Yang, X., Pan, H., Shaheen, S. M. & Wang, H. (2021). Immobilization of cadmium and lead using phosphorus-rich animal-derived and iron-modified plant-derived biochars under dynamic redox conditions in a paddy soil. 156. https://doi.org/10.1016/j.envint.2021.106628

[538] Yang, X.-E., Peng, H.-Y., Jiang, L.-Y. & Zhen-Li, H. (2005). Phytoextraction of copper from contaminated soil by elsholtzia splendens as affected by EDTA, citric acid, and compost. International Journal of Phytoremediation, 7(1), 69–83.

[539] Yao, L. & Wilding, L. P. (1995). Micromorphological study of compacted mine soil in east Texas. Intl. J. Rock Mech. Min. Sci. Geomech, 32, 219. http://dx.doi.org/10.1016/0148-9062(95)93274-S.

[540] Yeboah, D., Adams, B. & Oteng, D. (2019). The impact of mining on the environment: Case study from Ghana. International Journal of Environmental Science and Technology, 16(2), 347–362.

[541] Yen, L. V. & Saibeh, K. (2013). Phytoremediation using typha angustifolia l. for mine water effluence treatment: Case study of ex-mamut copper mine, ranau, sabah. Borneo Science, 33, 16–22.

[542] Yevugah, L. L., Darko, G. & Bak, J. (2021). Does mercury emission from small-scale gold mining cause widespread soil pollution in Ghana?. Environmental Pollution, 284, 116945.

[543] Yin, D., Wang, X., Chen, C., Peng, B., Tan, C. & Li, H. (2016). Varying effect of biochar on Cd, Pb and As mobility in a multi-metal contaminated paddy soil. Chemosphere, 152, 196–206. https://doi.org/10.1016/j.chemosphere.2016.01.044.

[544] Yoon, J., Cao, X., Zhou, Q. & Ma, L. Q. (2006). Accumulation of Pb, Cu, and Zn in native plants growing on a contaminated Florida site. Science of the Total Environment, 368, 456–464.

[545] Young, A. (1989). Agroforestry for Soil Conservation. CAB International, ICRAF, 93–103.

[546] Yu, K. & Rinklebe, J. (2011). Advancement in soil microcosm apparatus for biogeochemical research. Ecological Engineering, 37(12), 2071–2075. https://doi.org/10.1016/j.ecoleng.2011.08.017.

[547] Yu, K., Böhme, F., Rinklebe, J., Neue, H.-U. & DeLaune, R. D. (2007). Major biogeochemical processes in soils-a microcosm incubation from reducing to oxidizing conditions. Soil Science Society of America Journal, 71(4), 1406–1417. https://doi.org/10.2136/sssaj2006.0155.

[548] Yu, X., Li, Y., Li, Y., Xu, C., Cui, Y., Xiang, Q. . . . & Chen, Q. (2017). Pongamia pinnata inoculated with Bradyrhizobium liaoningense PZHK1 shows potential for phytoremediation of mine tailings. Applied Microbiology and Biotechnology, 101, 1739–1751.

[549] Zhang, L., Zhang, P., Yoza, B., Liu, W. & Liang, H. (2020). Phytoremediation of metal-contaminated rare-earth mining sites using Paspalum conjugatum. Chemosphere, 259, 127280.

[550] Zimmerman, A. J., Gutierrez, D. G., Campos, V. M., Weindorf, D. C., Deb, S. K., Chacón, S. U., Landrot, G., Flores, N. G. G. & Siebecker, M. G. (2021). Arsenic speciation in titanium dioxide (TiO2) waste produced via drinking water filtration: Potential environmental implications for soils, sediments, and human health. Environmental Advances, https://doi.org/10.1016/j.envadv.2021.100036.

[551] Zou, Q., An, W., Wu, C., Li, W., Fu, A., Xiao, R., Chen, H. & Xue, S. (2018). Red mud-modified biochar reduces soil arsenic availability and changes bacterial composition. Environmental Chemistry Letters, 16(2), 615–622. https://doi.org/10.1007/s10311-017-0688-1.

[552] Bradshaw, A. D. (1997). Restoration of mined lands—using natural processes. Ecological Engineering, 8(4), 255–269. https://doi.org/10.1016/S0925-8574(97)00022-0.

[553] Perrow, M. R. & Davy, A. J. (2002). Handbook of Ecological Restoration: Principles of Restoration (Vol. 1). UK: Cambridge University Press.

[554] Office of Surface Mining Reclamation and Enforcement. (2019). Annual report 2019. https://www.osmre.gov.

[555] Assel, P. G. (2006). Evaluating the usefulness of Acacia auriculiformis in ameliorating surface mine degraded lands. Unpublished B.Sc. dissertation, Department of Agroforestry, Kwame Nkrumah University of Science and Technology, Kumasi, Ghana. 1–24.

[556] Bonsu, M. & Quansah, C. (1992). The importance of soil conservation for agriculture and economic development of Ghana. In: Soil Resources Management towards Sustainable Agriculture in Ghana: The Role of the Soil Scientist: Proceedings of the 13th Annual General Meeting of the Soil Science Society of Ghana. 77–80.

[557] Asiamah, R. D., Asiedu, E. K. & Kwakye, P. K. (2001). Management and Rehabilitation of fertility declined soils: Case Study. Consultancy Report to FAO, Rome. 1–13.

[558] Appiah-Opoku, S. & Mulamoottil, G. (1997). Indigenous institutions and environmental assessment: The case of Ghana. Environmental Management, 21(2), 159–171.

[559] Kundu, N. K & Ghose, M. K. (1997). Soil profile characteristic in Rajmahal coalfield area. Indian Journal of Soil and Water Conservation, 25(1), 28–32.

[560] Wong, M. H. (2003). Ecological restoration of mine degraded soils, with emphasis on metal contaminated soils. Chemosphere, 50(6), 775–780.

[561] Sheoran, A. S., Sheoran, V. & Poonia, P. (2008). Rehabilitation of mine degraded land by metallophytes. Mining Engineers Journal, 10(3), 11–16.

[562] Grigg, A. H., Sheridan, G. J., Pearce, A. B. & Mulligan, D. R. (2006). The effect of organic mulch amendments on the physical and chemical properties and revegetation success of a saline-sodic minespoil from central Queensland, Australia. Soil Research, 44(2), 97–105.

[563] Viventsova, R. E., Kumpiene, J., Gunneriusson, L. & Holmgren, A. (2005). Changes in soil organic matter composition and quantity with distance to a nickel smelter – A case study on the Kola. Peninsula, NW Russia. Geoderma, 127(3–4), 216.

[564] Young, A. (1997). Agroforestry for Soil Management, Effects of Trees on Soils (pp. viii+-320). CAB International.

[565] Guo, T., Li, L., Zhai, W., Xu, B., Yin, X., He, Y., Xu, J., Zhang, T. & Tang, X. (2019). Distribution of arsenic and its biotransformation genes in sediments from the East China Sea. Environmental Pollution, 253, 949–958. https://doi.org/10.1016/jenvpol.2019.07.091.

[566] Zobeck, T. M. (2004). Rapid soil particle size analyses using laser diffraction. Applied Engineering in Agriculture, 20(5), 633.

[567] Isinkaye, M. O., Jibiri, N. N., Bamidele, S. I. & Najam, L. A. (2018). Evaluation of radiological hazards due to natural radioactivity in bituminous soils from tar-sand belt of southwest Nigeria using HpGe-Detector. International Journal of Radiation Research, 16(3), 351–362.

[568] Mensah, A. K., Marschner, B., Antoniadis, V., Stemn, E., Shaheen, S. M. & Rinklebe, J. (2021). Science of the total environment human health risk via soil ingestion of potentially toxic elements and remediation potential of native plants near an abandoned mine spoil in Ghana. Science of the Total Environment, 798, 149272. https://doi.org/10.1016/j.scitotenv.2021.149272.

[569] U.S. Environmental Protection Agency, 2002. Supplemental Guidance for Developing Soil Screening Levels for Superfund sites, pp. 1–184. https://nepis.epa.gov/Exe/ZyPDF.cgi/91003IJK.PDF?Dockey=91003IJK.PDF. (Accessed 7 February 2021).

[570] Taylor, A. A., Tsuji, J. S., Garry, M. S., McArdle, M. E., Goodfellow, W. L. Jr., Adams, W. J. & Menzie, C. A. (2020). Critical review of exposure and effects: Implications for setting regulatory health criteria for ingested copper. Environmental Management, 65, 131e159.

[571] Narh, C. T., Dzamalala, C. P., Mmbaga, B. T., Menya, D., Mlombe, Y., Finch, P. & Carreira, C. (2021). Geophagia and risk of squamous cell esophageal cancer in the African esopha- geal cancer corridor-findings from the ESCCAPE multi-country case-control studies. International Journal of Cancer, 149, 1274–1283.

[572] Duker, A. A., Carranza, E. J. & Hale, M. (2004). Spatial dependency of Buruli ulcer prevalence on arsenic-enriched domains in Amansie West District, Ghana: Implications for arsenic mediation in Mycobacterium ulcerans infection. International Journal of Health Geographics, 3, 1–10.

[573] Sparks, D. L., Page, A. L., Helmke, P. A., Loppert, R. H., Soltanpour, P. N., Tabatabai, M. A., Johnston, C. T. & Summner, M. E. (1996). Methods of Soil Analysis: Chemical Methods, Part 3. Madison, WI: ASA and SSSA.

[574] U.S. Environmental Protection Agency, 2007. Microwave Assisted Acid Digestion of Sedi- ments, Sludges, Soils, and Oils. 3051A. U.S.

[575] Beckers, F., Mothes, S., Abrigata, J., Zhao, J., Gao, Y., & Rinklebe, J. (2019). Mobilization of mercury species under dynamic laboratory redox conditions in a contaminated floodplain soil as affected by biochar and sugar beet factory lime. Science of the Total Environment, 672, 604–617. https://doi.org/10.1016/j.scitotenv.2019.03.401.

[576] Niane, B., Guédron, S., Feder, F., Legros, S., Ngom, P. M., & Moritz, R. (2019). Impact of recent artisanal small-scale gold mining in Senegal: Mercury and methylmercury contamination of terrestrial and aquatic ecosystems. Science of the Total Environment, 669, 185–193. https://doi.org/10.1016/j.scitotenv.2019.03.108.

[577] Renock, D., & Voorhis, J. (2017). Electrochemical investigation of arsenic redox processes on pyrite. Environmental Science and Technology, 51(7), 3733–3741. https://doi.org/10.1021/acs.est.6b06018.

[578] Shaheen, S. M., Rinklebe, J., Frohne, T., White, J. R. & DeLaune, R. D. (2016). Redox effects on release kinetics of arsenic, cadmium, cobalt, and vanadium in Wax Lake Deltaic freshwater marsh soils. Chemosphere, 150, 740–748.

[579] Rinklebe, J. & Shaheen, S. M. (2017). Redox chemistry of nickel in soils and sedimentsA review. Chemosphere, 179, 265–278. https://doi.org/10.1016/j.chemosphere.2017.02.153.

[580] Ye, X. X., Sun, B. & Yin, Y. L. (2012). Variation of As concentration between soil types and rice genotypes and the selection of cultivars for reducing As in the diet. Chemosphere, 87(4), 384–389. https://doi.org/10.1016/j.chemosphere.2011.12.028.

[581] Cancès, B., Juillot, F., Morin, G., Laperche, V., Polya, D., Vaughan, D. J. & Calas, G. (2008). Changes in arsenic speciation through a contaminated soil profile: A XAS based study. Science of the Total Environment, 397(1e3), 178–189.

[582] Oberthuer, T., Weiser, J. A., Amanor, & Chryssoulis, S. L. (1997). Mineralogical siting and distribution of gold in quartz veins and sulfide ores of the Ashanti mine and other deposits in the Ashanti belt of Ghana: Genetic implications. Mineralium Deposita, 32, 2e15.

[583] Hayford, E. K., Amin, A., Osae, E. K. & Kutu, J. (2009). Impact of gold mining on soil and some staple foods collected from selected mining communities in and around Tarkwa-Prestea area. West African Journal of Applied Ecology, 14(1).

[584] Chen, P., Li, J., Wang, H. Y., Zheng, R. L. & Sun, G. X. (2017). Evaluation of bioaugmentation and biostimulation on arsenic remediation in soil through biovolatilization. Environmental Science and Pollution Research, 24, 21739–21749.

[585] Gitt, M. J. & Dollhopf, D. J. (1991). Coal waste reclamation using automated weathering to predict lime requirement. Journal of Environmental Quality, 20, 285–288.

[586] Gould, A. B., Hendrix, J. W. & Ferriss, R. S. (1996). Relationship of mycorrhizal activity to time following reclamation of surface mine land in western Kentucky. Propagule and spore population densities. Canadian Journal of Botany, 74, 247–261.

[587] Maiti, S. K. & Ghose, M. K. (2005). Ecological restoration of acidic coal mine overburden dumps-an Indian case study. Land Contamination and Reclamation, 13(4), 361–369.

[588] Maiti, S. K. & Ghose, M. K. (2005). Ecological restoration of acidic coal mine overburden dumps-an Indian case study. Land Contamination and Reclamation, 13(4), 361–369.

[589] Sheoran, A. S., Sheoran, V. & Poonia, P. (2008). Rehabilitation of mine degraded land by metallophytes. Journal of Minerals Engineering, 10(3), 11–16.

[590] Barcelo, J. & Poschenrieder, C. (2003). Phytoremediation: Principles and perspectives. Contribution to Science, 2(3), 333–344.

[591] Das, M. & Maiti, S. K. (2005). Metal mine waste and phytoremediation. Asian Journal of Water, Environment and Pollution, 4(1), 169–176.

[592] Nicolau, J. M. (2002). Runoff generation and routing in a Mediterranean-continental environment: The Teruel coalfield, Spain. Hydrological Processes, 16, 631–647.

[593] Nicolau, J. M. (2002). Runoff generation and routing in a Mediterranean-continental environment: The Teruel coalfield, Spain. Hydrological Processes, 16, 631–647.

[594] Moreno-de Lasheras, M., Nicolau, J. M. & Espigares, M. T. (2008). Vegetation succession in reclaimed coal mining sloped in a Mediterranean-dry environment. Ecological Engineering, 34, 168–178.

[595] Hu, Z., Caudle, R. D. & Chong, S. K. (1992). Evaluation of firm land reclamation effectiveness based on reclaimed mine properties. International Journal of Mining, Reclamation and Environment., 6, 129–135.

[596] Maiti, S. K. & Saxena, N. C. (1998). Biological reclamation of coal mine spoils without topsoil: An amendment study with domestic raw sewage and grass-legumes mixture. International Journal of Mining, Reclamation and Environment, 12, 87–90.

[597] Singh, A. N., Raghubanahi, A. S. & Singh, J. S. (2004). Impact of native tree plantations on mine spoil in a dry tropical environment. Forest Ecology and Management, 187, 49–60.

[598] Singh, A. N. (2006). Experiments on ecological restoration of coal mine spoil using native trees in a dry tropical environment, India: A synthesis. New Forests, 31, 25–39.

[599] Heras, M. M. L. (2009). Development of soil physical structure and biological functionality in mining spoils affected by soil erosion in a Mediterranean-Continental environment. Geoderma, 149, 249–256.

[600] Six, J., Bossuyt, H., Degryze, S. & Denef, K. (2004). A history of research on the link between aggregates, soil biota, and soil organic matter dynamics. Soil and Tillage Research, 79, 7–31.

[601] Donahue, R. L., Miller, R. W. & Shickluna, J. C. (1990). Soils: An Introduction to Soils and Plant Growth (5th ed. p.234). Prentice-Hall.

[602] Maiti, S. K., Karmakar, N. C. & Sinha, I. N. (2002). Studies into some physical parameters aiding biological reclamation of mine spoil dump – A case study from Jharia coal field. Indian Mining & Engineering Journal, 41, 20–23.

[603] Rimmer, L. D. & Younger, A. (1997). Land reclamation after coal- mining operations. In Hester, R. E. & Harrison, R. M. (eds.). Contaminated Land and Its Reclamation (pp. 73–90). London: Thomas Telford.

[604] Daniels, W. L. (1999). Creation and Management of Productive Mine Soils, Powell River Project Reclamation Guidelines for Surface-Mined Land in Southwest Virginia. http://www.ext.vt.edu/pubs/mines/460-121/460-121.html.

[605] Singh, A. N. & Singh, J. S. (1999). Biomass and net primary production and impact of bamboo plantation on soil re-development in a dry tropical region. Ecology and Management, 119, 195–207.

[606] Singh, A. N. (2006). Experiments on ecological restoration of coal mine spoil using native trees in a dry tropical environment, India: A synthesis. New Forests, 31, 25–39.

[607] Prach, K. & Pysek, P. (2001). Using spontaneous establishment of woody plants in Central European derelicts sites and their potential for reclamation. Restoration Ecology, 2, 190–197.

[608] Bradshaw, A. D. (1997). Restoration of mined lands using natural processes. Ecological Engineering, 8, 255–269.

[609] Gathuru, G. (2011). The Performance of Selected Tree Species in the Rehabilitation of a Limestone Quarry at East African Portland Cement Company Land, Athi River, Kenya. PHD Thesis. Kenyatta University, Nairobi, Kenya.

[610] Edgerton, D. L., Harris, J. A., Birch, P. & Bullock, P. (1995). Linear relationship between aggregate stability and microbial biomass in three restored soils. Soil Biology and Biochemistry, 27, 1499–1501.

[611] Harris, J. P., Birch, P. & Short, K. C. (1989). Changes in the microbial community and physico-chemical characteristics of top soils stockpiled during opencast mining. Soil Use and Management, 5, 161–168.

[612] Visser, S., Fujikawa, J., Griffiths, C. L. & Parkinson, D. (1984). Effect of topsoil storage on microbial activity, primary production and decomposition potential. Plant and Soil, 82, 41–50.

[613] Gil-Sotres, F., Trasr-Cepeda, C., Leiros, M. C. & Seoane, S. (2005). Different approaches to evaluating soil quality using biochemical properties. Soil Biology and Biochemistry, 37, 877–887.

[614] Glick, B. R., Patten, C. L., Holguin, G. & Penrose, D. M. (1999). Biochemical and Genetic Mechanisms Used by Plant Growth-Promoting Bacteria (pp. 280). London, UK: Imperial College Press.

[615] Khan, A. G. (2005). Role of soil microbes in the rhizospheres of plants growing on trace element contaminated soils in phytoremediation. Journal of Trace Elements in Medicine and Biology, 18(4), 355–364.

[616] Marschner, H. (1995). Mineral Nutrition of Higher Plants (2nd ed. pp.889). New York: Academic Press. ISBN 0-12-473542- 8.

[617] Arriagada, C. A., Herrera, M. A., Garcia-Romera, I. & Ocampo, J. A. (2004). Tolerance of cadmium of soybean (Glycine max) and Eucalyptus (Eucalyptus globules) inoculated with Arbuscular mycorrhiza and saprobe fungi. Symbiosis, 36(3), 285–299.

[618] Arriagada, C. A., Herrera, M. A. & Ocampo, J. A. (2005). Contribution of Arbuscular mycorrhiza and saprobe fungi to the tolerance of Eucalyptus globules to Pb. Water, Air, & Soil Pollution, 166, 31–47.

[619] Miller, R. M., Carnes, B. A. & Moorman, T. B. (1985). Factors influencing survival of vesicular-arbuscular mycorrhiza propagules during topsoil storage. Journal of Applied Ecology, 22, 259–266.

[620] Rives, C. S., Bajwa, M. I. & Liberta, A. E. (1980). Effects of topsoil storage during surface mining on the viability of VA mycorrhiza. Soil Science, 129, 253–257.

[621] Fliessbach, A., Martens, R. & Reber, H. H. (1994). Soil microbial biomass and microbial activity in soils treated with heavy metal contaminated sewage sludge. Soil Biology & Biochemistry, 26(9), 1201–1205.

[622] Chander, K. P. C. B., Brookes, P. C. & Harding, S. A. (1995). Microbial biomass dynamics following addition of metal-enriched sewage sludges to a sandy loam. Soil Biology & Biochemistry, 27(11), 1409–1421.

[623] Koo, N., Lee, S. H. & Kim, J. G. (2012). Arsenic mobility in the amended mine tailings and its impact on soil enzyme activity. Environmental Geochemistry and Health, 34, 337–348.

[624] Amegbey, N. (2001). Lecture Notes on Environmental Engineering in Mining (pp. 117–118). Tarkwa, Ghana: University of Science and Technology, School of Mines.

[625] Addiscott, T. M., Whitmore, A. P. & Powlson, D. S. (1991). Farming Fertilizers and the Nitrate Problem (pp. 176). Wallingford, UK: CAB International.

[626] Johnson, D. B. & Williamson, J. C. (1994). Conservation of mineral nitrogen in restored soils at opencast coal mine sites: I. Results from field studies of nitrogen transformations following restoration. European Journal of Soil Science, 45(3), 311–317.

[627]    Isermann, K. (1994). Agriculture's share in the emission of trace gases affecting the climate and some cause-oriented proposals for sufficiently reducing this share. Environmental Pollution, 83(1–2), 95–111.

[628]    Davies, R., Hodgkinson, R., Younger, A. & Chapman, R. (1995). Nitrogen loss from a soil restored after surface mining. Journal of Environmental Quality, 24, 1215–1222.

[629]    Tibbett, M. (2010). Large-scale mine site restoration of Australian eucalypt forests after bauxite mining: Soil management and ecosystem development. Ecology of Industrial Pollution, 309–326.

[630]    Kundu, N. K. & Ghose, M. K. (1997). Soil profile Characteristic in Rajmahal Coalfield area. Indian Journal of Soil and Water Conservation, 25(1), 28–32.

[631]    Sendlein, L. V., Yazicigil, H. & Carlson, C. L. (1983). Surface Mining Environmental Monitoring and Reclamation Handbook.

[632]    Williams, M. (2001). Arsenic in mine waters: An international study. Environmental Geology, 40, 267–278.

[633]    Duker, A. A., Carranza, E. J. & Hale, M. (2004). Spatial dependency of Buruli ulcer prevalence on arsenic-enriched domains in Amansie West District, Ghana: Implications for arsenic mediation in Mycobacterium ulcerans infection. International Journal of Health Geographics, 3, 1–10.

[634]    Catalano, J. G., Park, C., Fenter, P. & Zhang, Z. (2008). Simultaneous inner-and-outer-sphere arsenate adsorption on corundum and hematite. Geochimica Et Cosmochimica Acta, 72(8), 1986–2004.

[635]    Kumah, A. (2006). Sustainability and gold mining in the developing world. Journal of Cleaner Production, 14(3–4), 315–323.

[636]    Dubiński, J. (2013). Sustainable development of mining mineral resources. Journal of Sustainable Mining, 12(1), 1–6.

[637]    Ghana Minerals Commission. (2012). Minerals and Mining (General) Regulations, 2012 (LI 2173).

[638]    Forkuor, G., Ullmann, T. & Griesbeck, M. (2020). Mapping and monitoring small-scale mining activities in Ghana using Sentinel-1 Time Series (2015–2019). Remote Sensing, 12(6), 911.

[639]    National Development Planning Commission (Ghana), & Ghana. Environmental Protection Agency. (2004). Strategic Environmental Assessment of the Ghana Poverty Reduction Strategy: Process Report. NDPC.

[640]    Damayanti, R. & Handayani, S. (2023, June). Rehabilitation plan for coal pit revegetation area East Kalimantan. In IOP Conference Series: Earth and Environmental Science (Vol. 1190, No. 1, p. 012016). IOP Publishing.

[641]    Coffie-Anum, E. & Bansah, K. J. (2016). Post-Mining Reclamation of Manganese Waste Dump. 4th UMaT Biennial International Mining and Mineral Conference, ES 64–73.

[642]    Smith, P., Gregory, P. J., Van Vuuren, D., Obersteiner, M., Havlik, P., Rounsevell, M. & Krug, T. (2013). The Role of Ecosystems and Their Management in Regulating Climate, and the Implementation of REDD+ Mechanisms. Oxford University Press.

[643]    Cao, X. & Harris, W. (2010). Properties of dairy-manure-derived biochar pertinent to its potential use in remediation. Bioresource Technology, 101(14), 5222–5228. https://doi.org/10.1016/j.biortech.2010.02.052.

[644]    Reever Morghan, K. J., Sheley, R. L. & Denny, M. K. (2006). Invasive annual grass and perennial forb suppression by revegetation across a northern big sagebrush ecological gradient. Ecological Restoration, 24(3), 251–257. https://doi.org/10.3368/er.24.3.251.

[645]    Wulder, M. A., Masek, J. G., Cohen, W. B., Loveland, T. R. & Woodcock, C. E. (2012). Opening the archive: How free data has enabled the science and monitoring promise of Landsat. Remote Sensing of Environment, 122, 2–10. https://doi.org/10.1016/j.rse.2012.01.010.

[646]    Gunningham, N. & Sinclair, D. (2002). Leaders and Laggards: Next-Generation Environmental Regulation. *Greenleaf Publishing*, https://doi.org/10.9774/GLEAF.9781783530228.

[647]  Hilderbrand, R. H., Watts, A. C. & Randle, A. M. (2005). The Myths of Restoration Ecology. Ecology and Society, 10(1), 19.

[648]  Laghlimi, M., Baghdad, B., El Hadi, H. & Bouabdli, A. (2015). Phytoremediation mechanisms of heavy metal contaminated soils: A review. Open Journal of Ecology, 5(8), 375–388.

[649]  Gregorowius, D., Lindemann-Matthies, P. & Huppenbauer, M. (2012). Ethical discourse on the use of genetically modified crops: A review of academic publications in the fields of ecology and environmental ethics. Journal of Agricultural and Environmental Ethics, 25, 265–293.

[650]  Munro, K., Hartt, C. M. & Pohlkamp, G. (2015). Social media discourse and genetically modified organisms. The Journal of Social Media in Society, 4(1).

[651]  Kohnke H, Anson Bertrand R (1959). Soil Conservation. McGraw-Hill Book Company, Inc. U.S.A. 256–257.

[652]  Ferguson, J. J., Rathinasabapathi, B. & Chase, C. A. (2013). Allelopathy: How plants suppress other plants. In Horticultural Sciences Department, Florida Cooperative Extension Service, Institute of Food and Agricultural Sciences. University of Florida, Available at:, http://edis.ifas.ufl.edu. Accessed on 5th June, 2015. 14:00 GMT.

[653]  Pandey, R. R., Sharma, G., Tripathi, S. K. & Singh, A. K. (2007). Litterfall, litter decomposition and nutrient dynamics in a subtropical natural Oak forest and managed plantation in northeastern India. Forest Ecology & Management, 240, 96–104.

[654]  Blum, W. E. H. (1988). Problems of soil conservation.nature and environment series 39, council of Europe, Strasbourg. In Towards Sustainable Land Use: Furthering Cooperation between People and Institutions,1(9), Selected papers of the Conference of ISCO. Catena Verlag, Reiskirchen, Germany. Advances in Geology, 31, 755–757.

[655]  Negri, M. C., Gatliff, E. G., Quinn, J. J. & Hinchman, R. R. (2003). Root development and rooting at depths. In McCutcheon, S. C. & Schnoor, J. L. (eds.). Phytoremediation:Transformation and Control of Contaminants (pp. 233–262). New York: Wiley.

[656]  Pulford, I. D. & Watson, C. (2003). Phytoremediation of heavy metal-contaminated land by trees – A review. Environment International, 29(4), 529–540.

[657]  Padmavathiamma, P. K. & Li, L. Y. (2007). Phytoremediation technology: Hyper-accumulation metals in plants. Water, Air, & Soil Pollution, 184, 105–126.

[658]  Shu, W. S., Xia, H. P., Zhang, Z. Q., Lan, C. Y. & Wong, M. H. (2002). Use of vetiver and three other grasses for revegetation of Pb/Zn mine tailings: Field experiment. International Journal of Phytoremediation, 4(1), 47–57.

[659]  Singh, A. N., Raghubanshi, A. S. & Singh, J. S. (2002). Plantations as a tool for mine spoil restoration. Current Science, 1436–1441.

[660]  Hao, X. Z., Zhou, D. M., Wang, Y. J. & Chen, H. M. (2004). Study of rye grass in copper mine tailing treated with peat and chemical fertilizer. Acta Pedologica Sinica, 41(4), 645–648.

[661]  Liao, L. P., Wang, S. L. & Chen, C. Y. (2000). Dynamics of litter fall in the mixed plantation of Cunninghamia lanceolata and Michelia macclurei: A ten-year's observation. China Journal of Applied Ecology, 11, Supp, 131–136.

[662]  Zhang, J. W., Liao, L. P., Li, J. F. & Su, Y. (1993). Litter dynamics of Pinusmassoniana and Micheliamacclurei mixed forest and its effect on soil nutrients. China Journal of Applied Ecology, 4(4), 359–363.

[663]  Sundarapandian, S. M. & Swamy, P. S. (1999). Litter production and leaf litter decomposition of selected tree species in tropical forests at Kodayar in the Western Ghats, India. Ecology and Management, 123, 231–244.

[664]  Yang, Y. S., Guo, J. F., Chen, G. S., Xie, J. S., Cai, L. P. & Lin, P. (2004). Litter fall, nutrient return, and leaf-litter decomposition in four plantations compared with a natural forest in subtropical China. Annals of Science, 61, 465–476.

[665]  Kelty, M. J. (2006). The role of species mixtures in plantation forestry. Ecology and Management, 233, 195–204.

[666]  Forrester, D. I., Bauhus, J. & Cowie, A. L. (2005). Nutrient cycling in a mixed-species plantation of Eucalyptus globulus and Acacia mearnsii. Canadian Journal of Forest Research, 35(12), 2942–2950.

[667]  Binkley, D., Dunkin, K. A., DeBell, D. & Ryan, M. G. (1992). Production and nutrient cycling in mixed plantations of Eucalyptus and Albizia in Hawaii. Science, 38, 393–408.

[668]  Binkley, D. (1992). Mixtures of nitrogen-fixing and non-nitrogen-fixing tree species. In Cannell, M. G. R., Malcolm, D. C. & Robertson, P. A. (eds.). The Ecology of Mixed Species Stands of Trees (pp. 99–123). London: Blackwell Scientific.

[669]  Binkley, D. & Ryan, M. G. (1998). Net primary production and nutrient cycling in replicated stands of Eucalyptus saligna and Albiziafacaltaria. Forest Ecology and Management, 112, 79–85.

[670]  Parrotta, J. A. (1999). Productivity, nutrient cycling, and succession in single and mixed-species plantations of Casuarina equisetifolia, Eucalyptus robusta, and Leucaena leucocephalain Puerto Rico. Forest Ecology and Management, 124, 45–77.

[671]  Li, Y. M., Chaney, R. L., Brewer, E. P., Roseberg, R. J., Angle, J. S., Baker, A. J. M., Reeves, R. D. & Nelkin, J. (2003). Development of a technology for commercial phytoextraction of nickel: Economic and technical considerations. Plant Soil, 249, 107–115.

[672]  Chaney, R. L., Angle, J. S., Broadhurst, C. L., Peters, C. A., Tappero, R. V. & Donald, L. S. (2007). Improved understanding of hyper-accumulation yields commercial phytoextraction and phytomining technologies. Journal of Environmental Quality, 36, 1429–14423.

[673]  Yang, B., Shu, W. S., Ye, Z. H., Lan, C. Y. & Wong, M. H. (2003). Growth and metal accumulation in vetivera and two Sesbania species on lead/zinc mine tailings. Chemosphere, 52(9), 1593–1600.

[674]  Song, S. Q., Zhou, X., Wu, H. & Zhou, Y. Z. (2004). Application of municipal garbage compost on revegetation of tin tailings dams. Rural Eco-Environment, 20(2), 59–61.

[675]  Zhang, J. W., Liao, L. P., Li, J. F. & Su, Y. (1993). Litter dynamics of Pinusmassoniana and Micheliamacclurei mixed forest and its effect on soil nutrients. China Journal of Applied Ecology, 4(4), 359–363.

[676]  Puga, A. P., Abreu, C. A., Melo, L. C. A. & Beesley, L. (2015). Biochar application to a contaminated soil reduces the availability and plant uptake of zinc, lead and cadmium. Journal of Environmental Management, 159, 86–93.

[677]  Sheoran, V. & Choudhary, R. P. (2021). Phytostabilization of mine tailings. In Phytorestoration of Abandoned Mining and Oil Drilling Sites (pp. 307–324). Elsevier.

[678]  Pathak, L. & Shah, K. (2024). Natural colonizers effectively restore heavy metal polluted wasteland. International Journal of Phytoremediation, 1–12.

[679]  Mendez, M. O. & Maier, R. M. (2008). Phytostabilization of mine tailings in arid and semiarid environments – An emerging remediation technology. Environmental Health Perspectives, 116(3), 278–283.

[680]  Hu, Y., Liu, X., Bai, J., Shih, K., Zeng, E. Y. & Cheng, H. (2013). Assessing heavy metal pollution in the surface soils of a region that had undergone three decades of intense industrialization and urbanization. Environmental Science and Pollution Research, 20, 6150–6159.

[681]  Hu, Z. & Xiao, W. (2013). Optimization of concurrent mining and reclamation plans for single coal seam: A case study in northern Anhui, China. Environmental Earth Sciences, 68, 1247–1254.

[682]  Kolay, A. K. (2000). Basic Concepts of Soil Science (2nd ed. pp. 78–80 90–91, 102–105, 138). New Delhi, India: New Age International Publishers.

[683]  Narh, C. T., Dzamalala, C. P., Mmbaga, B. T., Menya, D., Mlombe, Y., Finch, P. & Carreira, C. (2021). Geophagia and risk of squamous cell esophageal cancer in the African esopha- geal cancer corridor-findings from the ESCCAPE multi-country case-control studies. International Journal of Cancer, 149, 1274–1283.

[684] Erbaugh, J. T. & Oldekop, J. A. (2018). Forest landscape restoration for livelihoods and well-being. Current Opinion in Environmental Sustainability, 32, 76–83.

[685] Strong, W. L. (2016). Biased richness and evenness relationships within Shannon–Wiener index values. Ecological Indicators, 67, 703–713.

[686] Burger, J. A. (2009). Sustainable mined land reclamation in the eastern U.S. coalfields: A case for an ecosystem reclamation approach. Reclamation and Sustainability, 10(2), 119–133. https://doi.org/10.1016/j.recsus.2009.03.006.

[687] Richardson, D. M., Pysek, P., Rejmanek, M., Barbour, M. G., Panetta, F. D. & West, C. J. (2008). Naturalization and invasion of alien plants: Concepts and definitions. Diversity and Distributions, 6(2), 93–107. https://doi.org/10.1046/j.1472-4642.2000.00083.x.

[688] Cunningham, R. A. (2001). Managing forest health in reclamation projects: Lessons from western Canada. Forest Ecology & Management, 10(1), 36–44.

[689] Smith, D. C. & Jones, J. H. (2010). Soil and crop responses to reclamation on abandoned mine lands. Reclamation & Soil Science, 25(1), 80–98.

[690] Osinakachukwu, M. E., Donatus, A. O., Ikevuje, A. H. & Edwin, B. G. (2024). Stakeholder engagement and influence: Strategies for successful energy projects. International Journal of Management & Entrepreneurship Research, 6(7), 2375–2395. Doi: 10.51594/ijmer.v6i7.1330.

[691] Reynolds, S. (2024). Stakeholder Engagement and its Impact on Supply Chain Sustainability in the Context of Renewable Energy. https://doi.org/10.20944/preprints202406.0080.v1

[692] African Union. (2015). Agenda 2063: The Africa We Want. Addis Ababa, Ethiopia: African Union Commission, Available at https://au.int/sites/default/files/documents/33126-doc-framework_document_book.pdf, Accessed 19.12.2024.

[693] United Nations Convention to Combat Desertification (UNCCD). (2022). The Global Land Outlook (2nd ed.). Bonn: UNCCD.

[694] Nsiah, P. K. & Schaaf, W. (2020). Subsoil amendment with poultry manure as topsoil substitute for promoting successful reclamation of degraded mine sites in Ghana. Journal of Natural Resources and Development, 10, 01–12. https://doi.org/10.5027/jnrd.v10i0.01.

[695] Singh, A. K., Sisodia, A., Sisodia, V. & Padhi, M. (2019). Role of microbes in restoration ecology and ecosystem services. In New and Future Developments in Microbial Biotechnology and Bioengineering (pp. 57–68). Elsevier, https://doi.org/10.1016/B978-0-444-64191-5.00004-3.

[696] Kumar, K. V., Raj, B. A., Sriraghul, A., Sadanish, K., Raj, N. R., Prajith, K. S. & Tamilselvan, M. (2023). Comparing the effect of organic and inorganic amendments on soil health. Bhartiya Krishi Anusandhan Patrika, https://doi.org/10.18805/BKAP599.

[697] Tao, C., Li, R., Xiong, W., Shen, Z., Liu, S., Wang, B., Ruan, Y., Geisen, S., Shen, Q. & Kowalchuk, G. A. (2020). Bio-organic fertilizers stimulate indigenous soil Pseudomonas populations to enhance plant disease suppression. Microbiome, 8(1), 137. https://doi.org/10.1186/s40168-020-00892-z.

[698] Burgess, A. J., Correa Cano, M. E. & Parkes, B. (2022). The deployment of intercropping and agroforestry as adaptation to climate change. Crop and Environment, 1(2), 145–160. https://doi.org/10.1016/j.crope.2022.05.001.

[699] Kuyah, S., Whitney, C. W., Jonsson, M., Sileshi, G. W., Öborn, I., Muthuri, C. W. & Luedeling, E. (2019). Agroforestry delivers a win-win solution for ecosystem services in sub-Saharan Africa. A meta-analysis. Agronomy for Sustainable Development, 39(5), 47. https://doi.org/10.1007/s13593-019-0589-8.

[700] Hayden, B., Greene, D. F. & Quesada, M. (2010). A field experiment to determine the effect of dry-season precipitation on annual ring formation and leaf phenology in a seasonally dry tropical forest. Journal of Tropical Ecology, 26(2), 237–242. https://doi.org/10.1017/S0266467409990563.

[701] Mkuhlani, S., Zinyengere, N., Kumi, N. & Crespo, O. (2022). Lessons from integrated seasonal forecast-crop modelling in Africa: A systematic review. Open Life Sciences, 17(1), 1398–1417. https://doi.org/10.1515/biol-2022-0507.

[702]  Holl, K. D. (2020). Primer of Ecological Restoration. Island Press, Retrieved from https://island press.org/books/primer-ecological-restoration.

[703]  Ruwanza, S. (2020). Topsoil transfer from natural renosterveld to degraded old fields facilitates native vegetation recovery. Sustainability, 12(9), 3833. https://doi.org/10.3390/su12093833.

[704]  Sanjay, M. A., Rai, S., Patil, A. A., Nengparmoi, T., Devi, K. B., Dora, H. S. V. & Sharma, Y. (2023). Environmental sustainability through soil conservation: An imperative for future generations. International Journal of Environment and Climate Change, 13(10), 1700–1707. https://doi.org/10.9734/ijecc/2023/v13i102826.

[705]  Galabuzi, C., Eilu, G., Mulugo, L., Kakudidi, E., Tabuti, J. R. S. & Sibelet, N. (2014). Strategies for empowering the local people to participate in forest restoration. Agroforestry Systems, 88(4), 719–734. https://doi.org/10.1007/s10457-014-9713-6.

[706]  Gholap, V. B., Benke, S. R. & Patil, S. N. (2022). Management and economics of ornamental nursery. International Journal of Agricultural Sciences, 18(1), 195–201. https://doi.org/10.15740/HAS/IJAS/18.1/195-201.

[707]  Putra, A. B., Rahman, F. K., Suheri Sastri, M. & Saswini, A. A. U. (2024). empowering women farmers through the Moengko nursery house program in Poso district. Sociality: Journal of Public Health Service, 153–160. https://doi.org/10.24252/sociality.v3i2.50792.

[708]  Behera, R. D. (2022). Soil and Plant Analysis. NIPA, https://doi.org/10.59317/9789394490895.

[709]  Holl, K. D. (2020). Primer of Ecological Restoration. Island Press, Retrieved from https://island press.org/books/primer-ecological-restoration.

[710]  Karimi, N., Ghaderian, S. M., Maroofi, H. & Schat, H. (2009). Analysis of arsenic in soil and vegetation of a contaminated area in Zarshuran, Iran. International Journal of Phytoremediation, 12(2), 159–173.

[711]  Clark, J. (2000). Viii. an International Overview of Legal Frameworks for mine Closure. 67–77.

[712]  Laghlimi, M., Baghdad, B., El Hadi, H. & Bouabdli, A. (2015). Phytoremediation mechanisms of heavy metal contaminated soils: A review. Open Journal of Ecology, 5(8), 375–388.

[713]  Gregorowius, D., Lindemann-Matthies, P. & Huppenbauer, M. (2012). Ethical discourse on the use of genetically modified crops: A review of academic publications in the fields of ecology and environmental ethics. Journal of Agricultural and Environmental Ethics, 25, 265–293.

[714]  Munro, K., Hartt, C. M. & Pohlkamp, G. (2015). Social media discourse and genetically modified organisms. The Journal of Social Media in Society, 4(1).

[715]  Vangronsveld, J., Herzig, R., Weyens, N., Boulet, J., Adriaensen, K., Ruttens, A. ... Mench, M. (2009). Phytoremediation of contaminated soils and groundwater: Lessons from the field. Environmental Science and Pollution Research, 16, 765–794.

[716]  Eapen, S., Singh, S. & D'Souza, S. (2007). Advances in development of transgenic plants for remediation of xenobiotic pollutants. Biotechnology Advances, 25(5), 442–451.

[717]  Shackira, A. M. & Puthur, J. T. (2019). Phytostabilization of Heavy Metals: Understanding of Principles and Practices. Plant-Metal Interactions (pp. 263–282). Cham: Springer.

[718]  Zeng, S., Ma, J., Yang, Y., Zhang, S., Liu, G. J. & Chen, F. (2019). Spatial assessment of farmland soil pollution and its potential human health risks in China. Science of the Total Environment, 687, 642–653.

[719]  Jiang, Y., Lei, M., Duan, L. & Longhurst, P. (2015). Integrating phytoremediation with biomass valorisation and critical element recovery: A UK contaminated land perspective. Biomass Bioenergy, 83, 328–339.

[720]  Placer Dome Inc. (1998). Sustainability policy.

[721]  Gibson, R. B. (2000). Favouring the higher test: Contribution to sustainability as the central criterion for reviews and decisions under the Canadian environmental assessment act. Journal of Environmental Law and Practice, 10(1), 39–54.

[722] Wright, A. F. & Bailey, J. S. (2001). Organic carbon, total carbon, and total nitrogen determinations in soils of variable calcium carbonate contents using a Leco CN-2000 dry combustion analyzer. Communications in Soil Science and Plant Analysis, 32(19–20), 3243–3258.

[723] Bagherifam, S., Brown, T. C., Fellows, C. M. & Naidu, R. (2019). Derivation methods of soils, water and sediments toxicity guidelines: A brief review with a focus on antimony. Journal of Geochemical Exploration, 205, 106348.

[724] Chang, S. H., Su, W. B., Jian, W. B., Chang, C. S., Chen, L. J. & Tsong, T. T. (2002). Electronic growth of Pb islands on Si (111) at low temperature. Physical Review B, 65(24), 245401.

[725] Zimmer, D., Kiersch, K., Baum, C., Meissner, R., Müller, R., Jandl, G., & Leinweber, P. (2011). Scale-Dependent Variability of As and Heavy Metals in a River Elbe Floodplain. Clean - Soil, Air, Water, 39 (4), 328–337. https://doi.org/10.1002/clen.201000295.

[726] Abrahams, P. W. (2002). Soils: their implications to human health. Science of the Total Environment, 291(1–3), 1–32.

[727] Addiscott, T. M., Whitmore, A. P. & Powlson, D. S. (1991). Farming Fertilizers and the Nitrate Problem (p. 176). Wallingford, UK: CAB International.

[728] Agboola, A. A. (1990). Organic matter and soil fertility management in the humidtropics of Africa. In Elliot, C. R. et al (ed.). Organic Matter Management and Tillage in Humid and Sub-humid Africa (pp. 231–243). Bangkok, Thailand: International Board for Soil Research and Management, 1990. IBSRAM Proceedings, 10.

[729] Agodzo, S. K. & Adama, I. (2003). Bulk density, core index and water content relations for some ghanaian soils. In Lecture Given at the College of Engineering on Soil Physics. Trieste: KNUST, 3–21 March 2003.

[730] Agodzo, S. K. & Adama, I. (2003). Bulk density, cone index and water content relations for some Ghanaian soils. Kwame Nkrumah Univ. *Sci. Technol. Kumasi-Ghana*.

[731] Akala, V. A. & Lal, R. (2001). Soil organic carbon pools and sequestration rates in reclaimed mine soils in Ohio. Journal of Environmental Quality, 30, 2098–2104.

[732] Alengebawy, A., Abdelkhalek, S. T., Qureshi, S. R. & Wang, M.-Q. (2021). Heavy Metals and Pesticides Toxicity in Agricultural Soil and Plants: Ecological Risks and Human Health Implications. Toxics, 9(3), 42. doi: https://doi.org/10.3390/toxics9030042.

[733] Alexander, M. J. (1990). Reclamation after tin mining on the Jos plateau, Nigeria. Geographical Journal, 156, 44–50.

[734] Ali, H. & Khan, E. (2018). Bioaccumulation of non-essential hazardous heavy metals and metalloids in freshwater fish. Risk to human health. Environmental Chemistry Letters, 16(3), 903–917.

[735] Ali, H., Khan, E. & Sajad, M. A. (2013). Phytoremediation of heavy metals – Concepts and applications. Chemosphere, 91(7), 869–881.

[736] Ali, M. M., Hossain, D., Al-Imran, A., Khan, M. S., Begum, M. & Osman, M. H. (2021). Environmental pollution with heavy metals: A public health concern. Heavy Metals-Their Environmental Impacts and Mitigation, 771–783.

[737] Amegbey, N. (2001). Lecture Notes on Environmental Engineering in Mining (pp. 139). Tarkwa, Ghana: University of Science and Technology, School of Mines, 117–118; 120–122; 124; 127–128, 135.

[738] Anane-Sakyi, C. (1995). Organic matter and soil fertility management in the North Eastern Savannah Zone of Ghana. In: Proceedings of Seminar on Organic and Sedentary Agriculture.

[739] Anderson, J. M. & Ingram, J. S. I. (1989). Tropical soil biology and fertility. In A Handbook of Method (pp. 37–38). C.A.B. International.

[740] Andoh, J. & Lee, Y. (2018). Forest transition through reforestation policy integration: A comparative study between Ghana and the Republic of Korea. Forest Policy & Economics, 90, 12–21. doi: 10.1016/j.forpol.2018.01.009.

[742] Antoniadis, V., Levizou, E., Shaheen, S. M., Ok, Y. S., Sebastian, A., Baum, C., Prasad, M. N. V., Wenzel, W. W. & Rinklebe, J. (2017). Trace elements in the soil-plant interface: phytoavailability,

translocation, and phytoremedi- ation–a review. Earth-Science Reviews, 171(October 2016), 621–645. doi: 10.1016/j.earscirev.2017.06.005.

[743] Antoniadis, V., Shaheen, S. M., Boersch, J., Frohne, T., Du Laing, G. & Rinkleble, J. (2017). Bioavailability and risk assessment of potentially toxic ele- ments in garden edible vegetables and soils around a highly contam- inated former mining area in Germany. Journal of Environmental Management, 186(Pt 2), 192–200. doi: 10.1016/j.jenvman.2016.04.036.

[744] Apostoli, P. & Catalani, S. (2015). Metal ions Affecting Reproduction and Development. Metal Ions in Toxicology: Effects, Interactions, Interdependencies (pp. 263–304, Vol. 11). De Gruyter.

[745] Appiah-Opoku, S. & Mulamoottil, G. (1997). Indigenous Institutions and Environmental Assessment. The Case of Ghana, 21(2), 159–171.

[746] Arriagada, C. A., Herrera, M. A. & Ocampo, J. A. (2005). Contribution of Arbuscular mycorrhiza and saprobe fungi to the tolerance of Eucalyptus globules to Pb. Water, Air, & Soil Pollution, 166, 31–47.

[747] Arriagada, C. A., Herrera, M. A., Garcia-Romera, I. & Ocampo, J. A. (2004). Tolerance of cadmium of soybean (Glycine max) and Eucalyptus (Eucalyptus globules) inoculated with Arbuscular mycorrhiza and saprobe fungi. Symbiosis, 36(3), 285–299.

[748] Arthur-Holmes, F., Busia, K. A., Vazquez-Brust, D. A. & Yakovleva, N. (2022). Graduate unemployment, artisanal and small-scale mining, and rural transformation in Ghana: What does the 'educated' youth involvement offer? Journal of Rural Studies, 95, 125–139.

[749] Asafu-Agyei, J. N. (1995). The role of legumes in soil fertility improvement-soil fertility status in Ghana. In Proceedings of Seminar on Organic and Sedentary Agriculture.

[750] Asiamah, R. D., Asiedu, E. K. & Kwakye, P. K. (2001). Management and Rehabilitation of fertility declined soils: Case Study. Consultancy Report to FAO, Rome. 1–13.

[751] Asiedu, J. B. K. (2013). Technical report on reclamation of small-scale surface mined lands in Ghana: a landscape perspective. American Journal of Environmental Protection, 1, 28–33.

[752] Assel, P. G. (2006). Evaluating the usefulness of Acacia auriculiformis in ameliorating surface mine degraded lands. Unpublished B.Sc. dissertation, Department of Agroforestry, Kwame Nkrumah University of Science and Technology, Kumasi, Ghana. 1–24.

[753] Aubynn, A. K. (1997). Economic Restructuring Dynamics And Environmental Problems In Africa: Empirical Examples from the Forestry and Mining sectors of Ghana. UNU/IAS Working Paper No. 34

[754] Awotwi, A., Anornu, G. K., Quaye-Ballard, J. A. & Annor, T. (2018). Monitoring land use and land cover changes due to extensive gold mining, urban expansion, and agriculture in the Pra River Basin of Ghana, 1986–2025. Land Degradation & Development, 29(10), 3331–3343. https://doi.org/10.1002/ldr.3093.

[755] Baker, A. J. M., McGrath, S. P., Reeves, R. D. & Smith, J. A. C. (2000). Metal hyperaccumulator plants: A review of the ecology and physiology of a biological resource for phytoremediation of metal-polluted soils. In Terry, N. & Banuelos, G. S. (eds.). Phytoremediation of Contaminants in Soil and Water (pp. 85–107). Boca Raton (FL): CRC Press.

[756] Baker, A. J. M. (1981). Accumulators and excluders-strategies in the response of plants to heavy metals. Journal of Plant Nutrition, 3, 643–654.

[757] Barber, S. A. (1995). Soil Nutrient Bio-Availability (pp. 35–90). John Wiley and Sons Inc.

[758] Barcelo, J. & Poschenrieder, C. (2003). Phytoremediation: principles and perspectives. Contribution to Science, 2(3), 333–344.

[759] Barkworth, H. & Bateson, M. (1964). An investigation into the bacteriology of top soil dumps. Plant Soil, 21(3), 345–353.

[760] BC EMLI. (2014, June 25). Release of "Lease Area" and "SRW Access Road" from Reclamation Permit M-74. Victoria, British Columbia, Canada: British Columbia Ministry of Energy and Mines.

[761] Beckers, F. & Rinklebe, J. (2017). Cycling of mercury in the environment: sources, fate, and human health implications: a review. Critical Reviews in Environmental Science and Technology, 47(9), 693–794. doi: 10.1080/10643389.2017.1326277.

[762] Beesley, L., Moreno-Jiménez, E. & Gomez-Eyles, J. L. (2010). Effects of biochar and greenwaste compost amendments on mobility, bioavailability and toxicity of inorganic and organic contaminants in a multi-element polluted soil. Environmental Pollution, 158(6), 2282–2287.

[763] Beesley, L., Moreno-Jiménez, E., Gomez-Eyles, J. L., Harris, E., Robinson, B. & Sizmur, T. (2011). A review of biochars' potential role in the remediation, revegetation and restoration of contaminated soils. Environmental Pollution, 159(12), 3269–3282. doi: https://doi.org/10.1016/j.envpol.2011.07.023.

[764] Belden, S. E., Schuman, G. E. & Depuit, E. J. S. (1990). Salinity and moisture responses in wood residue amended bentonite mine spoil. Soil Science, 150, 874–882.

[765] Bell, L. C., Mulligan, D. R. & Mitchell, R. J. (1991). Definition of rehabilitation strategies for pre-strip tertiary overburden spoil at the Saraji open-cast coal mine. Progress report to BHP-Utah Coal Limited (2). February 1991, Department of Agriculture, University of Queensland.

[766] Bell, L. C., Mulligan, D. R., Mitchell, R. J. & Philp, M. W. (1992). Definition of rehabilitation strategies for pre-strip tertiary overburden spoil at the Saraji open-cut coal mine. Progress report to BHP Australia Coal Limited (5). October 1992, Department of Agriculture, University of Queensland.

[767] Bewley, J. D. & Black, N. (1978). Physiology and Biochemistry of Seeds in Relation to Germination (p. 2). Springer-Verlag, Berlin.

[768] Bhattacharjee, K. & Behera, B. (2018). Determinants of household vulnerability and adaptation to floods: Empirical evidence from the Indian State of West Bengal. International Journal of Disaster Risk Reduction, 31, 758–769.

[769] Binkley, D. (1992). Mixtures of nitrogen-fixing and non-nitrogen-fixing tree species In Cannell, M. G. R., Malcolm, D. C. & Robertson, P. A. (eds. ). The Ecology of Mixed Species Stands of Trees (pp. 99–123). London: Blackwell Scientific.

[770] Binkley, D., Dunkin, K. A., DeBell, D. & Ryan, M. G. (1992). Production and nutrient cycling in mixed plantations of Eucalyptus and Albizia in Hawaii. Science, 38, 393–408.

[771] Binkley, D. & Ryan, M. G. (1998). Net primary production and nutrient cycling in replicated stands of Eucalyptus saligna and Albiziafacaltaria. Forest Ecology and Management, 112, 79–85.

[772] Birch, H. F. (1958). The effect of soil drying on humus decomposition and nitrogen availability. Plant & Soil, 10, 9–31.

[773] Biswas, M. & Mukherjee, A. (1994). Synthesis and evaluation of metal-containing polymers. Photoconducting Polymers/Metal-Containing Polymers, 89–123.

[774] Blaylock, M. J., Salt, D. E., Dushenkov, S., Zakharova, O., Gussman, C., Kapulnik, Y., Ensley, B. D. & Raskin, I. (1997). Enhanced accumulation of Pb in Indian mustard by soil-applied chelating agents. Environmental Science & Technology, 35(9).

[775] Blaylock, M. J. & Huang, J. W. (2000). Phytoextraction of metals. In Raskin, I. & Ensley, B. D. (eds. ). Phytoremediation of Toxic Metals: Using Plants to Clean up the Environment (pp. 53–70). New York: Wiley.

[776] Blum, W. E. H. (1988). Problems of Soil Conservation. Nature and Environment Series 39, Council of Europe, Strasbourg. In Towards Sustainable Land Use: Furthering Cooperation between People and Institutions. 1 (9). Selected papers of the Conference of ISCO. Catena Verlag, Reiskirchen, Germany.Adv. Geoecol. (31), 755–757.

[777] Bonsu, M. & Quansah, C. (1992). Soil resources management towards sustainable agriculture in ghana: The role of the soil scientist: proceedings of the 13th annual general meeting of the soil science society of Ghana. In The Importance of Soil Conservation for Agriculture and Economic Development of Ghana (pp. 77–80).

[778] Bonsu, M., Quansah, C. & Agyemang, A. (1996). Interventions for land and water conservation in Ghana. In Paper Prepared for National Workshop on Soil Fertility Management Action Plan for Ghana (p. 31). Sasakawa Center, University of Cape Coast, Cape Coast.

[779] Bradshaw, A. D. (1997). Restoration of mined lands – Using natural processes. Ecological Engineering, 8(4), 255–269.

[780] Branco, A. 2008. Mobility and bioavailability ofarsenic in italian polluted soils. PhD thesis.

[781] Caravaca, F., Hernandez, M. T., Garia, C. & Roldan, A. (2002). Improvement of rhizosphere aggregates stacbility of afforested semi- arid, plant species subjected to mycorrhizal inoculation and compost addition. Geoderma, 108, 133–144.

[782] Cargnelutti, D., Tabaldi, L. A., Spanevello, R. M., De oliveira jucoski, G., Battisti, V., Redin, M. ... Schetinger, M. R. C. (2006). Mercury toxicity induces oxidative stress in growing cucumber seedlings. Chemosphere, 65(6), 999–1006.

[783] Carroll, C., Merton, L. & Berger, P. (2000). Impact of vegetative cover and slope on runoff, erosion, and water quality for field plots on a range of soil and spoil materials on central Queensland coal mines. Australian Journal of Soil Research, 38, 313–327, doi: 10.1071/SR99052.

[784] Catalano, J. G., Park, C., Fenter, P. & Zhang, Z. (2008). Simultaneous inner- and outer-sphere arsenate sorption on corundum and hematite. Geochimica et Cosmochimica Acta, 72(8), 1986–2004.

[785] Chaney, R. L., Angle, J. S., Broadhurst, C. L., Peters, C. A., Tappero, R. V. & Donald, L. S. (2007). Improved understanding of hyper-accumulation yields commercial phytoextraction and phytomining technologies. Journal of Environmental Quality, 36, 1429–14423.

[786] Choi, Y. J., Lee, K. S. & Oh, J. W. (2021). The impact of climate change on pollen season and allergic sensitization to pollens. Immunology and Allergy Clinics, 41(1), 97–109.

[787] Clements, F. E. (1916). Plant Succession: An Analysis of the Development of Vegetation. Carnegie Institution of Washington.

[788] Coates, W. (2005). Tree species selection for a mine tailings bioremediation project in Peru. Biomass Bioenergy, 28(4), 418–423.

[789] Conesa, H. M., Garcia, G., Faz, A. & Arnaldos, R. (2007b). Dynamics of metal tolerant plant communities' development in mine tailings from the Cartagena-La Union Mining District (SE Spain) and their interest for further revegetation purposes. Chemosphere, 68, 1180–1185.

[790] Conesa, H. M., Schulin, R. & Nowack, B. (2007a). A laboratory study on revegetation and metal uptake in native plant species from neutral mine tailings. Water, Air, and Soil Pollution, 183, 1–4, 201–212.

[791] Cooper, R. T. M., Leakey, R. R. B., Rao, M. R. & Reynolds, L. (1996). Agroforestry and Mitigation of Land Degradation in the Humid and Sub-humid Tropics of Africa. Experimental Agriculture, 32(3), July 1996.

[792] Coppin, N. J. & Bradshaw, A. D. (1982). The Establishment Of Vegetation in Quarries and Open-Pit Non-Metal Mines. London: Min J. Books, 112.

[793] Coppin, P., Hermy, M. & Honnay, O. (2000). Impact of Habitat Quality of Forest Plant Species Colonization. Forest Ecology and Management, 115, 157–170.

[794] Costanza, R., d'Arge, R., De groot, R., Farber, S., Grasso, M., Hannon, B. ... Van Den Belt, M. (1997). The value of the world's ecosystem services and natural capital. Nature, 387(6630), 253–260.

[795] Cox, R. M. & Hutchinson, T. C. (1981). Environmental factors influencing the rate of spread of the grass Deschampsia cespitosa invading areas around the Sudbury nickel-copper smelters. Water, Air & Soil Pollution, 16, 83–106.

[796] Cunningham, S. D. & Ow, D. W. (1996). Promises and prospects of phytoremediation. Plant Physiology, 110, 715–719.

[797]   Daniel, P., Kovacs, B., Prokisch, J. & Gyori, Z. (1997). Heavy metal dispersion detected in soils and plants alongside roads in Hungary. Heavy metal dispersion detected in soils and plants alongside roads in Hungary. Chem Speciat Bioavailab, 9, 83–93.

[798]   Daniels, W. L. (1999). Creation and Management of Productive Mine Soils, Powell River Project Reclamation Guidelines for Surface-Mined Land in Southwest Virginia. http://www.ext.vt.edu/pubs/mines/460-121/460-121.html.

[799]   Das, M. & Maiti, S. K. (2005). Metal mine waste and phytoremediation. Asian Journal of Water, Environment and Pollution, 4(1), 169–176.

[800]   Davies, R., Hodgkinson, R., Younger, A. & Chapman, R. (1995). Nitrogen loss from a soil restored after surface mining. Journal of Environmental Quality, 24, 1215–1222.

[801]   Davies, R. & Younger, A. (1994). The effect of different post- restoration cropping regimes on some physical properties of a restored soil. Soil Use and Management, 10, 55–60.

[802]   De Gryze, S., Cullen, M., Durschinger, L., Lehmann, J., Bluhm, D., Six, J. & Suddick, E. (2010). Evaluation of the opportunities for generating carbon offsets from soil sequestration of biochar. An Issues Paper Commissioned by the Climate Action Reserve, Final Version, 1–99.

[803]   Dexter, A. R. (1988). Advances in the characterization of soil structure. Soil and Tillage Research, 11, 199–238.

[804]   Diogo, R. V. C., Bizimana, M., Nieder, R., Rukazambuga Ntirushwa, D. T., Naramabuye, F. X. & Buerkert, A. (2017). Effects of compost type and storage conditions on climbing bean on Technosols of Tantalum mining sites in Western Rwanda. Journal of Plant Nutrition and Soil Science, 180(4), 482–490.

[805]   Donahue, R. L., Miller, R. W. & Shickluna, J. C. (1990). Soils: An Introduction to Soils and Plant Growth (p. 234, 5th Edition). Prentice-Hall.

[806]   Donkor, A. K., Bonzongo, J. C., Nartey, V. K. & Adotey, D. K. (2006). Mercury in different environmental compartments of the Pra River Basin, Ghana. Science of the Total Environment, 368 (1), 164–176. doi: 10.1016/j.scito-tenv.2005.09.046.

[807]   Doran, J. W. & Zeiss, M. R. (2000). Soil health and sustainability: Managing the biotic component of soil quality. Applied Soil Ecology, 15.

[808]   Duku, M. H., Gu, S. & Hagan, E. B. (2011). Biochar production potential in Ghana – a review. Renewable and Sustainable Energy Reviews, *15*(8), 3539–3551.

[809]   Dutta, R. K. (1999). Performance and Impact of Selected Exotic Plant Species on Coal Mine Spoil. Ph.D. Thesis. Banaras Hindu University, Varanasi, India.

[810]   Dutta, R. K. & Agrawal, M. (2002). Effect of tree plantations on the soil characteristics and microbial activity of coal mine spoil land. International Society for Tropical Ecol, 43(2), 315–324.

[811]   Eckermann, A., Dowd, T., Chong, E., Nixon, L., Gray, R. & Johnson, S. (2006). Binan Goonj: Bridging Cultures in Aboriginal Health (2nd edition). Sydney: Elsevier.

[812]   Edgerton, D. L., Harris, J. A., Birch, P. & Bullock, P. (1995). Linear relationship between aggregate stability and microbial biomass in three restored soils. Soil Biology and Biochemistry, 27, 1499–1501.

[813]   Elkins, N. Z., Parker, L. W., Aldon, E. & Whitford, W. G. (1984). Responses of soil biota to organic amendments in strip-mine spoil in northwestern New Mexico. Journal of Environmental Quality, 13, 215–219.

[814]   Emmerton, B. R. (1984). Delineation of the factors affecting pasture establishment on mined land at the Gregory open-cut coal mine, central Queensland. M.Sc. thesis, Department of Agriculture, University of Queensland, Australia.

[815]   Ensley, B. D. (2000). Rationale for use of phytoremediation. In Raskin, I. & Ensley, B. D. (eds. ). Phytoremediation of toxic metals: using plants to clean up the environment (pp. 3–11). New York: John Wiley and Sons.

[816]   EPA. (2004). Reclamation Security Agreement between Ghanaian – Australian Goldfields Ltd Iduapriem and EPA. Ghana.

[817]   Erbaugh, J. T. & Oldekop, J. A. (2018). Forest landscape restoration for livelihoods and well-being. Current Opinion in Environmental Sustainability, *32*, 76–83.

[818]   European Commission. (2019). Review of progress on implementation of the EU green infrastructure strategy. In Report from the Commission to the European Parliament, the Council, the European Economic and Social Committee and the Committee of the Regions. 2019. European Commission: Brussels.

[819]   FAO, SER & IUCN CEM (2023). Standards of practice to guide ecosystem restoration. A contribution to the United Nations Decade on Ecosystem Restoration. Summary report. Rome, FAO. https://doi.org/10.4060/cc5223en.

[820]   Fasinu, P. S. & Orisakwe, O. E. (2013). Heavy metal pollution in sub-saharan Africa and possible implications in cancer epidemiology. Asian Pacific Journal of Cancer Prevention, 14(6), 3393–3402.

[821]   Favorito, J. E., Luxton, T. P., Eick, M. J. & Grossl, P. R. (2017). Selenium speciation in phosphate mine soils and evaluation of a sequential extraction procedure using XAFS. Environmental Pollution, 229, 911–921. doi: 10.1016/j.envpol.2017.07.071.

[822]   Felker, P. (1978). State of the Art: Acacia albida as a Complementary Permanent Intercrops with Annual Crops. Report to USAID. River side, California, USA: University of California. 133.

[823]   Felker, P. (1981). Uses of tree legumes in semiarid regions. Economic Botany, 35(2), 174–186.

[824]   Festin, E. S., Tigabu, M., Chileshe, M. N., Syampungani, S. & Odén, P. C. (2019). Progresses in restoration of post-mining landscape in Africa. Journal of Forestry Research, 30(2), 381–396.

[825]   Festin, E. S., Tigabu, M., Chileshe, M. N., Syampungani, S. & Oden, P. C. (2019). Progresses in restoration of post-mining landscape in Africa. Journal of Forestry Research, 30(2), 381–396. https://doi.org/10.1007/s11676-018-0621-x.

[826]   Forkuor, G., Ullmann, T. & Griesbeck, M. (2020). Mapping and monitoring small-scalemining activities in Ghana using sentinel-1 time series2015–2019. Remote Sensing, 12, 6, https://doi.org/10.3390/rs12060911.

[827]   Formosa, M. L. & Kelly, E. C. (2020). Socioeconomic benefits of a restoration economy in the Mattole River Watershed, USA. *Society & Natural Resources, 33*(9), 1111–1128.

[828]   Forrester, D. I., Bauhus, J. & Cowie, A. L. (2005). Nutrient cycling in a mixed-species plantation of Eucalyptus globulus and Acacia mearnsii. Canadian Journal of Forest Research, 35(12), 2942–2950.

[829]   Franco, A. A. & Faria de, S. M. (1996). The contribution of N2-fixing tree legumes to land reclamation and sustainability in the tropics. Soil Biol Biochem, 29(5/6), 897–903.

[830]   Fu, W. G. & Wang, F. K. (2015). Effects of high soil lead concentration on photosynthetic gas exchange and chlorophyll fluorescence in Brassica chinensis L. Plant Soil and Environment, 61(7), 316–321.

[831]   Gadd, G. M. (2010). Metals, minerals and microbes: Geomicrobiology and bioremediation. Microbiology, 156(3), 609–643.

[832]   Gann, G. D., McDonald, T., Walder, B., Aronson, J., Nelson, C. R., Jonson, J., Hallett, J. G., Eisenberg, C., Guariguata, M. R., Liu, J., Hua, F., Echeverría, C., Gonzales, E., Shaw, N., Decleer, K. & Dixon, K. W. (2019). International principles and standards for the practice of ecological restoration. Second edition. Restoration Ecology, 27, S1, https://doi.org/10.1111/rec.13035.

[833]   Gathuru, G. (2011). The Performance of Selected Tree Species in the Rehabilitation of a Limestone Quarry at East African Portland Cement Company Land, Athi River, Kenya. PHD Thesis. Kenyatta University, Nairobi, Kenya.

[834]   Ghose, M. K. (2004). Effect of Opencast Mining on Soil Fertility. Journal of Scientific and Industrial Research, 63, 1006–1009.

[835]   Ghose, M. K. (2005). Soil conservation for rehabilitation and revegetation of mine-degraded land. TIDEE – TERI Information Digest on Energy and Environment, 4(2), 137–150.

[836] Ghosh, A. B., Bajaj, J. C., Hassan, R. & Singh, D. (1983). Soil and Water Testing Methods- Laboratory Manual (pp. 31–36). IARI, New Delhi.

[837] Gibson, R. B. (2012). Why sustainability assessment? In Sustainability Assessment (pp. 3–17). Routledge.

[838] Gill, H. S. & Abrol, I. P. (1986). Salt-affected soils and their amelioration through afforestation. In Prinsley, R. T. & Swift, M. J. (eds.). Amelioration of Soil by Trees: A Review of Current Concepts and Practices (pp. 43–56). London: Commonwealth Science Council.

[839] Gil-Sotres, F., Trasr-Cepeda, C., Leiros, M. C. & Seoane, S. (2005). Different approaches to evaluating soil quality using biochemical properties. Soil Biology & Biochemistry, 37, 877–887.

[840] Gitt, M. J. & Dollhopf, D. J. (1991). Coal waste reclamation using automated weathering to predict lime requirement. Journal of Environmental Quality, 20, 285–288.

[841] Glick, B. R., Patten, C. L., Holguin, G. & Penrose, D. M. (1999). Biochemical and Genetic Mechanisms Used by Plant Growth-promoting Bacteria (pp. 280). London, UK: Imperial College Press.

[842] Gong, S. Y. & Jiang, D. Q. (1977). Soil Failure and its Control in Small Gully Watersheds in the Loess Area on the Middle Reaches of the Yellow River, Paris Symp. Erosion and Soil Matter Transport in Inland Waters preprint.

[843] Gould, A. B., Hendrix, J. W. & Ferriss, R. S. (1996). Relationship of mycorrhizal activity to time following reclamation of surface mine land in western Kentucky. I Propagule and spore population densities. Canadian Journal of Botany, 74, 247–261.

[844] Gratani, M., Sutton, S. G., Butler, J. R. A., Bohensky, E. L. & Foale, S. (2016). Indigenous environmental values as human values. Cogent Social Sciences, 2, 1. https://doi.org/10.1080/23311886.2016.1185811

[845] Gregorich, E. G., Kachanoski, R. G. & Voroney, R. P. (1989). Carbon mineralization in soil size fractions after various amounts of aggregate disruption. Journal of Soil Science, 40, 649–659.

[846] Grewal, S. S. & Abrol, I. P. (1986). Agroforestry on Alkali Soils: Effect of some management practices on initial growth, biomass accumulation and chemical composition of selected tree species. 4, 221–232

[847] Grigg, A. H. & Catchpoole, S. (1999). Rehabilitation of pre-strip Tertiary spoils at Saraji mine. Assessment of soil and vegetation development 8 years after establishment, June 1999. Report to BHP Saraji mine, December 1999. Centre for Mined Land Rehabilitation, University of Queensland.

[848] Grigg, A. H., Sheridan, G. J., Pearce, A. B. & Mulligan, D. R. (2006). The Effect of Organic Mulch Amendments on the Physical and Chemical Properties and Revegetation Success of a Saline-sodic Mine Spoil from Central Queensland (pp. 1–9). Australia.

[849] Gunn, R. H. (1967). A soil catena on denuded laterite profiles in Queensland. Soil Research, 5(1), 117–132.

[850] Guo, T., Li, L., Zhai, W., Xu, B., Yin, X., He, Y., Xu, J., Zhang, T. & Tang, X. (2019). Distribution of arsenic and its biotransformation genes in sediments from the East China Sea. Environmental Pollution, 253,949–958. doi: 10.1016/j.envpol.2019.07.091.

[851] Gustin, M. S., Hou, D. Y. & Tack, F. M. (2021). The term" heavy metal (s)": History, current debate, and future use.

[852] Guuroh, R. T., Foli, E. G., Addo-Danso, S. D., Stanturf, J., Kleine, M. & Burns, J. (2021). Restoration of degraded forest reserves in Ghana. Reforesta, 12, 35–55.

[853] Haider, F. U., Wang, X., Zulfiqar, U., Farooq, M., Hussain, S., Mehmood, T., Naveed, M., Li, Y., Liqun, C., Saeed, Q., Ahmad, I. & Mustafa, A. (2022). Biochar application for remediation of organic toxic pollutants in contaminated soils; An update. Ecotoxicology and Environmental Safety, 248, 114322. https://doi.org/10.1016/j.ecoenv.2022.114322

[854] Hao, X. Z., Zhou, D. M., Wang, Y. J. & Chen, H. M. (2004). Study of rye grass in copper mine tailing treated with peat and chemical fertilizer. Acta Pedol Sin, 41(4), 645–648.

[855]  Harper, K. A., Drapeau, P., Lesieur, D. & Bergeron, Y. (2014). Forest structure and composition at fire edges of different ages: Evidence of persistent structural features on the landscape. Forest Ecology & Management, 314, 131–140.

[856]  Harris, J. P., Birch, P. & Short, K. C. (1989). Changes in the microbial community and physico-chemical characteristics of top soils stockpiled during opencast mining. Soil Use and Management, 5, 161–168.

[857]  Hartley, W. & Lepp, N. W. (2008). Remediation of arsenic contaminated soils by iron-oxide application, evaluated in terms of plant productivity, arsenic and phytotoxic metal uptake. Science of the Total Environment, 390(1), 35–44. https://doi.org/10.1016/j.scitotenv.2007.09.021.

[858]  Harwood, M. R. (1998). The selection of new pasture grasses for the revegetation of open-cut coal mines in central Queensland. PhD thesis, Department of Agriculture, University of Australia.

[859]  Harwood, M. R., Hacker, J. B. & Mott, J. J. (1999). Field evaluation of seven grasses for use in the revegetation of lands disturbed by coal mining in Central Queensland. Australian Journal of Experimental Agriculture, 39,307–316. doi: 10.1071/EA98119.

[860]  Hayford, E. K., Amin, A., Osae, E. K. & Kutu, J. (2009). Impact of gold mining on soil and some staple foods collected from selected mining communities in and around Tarkwa- Prestea area. West African Journal of Applied Ecology, 14, 1–12. https://doi.org/10.4314/wajae.v14i1.44708

[861]  He, D., Dong, Z. & Zhu, B. (2024). An optimal global biochar application strategy based on matching biochar and soil properties to reduce global cropland greenhouse gas emissions: Findings from a global meta-analysis and density functional theory calculation. Biochar, 6(1), 92. https://doi.org/10.1007/s42773-024-00383-6.

[862]  Heras, M. M. L. (2009). Development of soil physical structure and biological functionality in mining spoils affected by soil erosion in a Mediterranean-Continental environment. Geoderma, 149, 249–256.

[863]  Hey, C., Evans, H. & Burke, P. (2023). Integrating renewable energy in post-mining land uses. Ausenco Engineering Canada Inc., 4515 Central Boulevard, 18th Floor Burnaby, British Columbia, V5H 0C6.

[864]  Hiller, E., Lalinská, B., Chovan, M., Jurkovic, L., Klimko, T., Jankulár, M., Hovoric, R., Šottník, P., Flaková, R., Ženišová, Z. & Ondrejková, I. (2012). Arsenic and antimony contamination ofwaters, streamsediments and soils in the vicinity ofabandoned antimonymines in the Western Carpathians, Slovakia. Applied Geochemistry, 27(3), 598–614. https://doi.org/10.1016/j.apgeochem.2011.12.005.

[865]  Hobbs, R. J. & Norton, D. A. (1996). Towards a Conceptual Framework for Restoration Ecology. Restoration Ecology, 4(2), 93–110. https://doi.org/10.1111/j.1526-100X.1996.tb00112.x.

[866]  Hodge, A. (2010). Farm Animal Welfare and Sustainability. University of Exeter (United Kingdom).

[867]  Holland, J. E., Bennett, A. E., Newton, A. C., White, P. J., McKenzie, B. M., George, T. S. & Hayes, R. C. (2018). Liming impacts on soils, crops and biodiversity in the UK: A review. Science of the Total Environment, 610, 316–332.

[868]  Holling, C. S. (1978). Adaptive Environmental Assessment and Management. John Wiley & Sons.

[869]  Houben, D., Pircar, J. & Sonnet, P. (2012). Heavy metal immobilization by cost-effective amendments in a contaminated soil: Effects on metal leaching and phytoavailability. Journal of Geochemical Exploration, 123,87–94. doi: 10.1016/j.gexplo.2011.10.004.

[870]  Howard, J. J. (1988). Leaf cutting and diet selection: Relative influence of leaf chemistry and physical features. Ecology, 69(1), 250–260.

[871]  Hu, Z., Caudle, R. D. & Chong, S. K. (1992). Evaluation of firm land reclamation effectiveness based on reclaimed mine properties. International Journal of Mining, Reclamation and Environment., 6 (129), 135.

[872]  Hu, L., Zhang, Z., Xiang, Z. & Yang, Z. (2016). Exogenous application of citric acid ame- liorates the adverse effect of heat stress in tall fescue (Lolium arundinaceum). Frontiers Plant Science, 7, 179.

[873]  Hua, J., Zhang, C., Yin, Y., Chen, R. & Wang, X. (2012). Phytoremediation potential of three aquatic macrophytes in manganese-contaminated water. Water and Environment Journal, 26(3), 335–342.

[874]  Huang, J. W., Chen, J., Berta, W. R. & Cunningham, S. D. (1997). Phytoremediation of lead contaminated soil: Role of synthetic chelates in lead phytoextraction. Environmental Science & Technology, 31(3), 800–805.

[875]  Hunter, F. & Currie, J. A. (1956). Structural changes during bulk soil storage. Journal of Soil Science, 7, 75–86.

[876]  Hutchinson, T. C. & Whitby, L. M. (1977). The effects of acid rainfall and heavy metal particulates on a boreal forest ecosystem near the Sudbury smelting region of Canada. Water, Air & Soil Pollution, 7, 421–438.

[877]  Ingram, J. (1990). The Role of Trees in Maintaining and Improving Soil Productivity.A Review of the Literature.Commonwealth Science Council (pp. 1–19). London.

[878]  IPCC. (2021). Climate Change 2021: The Physical Science Basis. Contribution of Working Group I to the Sixth Assessment Report of the Intergovernmental Panel on Climate Change. Cambridge University Press.

[879]  Isermann, K. (1994). Agriculture's share in emission of trace gases affecting the climate and some cause oriented proposals for sufficiently reducing this share. Environmental Pollution, 83, 95–111.

[880]  Isinkaye, O. M. (2018). Distribution and multivariate pollution risks assessment of heavy metals and natural radionuclides around abandoned iron-ore mines in North Central Nigeria. Earth Systems and Environment, 2(2), 331–343. doi: 10.1007/s41748-018-0035-0.

[881]  Issaka, S. & Ashraf, M. A. (2021). Phytorestoration of mine spoiled:"Evaluation of natural phytoremediation process occurring at ex-tin mining catchment. In Phytorestoration of Abandoned Mining and Oil Drilling Sites (pp. 219–248). Elsevier.

[882]  Jenny, H. (1941). Factors of Soil Formation: A System of Quantitative Pedology. McGraw-Hill.

[883]  Jha, A. K. & Singh, J. S. (1991). Spoil Characteristics and vegetation development of an age series of mine spoils in a dry tropical environment. Vegetation, 97, 63–76.

[884]  Jha, A. K. & Singh, J. S. (1993). Growth performance of certain directly seeded plants on mine spoil in a dry tropical environment. India Forest, 119, 920–927.

[885]  Jia, Y., Maurice, C. & Öhlander, B. (2016). Mobility of as, Cu, Cr, and Zn from tailings covered with sealing materials using alkaline industrial residues: A comparison between two leaching methods. Environmental Science & Pollution Research International, 23(1), 648–660. doi: 10.1007/s11356-015-5300-2.

[886]  Jiang, J., Yuan, M., Xu, R. & Bish, D. L. (2015). Mobilization of phosphate in variable-charge soils amended with biochars derived from crop straws. Soil and Tillage Research, 146, 139–147. https://doi.org/10.1016/j.still.2014.10.009

[887]  Johnson, D. B. & Williamson, J. C. (1994). Conservation of mineral nitrogen in restored soils at opencast mines sites: I. Result from field studies of nitrogen transformations following restoration. European Journal of Soil Science, 45, 311–317.

[888]  Johnson, D. B., Williamson, J. C. & Bailey, A. J. (1991). Microbiology of soils at opencast sites. Short and Long- term transformation in stockpiled soils. Journal of Soil Science, 42, 1–8.

[889]  Kavamura, V. N. & Esposito, E. (2010). Biotechnological strategies applied to the decontamination of soils polluted with heavy metals. Biotechnology Advances, 28(1), 61–69.

[890]  Kelty, M. J. (2006). The role of species mixtures in plantation forestry. Ecology and Management, 233, 195–204.

[891]  Kemp, W., Bellairs, S. M., Joyce, J. & Henderson, J. (2023). Consideration of First Nations' cultural values in mine site rehabilitation by environmental professionals. Environmental Challenges, 13 (100757).

[892]  Khalil, S. & Hassan, N. E. (2024). Review of Heavy Metal Removal from Soil: Methods and Technologies. Global Academic Journal of Agriculture and Biosciences, 6.

[893]  Khamkhash, A., Srivastava, V., Ghosh, T., Akdogan, G., Ganguli, R. & Aggarwal, S. (2017). Mining-related selenium contamination in Alaska, and the state of current knowledge. Minerals, 7(3), 46. doi: 10.3390/min7030046.

[894]  Khan, A. G. (2005). Role of soil microbes in the rhizospheres of plants growing on trace element contaminated soils in phytoremediation. Journal of Trace Elements in Medicine and Biology, 18(4), 355–364.

[895]  Kim, D. M., Yun, S. T., Yoon, S. & Mayer, B. (2019). Signature of oxygen and sulfur isotopes of sulfate in ground and surface water reflecting enhanced sulfide oxidation in mine areas. Applied Geochemistry, 100, 143–151.

[896]  Kim, H., Yang, S., Rao, S. R., Narayanan, S., Kapustin, E. A., Furukawa, H. & Wang, E. N. (2017). Water harvesting from air with metal-organic frameworks powered by natural sunlight. Science, 356(6336), 430–434.

[897]  Klubi, E., Abril, J. M., Mantero, J., García-Tenorio, R. & Nyarko, E. (2020). Environmental radioactivity and trace metals in surficial sediments from estuarine systems in Ghana (Equatorial Africa), impacted by artisanal gold-mining. Journal of Environmental Radioactivity, 218, 106260.

[898]  Kobayashi, S. (2004). Landscape rehabilitation of degraded tropical forest ecosystems. A Case Study of the CIFOR/Japan Project in Indonesia and Peru, 1–2.

[899]  Koe, T., Assuncao, A. & Bleeker, P. (1999). Phytoremediation of the Jales mine spoils with arsenate tolerant genotypes of HolcuslanatusL. Rev Bio Lisb, 17, 23–32.

[900]  Kohnke, H. & Anson Bertrand, R. (1959). Soil Conservation (pp. 256–257). McGraw-Hill Book Company, Inc. U.S.A.

[901]  Kolay, A. K. (2000). Basic Concepts of Soil Science (pp. 138, 2nd edition). New Delhi, India: New Age International Publishers, Vols. 78-80, 90–91, 102–105.

[902]  Kundu, N. K. & Ghose, M. K. (1997). Soil profile Characteristic in Rajmahal Coalfield area. Indian Journal of Soil and Water Conservation, 25(1), 28–32.

[903]  Kunhikrishnan, A., Choppala, G., Seshadri, B., Wijesekara, H., Bolan, N. S., Mbene, K. & Kim, W. I. (2017). Impact of wastewater derived dissolved organic carbon on reduction, mobility, and bioavailability of As (V) and Cr (VI) in contaminated soils. Journal of Environmental Management, 186, 183–191.

[904]  Kuyah, S., Whitney, C. W., Jonsson, M., Sileshi, G. W., Öborn, I., Muthuri, C. W. & Luedeling, E. (2019). Agroforestry delivers a win-win solution for ecosystem services in sub-Saharan Africa. A meta-analysis. Agronomy for Sustainable Development, 39(5), 47. https://doi.org/10.1007/s13593-019-0589-8.

[905]  Lal, R. (2006). Enhancing crop yields in the developing countries through restoration of the soil organic carbon pool in agricultural lands. Land Degradation & Development, 17(2), 197–209.

[906]  Lam, E. J., Cánovas, M., Gálvez, M. E., Montofré, Í. L., Keith, B. F. & Faz, Á. (2017). Evaluation of the phytoremediation potential of native plants growing on a copper mine tailing in northern Chile. Journal of Geochemical Exploration, 182, 210–217.

[907]  Lasat, M. M., Baker, A. J. M. & Kochian, L. V. (1998). Altered Zn compartmentation in the root symplasm and stimulated Zn absorption into the leaf as mechanisms involved in Zn hyper-accumulation in Thlaspi caerulescens. Plant Physiology, 118(875), 883.

[908]  Leach, M. & Fairhead, J. (2000). Challenging neo-Malthusian deforestation analyses in West Africa's dynamic forest landscapes. Population and Development Review, 26(1), 17–43.

[909]  Lehmann, J. & Joseph, S. (2015). Biochar for Environmental Management: Science, Technology and Implementation. Routledge.

[910]  Li, M. S. (2006). Ecological restoration of mine land with particular reference to the metalliferous mine wasteland in China: A review of research and practice. Soil Total Environment, 357, 38–53.

[911]    Li, Y. M., Chaney, R. L., Brewer, E. P., Roseberg, R. J., Angle, J. S., Baker, A. J. M., Reeves, R. D. & Nelkin, J. (2003). Development of a technology for commercial phytoextraction of nickel: Economic and technical considerations. Plant Soil, 249, 107–115.

[912]    Liao, L. P., Wang, S. L. & Chen, C. Y. (2000). Dynamics of litter fall in the mixed plantation of Cunninghamia lanceolata and Michelia macclurei: A ten-year's observation. China Journal of Applied Ecology, 11(Supp), 131–136.

[913]    Lim, C. H. & Kim, H. J. (2022). Can forest-related adaptive capacity reduce landslide risk attributable to climate change? – Case of Republic of Korea. Forests, 13(1), 49.

[914]    Lin, C., Tong, X., Lu, W., Yan, L., Wu, Y., Nie, C. & Long, J. (2005). Environmental impacts of surface mining on mined lands, affected streams and agricultural lands in the Dabaoshan Mine region, southern China. Land Degradation & Development, 16(5), 463–474.

[915]    Lindemann, W. C., Lindsey, D. L. & Fresquez, P. R. (1984). Amendment of mine spoils to increase the number and activity of microorganisms. America Journal of Soil Science Society, 48, 574–578.

[916]    Loska, K., Wiechuła, D. & Korus, I. (2004). Metal contamination offarming soils affected by industry. Environment International, 30(2), 159–165. doi: 10.1016/S0160-4120(03)00157-0.

[917]    Maboeta, M. S. & Van Rensburg, L. (2003). Earthworm assessment of a bio-solids application to mine tailings. Ecotoxicology and Environmental Safety, 56(2), 236–244.

[918]    Madejon, E., De Mora, A. P., Felipe, E., Burgos, P. & Cabrera, F. (2006). Soil amendments reduce trace element solubility in a contaminated soil and allow re-growth of natural vegetation. Environmental Pollution, 139, 40–52.

[919]    Maiti, S. K. (2003). Moef report, an assessment of overburden dump rehabilitation technologies adopted in CCL, NCL, MCL, and SECL mines (Grant no. J-15012/38/98-IA IIM).

[920]    Maiti, S. K. & Ghose, M. K. (2005). Ecological restoration of acidic coal mine overburden dumps-an Indian case study. Land Contamination and Reclamation, 13(4), 361–369.

[921]    Maiti, S. K., Karmakar, N. C. & Sinha, I. N. (2002). Studies into some physical parameters aiding biological reclamation of mine spoil dump – A case study from Jharia coal field. Indian Mining & Engineering Journal, 41, 20–23.

[922]    Maiti, S. K. & Saxena, N. C. (1998). Biological reclamation of coal mine spoils without topsoil: An amendment study with domestic raw sewage and grass-legumes mixture. International Journal of Mining, Reclamation and Environment, 12, 87–90.

[923]    Marschner, H. (1995). Mineral Nutrition of Higher Plants (p. 889, 2nd edition). New York: Academic Press. ISBN 0-12-473542-8.

[924]    Martin, M., Celi, L., Barberis, E., Violante, A., Kozac, L. M. & Huang, P. M. (2009 Mar). Effect of humic acid coating on arsenic adsorption on ferrihydrite-kaolinite mixed systems. Canadian Journal of Soil Science, 89, 421–434.

[925]    Mbagwu, J. S. C., Unamba-Oparah, I. & Nevoh, G. O. (1994). Physico-chemical properties and productivity of two tropical soils amended with dehydrated swine waste. Bioresource Technology, 49(163), -171.

[926]    McGowen, S. L., Basta, N. T. & Brown, G. O. (2001). Use of diammonium phosphate to reduce heavy metal solubility and transport in smelter-contaminated soil. Journal of Environmental Quality, 30, 493–500.

[927]    McIntyre, T. (2003). Phytoremediation of heavy metals from soils. Advances in Biochemical Engineering Biotechnology, 78, 97–123.

[928]    McLaughlin, M. J., Tiller, K. G., Beech, T. A. & Smart, M. K. (1994). Soil salinity causes elevated cadmium concentrations in field-grown potato tubers. Journal of Environmental Quality, 23, 1013–1018.

[929]    McMurray, A. & Param, R. (2008). Culture-specific care for indigenous people: A primary health care perspective. Contemporary Nurse, 28(1–2), 165–172.

[930]  Mendez, M. O. & Maier, R. M. (2008a). Phytoremediation of mine tailings in temperate and arid environments. Reviews in Environmental Science and Biotechnology, 7, 47–59.

[931]  Mendez, M. O. & Maier, R. M. (2008b). Phytostabilization of mine tailings in arid and semi-arid environments- An emerging remediation technology. Environmental Health Perspectives, 116(3), 278–283.

[932]  Mendez, M. O. & Maier, R. M. (2008). Phytostabilization of mine tailings in arid and semiarid environments-an emerging remediation technology. Environmental Health Perspectives, 116(3), 278–283. doi: 10.1289/ehp.10608.

[933]  Meng, J., Wang, L., Liu, X., Wu, J., Brookes, P. C. & Xu, J. (2013). Physicochemical properties of biochar produced from aerobically composted swine manure and its potential use as an environmental amendment. Bioresource Technology, 142,641–646. doi: 10.1016/j.biortech.2013.05.086.

[934]  Mensah, A. K. (2015). Role of revegetation in restoring fertility of degraded mined soils in Ghana: A review. International Journal of Biodiversity and Conservation, 7(2), 57–80. https://doi.org/10.5897/ijbc2014.0775.

[935]  Mensah, A. K., Marschner, B., Shaheen, S. M. & Rinklebe, J. (2022). Biochar, compost, iron oxide, manure, and inorganic fertilizer affect bioavailability of arsenic and improve soil quality of an abandoned arsenic-contaminated gold mine spoil. Ecotoxicology and Environmental Safety, 234, 113358. https://doi.org/10.1016/j.ecoenv.2022.113358

[936]  Mensah, A. K., Marschner, B., Wang, J., Bundschuh, J., Wang, S. L., Yang, P. T. & Rinklebe, J. (2022). Reducing conditions increased the mobilisation and hazardous effects of arsenic in a highly contaminated gold mine spoil. Journal of Hazardous Materials, 436, 129238.

[937]  Mensah, A. K., Samuel Obeng, A., Addai, P., Owusu-Ansah, A. & Owusu-Ansah, D.-G. E. J. (2024). Potentially harmful elements in mining sites in Ghana: Assessment of their carcinogenic and non-carcinogenic health risks for children and adults. Management of Environmental Quality: An International Journal, ahead-of-print(ahead-of-print). https://doi.org/10.1108/MEQ-03-2024-0118.

[938]  Mensah, A. K., Sekyi-Annan, E., Addai, P., Ulzen, O. O., Salifu, M. & Adam, S. (2024). Manure and iron oxide show potential for reducing uptake of arsenic and mercury in lettuce grown in a contaminated mining site. Journal of Hazardous Materials Advances, 100545, https://doi.org/10.1016/j.hazadv.2024.100545.

[939]  Mensah, A. K., Shaheen, S. M., Rinklebe, J., Heinze, S. & Marschner, B. (2022). Phytoavailability and uptake of arsenic in ryegrass affected by various amendments in soil of an abandoned gold mining site. Environmental Research, 214, 113729. https://doi.org/10.1016/j.envres.2022.113729

[940]  Mertens, J., Van Nevel, L., De Schrijver, A., Piesschaert, F., Oosterbean, A., Tack, F. M. G. & Verheyen, K. (2007). Tree species effect on the redistribution of soil metals. Environmental Pollution, 149(2), 173–181.

[941]  Ezeh, M. O., Ogbu, A. D., Ikevuje, A. H. & George, E. P-. E. (2024). Stakeholder engagement and influence: Strategies for successful energy projects. International Journal of Management & Entrepreneurship Research, 6(7), 2375–2395. https://doi.org/10.51594/ijmer.v6i7.1330.

[942]  Mignone, J. & O'Neil, J. (2005). Social capital and youth suicide risk factors in first Nations communities. Canadian Journal of Public Health, 96(S1), S51–S54.

[943]  Miller, R. M., Carnes, B. A. & Moorman, T. B. (1985). Factors influencing survival of vesicular-arbuscular mycorrhiza propagules during topsoil storage. Journal of Applied Ecology, 22, 259–266.

[944]  Mirsal, I. A. (2008). Planning and Realisation of Soil Remediation. In Soil Pollution (pp. 265–281). Berlin Heidelberg: Springer. https://doi.org/10.1007/978-3-540-70777-6_13.

[945]  Mkandawire, M. & Dudel, E. G. (2005). Accumulation of arsenic in Lemna gibba L. (duckweed) in tailing waters of two abandoned uranium mining sites in Saxony, Germany. Science of the Total Environment, 336, 81–89.

[946] Mleczek, P., Borowiak, K., Budka, A. & Niedzielski, P. (2018). Relationship between concentration of rare earth elements in soil and their dis- tribution in plants growing near a frequented road. Environmental Science & Pollution Research International, 25(24), 23695–23711. doi: 10.1007/s11356-018-2428-x.

[947] Moffat, A. J. & McNeill, J. (1994). Reclaiming Disturbed Land for Forestry (pp. xii+-103, No. 110). HM Stationery Office.

[948] Montreemuk, J., Stewart, T. N. & Prapagdee, B. (2024). Bacterial-assisted phytoremediation of heavy metals: Concepts, current knowledge, and future directions. Environmental Technology & Innovation, 33, 103488. https://doi.org/10.1016/j.eti.2023.103488

[949] Moreno, F. N., Anderson, C. W., Stewart, R. B., Robinson, B. H., Ghomshei, M. & Meech, J. A. (2005). Induced plant uptake and transport of mercury in the presence of sulphur-containing ligands and humic acid. New Phytologist, 166(2), 445–454.

[950] Moreno-de Lasheras, M., Nicolau, J. M. & Espigares, M. T. (2008). Vegetation succession in reclaimed coal mining sloped in a Mediterranean-dry environment. Ecological Engineering, 34, 168–178.

[951] Morgan, R. P. C. (2005). Soil Erosion and Conservation (pp. 190–191, 3rd edition). U.K: Blackwell Publishing Company Ltd, Vol. 59. 169–170.

[952] Mukhopadhyay, S., Masto, R. E., Yadav, A., George, J., Ram, L. C. & Shukla, S. P. (2016). Soil quality index for evaluation of reclaimed coal mine spoil. Science of the Total Environment, 542, 540–550.

[953] Mulligan, C. N., Yong, R. N. & Gibbs, B. F. (2001). Remediation technologies for metal-contaminated soils and groundwater: An evaluation. Engineering Geology, 60(1–4), 193–207. https://doi.org/10.1016/S0013-7952(00)00101-0.

[954] Murphy, H., Valiyev, T., Astudillo Paredes, D., De Leon, M. D. & Edwards, E. (2020). Community liaison officers – generating pride and enabling engagement. In SPE International Conference and Exhibition on Health, Safety, Environment, and Sustainability (pp. D031S009R001). https://doi.org/10.2118/199399-MS.

[955] Naidu, R. & Biswas, B. (2024). Introduction to inorganic contaminants and radionuclides: Global issues and challenges. In Inorganic Contaminants and Radionuclides (pp. 1–10). Elsevier.

[956] Negri, M. C., Gatliff, E. G., Quinn, J. J. & Hinchman, R. R. (2003). Root development and rooting at depths. In McCutcheon, S. C. & Schnoor, J. L. (eds.). Phytoremediation:Transformation and Control of Contaminants (pp. 233–262). New York: Wiley.

[957] Ngaba, M. J. Y., Uwiragiye, Y., Miao, H., Li, Z., Pereira, P. & Zhou, J. (2022). Ecological restoration stimulates environmental outcomes but exacerbates water shortage in the Loess Plateau. Peer, Journal, 10, e13658.

[958] Niane, B., Guédron, S., Moritz, R., Cosio, C., Malick, N. P., Deverajan, N., Pfeifer, H. R. & Poté, J. (2015). Human exposure to mercury in artisanal small-scale gold mining areas of Kedougou region, Senegal as a function of occupational activity and fish consumption. Environmental Science & Pollution Research International, 22(9), 7101–7111. doi: 10.1007/s11356-014-3913-5.

[959] Nicolau, J. M. (2002). Runoff generation and routing in a Mediterranean-continental environment: The Teruel coalfield, Spain. Hydrological Processes, 16, 631–647.

[960] Nriagu, J. O. & Pacyna, J. M. (1988). Quantitative assessment of worldwide contamination of air, water and soils by trace metals. Nature, 333(6169), 134–139.

[961] O'Reilly, S. E., Strawn, D. G. & Sparks, D. L. (2001). Residence time effects on arsenate adsorp- tion/desorption mechanisms on goethite. Soil Science Society of America Journal, 65, 67–77.

[962] Oades, J. M. (1984). Soil organic matter and structural stability: Mechanisms and implications for management. Plant & Soil, 76, 319–337.

[963] Obeng, E. A., Shrestha & Lal, 2006, Oduro, K. A., Obiri, B. D., Abukari, H., Guuroh, R. T., Djagbletey, G. D. ... Appiah, M. (2019). Impact of illegal mining activities on forest ecosystem services: Local

communities' attitudes and willingness to participate in restoration activities in Ghana. Heliyon, 5 (10). https://doi.org/10.1016/j.heliyon.2019.e02617.

[964] Odum, E. P. (1969). The strategy of ecosystem development. Science, 164(3877), 262–270.

[965] Ontl, T., Janowiak, M., Swanston, C., Daley, J., Handler, S., Cornett, M., Hagenbuch, S., Handrick, C., Mccarthy, L. & Patch, N. (2020). Forest management for carbon sequestration and climate adaptation. Journal of Forestry, 118, 86–101. https://doi.org/10.1093/jofore/fvz062.

[966] Owen, O. S., Chiras, D. D. & Reganold. (1998). Natural Resource Conservation for Sustainable Future (pp. 503–504, 7th edition).

[967] Owusu-Nimo, F., Mantey, J., Nyarko, K. B., Appiah-Effah, E. & Aubynn, A. (2018). Spatial distribution patterns of illegal artisanal small scale gold mining (galamsey) operations in Ghana: A focus on the Western Region. Heliyon, 4(2).

[968] Padmavathiamma, P. K. & Li, L. Y. (2007). Phytoremediation technology: Hyper-accumulation metals in plants. Water, Air, Soil Pollution, 184, 1–4, 105–126.

[969] Pagouni, C., Pavloudakis, F., Kapageridis, I. & Yiannakou, A. (2024). Transitional and post-mining land uses: A global review of regulatory frameworks, decision-making criteria, and methods. Land, 13(7), 1051. https://doi.org/10.3390/land13071051.

[970] Palmer, M. A., Bernhardt, E. S., Schlesinger, W. H., Eshleman, K. N., Foufoula-Georgiou, E., Hendryx, M. S., Lemly, A. D., Likens, G. E., Loucks, O. L., Power, M. E. et al. (2010). Mountaintop mining consequences. Science, 327(5962), 148–149. doi: 10.1126/science.1180543.

[971] Parkinson, D. (1979). Soil microorganisms and plant roots. In Burges, A. & Raw, F. (eds. ). Soil Biol (pp. 449–478). New York: Academic Press.

[972] Parren, M. P. E. & De Graaf, N. R. (1995). The quest for natural forest management in Ghana, Côte d'Ivoire and Liberia.

[973] Parrotta, J. A. (1999). Productivity, nutrient cycling, and succession in single and mixed-species plantations of Casuarina equisetifolia, Eucalyptus robusta, and Leucaena leucocephalain Puerto Rico. Forest Ecology and Management, 124, 45–77.

[974] Pathak, L. & Shah, K. (2021). Phytoremediation of abandoned mining areas for land restoration: Approaches and technology. In Phytorestoration of Abandoned Mining and Oil Drilling Sites (pp. 33–56). Elsevier.

[975] Philp, M. W. (1992). Investigation of strategies for establishment of native shrub and tree species on saline-sodic spoil at the Saraji open-cut coal mine, Bowen Basin. In Proceedings of the 17th Annual Environmental Workshop (pp. 139–154). Yeppoon. Canberra: Australian Mining Industry Council.

[976] Plaster, J. E. (2009). Soil Science and Management (pp. 124–130, 5th edition, Vol. 495). U.S.A.: Delmar, Concave learning.

[977] Pourret, O. & Bollinger, J. C. (2017). Heavy metals'-what to do now: To use or not to use? Science of the Total Environment, 610, 419–420.

[978] Pourret, O. & Hursthouse, A. (2019). It's time to replace the term "heavy metals" with "potentially toxic elements" when reporting environmental research. International Journal of Environmental Research and Public Health, 16(22), 4446.

[979] Prach, K. & Pysek, P. (2001). Using spontaneous establishment of woody plants in Central European derelicts sites and their potential for reclamation. Restoration Ecology, 2, 190–197.

[980] Prasad, M. N. V. (2006). Plants that accumulate and/or exclude toxic trace elements play an important role in phytoremediation. In Prasad, M. N. V., Sajwan, K. S. & Naidu, R. (eds.). Trace Elements in the Environment: Biogeochemistry, Biotechnology and Bioremediation (pp. 523–547). Boca Raton, FL: CRC Press.

[981] Pueyo, M., López-Sanchez, J. F. & Rauret, G. (2004). Assessment of CaCl2, NaNO3 and NH4NO3 extraction procedures for the study of Cd, Cu, Pb and Zn extractability in contaminated soils. Analytica Chimica Acta, 504, 217–226.

[982]   Puga, A. P., Abreu, C. A., Melo, L. C. A. & Beesley, L. (2015). Biochar application to a contami- nated soil reduces the availability and plant uptake of zinc, lead and cadmium. Journal of Environmental Management, 159, 86–93. https://doi.org/10.1016/j.jenvman.2015.05.036

[983]   Pulford, I. D. & Watson, C. (2003). Phytoremediation of heavy metal-contaminated land trees-a review. In Environ. Int (pp. 529–540, Vol. 29). Queensland, Australia.

[984]   Ramírez, O., de la Campa, A. M. S., Sánchez-Rodas, D. & Jesús, D. (2020). Hazardous trace elements in thoracic fraction ofairborne particulate matter: Assessment oftemporal var- iations, sources, and health risks in a megacity. Science of the Total Environment, 710, 136344.

[985]   Redman, A. D., Macalady, D. L. & Ahmann, D. (2002). Natural organic matter affects arsenic speciation and sorption onto hematite. Environmental Science & Technology, 36, 2889e2896.

[986]   Ren, H. & Yu, Z. (2008). Biomass changes of an Acacia mangiumplantation in Southern China. Tropical Forest Science, 20(2), 105–110.

[987]   Ribet, J. & Drevon, J. J. (1996). The phosphorus requirement of N2-fixing and urea-fed Acacia mangium. New Phytologist, 132, 383–390.

[988]   Richart, S. I., Nancy, J. H., David, T., John, R. T., Mark, S. & Kothleanc, Z. (1987). Old field succession on Minnesota sand plain. Ecology, 68, 12–26.

[989]   Richter, D. D. & Markewitz, D. (2001). Understanding Soil Change: Soil Sustainability over Millennia, Centuries, and Decades. Cambridge University Press.

[990]   Rimmer, L. D. & Younger, A. (1997). Land reclamation after coal- mining operations. In Hester, R. E. & Harrison, R. M. (eds.). Contaminated Land and Its Reclamation (pp. 73–90). London: Thomas Telford.

[991]   Rives, C. S., Bajwa, M. I. & Liberta, A. E. (1980). Effects of topsoil storage during surface mining on the viability of VA mycorrhiza. Soil Science, 129, 253–257.

[992]   Russell, W. E. (1973). Soil Conditions and Plant Growth. New York: Longman Inc.

[993]   Saha, L. & Bauddh, K. (2021). Characteristics of mining spoiled and oil drilling sites and adverse impacts of these activities on the environment and human health. In Phytorestoration of Abandoned Mining and Oil Drilling Sites (pp. 87–101). Elsevier.

[994]   Saha, L., Tiwari, J., Bauddh, K. & Ma, Y. (2021). Recent developments in microbe–plant-based bioremediation for tackling heavy metal-polluted soils. Frontiers in Microbiology, 12, 731723.

[995]   Said, A. S. (2009). Influence of Revegetation on Fertility of Degraded Gold-mined land: A Case Study at AngloGold Ashanti Concession at Obuasi. MPHIL Thesis. Department of Soil Science, University of Cape Coast, Ghana. 2–3;9–11; 14; 20–25; 27; 48.

[996]   Sapkota, A., White, J. D. & Campbell, J. E. (2021). Applications of GIS and remote sensing in reclamation. Environmental Monitoring and Assessment, 193(8), 1–16.

[997]   Schnoor, J. L. (1997). Phytoremediation: Technology overview report. Ground-Water Remediation Technologies Analysis Center.

[998]   Schnurer, J., Clarholm, M. & Rosswal, T. (1985). Microbial biomass and activity in agricultural soil with different organic matter contents. Soil Biology & Biochemistry, 17, 611–618.

[999]   Shaheen, S. M., Ali, R. A., Abowaly, M. E., Rabie, A. E. M. A., El Abbasy, N. E. & Rinklebe, J. (2018). Assessing the mobilization of As, Cr, Mo, and Se in Egyptian lacustrine and calcareous soils using sequential extraction and biogeochemical microcosm techniques. Journal of Geochemical Exploration, 191(May), 28–42. https://doi.org/10.1016/j.gexplo.2018.05.003.

[1000]  Shen, X., Dai, M., Yang, J., Sun, L., Tan, X., Peng, C., Ali, I. & Naz, I. (2022). A critical review on the phytoremediation of heavy metals from environment: Performance and challenges. Chemosphere, 291, 132979. https://doi.org/10.1016/j.chemosphere.2021.132979

[1001]  Sheoran, A. S., Sheoran, V. & Poonia, P. (2008). Rehabilitation of mine degraded land by metallophytes. Journal of Minerals Engineering, 10(3), 11–16.

[1002] Sheoran, V., Sheoran, A. S. & Poonia, P. (2010). Soil reclamation of abandoned mine land by revegetation: A review. International Journal of Ground Sediment & Water, 3, 1–21. Available at: http://scholarworks.umass.edu/intljssw/vol3/iss2/13.

[1003] Sheoran, V. & Choudhary, R. P. (2021). Phytostabilization Ofmine Tailings. Phytorestoration of Abandoned Mining and Oil Drilling Sites (pp. 307–324). Elsevier.

[1004] Shrestha, R. K. & Lal, R. (2011). Changes in physical and chemical properties of soil after surface mining and reclamation. Geoderma, 161(3–4), 168–176. https://doi.org/10.1016/j.geoderma.2010.12.001.

[1005] Shu, W. S., Xia, H. P., Zhang, Z. Q. & Wong, M. H. (2002). Use of vetivera and other three grasses for revegetation of Pb/Zn mine tailings: Field experiment. International Journal of Phytomedicine, 4(1), 47–57.

[1006] Siachoono, S. M. (2010). Land reclamation efforts in Haller Park, Mombasa. International Journal of Biodiversity and Conservation, 2(2), 19–25.

[1007] Sillanpää, M. & Jansson, H. (1992). Status of Cadmium, Lead, Cobalt and Selenium in Soils and Plants of Thirty Countries (No. 65). Food & Agriculture Org.

[1008] Singh, A. N., Raghubanahi, A. S. & Singh, J. S. (2004). Impact of native tree plantations on mine spoil in a dry tropical environment. Forest Ecology and Management, 187, 49–60.

[1009] Singh, A. N. & Singh, A. N. (2006). Experiments on ecological restoration of coal mine spoil using native trees in a dry tropical environment, India: A synthesis. New Forests, 31, 25–39.

[1010] Singh, A. N. & Singh, J. S. (1999). Biomass and net primary production and impact of bamboo plantation on soil re-development in a dry tropical region. Ecology and Management, 119, 195–207.

[1011] Singh, J. S., Raghubanshi, A. S., Singh, R. S. & Srivasta, S. C. (1989). Microbial biomass acts as a source of plant nutrients in dry tropical forest and savanna. Nature, 338, 499–500.

[1012] Singh, J., Ardian, A. & Kumral, M. (2021). Gold-copper mining investment evaluation through multivariate copula-innovated simulations. Mining, Metallurgy & Exploration, 38(3), 1421–1433.

[1013] Singh, A. N., Raghubanshi, A. S. & Singh, J. S. (2002). Plantations as a tool for mine spoil restoration. Current Science, 82(12), 1436–1441.

[1014] Six, J., Bossuyt, H., Degryze, S. & Denef, K. (2004). A history of research on the link between aggregates, soil biota, and soil organic matter dynamics. Soil and Tillage Research, 79, 7–31.

[1015] Smith, J. A., Schuman, G. E., Depuit, E. J. & Sedbrook, T. A. (1985). Wood residue and fertilizer amendment of bentonite mine spoils: I. Spoil and general vegetation responses. Journal of Environmental Quality, 14, 575–580.

[1016] Song, S. Q., Zhou, X., Wu, H. & Zhou, Y. Z. (2004). Application of municipal garbage compost on revegetation of tin tailings dams. Rural Eco-Environment, 20(2), 59–61.

[1017] Stark, N. M. (1977). Fire and nutrient cycling in a douglas fir/larch forest. Ecology, 58, 16–30.

[1018] Sundarapandian, S. M. & Swamy, P. S. (1999). Litter production and leaf litter decomposition of selected tree species in tropical forests at Kodayar in the Western Ghats, India. Ecology and Management, 123, 231–244.

[1019] Svobodova, K. (2022). Turning Mined Lands into Beautiful Places: The Aesthetics of Ecological Restoration. People, Nature, Landscapes. https://medium.com/people-nature-landscapes/turning-mined-lands-into-beautiful-places-the-aesthetics-of-ecological-restoration-7176a160f4c4.

[1020] Tack, F. M. G., Van Ranst, E., Lievens, C. & Vandenberghe, R. E. (2006). Soil solu- tion Cd, Cu and Zn concentrations as affected by short-time drying or wetting: The role of hydrous oxides of Fe and Mn. Geoderma, 137(1–2), 83–e89. doi: 10.1016/j.geoderma.2006.07.003.

[1021] Tang, J., Sun, Q. & Wang, R. (2009). Management on the bio remediation of petroleum contaminated soil and its cost analysis. In 2009 16th International Conference on Industrial Engineering and Engineering Management (pp. 454–458). https://doi.org/10.1109/ICIEEM.2009.5344551

[1022] Tchounwou, P. B., Yedjou, C. G., Patlolla, A. K. & Sutton, D. J. (2012). Heavy metal toxicity and the environment. Experientia Supplementum, 101, 133–164. https://doi.org/10.1007/978-3-7643-8340-4_6

[1023] Tibbett, M. (2010). Large-scale mine site restoration of Australian eucalypt forests after bauxite mining: Soil management and ecosystem development. Ecology of Industrial Pollution, 309–326.

[1024] Tordoff, G. M., Baker, A. J. M. & Willis, A. J. (2000). Current approaches to the revegetation and reclamation of metalliferous mine wastes. Chemosphere, 41, 219–228.

[1025] Toussaint, S., Liam-Puttong, P. & Gardner, H. (2003). Our shame, blacks live poor, die young': Indigenous health practice and ethical possibilities for reform. Health, Social Change Communities, 241–256.

[1026] Troeh, F. R., Hobbs, J. A. & Donahue, R. L. (1980). Soil and Water Conservation for Production and Environmental Protection. Eaglewood Cliffs NJ: Prentice- Hall, Inc.

[1027] U.S. Environmental Protection Agency. (1998). A Citizen's Guide to Phytoremediation. Tech. Fact Sheet, NCEPI. Washington, DC. EPA/542/F-98/011.

[1028] U.S. EPA. (2009, January). Re-Powering America's Land: Siting Renewable Energy on Potentially Contaminated Land and Mine Sites – Summitville Mine, Rio Grande County, Colorado Success Story Hydroeleftric Plant Powers Contaminated Water Treatment at Former Gold Mine. Retrieved from United States Environmental Protection Agency: https://www.epa.gov/sites/default/files/2015-04/documents/success_summitvillemine_co.pdf.

[1029] University of Naples (Italy) Federico II. p. 122.

[1030] USEPA. (2002). Supplemental guidance for developing soil screening levels for superfund sites. U. S. Environmental Protection Agency, Office of Emergency and Remedial Response.

[1031] Van der Lelie, D., Schwitduebel, J., Glass, D. J., Vangronsveld, J. & Baker, A. (2001). Assessing phytoremediation progress in the United States and Europe. Environmental Science & Technology, 35(21), 447A–452A.

[1032] Vanlauwe, B., Aihou, K., Tossah, B. K., Diels, J., Sanginga, N. & Merckx, R. (2005). Sennasiamea trees recycle Ca from Ca-rich subsoil and increase the topsoil pH in Agroforestry systems in the West African derived savanna zone. Journal of Plants and Soils, 269(1–2), 285–296.

[1033] Violante, A., Krishnamurti, G. S. R. & Pigna, M. (2008). Mobility of trace elements in soil environments. In Violante, A., Huang, P. M. & Gadd, G. M. (eds. ). Biophysico-chemical Processes of Metals and Metalloids in Soil Environments (pp. 169–213). Hoboken, NJ: John Wiley & Sons.

[1034] Visser, S., Fujikawa, J., Griffiths, C. L. & Parkinson, D. (1984). Effect of topsoil storage on microbial activity, primary production and decomposition potential. Plant & Soil, 82, 41–50.

[1035] Viventsova, R. E., Kumpiene, J., Gunneriusson, L. & Holmgren, A. (2005). Changes in Soil Organic Matter Composition and Quantity with Distance to a Nickel Smelter – A Case Study on the Kola. Peninsula, NW Russia. Geoderma, 127(3–4), 216.

[1036] Von der Heyden, B. P., Benoit, J., Fernandez, V. & Roychoudhury, A. N. (2020). Synchrotron X-ray radiation and the African earth sciences: A critical review. Journal of African Earth Sciences, 104012. https://doi.org/10.1016/j.jafrearsci.2020.104012.

[1037] Walker, B. H., Holling, C. S., Carpenter, S. R. & Kinzig, A. (2004). Resilience, adaptability and transformability in social–ecological systems. Ecology and Society, 9(2), 5.

[1038] Wang, S. L., Liao, L. P. & Ma, Y. Q. (1997). Nutrient return and productivity of mixed. Cunninghamia lanceolata and Micheliamacclurei plantations. China Journal of Applied Ecology, 8(4), 347–352.

[1039] Wardle, D. A., Bardgett, R. D., Klironomos, J. N., Setälä, H., Van Der Putten, W. H. & Wall, D. H. (2004). Ecological linkages between aboveground and belowground biota. Science, 304(5677), 1629–1633.

[1040] Watanabe, M. E. (1997). Phytoremediation on the brink of commercialization. Environmental Science & Technology, 31, 182–186.

[1041]  Wei, B. & Yang, L. (2010). A review of heavy metal contaminations in urban soils, urban road dusts and agricultural soils from China. Microchemical Journal, 94(2), 99–107.

[1042]  Williams, H. T. & Bellito, M. W. (1998). Revegetation of historic high altitude mining wastes in the western United States. In Fox, M. & McIntosh, (eds. ). Land Reclamation: Achieving Sustainable Benefits (pp. 193–194).

[1043]  Williamson, J. C. & Johnson, D. B. (1991). Microbiology of soils at opencast sites: II. Population transformations occurring following land restoration and the influence of rye grass/ fertilizer amendments. Journal of Soil Science, 42, 9–16.

[1044]  Wong, M. H. (2003). Ecological restoration of mine degraded soils, with emphasis on metal contaminated soils. Chemosphere, 50, 775–780.

[1045]  Xing, K., Zhou, S., Wu, X., Zhu, Y., Kong, J., Shao, T. & Tao, X. (2015). Concentrations and characteristics of selenium in soil samples from Dashan Region, a selenium-enriched area in China. Soil Science and Plant Nutrition, 61(6), 889–897. doi: 10.1080/00380768.2015.1075363.

[1046]  Xu, D., Yang, Z. & He, Q. (1998). Above the ground biomass production and nutrient cycling of middle-age plantation of Acacia mangium. Forest Res, 11, 592–598.

[1047]  Yang, B., Shu, W. S., Ye, Z. H., Lan, C. Y. & Wong, M. H. (2003). Growth and metal accumulation in vetivera and two Sesbania species on lead/zinc mine tailings. Chemosphere, 52(9), 1593–1600.

[1048]  Yang, S., He, M., Zhi, Y., Chang, S. X., Gu, B., Liu, X. & Xu, J. (2019). An integrat- ed analysis on source-exposure risk of heavy metals in agricultural soils near intense electronic waste recycling activities. Environment International, 133(Pt B), 105239. doi: 10.1016/j.envint.2019.105239.

[1049]  Yang, Y. S., Guo, J. F., Chen, G. S., Xie, J. S., Cai, L. P. & Lin, P. (2004). Litter fall, nutrient return, and leaf-litter decomposition in four plantations compared with a natural forest in subtropical China. Annals of Science, 61, 465–476.

[1050]  Yang, P.-T. & Wang, S.-L. (2025). Chemical speciation and availability of molybdenum in soils to wheat uptake. Journal of Environmental Management, 374, 124097. https://doi.org/10.1016/j.jenv man.2025.124097

[1051]  Yao, L. & Wilding, L. P. (1995). Micromorphological study of compacted mine soil in east Texas. International Journal of Rock Mechanics and Mining Sciences and Geomechanics, 32, 219.

[1052]  Yeo, H. C. & Lim, C. H. (2022). Can forest restoration enhance the water supply to respond to climate change? – The Case of North Korea. Forests, 13(10), 1533.

[1053]  Young, A. (1989). Agroforestry for Soil Conservation (pp. 93–103). CAB International, ICRAF.

[1054]  Young, A. (1997). Agroforestry for Soil Management, Effects of Trees on Soils (pp. 23). CAB International.

[1055]  Young, R. E., Gann, G. D., Walder, B., Liu, J., Cui, W., Newton, V., Nelson, C. R., Tashe, N., Jasper, D., Silveira, F. A. O., Carrick, P. J., Hägglund, T., Carlsén, S. & Dixon, K. (2022). International principles and standards for the ecological restoration and recovery of mine sites. Restoration Ecology, 30 (S2), e13771. https://doi.org/10.1111/rec.13771.

[1056]  Yu, X., Kang, X., Li, Y., Cui, Y., Tu, W., Shen, T., Yan, M., Gu, Y., Zou, L., Ma, M. et al. (2019). Rhizobia population was favoured during in situ phy- toremediation of vanadium-titanium magnetite mine tailings dam using Pongamia pinnata. Environmental Pollution, 255(Pt 1), 113167. doi: 10.1016/j. envpol.2019.113167.

[1057]  Yu, X., Li, Y., Li, Y., Xu, C., Cui, Y., Xiang, Q. ... Chen, Q. (2017). Pongamia pinnata inoculated with Bradyrhizobium liaoningense PZHK1 shows potential for phytoremediation of mine tailings. Applied Microbiology and Biotechnology, 101, 1739–1751.

[1058]  Zak, D., Gelbrecht, J. & Steinberg, C. E. W. (2004). Phosphorus retention at the redox interface of peatlands adjacent to surface waters in Northeast Germany. Biogeochemistry, 70(3), 357–368. https://doi.org/10.1007/s10533-003-0895-7.

[1059] Zhang, J. & Liu, C. L. (2002). Riverine composition and estuarine geochemis- try of particulate metals in China – Weathering features, anthropo- genic impact and chemical fluxes. Estuarine, Coastal and Shelf Science, 54(6), 1051–1070. doi: 10.1006/ecss.2001.0879.

[1060] Zhang, J. W., Liao, L. P., Li, J. F. & Su, Y. (1993). Litter dynamics of Pinusmassoniana and Micheliamacclurei mixed forest and its effect on soil nutrients. China Journal of Applied Ecology, 4 (4), 359–363.

[1061] Oduro, K. A., Arts, B., Hoogstra-Klein, M. A., Kyereh, B., & Mohren, G. M. J. (2014). Exploring the future of timber resources in the high forest zone of Ghana. International Forestry Review, 16(6), 573–585.

[1062] Mulligan, C., & Bronstein, J. M. (2020). Wilson disease: an overview and approach to management. Neurologic Clinics, 38(2), 417–432.

[1063] Ashraf, M. A., Maah, M. J., & Yusoff, I. (2013). Evaluation of natural phytoremediation process occurring at ex-tin mining catchment. Chiang Mai Journal of Science, 40(2), 198–213.

[1064] Yu, X., Li, Y., Zhang, C., Liu, H., Liu, J., Zheng, W., ... & Chen, Q. (2014). Culturable heavy metal-resistant and plant growth promoting bacteria in V-Ti magnetite mine tailing soil from Panzhihua, China. PloS one, 9(9), e106618.

[1065] Pankhurst, C. E. (1997). Biodiversity of soil organisms as an indicator of soil health. Pp. 297–324.

[1066] Kundu, N. K., & Ghose, M. K. (1997). Shelf life of stock-piled topsoil of an opencast coal mine. *Environmental Conservation, 24*(1), 24–30.

[1067] Reeves, R. D., Baker, A. J., Jaffré, T., Erskine, P. D., Echevarria, G., & van Der Ent, A. (2018). A global database for plants that hyperaccumulate metal and metalloid trace elements. *New Phytologist, 218* (2), 407–411.

[1068] Bowman, B., & Baker, D. (1998). Mine reclamation planning in the Canadian North. *Yellowknife: Canadian Arctic Resources Committee.*

[1069] Bawua, S. A., & Owusu, R. (2018). Analyzing the effect of Akoben programme on the environmental performance of mining in Ghana: a case study of a gold mining company. *Journal of Sustainable\ Mining, 17*(1), 11–19.

[1070] Liu, S., Ali, S., Yang, R., Tao, J., & Ren, B. (2019). A newly discovered Cd-hyperaccumulator Lantana camara L. *Journal of hazardous materials, 371*, 233–242.

[1071] Littlefield, T., Barton, C., Arthur, M., 2006. Carbon and nutrient dynamics in refor- ested mine sites within the eastern Kentucky coal fields. In: Barnhisel, R.I. (Ed.), Paper presented at the Seventh International Conference on Acid Rock Drainage (ICARD). March 26–30, 2006, St. Louis, MO. American Society of Mining and Reclamation (ASMR), Lexington, KY.

[1072] Sencindiver, J.C., Ammons, J.T., 2000. Minesoil genesis and classification. In: Barnhisel, R., et al. (Eds.), Reclamation of Drastically Disturbed Lands, 2nd ed. Agron. Monogr. 41, ASA, CSSA, and SSSA, Madison, WI, pp. 595–613.

[1073] Arshad, M.A., Martin, S., 2002. Identifying critical limits for soil quality indicators in agro-ecosystems. Agric. Ecosyst. Environ. 88, 153–160.

[1074] Rodrigue, J. A., & Burger, J. A. (2004). Forest soil productivity of mined land in the midwestern and eastern coalfield regions. Soil Science Society of America Journal, 68(3), 833–844.

[1075] Dickinson, N.M., Hartly, W., Uffindell, L.A., Plumb, A.N., Rawlinson, H., Putwain, P., 2005. Robust biological descriptors of soil health for use in reclamation of brownfield land. Land Contam. Reclam. 13(4), 317–326.

[1076] Andrews SS, Carroll CR 2001: Designing a soil quality assessment tool for sustainable agroecosystem management. Ecol. Appl., 11, 1573–1585

[1077] Mukhopadhyay, S., Maiti, S.K., Masto, R.E., 2014. Use of reclaimed mine soil index (RMSI) for screening of tree species for reclamation of coal mine degraded land. Ecol. Eng. 57, 133–142.

[1078]  Masto, R.E., Chhonkar, P.K., Singh, D., Patra, A.K., 2008. Alternative soil quality indices for evaluating the effect of intensive cropping, fertilization and manuring for 31 years in the semi-arid soils of India. Environ. Monit. Assess. 136, 419–435.

[1079]  Andrews, S.S., Karlen, D.L., Mitchell, J.P., 2002. A comparison of soil quality indexing methods for vegetable production systems in Northern California. Agric. Ecosyst. Environ. 90, 25–45.

[1080]  Jia Y, Maurice C, Öhlander B. 2016. Mobility of as, Cu, Cr, and Zn from tailings covered with sealing materials using alkaline industrial residues: a comparison between two leaching methods. Environ Sci Pollut Res Int. 23(1):648–660. doi: 10.1007/s11356-015-5300-2

[1081]  Yu X, Kang X, Li Y, Cui Y, Tu W, Shen T, Yan M, Gu Y, Zou L, Ma M, et al. 2019. Rhizobia population was favoured during in situ phytoremediation of vanadium-titanium magnetite mine tailings dam using Pongamia pinnata. Environ Pollut. 255(Pt 1):113167. doi: 10.1016/j.envpol.2019.113167

[1082]  Cobbett, F. (2007). *Mercury Fluxes and speciated concentrations above terrestrial surfaces in Canada during colder periods* (Doctoral dissertation, University of Guelph).

[1083]  Shackira, A.M., Puthur, J.T., 2017. Enhanced phytostabilization of cadmium by a halophyte Acanthus ilicifolius L. Int. J. Phytoremediat. 19, 319–326. https://doi.org/10.1080/15226514.2016.1225284.

[1084]  Ofosu-Mensah, E. A. (2017). Historical and modern artisanal small-scale mining in Akyem Abuakwa, Ghana. Africa Today, 64(2), 69–91.

# Index

https://doi.org/10.1515/9783111662046-025

www.ingramcontent.com/pod-product-compliance
Lightning Source LLC
Chambersburg PA
CBHW080130220326
41598CB00032B/5020